Arsenic in Groundwater
Poisoning and Risk Assessment

Arsenic in Groundwater
Poisoning and Risk Assessment

M. Manzurul Hassan

CRC Press is an imprint of the
Taylor & Francis Group, an **informa** business

CRC Press
Taylor & Francis Group
6000 Broken Sound Parkway NW, Suite 300
Boca Raton, FL 33487-2742

© 2018 by Taylor & Francis Group, LLC
CRC Press is an imprint of Taylor & Francis Group, an Informa business

No claim to original U.S. Government works

Printed on acid-free paper

International Standard Book Number-13: 978-1-4398-3927-0 (Hardback)

This book contains information obtained from authentic and highly regarded sources. Reasonable efforts have been made to publish reliable data and information, but the author and publisher cannot assume responsibility for the validity of all materials or the consequences of their use. The authors and publishers have attempted to trace the copyright holders of all material reproduced in this publication and apologize to copyright holders if permission to publish in this form has not been obtained. If any copyright material has not been acknowledged please write and let us know so we may rectify in any future reprint.

Except as permitted under U.S. Copyright Law, no part of this book may be reprinted, reproduced, transmitted, or utilized in any form by any electronic, mechanical, or other means, now known or hereafter invented, including photocopying, microfilming, and recording, or in any information storage or retrieval system, without written permission from the publishers.

For permission to photocopy or use material electronically from this work, please access www.copyright.com (http://www.copyright.com/) or contact the Copyright Clearance Center, Inc. (CCC), 222 Rosewood Drive, Danvers, MA 01923, 978-750-8400. CCC is a not-for-profit organization that provides licenses and registration for a variety of users. For organizations that have been granted a photocopy license by the CCC, a separate system of payment has been arranged.

Trademark Notice: Product or corporate names may be trademarks or registered trademarks, and are used only for identification and explanation without intent to infringe.

Library of Congress Cataloging-in-Publication Data

Names: Hassan, M. Manzurul, author.
Title: Arsenic in groundwater : poisoning and risk assessment / M. Manzurul Hassan.
Description: Boca Raton : Taylor & Francis, a CRC title, part of the Taylor & Francis imprint, a member of the Taylor & Francis Group, the academic division of T&F Informa, plc, 2018. | Includes bibliographical references and index.
Identifiers: LCCN 2017051747| ISBN 9781439839270 (hardback : acid-free paper) | ISBN 9781315117034 (ebook)
Subjects: LCSH: Groundwater--Pollution. | Arsenic--Environmental aspects. | Arsenic--Toxicology.
Classification: LCC TD427.A77 H37 2018 | DDC 628.1/14--dc23
LC record available at https://lccn.loc.gov/2017051747

Visit the Taylor & Francis Web site at
http://www.taylorandfrancis.com

and the CRC Press Web site at
http://www.crcpress.com

This book is dedicated to those who died with incurable groundwater arsenic poisoning.

Contents

List of Figures xiii
List of Tables xvii
Preface xix
Acknowledgment xxiii
Author xxv

Chapter 1 Arsenic Poisoning through Ages: Victims of Venom 1

    1.1 Arsenic: Terminology and History 2
    1.2 Arsenic Chemistry 3
        1.2.1 Principal Arsenic Compounds 3
        1.2.2 Properties of Arsenic 6
        1.2.3 Arsenic Analysis: Geochemical Speciation 7
    1.3 Arsenic Applications 8
        1.3.1 Homicidal and Suicidal Arsenic 8
        1.3.2 Medicinal Arsenic 9
        1.3.3 Arsenic in Industries 11
        1.3.4 Arsenic in Agriculture 11
    1.4 Arsenic in Environment 12
        1.4.1 Arsenic in Water 12
        1.4.2 Arsenic in Rocks and Soils 14
        1.4.3 Arsenic in Atmosphere 15
        1.4.4 Dietary Sources of Arsenic 16
    1.5 Mechanism of Arsenic in Groundwater 17
    1.6 Health Impacts of Arsenic Toxicity 20
    1.7 Regulation for Arsenic Exposure 22
    1.8 Concluding Remarks 24

Chapter 2 Groundwater Arsenic Catastrophe: The Global Scenario 25

    2.1 Worldwide Arsenic Catastrophe 26
    2.2 Arsenic in Asian Countries 26
        2.2.1 Alluvial and Deltaic Aquifers, Bangladesh 26
        2.2.2 Irrawaddy Delta, Burma 38
        2.2.3 Mekong Delta, Cambodia 39
        2.2.4 Alluvial Plains, China 40
        2.2.5 Ganges-Brahmaputra Plain, India 41
        2.2.6 Kurdistan Province, Iran 44
        2.2.7 Japan 45
        2.2.8 Terai Alluvial Plain, Nepal 45
        2.2.9 Indus Alluvial Plain, Pakistan 46
        2.2.10 Quaternary Aquifer, Taiwan 47
        2.2.11 Ron Phibun, Thailand 48
        2.2.12 Western Anatolia, Turkey 49
        2.2.13 Red River Basin and Mekong Delta, Vietnam 50
    2.3 Arsenic in North America 51
        2.3.1 Canada 51
        2.3.2 Mexico 52

		2.3.3	The USA	54
	2.4	Arsenic in South America		55
		2.4.1	Chaco-Pampean Plain, Argentina	55
		2.4.2	Brazil	57
		2.4.3	Chile	57
	2.5	Arsenic in Europe		58
		2.5.1	France	58
		2.5.2	Germany	59
		2.5.3	Greece	59
		2.5.4	The Great Hungarian Plain	59
		2.5.5	Italy	60
		2.5.6	The Danube Basin: Romania, Slovakia, and Croatia	60
		2.5.7	Scandinavia	60
		2.5.8	Madrid and Duero Basin, Spain	61
		2.5.9	Switzerland	61
		2.5.10	United Kingdom	62
	2.6	Arsenic in Africa		62
		2.6.1	Burkina Faso	62
		2.6.2	Cameroon	62
		2.6.3	Ethiopia	63
		2.6.4	Ghana	63
		2.6.5	Nigeria	64
	2.7	Arsenic in Oceania		64
		2.7.1	Australia	64
		2.7.2	New Zealand	64
	2.8	Other Areas with Arsenic Contamination		65
	2.9	Natural Geochemical Processes of Arsenic Mechanism		65
	2.10	Concluding Remarks		67

Chapter 3 Groundwater Arsenic Discontinuity: Spatial Mapping, Spatial Planning and Public Participation 69

	3.1	Spatial Characteristics: What Is Special about Spatial Analysis?		70
	3.2	Spatial Variability: Water Quality Investigation		72
	3.3	Spatial Continuity: Theories of Estimation		74
		3.3.1	Geostatistical Estimation	74
		3.3.2	Semivariogram Estimation	75
		3.3.3	Kriging Estimation	77
		3.3.4	Cross-Validation and Prediction Error	82
	3.4	Descriptive Statistics and Regression		82
	3.5	Attributes for Spatial Arsenic Concentration		84
	3.6	Groundwater Arsenic Concentrations: Pattern and Array		85
		3.6.1	Probabilistic Scale: How to Quantify the Pattern?	85
		3.6.2	Spatial Scale: What Are the Areal Discontinuity and at What Extent?	85
		3.6.3	Depth Scale: Which Aquifer Is Contaminated at Which Level?	89
		3.6.4	Spatial Variation: What Is the Possible Mechanism?	92
	3.7	Public Participation GIS: Theoretical Framework		93
	3.8	GIS and Spatial Arsenic Mitigation Planning		94
		3.8.1	Suitable Location for Arsenic-Safe Tubewell Installation	94

Contents   ix

		3.8.2	Spatial Deep Tubewell Planning with PPGIS	94
	3.9	Concluding Remarks		97

Chapter 4 Chronic Arsenic Exposure to Drinking Water: An Environmental Health Concern 99

	4.1	Toxic Effects of Inorganic Arsenic	100
	4.2	Noncarcinogenic Effects of Chronic Arsenic Exposure	102
		4.2.1 Integumentary System	102
		4.2.2 Vascular System	123
		4.2.3 Pulmonary System	125
		4.2.4 Endocrine System	127
		4.2.5 Nervous System	129
		4.2.6 Hepatic System	131
		4.2.7 Hematological System	132
		4.2.8 Renal System	133
		4.2.9 Reproductive System	134
	4.3	Carcinogenic Effects	137
		4.3.1 Skin Cancer	137
		4.3.2 Lung Cancer	145
		4.3.3 Liver Cancer	146
		4.3.4 Bladder and Kidney Cancer	147
		4.3.5 Prostate Cancer	148
	4.4	Susceptibility and Latency Period	149
	4.5	Mechanisms of Arsenic Toxicity	149
	4.6	Concluding Remarks	154

Chapter 5 Risk from Groundwater Arsenic Exposure: Epidemiological and Spatial Assessment 155

	5.1	Chronological Development of Risk	156
	5.2	Conceptual Focus of Risk	158
		5.2.1 Defining Risk	158
		5.2.2 Expression of Risk	160
		5.2.3 Risk Analysis, Risk Assessment, and Risk Management	161
	5.3	Relevant Risk Terminologies	164
		5.3.1 Risk vs. Hazard and Disaster	164
		5.3.2 Risk vs. Threat	165
		5.3.3 Risk vs. Uncertainty	166
	5.4	Risk Assessment Theories: Arsenic through Drinking Water	166
		5.4.1 Hazard Identification	167
		5.4.2 Toxicity (Dose-Response) Assessment	170
		5.4.3 Exposure Assessment	173
		5.4.4 Risk Characterization	176
	5.5	Human Exposure to Arsenic and Spatial Risk	179
		5.5.1 GIS and Human Exposure	179
		5.5.2 GIS and Spatial Risk Assessment	180
	5.6	Groundwater Arsenic Dose and Risk Response	181
		5.6.1 Noncancerous Responses	181
		5.6.2 Cancerous Responses	183

	5.7	Groundwater Arsenic Exposure and Risk: A Bangladesh Case Study 185
		5.7.1 Epidemiological Risk 185
		5.7.2 Spatial Risk and Mapping 190
	5.8	Concluding Remarks 194

Chapter 6 Arsenic-Induced Health and Social Hazard and Survival Strategies: Experiences from Arsenicosis Patients 195

    6.1 Theories of Qualitative Research 196
        6.1.1 Qualitative Research Philosophy 196
        6.1.2 Interpretive Research Paradigm 197
        6.1.3 Hermeneutic Phenomenology: Philosophical Underpinnings 198
        6.1.4 Meta-Synthesis: Qualitative Foundations 200
    6.2 Verbatim and Narratives 201
        6.2.1 Qualitative Transcripts 201
        6.2.2 Rich Descriptive Narratives 203
    6.3 Rigor and Trustworthiness 204
    6.4 Terminological Issues and People's Understandings 206
    6.5 Arsenic Exposure and Health Effects 206
        6.5.1 Symptom Recognition and Health Conditions 207
        6.5.2 Health within Illness 208
    6.6 Arsenic Exposure and Social Implications 208
        6.6.1 Ostracism 209
        6.6.2 In-Family Situation 210
        6.6.3 Marriage and Conjugal Life 211
        6.6.4 Difficulties in Daily Activities 212
        6.6.5 Schooling Children 213
    6.7 Attitudes of Local Leaders and Service Providers 214
    6.8 Arsenic Poisoning and Survival Strategies 214
        6.8.1 Coping Strategies 215
        6.8.2 Adaptation Strategies 216
    6.9 Health and Social Hazards 217
    6.10 Concluding Remarks 219

Chapter 7 Policy Response and Arsenic Mitigation in Bangladesh 221

    7.1 Arsenic Toxicity: Victims of Venom 222
    7.2 Arsenic Policies in Bangladesh 223
        7.2.1 Relevant Policies and Strategies 223
        7.2.2 National Institutional Framework 226
    7.3 Arsenic Projects: Situation and Mitigation 228
    7.4 Existing Mitigation Options: Effectiveness 233
        7.4.1 Deep Tubewells 233
        7.4.2 Rainwater Harvesting 238
        7.4.3 Pond Sand Filter 239
        7.4.4 Dug-Wells 239
        7.4.5 Sharing Arsenic-Safe Tubewells 240
        7.4.6 Surface Water Usage: Digging Ponds and Boiling Water 240
        7.4.7 Reflexive Sedimentation 240
        7.4.8 Technology Options 241
        7.4.9 Piped-Water Systems 241

Contents

	7.5	Natural Arsenic Mitigation	242
		7.5.1 Irrigation with Surface Water	242
		7.5.2 Switching Drinking Water	243
	7.6	Policy Issues: Vigor or Lethargy?	243
		7.6.1 Government Vision	243
		7.6.2 Safety Standards: 50 µg/L or 10 µg/L?	244
		7.6.3 Opinion of Things: Arsenic Data Accuracy	246
		7.6.4 Arsenic Awareness: Shortcomings	247
	7.7	Mitigation Actions: Justification and Validation	249
		7.7.1 Government Actions	249
		7.7.2 DPHE Activities	249
		7.7.3 Health and Family Planning Department	250
		7.7.4 Local Elected Administrators	250
		7.7.5 NGO Activities	251
		7.7.6 Donors and International Organizations	251
	7.8	Potential Policy Strategy in Bangladesh	252
		7.8.1 Short-Term Strategy	252
		7.8.2 Long-Term Strategy	253
	7.9	Concluding Remarks	255
Chapter 8	Arsenic Poisoning in Bangladesh and Legal Issues of Responsibility		257
	8.1	Sutradhar v. the Natural Environment Research Council	257
	8.2	What Space for Environmental Justice?	258
	8.3	The Body-Environment Nexus	258
	8.4	Toxic Torts: *Sutradhar v. NERC*	259
	8.5	House of Lords	261
	8.6	Concluding Remarks	261
Chapter 9	Epilogue and Way Forward		263
	9.1	Précis of Venomous Arsenic: Policy Mapping and Legal Issues	263
		9.1.1 Arsenic Scenario and Mechanism	263
		9.1.2 Arsenic-Induced Health	264
		9.1.3 Assessing Risk of Arsenic Toxicity	265
		9.1.4 Social Implications of Arsenic Toxicity	265
		9.1.5 Spatial Arsenic Mapping and Planning	266
		9.1.6 Arsenic and Legal Responsibility	266
	9.2	Way Forward	267
	9.3	Concluding Remarks	268
Bibliography			269
Index			347

List of Figures

Figure 1.1 Usage of arsenic in agriculture and as an antidote: (a) as pesticide and insecticide sprays in agriculture and (b) as medicinal for several diseases. Its application has been banned in agriculture due to the toxic effects of arsenic for food safety and in environmental contamination. 3

Figure 1.2 Comprehensive transfer and mobilization of arsenic in the environment in terms of lithosphere, hydrosphere, atmosphere, and biosphere. 13

Figure 1.3 Arsenic transformation in groundwater through oxidation process. 18

Figure 1.4 Modeled global probability of geogenic arsenic contamination in groundwater for (a) reducing groundwater conditions and (b) high-pH/oxidizing conditions where arsenic is soluble in its oxidized state. 19

Figure 1.5 Lithology and arsenic concentrations with sediment depths in different boreholes: (a) three boreholes in Jianghan Plain of China and (b) Ghona union of Satkhira district of Bangladesh. 19

Figure 1.6 Chronic arsenic impacts on human health: keratosis to cancers. 21

Figure 2.1 Geographical location of worldwide arsenic catastrophe. More than one-third of the countries in this globe are facing arsenic poisoning from their groundwater. 27

Figure 2.2 Groundwater arsenic concentrations and arsenic-affected patients in Bangladesh: (a) spatial pattern of arsenic concentrations—most of the areas in southern part of Bangladesh and some areas in north-eastern Bangladesh are found to be severely contaminated with arsenic, while the northern part of Bangladesh and hilly districts in south-eastern Bangladesh are found to be safe; (b) percent of contaminated tubewells—a significant number of contaminated tubewells are located in the central part of Bangladesh; and (c) arsenic-affected patient diagnosed—the maximum number of patients diagnosed with arsenic poisoning are in the areas of lower-right corner of central Bangladesh and areas in western part of Bangladesh 34

Figure 2.3 Groundwater arsenic concentrations in some regions of Asian countries: (a) West Bengal, India; (b) China; (c) Taiyuan Basin, China; (d) Part of Inner Mongolia, China; (e) Maydavood area of Southeastern Iran; (f) Hanoi, Vietnam; and (g) Katmandu, Nepal..42

Figure 2.4 Groundwater arsenic concentrations in some regions of North American and South American countries: (a) Alaska, USA; (b) Upper Midwest, USA; (c) New England, USA; (d) Chaco-Pampa Plain, Argentina; (e) Robles area, Argentina; (f) Region Lagunera, Mexico; and (g) Chinchorro, Chile......................53

Figure 2.5 Natural geochemical processes that release arsenic into groundwater in different arsenic-contaminated countries of the world...66

Figure 3.1 Different theoretical models of experimental semivariogram. The shapes of a number of models (e.g. spherical, exponential, Gaussian, circular, pentaspherical) were developed from arsenic attributes collected from Ghona union in Satkhira district of southwest Bangladesh. The shape of a spherical model shows a steady increase up to a distance equal to the

	"range" and then reaches a plateau, the exponential model reaches the threshold value with an asymptotic trend, a graph of a Gaussian model has a parabolic form near the origin, in a pentaspherical model, experimental semivariogram increases with distance until it reaches the sill value at the distance equal to the model range, and a circular model reaches the threshold value with a steady increase up to a distance equal to the model range and then reaches a plateau.	78
Figure 3.2	Properties of cross-validation and prediction-error for IK prediction for spatial arsenic discontinuity in Ghona union of Satkhira district of southwest Bangladesh.	83
Figure 3.3	The pattern of groundwater arsenic concentrations and spatial discontinuity of groundwater arsenic in Bangladesh with different GIS and geostatistical methods: (a) arsenic concentration in each tubewell with point proportion following the low concentration with a small dot and high concentration with a bigger circle; (b) the pattern of spatial arsenic concentration with the ordinary kriging prediction method; (c) spatial variation of arsenic concentrations with the IDW prediction method; and (d) spatial arsenic variation with the IK prediction method. It is noted that this figure is completely different from Figure 2.2 of Chapter 2, which was prepared based on the choropleth mapping concept.	87
Figure 3.4	The pattern of spatial arsenic concentrations and its variations at three different threshold limits of IK prediction method. Arsenic-safe zones as well as arsenic-contaminated zones are found in different areas with different degrees in an area (Ghona union of Satkhira district) of Southwest Bangladesh.	88
Figure 3.5	Relationships between arsenic concentrations and aquifer depth: (a) concentrations in different aquifer depths and (b) three-dimensional view of arsenic concentrations with depth and time.	90
Figure 3.6	The pattern of arsenic concentrations at certain depths. The symbology shows the qualitative variation of arsenic concentrations at very close distances within a certain depth in an aquifer. In shallow aquifers (e.g. a depth of 42, 49, 62 and 74 m), a significant number of tubewells were found to be located within few meters from each other, but extensive variations of arsenic concentrations are noticeable in the tubewells. In deep aquifers (e.g. 180 m depth), arsenic concentrations in tubewells were almost within the safe level, mainly within 50 µg/L of arsenic.	91
Figure 3.7	Suitable arsenic-safe aquifer at different depths: (a) at the depth below 20 meters and the southern part covers almost all of the safe water table; (b) at the depth between 20 and 24 meters and the northern, central, and southern parts cover the safe zones and (c) at the depth more than 24 meters and northern, central, and southern parts cover the safe zones. People can get their access to arsenic-safe water with shallow tubewells.	95
Figure 3.8	Participatory views of spatial deep tubewell planning: views (a) from farmers; (b) from school teachers; (c) from local political leaders; (d) combining all the participatory views and (e) transformation of participatory planning into GIS for installing deep tubewells for arsenic-safe water for the unserved areas of Ghona union.	97

List of Figures

Figure 4.1	Significant effects of chronic arsenic poisoning on major human organ systems.	101
Figure 4.2	Impacts of chronic arsenic exposure from drinking water on human health. Arsenic can be responsible for both the malignant and nonmalignant effects on human health.	102
Figure 4.3	Arsenical skin lesions: (a) nonpitting edema; (b) diffuse melanosis; (c) spotted melanosis; (d) leuco-melanosis; (e) diffused and nodular keratosis on palm; (f) spotted keratosis on sole; (g) dorsal keratosis; (h) an arsenicosis patient with severe keratosis who died of lung cancer.	121
Figure 4.4	Schematic representation of arsenic-induced carcinogenic mechanisms and various disease outcomes that are mostly generated due to its biotransformation process, having effects at genetic and epigenetic levels.	152
Figure 4.5	Genomic integrity in normal cell and its disruption are due to arsenic exposure, leading to MIN (microsatellite instability) and CIN (chromosomal instability).	153
Figure 4.6	Mechanism of arsenic toxicity and carcinogenesis with ROS.	154
Figure 5.1	Risk is possible when three basic elements (vulnerability, threat, and opportunity) exist at the same time.	158
Figure 5.2	Theoretical relationships between the severity of environmental hazard, probability, and risk.	159
Figure 5.3	Risk assessment and risk management process. Dose-response assessment is the main process for risk characterization.	162
Figure 5.4	Graphical position of risk with relation to hazard and vulnerability.	165
Figure 5.5	Sequence of exposure and risk as well as an environmental health paradigm with relationship to risk assessment and risk management framework.	168
Figure 5.6	Dose-response relationship in epidemiology—the greater the toxic chemical exposure (dose) an individual has, the greater their response to the poison will be.	171
Figure 5.7	The domain of exposure assessment in relation to an environmental health paradigm.	175
Figure 5.8	Basic elements in the estimation and prioritization of environmental health risks.	177
Figure 5.9	Theoretical categorization of chronic arsenic risk.	179
Figure 5.10	Spatial risk mapping with methodological concern: (a) buffer generation of tubewells; (b) extract operation for spatial risk zoning; (c) arsenic concentrations with the buffer areas; and (d) spatial risk zones of arsenic exposure.	191
Figure 6.1	FGD with people in different areas of Bangladesh: (a) Political leaders, Ghona, Satkhira; (b) Community people, Galachipa, Patuakhali; (c) Community people, Shivalaya, Manikganj; (d) Community people, Mongla, Bagerhat; and (e) NGO people, Kotalipara, Gopalganj.	202
Figure 6.2	Health and social hazard posed by groundwater arsenic poisoning in arsenic-affected areas of Bangladesh.	218

Figure 7.1 Activities of different stakeholders and their roles in groundwater arsenic mitigation in Bangladesh. 227

Figure 7.2 Arsenic-safe water points in different areas of Bangladesh and sharing the existing arsenic-safe tubewells as a preventive measure to reduce arsenic poisoning. 237

List of Tables

Table 1.1	Timeline of Some Historic Events of Arsenic Toxicity	4
Table 1.2	Significant Arsenic Minerals and Their Characteristics with Geological Occurrences	5
Table 1.3	Physical and Electronic Properties of Elemental Arsenic	6
Table 1.4	Current National Standards for Arsenic in Drinking Water in Some Countries	23
Table 2.1	Groundwater Arsenic Distribution and Problems in Some Countries Around the World	28
Table 2.2	Arsenic Concentrations in Bangladesh in Different Research as well as Impacts of Arsenic on Human Health and Its Mitigation	35
Table 3.1	Descriptive Statistics of Arsenic Concentrations in Different Study Sites at the Coastal Belts of Bangladesh	86
Table 3.2	Arsenic Concentration Pattern with IK Prediction Method at Three Different Threshold Limits	89
Table 3.3	Suitable Area for Arsenic-Safe Tubewell Installation in Dhopakhali Union	96
Table 4.1	Epidemiology of Arsenic Exposure to Drinking Water and Prevalence of Non-carcinogenic Effects on Human Health	104
Table 4.2	Epidemiological Analysis of Arsenic Concentrations in Drinking Water and Its Ingestion with Different Cancers	138
Table 4.3	Levels of Significant Exposure to Inorganic Arsenic Ingestion and Exposure Duration and Its Impacts on Human Health	150
Table 5.1	Consecutive Steps in Arsenic-Induced Health Risk Characterization	169
Table 5.2	Dose-Response Relationship of Arsenic Poisoning	182
Table 5.3	Relevant Parameters Used in Groundwater Arsenic-Induced Health Risk Model	187
Table 5.4	Estimation of Lifetime Cancer Risk	187
Table 5.5	Estimation of Risk Ratio for Arsenic Poisoning	188
Table 5.6	Buffer Generation for Assessing Spatial Risk of Arsenic Poisoning	192
Table 5.7	Spatial Risk Zones and Population Are at Risk	193
Table 7.1	A Timeline of Arsenic Issues in Bangladesh	225
Table 7.2	Suitability and Affordability Analysis of Existing Water Technology for Arsenic-Safe Portable Water	234

Preface

> No one can avoid a challenge in life without breeding regret, and regret is the arsenic of life.
>
> **Esther J. Williams**
> *American competitive swimmer and actress, 1921–2013*

Groundwater is the main source of safe drinking water in many countries of the world, but much of that groundwater has been found to be contaminated with toxic levels of arsenic. It is ironic that so many tubewells have been installed for drinking water that are safe from water-borne diseases but contaminated with toxic levels of arsenic. It is estimated that more than 300 million people in 70 countries worldwide are at risk of groundwater arsenic poisoning. Apart from Bangladesh and the neighboring Indian state of West Bengal, which between them have the largest problem, there have been warnings from Argentina, Chile, Taiwan, Vietnam, China, Pakistan, Thailand, and even the southwestern part of the USA. The sheer scale of this collective hazard deserves greater publicity and certainly greater attention from researchers. There is a substantial literature on environment-induced diseases, but less in monograph format. This book will make an important contribution by focusing on the arsenic tragedy and its possible mitigation.

The situation of arsenic poisoning in Bangladesh will be addressed in this book. Groundwater arsenic toxicity presents a new dimension of hazard in Bangladesh above and beyond the well-known calamities, such as floods, cyclones, tidal surges, famine, and infectious disease. It has been estimated that as many as 85 million people in Bangladesh are exposed to toxic levels of arsenic in drinking water, making it presently the world's largest environmental health hazard. This situation has been described as the "worst mass poisoning in human history" and the scale is well beyond that of the accident in Bhopal, India, in 1984, or Chernobyl, Ukraine, in 1986.

The book deals with the methodological issues of spatial, quantitative, and qualitative enquiries on arsenic poisoning, for instance, using Geographical Information Systems (GIS) to investigate the distribution of arsenic-laced water in space-time to uncover the pattern of variations over scales from meters to kilometers. Spatial risk mapping can provide an indication to researchers, policy makers, and politicians of possible long-term strategies for arsenic mitigation. Qualitative methodological approaches uncover the hidden issues of arsenic poisoning on human health and their social implications as well as their coping strategies and adaptation in the face of community and in-family ostracism. There are very few books on arsenic issues, and almost all the books are mainly concentrated on geology, geochemistry, and health issues. Arsenic risk mapping with spatial modelling, social implications, and legal issues of arsenic toxicity are the areas where serious methodological limitations are evident. Therefore, this book is a departure for health geography with a social science and legal context.

The book is particularly relevant to the advanced undergraduate and masters level. It can contribute to modules on public health, environmental studies, water quality, GIS methods and applications, human rights, and legal issues. The book also discusses a wide range of methodological issues for arsenic poisoning and its policy response to mitigation in the social science/geography perspective, so readers in arsenic-contaminated countries will be interested in this book.

The book contains a total of nine chapters covering diversified issues of arsenic. The introductory chapter sets the scene concerning arsenic and its poisonous nature with historical events, physical and chemical properties of arsenic, usages of arsenic in medicines and industries, concentration patterns of arsenic in the environment, toxic effects of arsenic on human health, mechanism of releasing arsenic into groundwater, and regulatory limits of inorganic arsenic exposure from drinking water. The intention is to describe arsenic issues concisely from the substantial literature that has been published in the last couple of decades. The history of arsenic dates back to several thousand years ago, and arsenic toxicity has been utilized in many societies from the Greek civilization to the present era. During the Middle Ages and Renaissance, arsenic was referred to as the "King of Poisons" and the "Poison of Kings" over the years. The physical and chemical properties of arsenic with geochemical speciation are important in the dose-response pattern upon which the regulatory limits of arsenic exposure from drinking water are based.

The second chapter describes the worldwide distribution of groundwater arsenic. High arsenic concentrations in groundwater have been detected in many countries around the world in the last few decades, and a few major incidents of groundwater arsenic contamination have been reported in Argentina, Bangladesh, Chile, Xinjiang and Inner Mongolia in China, West Bengal in India, Mexico, and Taiwan. In addition, about one-third of countries around the world have arsenic contamination in their groundwater. Wide-ranging information with maps and figures is incorporated in this chapter, which introduces the sources and mobilization of arsenic in different countries.

The micro-level spatial distribution and concentration scenario of groundwater arsenic is discussed in the third chapter. Groundwater contamination is a multi-dimensional phenomenon which has a spatial component, and GIS is well suited to groundwater applications. There is a complex pattern of spatial variability of arsenic concentrations in groundwater with differences between neighboring wells at different scales. Geostatistics and GIS technologies can be used as a management and decision-support tool in spatial discontinuity for groundwater arsenic concentration. Public participation GIS (PPGIS), in recent times, is an important issue in spatial planning for arsenic mitigation. A PPGIS can be used as a new window to view the whole of GIS practices in social settings within the domain of 'information-democracy'. This chapter is mainly focused on the Bangladesh issue, with some examples from other arsenic-affected countries. Several spatial maps have been presented in this chapter to analyze the pattern of micro-level spatial discontinuity of groundwater arsenic concentrations. GIS methodological approaches in terms of spatial analysis and geostatistical analysis were adopted for mapping the 'spatial arsenic concentrations'. Furthermore, PPGIS applications were incorporated in spatial decision-support systems for groundwater arsenic mitigation planning.

Arsenic is a known carcinogen and only a small quantity can constitute a serious health hazard. The impacts of arsenic exposure on human health at different dose levels are pointed out in the fourth chapter. The chronic deleterious effects of arsenic have been well known for the last couple of centuries. Chronic exposure to low-level environmental arsenic compounds at parts per billion to a few parts per million has been linked to morbidity and mortality associated with diseases affecting millions of people worldwide. Inorganic arsenic compounds are known internal and skin carcinogens. This chapter includes a state-of-the-art review of the literature and cutting-edge scientific evidence for arsenic-related health effects. The non-carcinogenic effects of chronic arsenic exposure in terms of the integumentary system, vascular system, pulmonary system, endocrine system, nervous system, hepatic system, renal system, and reproductive system have been discussed in this chapter. Moreover, carcinogenic effects of high levels of arsenic ingestion from drinking water including skin cancer, lung cancer, liver cancer, bladder and kidney cancer, and prostate cancer were reviewed in this chapter with susceptibility and latency period. Furthermore, acute impacts of arsenic exposure on human health were examined. Moreover, the mechanism of arsenic toxicity in terms of genotoxicity, carcinogenesis, and cytotoxicity has been illustrated here. This chapter is a source of valuable information for health professionals, environmental scientists, and epidemiologists, as well as public health practitioners.

Risk assessment has become an important priority issue in recent years. Exploring risk characterization in terms of probabilistic environmental health risk and spatial risk zoning of arsenic toxicity is the purpose of the fifth chapter. Risk is the probability of some adverse health effect that results from exposure to some contaminant in the environment. It is the potential for realization of unwanted negative consequences of groundwater arsenic exposure. In risk assessment, an estimate is made of the magnitude of risk to human health posed by exposure to an environmental hazard. Spatial risk mapping is the process of estimating the spatial magnitude of risk to human health posed by exposure to hazardous arsenic. The assessment of arsenic risk is based on a combination of information on the amount of arsenic that people are exposed to and its toxicity as well as the arsenic exposure period. Proper risk assessment can mitigate possible arsenic poisoning. The chapter provides risk assessment theories of arsenic through drinking water pathways including hazard identification, toxicity assessment, exposure assessment, and risk characterization. This chapter also stipulates a direction for assessing spatial risk zoning with spatial methodologies.

Apart from its health impact, arsenic toxicity is also creating widespread social problems for its victims and their families as well as panic among unaffected people in arsenic-prone areas. There has been a lot of works on arsenic poisoning on human health, but the social issues have so far been underplayed. This issue has been introduced in the sixth chapter. There is a lack of information about the pain of arsenic-affected people concerning their health situation as well as the social problems created with visible arsenicosis symptoms on their bodies. There is even a lack of literature about the survival strategies of arsenic-affected people, i.e. how they manage their health situation as well as their regular social life once challenged by arsenic poisoning. This chapter focuses on the inherent health situation and social problems as well as the social stigma that arsenicosis patients experience during their illness. As a social consequence of arsenic poisoning, there is, for instance, a tendency to ostracize arsenic-affected people since arsenicosis is thought of as a contagious disease. The coping and survival strategies of arsenic-affected people are incorporated in this chapter. Qualitative methodological approaches have mainly been explored for applicability and functionality in identifying health issues of arsenicosis patients as well as their social implications.

The seventh chapter deals with the policy response of governments and the different stakeholders for arsenic mitigation. The water policies and strategies concerning groundwater arsenic issues can focus on arsenic mitigation from drinking water. A number of policies have been promulgated over the last two decades for groundwater arsenic mitigation in Bangladesh. The prominent feature of institutional landscape in arsenic-related interventions is that of engaging a large number of agencies (governmental and non-governmental), but there is poor coordination in current institutional arrangements at the national level for facing the challenges of the sector. Numerous arsenic projects funded by donor agencies, international organizations, and government with mitigation opportunities and their success stories are very unsatisfactory. The existing mitigation options that have been proposed by government, donor agencies, and stakeholders, along with their strengths and weaknesses, have been considered. Moreover, mitigation actions implemented by government institutions, NGOs, donors, national and international organizations were analyzed with their shortcomings. Furthermore, it was identified several proposed short-term and long-term mitigation strategies that may be significant to minimize arsenic toxicity in Bangladesh. Such strategies may also be applicable in other arsenic-affected countries.

The eighth chapter looks at the environmental justice and legal issues of drinking water and its quality. The identification and consequences of arsenic concentrations in groundwater have been a matter of international legal dispute in recent years. A case study from Bangladesh (*Binod Sutradhar v. Natural Environment Research Council-NERC*) is an important issue regarding environmental justice. Mr. Sutradhar, a carpenter from Ramrail in Brahmanbaria district of Bangladesh, suffered from painful arsenicosis on his hands and feet, which he claimed were the result of consuming water contaminated with minute traces of arsenic. Mr. Sutradhar alleged a "tort", legally defined as damage caused by someone's action or inaction. The point of law at stake was the controversial notion of "proximity": the nature of the relationship between the plaintiff and defendant in terms of geography and "duty of care". The chapter presents the background and investigates a number of philosophical and legal principles that are helpful in understanding the "rights" context of environmental health challenges in general terms.

Finally, the concluding chapter summarizes all the arsenic issues presented in this book and proposes an agenda of further research. The main findings from different issues about groundwater arsenic toxicity have been organized in this closing chapter. As a documented carcinogen, groundwater arsenic poisoning is considered as a global catastrophe. Therefore, future research on a range of different issues and employing state-of-the-art technologies for mitigation measures is vital for protecting millions of people who are still suffering from arsenic-induced health and social problems.

M. Manzurul Hassan
Jahangirnagar University
Savar, Dhaka, Bangladesh

Acknowledgment

I am particularly grateful to Em. Professor Peter J. Atkins (Department of Geography, Durham University, UK), who has read every chapter and has provided many ideas and impressions regarding several arsenic-related issues, as well as helped to polish the English. He supervised my doctoral thesis and has been a co-author of many arsenic-related papers.

I am grateful to Professor Dipankar Chakraborti (School of Environmental Studies, Jadavpur University, Kolkata, India) for his heartfelt support in analyzing the water samples in the SOES laboratory. I would like to express my sincere thanks to Dr. Nicholas J Cox of Durham Geography for his guidance and support in analyzing the arsenic data with STATA.

I would like to express my sincere thanks to the ICCO Cooperation, the Netherlands; the World Bank, NGO Forum for Public Health; and Caritas Bangladesh for their support in relevant arsenic and water quality information. They provided me with some information regarding the policy issues of water quality and arsenic poisoning. Special thanks will go to F. M. Sarwar Hossain, ICCO Cooperation, Bangladesh, for his support in completion of a WASH project that I was awarded. A significant amount of information from the WASH project was utilized for this book.

Again, special thanks will go to my younger brother (Md Munibur Rahman) for his continuous assistance in collecting the primary data. Without his constant support, it would have been difficult to collect the relevant data properly and on time. In addition, I am indebted to Shahidul Islam, Nurul Islam, and Atiur Rahman for providing their valuable support during the fieldworks. Thanks are also due to Mr. Peshkar for assisting me in the primary data collection from Ghona of southwest Bangladesh. Moreover, the UP Chairmen and all members from Ghona Union of Satkhira district and Dhopakhali Union of Bagerhat district contributed in collecting the relevant information as well as provided me with some valuable information. Therefore, I am grateful to them.

I would like to express my sincere thanks to Irene Wilkinson, Alan Lumsdon, and Alison Lumsdon for their constant inspiration and support. Thanks are also due to many of my family members (e.g., Shamsunnahar, Mahbubul Alam, Taslima Jahan, Mahfuzul Alam Russel, and Mahmudul Alam Tareq) and friends and well-wishers (Mozammel Haque, Mohammad Hemayatul Islam, and Raihan Ahamed) who contributed to the completion of the manuscript. In addition, I am grateful to Md. Jainal Abedin, for his inspiration in developing the manuscript and to Bobby Akhter for her unlimited support during the preparation of the manuscript.

I am deeply indebted to my parents and parents-in-law who gave me inspiration and moral support for this book. Finally, my deepest thanks go to my wife (Salma Jahan) and son (Saquib M. Hassan) whose constant and consistent support and blessings have made it possible to develop the book manuscript.

Author

Dr. M. Manzurul Hassan is a professor of Geography and Environment, Jahangirnagar University, Dhaka, Bangladesh. He is also a part-time faculty member in the Department of Public Health and in the Department of Environmental Science and Management, North South University, Dhaka, Bangladesh. His research interests include groundwater arsenic poisoning, medical waste management, climate change and human health, water supply and sanitation, and Geographical Information Systems. Furthermore, he has a wide range of consulting experience in the above fields as well as monitoring and evaluation, air quality monitoring, and environmental assessment with many national and international organizations as well as United Nations sister organizations. More information about the author can be available at <http://www.manzurul.net> and <http://lnkd.in/bhymQgN>.

1 Arsenic Poisoning through Ages
Victims of Venom

Arsenic sticks around and today it's easily found after death if somebody thinks of looking for it, because the problem with arsenic, it isn't looked for in the common tests for drugs.

Michael M. Baden
Physician and Forensic Pathologist, 1934

Arsenic is a well-known toxic element and a sinister poison that is ubiquitous in the environment (Garelick et al., 2008). It is a metalloid chemical element that is best known as a terminator of life. Arsenic exists naturally in the earth's crust at low levels, mainly at an average concentration of 1.8 ppm by weight (Ravenscroft et al., 2009: 26). Arsenic ranks as the 20th most common element in natural abundance comprising about 0.00015% (1.5 ppm) of the earth's crust, 14th in the sea, and 12th in the human body (Flora, 2015: 2; Mandal and Suzuki, 2002). Arsenic is mobilized through a combination of natural processes such as weathering reactions, biological activity, and volcanic emissions as well as through a range of anthropogenic activities such as gold mining, nonferrous smelting, petroleum-refining, combustion of fossil fuel in power plants, and the use of arsenical pesticides and herbicides (Biswas et al., 2008; Smedley and Kinniburgh, 2002; Violante et al., 2006).

Humans have been using arsenic since ancient times both as a poison and a curative. It has also been used in pyrotechnics, metallurgy, warfare, and pigmentation and for decoration. Tasteless and odorless, arsenic trioxide (white powder) was used in the past as a chemical warfare agent, and green colored copper acetoarsenate was traditionally used in wallpapers as a pigment (Nriagu, 2002). However, arsenic has been used as a medicinal agent for a number of fatal diseases, and arsenic trioxide is the best example of this paradox. This potent and lethal inorganic arsenical has not only been commonly used to commit homicide but also been used more recently as an effective cancer chemotherapeutic agent (Hughes, 2016: 3). Arsenic compounds are used in wood preservation and insecticides. Small amounts of arsenic (less than 2%) can be used in lead alloys for ammunition. Despite its potential toxicity, arsenic is also an essential element, necessary to our physiology. A level of 0.00001% is needed for growth and for a healthy nervous system. Arsenic, in very small quantities, is necessary as a nutrient to humans, but ingesting excessive amounts can be poisonous (Harding, 1983). Arsenic is immediately dangerous to life or health at 5.0 mg/m^{-3}. The human body does not readily absorb the element itself; hence, pure arsenic is much less dangerous than arsenite (As^{+3}) compounds such as arsenic trihydride or arsine (AsH$_3$) and arsenic trioxide (As$_2$O$_3$), which are absorbed easily and are carcinogenic at high toxicity.

Humans are exposed to arsenic, mainly inorganic arsenic and predominantly through contaminated drinking water, whereas inhalation and skin absorption are minor routes of exposure (Shi et al., 2004). Chronic arsenic exposure through drinking water leads to carcinogenesis of almost all organs, skin diseases (e.g., hyper-pigmentation, hyperkeratosis) leading to different internal cancers, and skin cancer (Bhattacharya, 2017). As a known documented carcinogen, arsenicosis leads to irreversible damage in several vital organs. Inorganic arsenic was first classified as a carcinogen in 1980 by the International Agency for Research on Cancer (IARC, 1980, 2012). Chronic arsenic toxicity from drinking of arsenic-contaminated groundwater is a major environmental health hazard throughout the world, especially affecting Bangladesh. Improvement of symptoms of arsenicosis patients in Bangladesh have been reported to occur following use of vitamins A, C, and E (Khandker et al., 2006). Vitamin E and selenium, either alone or in combination, slightly improved arsenic-induced skin lesions (Verret et al., 2005). Moreover, vitamins C and E significantly improve arsenic-induced keratotic skin lesions in arsenicosis patients in Bangladesh (Melkonian et al., 2012). However, apart from avoiding arsenic-contaminated drinking water and certain symptomatic treatments, there are no evidence-based definitive treatment regimens to treat chronic arsenic toxicity in humans (Bhattacharya, 2017).

Chronic arsenicism has affected more than 300 million people worldwide, approximately 85 million of whom reside in Bangladesh. The major arsenic-prone areas have been reported in large deltas and/or along major river basins across the world (Fendorf et al., 2010; Shankar et al., 2014). A number of regions with deltas and river basins including alluvial and deltaic aquifers in Bangladesh (Chakraborti et al., 2015; Smith et al., 2000a), the Paraiba do Sul delta in Brazil (Mirlean et al., 2014), the Ganges-Brahmaputra Plain in India (Das et al., 2009), the Red river basin and Mekong Delta in Vietnam (Nguyen et al., 2009), the lower alluvial delta of the Mekong River in Cambodia (Benner et al., 2008; Luu et al., 2009; Nguyen and Itoi, 2009; Sthiannopkao et al., 2008), the Huhhot Basin and Hetao river basin in Mongolia (He et al., 2009), the Great Hungarian Plain (Varsányi and Kovács, 2006), and the Danube river basin in Hungary (Nriagu et al., 2007) are severe arsenic-prone areas. Since drinking water is the main route through which arsenic enters the human body (Chen et al., 2009), the understanding of the processes of arsenic contamination in groundwater, associated health risks, and mitigation measures of arsenic is a prerequisite to prevent arsenic poisoning. This chapter provides information regarding physical and chemical properties of arsenic, historical evidence of usage of arsenic, toxic effects of arsenic on human health, mechanism of releasing arsenic into groundwater, and regulatory limits of inorganic arsenic exposure from drinking water.

1.1 ARSENIC: TERMINOLOGY AND HISTORY

Arsenic was known to the ancients in its sulfide compound. The term "arsenic" has its origin in the Syriac word "(al) zarniqa" from the Persian "zarnikh" meaning "yellow" and hence "yellow orpiment". The terminology was adopted into Greek as "arsenikon" and later adopted into Latin as "arsenicum" meaning "yellow orpiment" (Swami et al., 2014). The elemental name of arsenic is believed to have come from the Greek word "arsenikos" meaning potent, an allusion to the ease with which it combines with many other elements (Cullen, 2008: 1; Flora, 2015: 2). Arsenic sulfides (orpiment and realgar) and oxides have been known and used since ancient times (Bentley and Chasteen, 2002). As early as the 4th century BC, Greek philosopher Aristotle wrote of a substance called "sandarache"—now believed to have been a sulfide of arsenic (Evans, 1998). The Greek philosopher Theophrastus (371–287 BC) knew about two arsenic minerals: bright yellow orpiment (As_2S_3) and red colored realgar (As_4S_4). Zosimos of Panopolis (about 300 AD), Egyptian or Greek alchemist, described roasting "sandarach" (realgar) to obtain "cloud of arsenic" (arsenic trioxide), which he then reduced to metallic arsenic. The Greek historian Olympiodorus of Thebes (5th century AD) was the first to obtain white arsenic (As_2O_3) by heating arsenic sulfide (Flora, 2015: 2).

Albertus Magnus (Albert the Great, 1193–1280), a German philosopher and theologian, was the first to discover elemental arsenic and to describe the metallic behavior of arsenic, and he is believed to have been the first to isolate the element from a compound in 1250 by heating soap together with orpiment (Cullen, 2008: 1). Arsenic trioxide (As_2O_3), also known as white arsenic, a by-product of copper refining, when mixed with olive oil and heated gives arsenic metal. Li Shih Chen (1518–1593), a Chinese scientist, developed a book, *Pen Tsao Kang Mu*, in which he studied toxicity of arsenic compounds in the 1500s during the Ming dynasty (1368–1644) and mentioned their use as pesticides in the rice fields (Flora, 2015: 3). Johann Schroeder (1600–1664), a German physician and pharmacologist, was the first person to recognize that arsenic was an element. In 1649, he prepared arsenic by heating its oxide with charcoal (Evans, 1998). Although groundwater arsenic is toxic to humans, through the ages arsenic has been used in medicine, the cosmetics industry, and agriculture and has also been used as an insecticide, desiccant, rodenticide, and herbicide (Azcue and Nriagu, 1994; DeSesso et al., 1998) (Figure 1.1).

Arsenic Poisoning through Ages

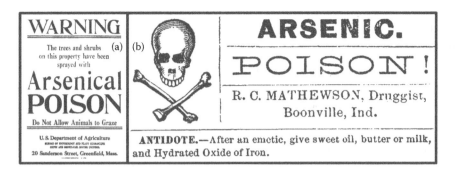

FIGURE 1.1 Usage of arsenic in agriculture and as an antidote: (a) as pesticide and insecticide sprays in agriculture and (b) as medicinal for several diseases. Its application has been banned in agriculture due to the toxic effects of arsenic for food safety and in environmental contamination. (After www.fs.fed.us/ne/morgantown/4557/otis/arsenic.html.)

By the 18th century, arsenic was well known as a unique semimetal. It was 1733 that this metal was examined by Georg Brandt (1694–1768), a Swedish chemist and mineralogist, who showed that white arsenic is an oxide of the metal (Gusenius, 1967). During the Renaissance, it was Theophrastus Paracelsus (1493–1541), a Swiss-German physician and the father of toxicology, who pioneered the use of arsenic compounds in medicine. Arsenic acid and arsenic hydride were discovered in 1775 by Carl Wilhelm Scheele (1742–86), a Swedish apothecary (Grund et al., 2005; Partington, 1962: 151). Arsenic trioxide has been produced in China for some 500 years. In the Victorian era, white arsenic was mixed with vinegar and chalk and rubbed into the faces and arms of women to improve the complexion of their faces (Murphy and Aucott, 1998). Table 1.1 highlights some of the historical aspects of arsenic over 300 years.

1.2 ARSENIC CHEMISTRY

1.2.1 Principal Arsenic Compounds

Arsenic is a metalloid that exhibits both metallic and nonmetallic chemical and physical properties. Arsenic does not often form in its elemental state, but more commonly it is found either in primary arsenic-bearing minerals or adsorbed on to various mineral phases such as iron and aluminum oxides, clays, and iron sulfides. Worldwide, most arsenic is produced as a by-product of copper and lead smelting. Native arsenic is found in the environment in small quantities, usually in ores containing gold, silver, cobalt, nickel, and antimony. The most common arsenic-bearing minerals are arsenopyrite (FeAsS), orpiment (As_2S_3), realgar (As_2S_2), scorodite ($FeAsO_4 \cdot 2H_2O$), löllingite ($FeAs_2$), and tennantite ($[Cu,Fe]_{12}As_4S_{13}$), of which arsenopyrite is the most common (Table 1.2). Other arsenic-bearing minerals include enargite (Cu_3AsS_4), cobaltite (CoAsS), niccolite (NiAs), arsenolite (As_4O_6), claudetite (As_2O_3), and erythrite ($Co_3(AsO_4)_2 \cdot 8H_2O$). However, these mostly occur as accessory minerals in ore deposits (Grund et al., 2005; Mason and Berry, 1978; Ravenscroft et al., 2009: 26) and are not known to be responsible for any cases of arsenic pollution.

Arsenic occurs in water in several different forms depending on the pH and oxidation potential of water (Kartinen and Martin, 1995). Arsenic occurs in the environment both in inorganic (trivalent or arsenite) and organic (pentavalent or arsenate) forms with different degrees of toxicity as well as different valence or oxidation states. The common species of arsenic are arsenite (As^{+3}), arsenate (As^{+5}), monomethyl arsenic acid (MMAA) and dimethyl arsenic acid (DMAA). The primary valence states or the oxidation states of arsenic include arsine (As^{-3}), elemental arsenic (As), arsenite, and arsenate. Arsenic with oxygen forms three oxides, namely arsenic trioxide (As_2O_3), arsenic pentoxide (As_2O_5), and arsenic dioxide (As_2O_4), which apparently contains both trivalent and pentavalent arsenic (Grund et al., 2005). The trivalent forms (arsenite) and the pentavalent

TABLE 1.1
Timeline of Some Historic Events of Arsenic Toxicity

Year	Events	References
1733	Georg Brandt discovered white arsenic that is the oxide of metal.	Gusenius (1967)
1775	Arsenic acid and arsenic hydride were discovered by Carl Wilhelm Scheele.	Partington (1962): 151
1786	Fowler's solution (1% solution of potassium arsenite).	Rohe (1897): 146
1832	Marsh Test for detection of arsenic developed.	Marsh (1837)
1842	Dimethylarsinic acid detected in the environment.	Hughes et al. (2011)
1867	Paris green (copper acetoarsenite) used as insecticide against the Colorado potato beetle in the USA.	Peryea (1998)
1887	Hutchinson proposes arsenic is a human skin carcinogen.	Hutchinson (1887)
1910	Nobel laureate Paul Ehrlich developed Salvarsan (arsphenamine: arsenic-based drug) used as a chemotherapeutic agent.	Aronson (1994)
1930	Reported to be effective in chronic myelogenous leukemia.	Forkner & Scott (1931)
1942	The USA arsenic drinking water interim standard set at 50 µg/L by the US Public Health Services.	USPHS (1943)
1945	British antilewisite (Dimercaprol), a medication, was developed to treat acute poisoning by arsenic.	Peters et al. (1945)
1963	WHO lowered the allowable concentration to 50 µg/L for drinking water arsenic.	WHO (1963)
1968	Prevalence of skin cancer in an arsenic-exposed Taiwanese population was investigated.	Tseng et al. (1968)
1975	USEPA adopts drinking water interim standard at 50 µg/L.	USEPA (1988)
1993	WHO recommends drinking water standard of 10 µg/L.	WHO (1993)
1995	Dimethylarsinic acid a tumor promoter in four rat organs.	Hughes et al. (2011)
1998	Dimethylarsinic acid a complete carcinogen in rat urinary bladder.	Hughes et al. (2011)
2000	US Food and Drug Administration (USFDA) approves arsenic trioxide for leukemia chemotherapy.	Antman (2001)
2001	USEPA lowers US arsenic drinking water standard to 10 µg/L.	USEPA (2001)
2001	Inorganic arsenic a complete carcinogen in adult mice after transplacental exposure.	Waalkes et al. (2007)
2011	Inorganic arsenic a complete carcinogen in adult mice after whole life exposure.	Tokar et al. (2011)

forms (arsenate) of inorganic arsenic tend to be more prevalent in water than the organic arsenic species (Clifford and Zhang, 1994).

Arsenic combined with oxygen, chlorine and sulfur is referred to as inorganic arsenic, while organic arsenic is combined with carbon and hydrogen (ATSDR, 2007). The form and valence state of arsenic is important in its potential toxic effects. In a general form, it is well documented that (a) inorganic arsenicals dissolved in groundwater are more potent and harmful than organic arsenicals present in food; (b) trivalent arsenicals are more potent than pentavalent arsenicals; and (c) trivalent organic arsenicals are equally or more potent than trivalent inorganic arsenicals (Hughes, 2016: 4). Inorganic arsenic is a documented carcinogen, and cancers occur chronically after a long-time exposure to even a small amount of daily arsenic intake (Kartinen and Martin, 1995).

The trivalent form of arsenic species is the dominant species under reducing conditions, and the pentavalent form of arsenic species is the dominant species under oxidizing conditions (Swami et al., 2014). Depending on the redox condition, pH range, presence of complexing ions and the microbial activity of the environment, the pentavalent form of arsenic species may exist as arsenic acid ($HaAsO_4$) and arsenate ions (AsO_4^{3-}). Similarly, the trivalent form of arsenic species may be present as arsenious acid (H_3AsO_3) and arsenite ions (AsO_3^{3-}) (Swami et al., 2014). Pentavalent arsenicals (arsenate) are more likely to occur in aerobic surface waters and trivalent arsenicals (arsenite) are more likely to occur in anaerobic ground waters (USEPA, 2000).

TABLE 1.2
Significant Arsenic Minerals and Their Characteristics with Geological Occurrences

Mineral	Chemical Formula	Arsenic Content (%)	Mohs Hardness	Characteristics	Geological Occurrence
Native Arsenic	As	90–100	3–4	It is light grey to dark grey in color. Metalloid chemical element. It occurs in many minerals, usually in combination with sulfur and metals, but also as a pure elemental crystal. Also, occurs in various organic forms in the environment.	Hydrothermal veins in crystalline rock. Arsenic comprises about 1.5 ppm (0.00015%) of the earth's crust, and is the 53rd most abundant element.
Orpiment	As_2S_3	61	1.5–2	Deep orange, yellow to yellowish brown colored arsenic sulfide mineral. Pearly or resinous, transparent prisms. It takes its name from the Latin "auripigmentum" because of its deep-yellow color.	Found in low temperature hydrothermal veins and mineralization and at hot springs. It is formed both by sublimation and as a by-product of the decay of another arsenic mineral, realgar.
Löllingite	$FeAs_2$	73	5–5.5	An iron arsenide mineral and often associated with arsenopyrite from which it is hard to distinguish. Cobalt, nickel, and sulfur substitute in the structure.	Occurs in mesothermal ore deposits associated with skutterudite, native bismuth, nickeline, nickel-skutterudite, siderite and calcite.
Realgar	AsS	70	1.5–2	An arsenic sulfide mineral, also known as "ruby sulfur" or "ruby of arsenic." Orange to red in color and resinous prisms. Transparent when fresh.	Minor constituent of hydrothermal sulfide veins. Occasionally found with limestone and clay in volcanic terrain.
Arsenopyrite	FeAsS	46	5.5–6	It is an iron arsenic sulfide. Silver-white to steel-grey in color and metallic prisms with rhombic cross-sections. Along with orpiment and 46% arsenic content, arsenopyrite is a principal ore of arsenic.	Formed at moderate to high temperature, associated with Tin, Gold, and Nickel-Cobalt-Silver ores. Also, found in high temperature hydrothermal veins, in pegmatites, and in areas of contact metamorphism.
Scorodite	$FeAsO_4 \cdot 2H_2O$	—	3.5–4	It is a common hydrated iron arsenate mineral. Yellowish-green to greenish brown, bluish green, or very dark green; may be fibrous, granular or earthy. Named from the Greek "Scorodion" means "garlicky". When heated, it smells of garlic.	Found in hydrothermal deposits and as a secondary mineral in gossans worldwide. An intensively oxidized rock capping a mineral deposit.

Source: Mason, B. and L. G. Berry. 1978. *Elements of Mineralogy.* New York; Gaines, R. V. et al. 1997. *Dana's New Mineralogy* (8th Edition). New York: John Wiley & Sons.; Grund, S. C. et al., 2005. *Ullmann's Encyclopedia of Industrial Chemistry,* ed. B. Elvers, 199–240. Vol 4. Weinheim: Wiley-VCH.

1.2.2 Properties of Arsenic

Arsenic is a metalloid chemical element in the nitrogen family having the atomic number 33 with an atomic weight of 74.921, placing it as heavier than iron, manganese, and nickel, but lighter than lead, gold, or silver. In its most stable free state, arsenic is a steel-grey, brittle solid with low thermal and electrical conductivity (Table 1.3). Arsenic mainly occurs in three distinct solid forms in the environment: grey arsenic, yellow arsenic, and black arsenic. Grey arsenic is the most common in the environment. Grey arsenic is a semimetal and is the most stable form. It is brittle and has a relatively low Mohs scale of mineral hardness of 3.5. It has a metallic luster and conducts electricity (ATSDR, 2007). Yellow arsenic is soft and waxy. It is metastable, is a poor electrical conductor, and does not have a metallic sheen. It is produced by rapid cooling grey arsenic vapor (As_4) in liquid air. It reverts to grey arsenic at room temperature. The yellow form of arsenic has a density of 1.97 g/cm^3. Black arsenic can be prepared by cooling arsenic vapor (As_4) at 100°C–220°C. It is glassy and brittle, and it is a poor electrical conductor (Flora, 2015: 5).

The electronic structure of the arsenic atom resembles that of nitrogen and phosphorus. The inorganic arsenic compounds are solids at normal temperatures and are not likely to volatilize. In water, they range from quite soluble (sodium arsenite and arsenic acid) to practically insoluble (arsenic trisulfide). Pure inorganic arsenic is a grey-colored metal and is usually found combined with other elements such as oxygen, chlorine, and sulfur. The luster of arsenic is metallic, and its transparency is crystal and opaque. Its hardness ranges between 3 and 4, its specific gravity between 5.4 and 5.9, and its streak is black. Arsenic does not evaporate, and most arsenic compounds can dissolve in water.

TABLE 1.3
Physical and Electronic Properties of Elemental Arsenic

Property	Value
Classification	A metalloid
Atomic number	33
Atomic mass (weight)	74.9216 g/mol
Melting point	814°C
Boiling point	615°C
Electrons	33
Protons	33
Neutrons in most abundant isotope	42
Electron shells	2,8,18,5
Electron configuration	[Ar] $3d^{10}\,4s^2\,4p^3$
Density	5.7 g/cm^3 at 14°C
Molar volume	13.08 $cm^3/mole$
Appearance (Color)	Grey, brittle, nonmetal flakes
Lustre of arsenic	Metallic
State	Solid
Streak of arsenic	Black
Hardness	Between 3 and 4
Specific gravity	Between 5.4 and 5.9
Transparency of arsenic	Crystal and opaque

Source: ATSDR. 2007. *Toxicological Profile for Arsenic*. Atlanta: Agency for Toxic Substances and Disease Registry, U.S. Department of Health and Human Services; Flora, S. J. S. 2015. *Handbook of Arsenic Toxicity*, ed. S. J. S. Flora, 1–49. Amsterdam: Elsevier and Academic Press.

Arsenic gets into the air when contaminated materials are burned, but it settles from the air to the ground and it does not break down, though it can change from one form to another (ATSDR, 2007).

1.2.3 ARSENIC ANALYSIS: GEOCHEMICAL SPECIATION

Analytical methods for determining different arsenic species have become increasingly important due to the different toxicity and chemical behavior of the various arsenic forms (Baig et al., 2010; Brahman et al., 2013; Rajaković et al., 2013; Sorg et al., 2014). Arsenic is measured mostly by Atomic Absorption Spectrometry (AAS) techniques, with samples prepared by digestion with nitric, sulfuric acid, and/ or perchloric acids (Altunay and Gürkan, 2017; Carrera et al., 2017; Hershey et al., 1988). The generation of hydrides, which is also successfully applied in Inductively Coupled Plasma-Atomic Emission Spectrometry (ICP-AES) (Fengzhou et al., 1991; Smedley et al., 2002), and successful investigation of arsenic at trace levels is possible by Nonflame Atomic Spectrometry without the use of commercial devices (Pesic and Srdanov, 1977). Other methods are employed, including spectrophotometric techniques such as Graphite Furnace-Atomic Absorption Spectrometry (GF-AAS) (Hagiwara et al., 2013; Shamsipur et al., 2014), Inductively Coupled Plasma-Mass Spectrometry (ICP-MS) (Chung et al., 2016; Musil et al., 2014), and X-ray Fluorescence (XRF) (Sbarato and Sánchez, 2001).

Well-established methods that involve the coupling of separation techniques, such as Ion Chromatography (IC) (Duarte et al., 2007; Pohl and Pruisisz, 2004) and High Performance Liquid Chromatography (HPLC) (Bednar et al., 2004; Komorowicz and Barałkiewicz, 2016; Liévremont et al., 2009; Ma et al., 2014), with a sensitive detection system, such as ICP-MS (Jabłońska-Czapla et al., 2014; Komorowicz and Barałkiewicz, 2011, 2014), Hydride Generation-Atomic Fluorescence Spectrometry (HG-AFS) (Farías et al., 2015; Yu et al., 2014), Hydride Generation Atomic Absorption Spectrometry (HG-AAS) (Affum et al., 2015; Jitmanee et al., 2005; Musil et al., 2014), and GF-AAS (Hagiwara et al., 2015; Peng et al., 2015; Vukašinović-Pešić et al., 2009), are the methods of choice for the routine determination of a large number of water samples. They are characterized by method detection limits ranging from 0.5 µg/L to 50 µg/L (Luong et al., 2007; Ma et al., 2014; USEPA, 2004).

Pazirandeh et al. (1998) measured the magnitude of arsenic in the scalp hair of people of a village in the west of Iran using neutron activation analysis. Chen and Jiang (1996) pointed out a simple and very inexpensive in-situ nebulizer/hydride generator with ICP-MS for the determination of arsenic, antimony (Sb), bismuth (Bi), and mercury (Hg) in water samples. Piñeiro et al. (2013) measured arsenic species in human scalp hair with ICP-MS and HPLC coupled to ICP-MS. Ding and Sturgeon (1996) pointed out a development of a continuous flow electrochemical hydride generation technique coupled with in situ concentration in a graphite furnace for determination of arsenic and selenium in seawater.

Roig-Navarro et al. (2001) conducted the simultaneous determination of arsenic species in drinking water by IC coupled to ICP-MS under the use of anion exchange. Vassileva et al. (2001) also found the applicability of the same method to determine arsenic and selenium species in groundwater. In addition, Grégoire and Ballinas (1997) pointed out the process of determination of arsenic from water by electrothermal vaporization ICP-MS. The XRF technique is an important analysis procedure of arsenic pollution in groundwater aquifers (Sbarato and Sánchez, 2001). By means of XRF and using an energy-dispersive spectrometer on 50 groundwater samples from La Francia, Córdoba in Argentina, a high percentage showed concentrations less than or equal to 50 µg/L. He et al. (1997) developed a rapid, simple, and sensitive fluorometric method for the determination of arsenite with fluorescein as the fluorogenic reagent.

Krishna et al. (2001) studied the functionality of an ICP-QMS and a HPLC-ICPMS procedure for speciation and determination of both arsenite and arsenate in water samples. Näykki et al. (2001) describe the optimization of a Flow Injection Hydride Generation (FI-HG) technique together with AAS for the determination of arsenic, antimony, and selenium in iron-based water treatment. Chakraborti et al. (2016a) determined arsenite and arsenate from groundwater by Flow Injection Hydride Generation Atomic Absorption Spectrometry (FI-HG-AAS). The FI-HG-AAS is a simple procedure for the direct determination of arsenite and arsenate in water samples (Burguera et al., 1998; Coelho et al., 2002; Luong et al., 2007; Samanta et al., 1999). The HPLC separation, which is followed by ICP-MS or HG-AFS, belongs to the most often used hyphenated techniques (Komorowicz and Barałkiewicz, 2016). The ICP-MS is commonly applied for arsenic speciation since it provides reliable quantitative data for arsenic species (Ma et al., 2014). The HPLC/ICP-MS can detect inorganic and organic arsenic with high accuracy.

Saad and Hassanien (2001) assessed arsenic levels in hair of the nonoccupational Egyptian population, which was measured by means of hydride atomic absorption spectrophotometry. Gong et al. (2001) studied the performance of a microwave plasma torch discharge AES system directly coupled with Hydride Generation for the determination of arsenic and antimony. Rasul et al. (2002) describe the development of an Anodic Stripping Voltammetric technique for speciation of arsenic in groundwater. The measurements are validated by atomic absorption, atomic emission, and other techniques (Rasul et al., 2002).

Kinniburgh and Kosmus (2002) describe some analytical options in identifying arsenic concentrations in Bangladesh groundwater, while Korngold et al. (2001) pointed out the mechanism of removal of arsenate from drinking water by anion exchangers. Van Elteren et al. (2002) describe the speciation of inorganic arsenic species (i.e., arsenite and arsenate) in some bottled mineral waters from the Radenska and Rogaška springs in Slovenia using existing speciation procedures. The hyphenated technique (HPLC-HGAFS) and a more conventional selective coprecipitation of arsenite combined with Flow-Injection Hydride Generation Atomic Fluorescence Spectrometry (FI-HG-AFS) were used for the speciation of inorganic arsenic. Semenova et al. (2002) developed a software-controlled time-based multi-syringe flow-injection system for total inorganic arsenic determination by HG-AFS. Ferreira and Barros (2002) describe a simple, fast, and quantitative method for determination of arsenite and total arsenic in drinking water using square wave cathodic stripping voltammetry at a hanging mercury drop electrode. The method is validated by the application of recovery and duplicate tests in the measurements of arsenite and total arsenic in natural water.

1.3 ARSENIC APPLICATIONS

1.3.1 Homicidal and Suicidal Arsenic

Arsenic is a popular murder weapon. As one of the most enigmatic elements, arsenic has been used as a deliberate human poison for centuries. Because of its potency and its involvement in many high-profile murders, mainly removing members of the ruling class during the Middle Ages and Renaissance, arsenic has often been referred to as the "King of Poisons" and the "Poison of Kings" over the years (Vahidnia et al., 2007). Many arsenic compounds resemble white sugar, and this apparent innocuousness is enhanced by being tasteless and odorless, which was publicized by Frank Capra's film "Arsenic and Old Lace" in which two elderly ladies used arsenic in elderberry wine to murder their male suitors (Ratnaike, 2003). There have been many suspicious poisonings by arsenic of some renowned persons in the history of several centuries. It has been suggested that Alexander the Great (356–323 BC) and Claudius Caesar Britannicus (41–55 AD, the son of Roman Emperor Claudius) were poisoned by arsenic (Bentley and Chasteen, 2002; Cullen, 2008: 167; Gorby, 1988). The Greek physician Pedanius Dioscorides (40–90 AD) included arsenic as a poison in his five-volume publication, "De Materia Medica" (Meek, 1955).

Because of its easy availability and undetectability in food or beverages, and its odorless and tasteless characteristics, arsenic remained a popular poison until the mid-1850s. Moreover, for a long time, there was no reliable analytical method for detecting arsenic in tissue or other media, although early tests for arsenic were introduced in the mid-1700s (Hughes et al., 2011). However, at present, owing to improved arsenic measurement, one cannot readily "get away with murder" by using arsenic anymore. However, incidents do still occur, for example, as in 2003, arsenic poisoning made headlines when arsenic was detected in coffee served at a church meeting in Maine (Zernike, 2003). Between 1972 and

1982, a total of 28 deaths were attributed to arsenic exposure in North Carolina in the USA; of them, 14 were declared homicides and 7 suicides (Massey et al., 1984). Four of the confirmed arsenic homicides were attributed to one woman, with the crimes occurring over a four-year period (Hughes, 2016: 6).

During the Middle Ages and Renaissance periods, murder by poisoning reached its zenith (Bentley and Chasteen, 2002; Cullen, 2008: 104). Some of the poisonings were politically motivated, particularly in the Catholic Church, as several senior clergymen were poisoned with arsenic over a 500-year period (Cullen, 2008: 168; Lucanie, 1998). It is recognized that arsenic was among the toxins used by the Medici and Borgia papal families to eradicate rivals (Cullen, 2008: 168). Poison accounted for the death of Pope John XIV (984); Clement II (1047); Benedict XI (1304); Leo X (Giovanni de' Medici) (1521); Adrian VI (1523); Clement VII (Giulio de' Medici) (1534), Alexander VI (Rodrigo Borgia) (1503) (Lucanie, 1998), Giulia Toffana, Hieronyma Spara, and the French woman Marie de Brinvillers (Bentley and Chasteen, 2002; Cullen, 2008: 168). As a notorious poison, arsenic was used in several other prominent murder cases, most famously in the death of Napoleon Bonaparte that some conspiracy theorists claim was a political assassination. Napoleon died on the 5th May 1821 on the island of Saint Helena. All four autopsy reports revealed the cause of death to be an extensive carcinoma of the stomach complicated by terminal bleeding, the same disease that took his father and several close relatives (Hindmarsh and Corso, 2008: 104). In the 1960s, when neutron and some activation techniques became available to analyze trace elements from human hairs, a series of literature reported increased arsenic concentrations that had been found in Napoleon's hair, leading to claims that he had been deliberately poisoned while on Saint Helena (Forshufvud, 1962: 256; Weider and Forshufvud, 1995: 555).

The detection of arsenic took a leap forward in 1832 when James Marsh (a chemist at the Royal British Arsenal in Woolwich) decided to investigate analytical methods to provide juries with more reliable evidence of "visible arsenic" (Cullen, 2008: 173). His test method was first used in the trial of Marie LaFarge in France in 1840, in which Marie LaFarge was charged with poisoning her husband with arsenic-laden cakes (Cullen, 2008: 173). The Marsh test represented a turning point in arsenic analytics, and it is effective enough to the present day. The Marsh test was the beginning of the end of undetected arsenic poisonings (Marsh, 1837; Newton, 2007: 7). Arsenic poisoning involved several elderly women of Nagyrev village in south-central Hungary (Gyorgyey, 1987). In 1929, four women were brought to trial, accused of murdering family members with arsenic. There were other suspicious deaths in this village during that time—some 46 of the deceased had arsenic levels high enough to be lethal. Other women were brought to trial later and charged with the murders of husbands, fathers, sons, mothers-in-law, and fathers-in-law with arsenic. Several of the women were found guilty of murder and punished (Hughes, 2016: 6).

1.3.2 Medicinal Arsenic

Despite its toxicity, arsenic has been used for many years for medicinal purposes. As a therapeutic agent, arsenic has been documented back 4000 years (Hyson, 2007). The Father of Medicine, Hippocrates of Kos (460–370 BC), is thought to have used an arsenic paste to treat ulcers and abscesses (Riethmiller, 2005). Hippocrates also used arsenic as a healing agent. Other pioneering physicians (e.g., Aristotle and Paracelsus) are also reported to have used arsenic medicinally (Cullen, 2008: 7). Although arsenic has been used throughout history, more detailed documentation of its use began in the late 18th century. During the last 300 years, a number of arsenic compounds were used as medicines, including arsenic trioxide by Thomas Fowler (1777–1843, an English scientist) and arsphenamine by Paul Ehrlich (1854–1915, a German physician and scientist) (Gibaud and Jaouen, 2010).

Fowler's solution, which was discovered in 1786, is a 1% solution of arsenic trioxide that was widely used in the treatment of various diseases during the 19th century (Scheindlin, 2005). In 1958 the listed indications for Fowler's solution were leukemia, malaria, syphilis, asthma, chorea, skin conditions (psoriasis, dermatitis herpetiformis, and eczema), stomatitis and gingivitis in infants, and

Vincent's angina (Kwong and Todd, 1997; Ratnaike, 2003). Fowler's solution was prescribed as a "health energizer". Paul Ehrlich introduced a new arsenic-based drug in 1910 called arsphenamine (Salvarsan), which became known as the "magic bullet" for treating syphilis and was used until the use of penicillin became more prevalent after World War II (Riethmiller, 2005). Arsphenamine (neoarsphenamine), a light-yellow compound containing 30% arsenic, was used intravenously to treat syphilis, yaws, and some protozoan infections (Ratnaike, 2003) until it was superseded by modern antibiotics.

Arsenic can target the product of a genetic lesion behind a specific type of leukemia, mainly the acute promyelocytic leukemia (APL). Zhang et al. (2017a) considered APL to be a malignancy of the bone marrow, which is characterized by abnormal accumulation of promyelocyte. Arsenic trioxide was previously described to be able to induce apoptosis and differentiation in APL cells. The therapeutic effect of arsenic trioxide in APL cells is strongly associated with the growth suppressor promyelocytic leukemia protein (PML) (Chen et al., 2015). Since 1970, as an active ingredient of traditional Chinese medicine, arsenic trioxide has been used to treat cancer (Sweeney et al., 2010). Arsenic trioxide can induce remission in more than 90% of patients, but it is ineffective in patients with non-APL acute myeloid leukemia (Coe and Schimmer, 2008). In 1878, it was found that Fowler's solution could be effective in lowering the white blood cell count in leukemia patients. Arsenic has also a rich history as a cancer chemotherapeutic. The use of arsenical pastes and arsenic trioxide was found to be effective for the treatment of skin and breast cancer (Antman, 2001). Arsenic trioxide emerged as an effective chemotherapeutic drug for treating APL based on its mechanism as an inducer of apoptosis (programmed cell death) (Fenaux et al., 2001; Rojewski et al., 2002; Zhang et al., 2001a, 2017a). Moderate to high doses (between 0.06 and 0.2 mg/kg/day) of arsenic trioxide given for a period of 30 days can induce remissions in patients with APL (Bode and Dong, 2002; Platanias, 2009; Soignet et al., 1998; Zhu et al., 1997).

The use of Fowler's solution eventually declined over time due to its overt toxicity. Chronic arsenic intoxication from the long-term use of Fowler's solution may cause some carcinoma, for example, haemangiosarcoma (Regelson et al., 1968), angiosarcoma of the liver (Lander et al., 1975; Neshiwat et al., 1992), and nasopharyngeal carcinoma (Zaw and Emett, 2002). Arsenic trioxides are widely used in the treatment of solid tumors such as liver (Alarifi et al., 2013; Wang et al., 2017; Zhang et al., 2012) and gastrointestinal (Lee et al., 2013) carcinomas. The antitumor effects of chemotherapeutic drugs' treatment modalities are enhanced when combined with arsenic trioxides, indicating that arsenic trioxide might contribute in treating tumors (Zhai et al., 2015). Arsenic trioxide enhances antitumor immunity in myeloma models (Ge et al., 2009) and breast cancer cells (Baj et al., 2002). Preclinical studies showed that arsenic trioxide affects peripheral blood leukocyte and platelet counts (Jiang et al., 2010) that may ultimately alter the immune system (Wang et al., 2017). Arsenic trioxide generates Reactive Oxygen Species (ROS), which activates c-Jun N-terminal Kinase (JNK), upregulates pro-apoptotic proteins, and downregulates anti-apoptotic proteins. Thus, the leukemic cells undergo arsenic-induced apoptosis and the patients enter a state of remission to the cancer (Hughes et al., 2011). Also, arsine (AsH_3) has long been recognized as a selected toxin that can be used as therapy for malaria (Parris, 2005).

Arsenic continues to be an essential constituent of many traditional medicine products. The Chinese began to practice herbal medicine around 2800 BC. This experience was eventually gathered into Shen Nong's Materia Medica (200 BC) around the time of the unification of the nation and the beginning of the Han dynasty (Chan, 2002: 71; Porter, 1997: 149). As early as 222 BC, some Chinese traditional medications contain realgar (arsenic sulfide) and are available as pills, tablets, and other preparations. They were used for asthma, hemorrhoids, psoriasis, rheumatism, syphilis, and so on. They were also prescribed as a health tonic, an analgesic, an anti-inflammatory agent, and a treatment for some malignant tumors (Ko, 1999; Wong et al., 1998). In India, herbal medicines containing arsenic are used in some homoeopathic preparations (Kew et al., 1993) and for treating hematological malignancies (Prakash, 1994). In Korea, arsenic is prescribed in herbal medicine for hemorrhoids (Mitchell-Heggs et al., 1990).

1.3.3 Arsenic in Industries

Arsenic and arsenic compounds are being used in industries. Arsenic with high purity is used in semiconductor applications, solar-cells, optoelectronic devices, and so on (Azcue and Nriagu, 1994). Arsenic compounds were employed in bronze alloys as early as 3000 BC (BGS/DPHE, 1999). The use of chromated copper arsenate and ammoniac copper arsenate as wood preservatives was very common until the recent past (Woolson, 1983). In industry, arsenic is used to manufacture paints, fungicides, insecticides, pesticides, herbicides, wood preservatives, and cotton desiccants (Ratnaike, 2003). Arsenic-based pigment (e.g., Paris green or copper acetoarsenite) was used in many consumer products (e.g., toys, candles, and fabric), but its use in wallpaper was particularly linked to widespread sickness and death during the 18th and 19th centuries (Scheindlin, 2005). Although the use of arsenic has been phased out from pigment products, it is still used in the production of glass and semiconductors (ATSDR, 2007).

Metallic arsenic can be used in alloys with lead. Lead components in car batteries are strengthened by the presence of a very small percentage of arsenic (Bagshaw, 1995). Refined arsenic trioxide is used in glassware production and tertiary arsines are used in polymerization of unsaturated compounds (BGS/DPHE, 1999). High-purity arsenic (at least 99.999%) is used in electronics in conjunction with gallium or indium. Gallium arsenide (GaAs) is an important semiconductor material used in integrated circuits, which are much faster than those made from silicon. GaAs can be used in laser diodes and light-emitting diodes (LEDs) to convert electrical energy directly into light (Grund et al., 2005). In addition, arsanilic acid is used in motor fuel, arsonic and arsenic acid are used in the steel industries, and roxarsone is used in feed additives (USEPA, 2000).

Arsenic compounds are present in weed killers, embalming fluids, paints, dyes, soaps, metals, wood preservations, automotive body solder, industrial battery grid, and electrophotography (Azcue and Nriagu, 1994). Gallium arsenide or aluminum gallium arsenide crystals are components of semiconductors, light emitting diodes, lasers, and a variety of transistors (Ratnaike, 2003). In addition, aromatic arsenic compounds are used in drugs, and phenylarsenic compounds are used in animal feed additives and disease prevention (BGS/DPHE, 1999). Moreover, the production of chromated copper arsenate (CCA), an inorganic arsenic compound and wood preservative, accounts for approximately 90% of the arsenic used annually by industry in the USA (USEPA, 2000). CCA is used to pressure treat timber, which is typically used for the construction of decks, fences, and other outdoor applications (Smedley and Kinniburgh, 2002).

1.3.4 Arsenic in Agriculture

Agricultural inputs such as pesticides, desiccants, and fertilizers are major sources of arsenic in soils (Jiang and Singh, 1994; Saxe et al., 2006: 281). The use of arsenic-containing fertilizers and pesticides represents an historic and continuing addition to background concentrations of arsenic in soils. From the late 1800s and until the introduction of dichlorodiphenyltrichloroethane (DDT), several arsenic compounds (e.g., lead arsenate, calcium arsenate, magnesium arsenate, zinc arsenate, zinc arsenite, and Paris green) were extensively used as pesticides in orchards (Merry et al., 1983; Smith et al., 1998a: 150). Early in the 20th century, pesticides including lead arsenate and calcium arsenate were commonly applied to turf grass (e.g., golf courses, sod farms) and agricultural crops (e.g., apple orchards, vegetable fields) (Alden, 1983; Welch et al., 2000). Arsenical pesticides were also widely used in livestock dips to control ticks, fleas, and lice (Vaughan, 1993). Paris green was used as an insecticide from 1867 to 1900, and it was effective in controlling Colorado potato beetles and mosquitoes (Cullen, 2008: 61; Peryea, 1998). Through the early 1900s, lead arsenate, another arsenic-based pesticide, was widely used as a pesticide for apple and cherry orchards. It is noted that arsenite of lime and arsenate of lead were used widely as insecticides until the discovery of DDT in 1942 (Murphy and Aucott, 1998).

Arsenic-based pesticide solutions were widely used in Queensland and northern New South Wales, Australia, from the early 1900s to 1955 to control ticks in cattle (Smith et al., 1998a: 150).

However, these applications have been banned due to the toxic effects of arsenic and later public awareness about food safety and environmental contamination (Smith et al., 1998a: 162). Lead hydrogen arsenate was the most extensively used of the arsenical insecticides, more specifically on fruit trees (Peryea, 1998). It was first prepared as an insecticide in 1892 for use against the gypsy moth in the USA. Lead hydrogen arsenate applied in foliar sprays was longer lasting for its pesticide effect, and farmers used to adopt lead arsenate for insecticidal use (Peryea, 1998). In addition, contact with the compound sometimes resulted in brain damage among those working the sprayers. By 1960, most uses of lead arsenate were phased out after it was recognized that its use was associated with health effects (Smith et al., 1998a: 165). The use of lead arsenate was banned in the USA in 1988 (Peryea, 1998).

Arsenical insecticides have been used in agriculture for centuries. The earliest available records indicate the use of arsenic sulfides in China as early as the 10th century AD. During the Middle Ages, arsenic compounds were widely used in agriculture and in herbicides. The organic arsenic compounds, e.g., herbicides monosodium methyl arsenate (MSMA), disodium methyl arsenate (DSMA), arsenic acid and dimethyl arsenic acid (DMAA), were used in most important pesticides (Wauchope and McDowell, 1984). In the second half of the 20th century, MSMA and DSMA—less toxic organic forms of arsenic—replaced lead arsenate in agriculture. The first insecticidal use of the copper acetoarsenite pigment commonly known as "Paris green" [$(CH_3COO)_2Cu \cdot 3Cu(AsO_2)_2$] appears to have been in 1867 on the Colorado potato beetle in the USA. Paris green sprays were soon adopted by fruit growers for control of codling moth in apple orchards, a use that continued through about 1900. Paris green was also used for mosquito abatement, as well as for control of other insects (Peryea, 1998).

Some modern uses of arsenic-based pesticides still exist. The toxicity of arsenic to insects, bacteria, and fungi led to its use as a wood preservative (Rahman et al., 2004). CCA (or Tanalith) has been registered for use in the USA since the 1950s as a wood preservative, protecting wood from insects and microbial agents. CCA is still approved for use in nonresidential applications (WPSC, 2008), and, for decades, this treatment was the most extensive industrial use of arsenic. The use of CCA in consumer products was banned in the European Union (EU) and the USA in 2004. However, CCA remains in heavy use in other countries (such as on Malaysian rubber plantations) (Grund et al., 2005). The use of organic arsenical pesticides began in the 1950s and has continued into the present time. Organic arsenicals in the pentavalent oxidation state are much less toxic than inorganic arsenicals (Cohen et al., 2006). Although both MSMA and DSMA are effective as selective grass suppressors, MSMA has been used almost exclusively in Australia (Smith et al., 1998a: 162). It was determined that DMAA had the potential to cause cancer in humans when doses are high enough to cause bladder cell cytotoxicity. Therefore, some uses of these compounds as herbicides were discontinued in 2009 (Hughes et al., 2011).

Arsenic is intentionally added to the feed of chickens raised for human consumption. Arsenic is used as a feed additive in poultry and swine production to increase weight gain, improve feed efficiency, and prevent diseases (Nachman et al., 2005). Organic arsenic compounds are less toxic than pure arsenic and promote the growth of chickens. Under some conditions, arsenic in chicken feed is converted to the toxic inorganic form (Gray, 2012). Roxarsone, for example, had been used as a broiler starter, but it was banned in industrial swine and poultry production since elevated levels of inorganic arsenic were found in treated chickens.

1.4 ARSENIC IN ENVIRONMENT

1.4.1 Arsenic in Water

Arsenic occurs at a very low concentration in nature, but higher concentrations are associated with anthropogenic sources that may introduce arsenic into drinking water and food. The primary natural sources include geological formations (e.g., rocks, soil, and sedimentary deposits), geothermal activity, and volcanic activity (Figure 1.2). Volcanic activity appears to be the largest natural source of arsenic emissions to the atmosphere (ATSDR, 2007). Arsenic compounds, both inorganic and organic, are also found in food (Gunderson, 1995). Arsenic accumulates via weathering, smelting and mining, hydrothermal and geothermal activities, and oxidative dissolution of arsenic-bearing mineral releases into the soil and water. Arsenic undergoes oxidation, reduction, methylation, demethylation, and precipitation reaction and finally returns to soil and forms arsenic-rich sediments, for example, arsenopyrite, orpiment, realgar, and so on. Industrial processes, fossil fuels, burning of coal, and use of pesticides release arsenic into the atmosphere. None of the various species of arsenic escapes from the environment; rather they continue to cycle from one species to another and from one medium to another (Flora, 2015: 14). Volcanic activity, erosion of rocks and minerals, and forest fires are natural sources of arsenic in the environment. The terrestrial abundance of arsenic is around 1.5–3.0 mg/kg (Mandal and Suzuki, 2002). Arsenic enters the human body mainly through drinking water, and this is the greatest menace to human health. Drinking water contains arsenic at various concentrations depending on the sources. However, groundwater holds the highest concentration of arsenic owing to various natural and human activities (Flora, 2015: 14).

Arsenic is found at low concentrations in natural water almost entirely in the inorganic form and can be stable as both arsenite and arsenate, trivalent and pentavalent inorganic arsenicals, respectively (Saxe et al., 2006: 280). Seawater ordinarily contains 1.0–8.0 µg/L of arsenic (Cutter

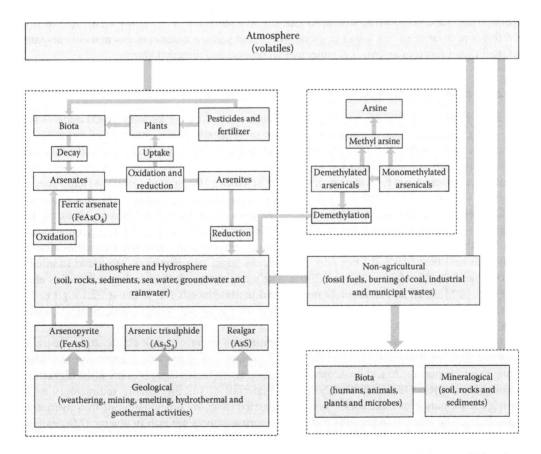

FIGURE 1.2 Comprehensive transfer and mobilization of arsenic in the environment in terms of lithosphere, hydrosphere, atmosphere, and biosphere. (Modified from Flora, S. J. S. 2015. Arsenic: Chemistry, occurrence, and exposure. In *Handbook of Arsenic Toxicity*, ed. S. J. S. Flora, 1–49. Amsterdam: Elsevier and Academic Press. With permission.)

et al., 2001). In oxic seawater, arsenic is typically dominated by arsenate, although some arsenite is invariably present and becomes of increasing importance in anoxic bottom waters (Peterson and Carpenter, 1983; Pettine et al., 1992). Relatively high proportions of arsenious acid (H_3AsO_3) are found in surface ocean waters (Cullen and Reimer, 1989; Cutter et al., 2001). These coincide with zones of primary productivity (Smedley and Kinniburgh, 2002).

The direct sources of arsenic pollution in surface water include domestic and industrial waste water, electric power plants, base metal mining and smelting, and atmospheric fallout of contaminated aerosols, while the indirect sources of pollution include the residues of pesticides and fungicides from soils (Flora, 2015: 14). It should be noted that monosodium methanoarsonate (MSMA) and cacodylic acid are highly soluble in water and can be washed away before they are stabilized in soils (BGS/DPHE, 1999). Arsenic predominantly presents in groundwater in the form of arsenite and arsenate. Methylation of inorganic arsenic to methyl and dimethyl arsenic acid is associated with biological activity in water (Azcue and Nriagu, 1994). The concentrations of arsenic in unpolluted fresh waters typically range between 1 µg/L and 10 µg/L, rising to 100 µg/L–5000 µg/L in areas of sulfide mineralization and mining (Smedley et al., 1996). Arsenic speciation was performed on groundwater samples from an area around Alaska containing high levels of arsenic; 3 to 39% contained arsenite, and the rest were arsenate (Harrington et al., 1978).

Geothermal water can be a source of inorganic arsenic in surface water and groundwater. Welch et al. (1988) identified 14 areas in the western USA where dissolved arsenic concentrations ranged between 80 µg/L and 15,000 µg/L. Geothermal water in Japan contains arsenic between 1800 µg/L and 6400 µg/L and neighboring streams about 2 µg/L (Nakahara et al., 1978). Generally, methylated forms of arsenic are not found in groundwater, but surface water contains arsenite, arsenate, and methylated forms of arsenic, i.e., monomethyl arsenic and dimethyl arsenic.

1.4.2 Arsenic in Rocks and Soils

The natural content of arsenic in soils globally ranges from 0.01 mg/kg to over 600 mg/kg (Yan-Chu, 1994), with an average of about 2 mg/kg, and subsequently it was estimated in various forms of rocks and soils (Flora, 2015: 17). Arsenic concentration in soil is almost entirely in the inorganic form, except in areas where higher levels of organic compounds can be found. In soils, pentavalent arsenic predominates due to oxidation of trivalent arsenicals (Gong et al., 2001). Arsenic concentrations in soils are mostly present in sulfide ores of metals including copper, lead, silver, and gold (BGS/DPHE, 1999). Arsenic in soils may originate from parent materials (Tanaka, 1988), but it is present in soils in higher concentrations than those in rocks (Peterson et al., 1981). Uncontaminated soils usually contain between 1.0 and 40.0 mg/kg of arsenic with the lowest concentrations in sandy soils and those derived from granites, whereas larger concentrations are found in alluvial and organic soils (Mandal and Suzuki, 2002). The mean arsenic content in Japanese soil is about 11 mg/kg, in Mexican soil is about 14 mg/kg, and in Bangladesh soil is about 22.1 mg/kg. The natural level of arsenic in sediments is usually below 10 mg/kg and varies considerably all over the world (Mandal and Suzuki, 2002).

The principal factor influencing arsenic concentrations in soils is rock composition. There are many arsenic-containing minerals, and the most important ores of arsenic are arsenopyrite, realgar, löllingite, and orpiment. Arsenic concentrations in soils enriched in these ores are often higher than in normal soil (BGS/DPHE, 1999). The parent materials of soil are usually sedimentary rocks. During the formation of these rocks, arsenic is carried down by precipitation of iron hydroxides and sulfides. Therefore, iron deposits and sedimentary iron ores are rich in arsenic (Maclean and Langille, 1981). The natural level of arsenic in sediments is usually below 10 mg/kg in dry weight (Mandal and Suzuki, 2002). Arsenic retention and release by sediments depends on the chemical properties of the sediments, especially on the amount of iron and aluminum oxides and hydroxides they contain (BGS/DPHE, 1999). The amount of sedimentary iron is an important factor that influences arsenic retention in sediments (Mandal and Suzuki, 2002). Arsenic concentrations vary from

1.5 to 5.9 mg/kg in igneous rocks, from 5.0 to 10 mg/kg in sedimentary rocks, and less than 5.0 mg/kg in metamorphic rocks (Flora, 2015: 17). Volcanic materials are the main contributor for generating high-arsenic levels in igneous rocks (Riedel and Eikmann, 1986). Argillaceous deposits (e.g., fine-grained clastic sediments) have a high arsenic concentration (13 mg/kg) owing to enhanced sulfide mineral, oxides, clays, and organic matter in sedimentary rocks. Iron-rich rocks, phosphorites, coal, and bituminous deposits are enriched with exceptionally high arsenic concentration ranging between 100 and 900 mg/kg (Belkin et al., 2000). Pelitic metamorphic rocks (e.g., a metamorphosed fine-grained sedimentary rock, i.e., mudstone or siltstone) contain high arsenic concentration of 18 mg/kg (Flora, 2015: 17).

Arsenic occurs mainly as inorganic species, but it can bind to organic material in soils. Arsenic may accumulate in soils through arsenical pesticides, herbicide, fertilizer, and so on. Inorganic arsenic may be converted to arsenic compounds by soil micro-organisms (Wei et al., 1991). Although soil is a source of arsenic in the environment, the main accumulations are in the topsoil layers (Adriano, 1986; Woolson, 1983: 395). A number of factors govern the transport and situation of arsenic in soil, including the chemical form of arsenic, the chemical and physical characteristics of soil such as soil pH and redox potential, and the presence of other soil constituents that may limit arsenic mobility by adsorption or precipitation (Saxe et al., 2006: 282). Arsenate is the most common form of inorganic arsenic in soil. Arsenite tends to exist under reducing redox conditions. Arsenate is typically less mobile in soil, and arsenite is more mobile; organic compounds that contain arsenic are generally not mobile in soil (ATSDR, 2007). The water solubility of arsenic is a key factor in determining its mobility. The solubility of the various chemical forms of arsenic depend in part on soil pH, which naturally varies between soils and within a soil over time (Saxe et al., 2006: 282).

The amount of oxygen in soil influences the water solubility of arsenic. If very little free oxygen is present in water, arsenite will transform to arsenate only at high pH values (alkaline conditions), but when more oxygen is present, the arsenite-to-arsenate transformation will take place at low pH values (acidic conditions). In general, arsenic is least mobile in soil when the soil's pH is near neutral or slightly acidic (pH 4–8) (Saxe et al., 2006: 283). Through the adsorption mechanism, arsenic is retained in soil, thus limiting its mobility. The quantity of arsenic bound to soil through adsorption depends upon adsorbing minerals present in soil (e.g., clays), soil pH, and other components that compete for adsorption sites (Matera and Le Hecho, 2001: 213). There is a significant correlation between the amount of iron or aluminum in soil and the amount of arsenic in soil (Chen et al., 2002). The total amount of arsenic in soils and its chemical forms have an important influence on plant, animal, and human health (Azcue and Nriagu, 1994).

1.4.3 Arsenic in Atmosphere

The concentrations of arsenic in the atmosphere are usually low, but are increased by inputs from smelting and other industrial operations, fossil-fuel combustion, and volcanic activity (Flora, 2015: 19; Smedley and Kinniburgh, 2002). Due to specific evaporation-condensation processes at high temperatures, arsenic partially or entirely evaporates into the atmosphere (Flora, 2015: 19). Much of the atmospheric arsenic is particulate, and it is usually present as a mixture of arsenite and arsenate. The human exposure to arsenic in the air, which is almost entirely as inorganic arsenic, is generally very low. Arsenic concentrations in the air generally range from 0.0004 to 0.030 $\mu g/m^{-3}$ (WHO, 1996). There is little evidence to suggest that atmospheric arsenic poses a real health threat, but atmospheric arsenic arising from coal burning has been invoked as a major cause of lung cancer, as found in Guizhou Province of China (Finkelman et al., 1999). Particles of less than 1.0 μm when inhaled may be deposited in the pulmonary tissue of the respiratory tract and enter the blood. The amount of arsenic inhaled per day is about ≤ 0.05 $\mu g/m^{-3}$ in unpolluted areas (WHO, 1981). Typical arsenic concentrations accounting to around 10^{-5}–10^{-3} $\mu g/m^{-3}$ have been recorded in unpolluted areas, 0.003–0.18 $\mu g/m^{-3}$ in urban areas, and more than 1.0 $\mu g/m^{-3}$ close to industrial plants (WHO, 2001).

Coal is a natural source of arsenic and is primarily responsible for the release of arsenic into the environment. Coal combustion for thermal power plants is one of the major sources of anthropogenic arsenic emission into the biosphere. Coal contains both inorganic and organic forms of arsenic, and arsenic-bearing sulfides contain high concentrations of arsenic, whereas pyrites contain very low arsenic concentrations (WHO, 2001; Yudovich and Ketris, 2005). Release of arsenic during coal combustion depends mainly on combustion temperature. During low temperature combustion (<1200°C) arsenic oxide (AsO) and arsenic trioxide (As_2O_3) escape in the gaseous phase, while at high temperature (>1200°C) only As_2O_3 is released (Flora, 2015: 20).

Release of arsenic from fly ash can also be responsible for environmental pollution. During the combustion of coal, the escaping arsenic in the gaseous phase is captured by fly ash and collected by an electrostatic precipitator. Arsenic associated with fly ash ends up in groundwater through leaching. Arsenic concentrations in fly ash generally range from 2 to 440 mg/kg and sometimes can be as high as 1000 mg/kg (Eary et al., 1990). The concentration of arsenic in fly ash depends upon sulfur content of the coal. Arsenic is captured by calcium-bearing minerals and hematite and forms a stable complex with calcium or iron in fly ash. Arsenic is mostly present in calcium-rich fly ash as calcium orthoarsenate ($Ca_3(AsO_4)_2$) and calcium pyroarsenate ($Ca_2As_2O_7$) because calcium can react with arsenic vapor and capture the metal in water-insoluble forms as arsenate (Sterling and Helble, 2003).

1.4.4 Dietary Sources of Arsenic

Since arsenic occurs naturally, everyone is exposed to low levels of arsenic through food, water, air, and contact with the soil. About one-tenth of inorganic arsenic can be found in fish and seafood, while other foods contain the maximum level of inorganic arsenic. Inorganic arsenic intake from food in the USA is 1.3 µg/day for infants under 1 year old, 10 µg/day for 25–30 year old males, and 12.5 µg/day for 60–65 year old males (NRC, 1999b). In addition, the mean inorganic arsenic consumption for adults is 10.22 µg/day, with a standard deviation of 6.54 µg/day (MacIntosh et al., 1997).

Arsenic in plants depends mainly on soil texture. Elevated level of arsenic can be found in coarse-textured soils with less colloidal material and little ion exchange capacity, while low levels of concentration can be in fine-textured soils that are high in clay, organic material, iron, calcium, and phosphate (Flora, 2015: 23). In arsenic-contaminated soil, the uptake of arsenic by plants, particularly in vegetables and edible crops, was found to increase significantly (Larsen et al., 1992; Shrivastava et al., 2017). Dietary intake of arsenic, particularly through yams and rice, is an important exposure route that can contribute to "adverse health effects" (Schoof et al., 1998). Many researchers have highlighted the importance of rice as an exposure route (Banerjee et al., 2013; Cano-Lamadrid et al., 2016; Ciminelli et al., 2017; Gundert-Remy et al., 2015; Mondal and Polya, 2008).

It is reported that nearly half of arsenic in the human body comes from the food chain, and arsenic intake via food varies from 17 to 291 µg/day in different countries (Flora, 2015: 23). Seafood accounts for 60%–96% of the total dietary intake of arsenic, mostly in the form of arsenobetaine and arsenosugars, relatively nontoxic forms of arsenic (Delgado-Andrade et al., 2003). Other food sources are vegetables, mushrooms, grains, milk, chicken, and beef, which account for inorganic arsenic consumption (Lasky et al., 2004). In nonarsenic-endemic regions, the principal sources of inorganic arsenic in the diet are rice and chicken, which result in the accumulation (55 to 97 ng/g) of methylated arsenic compounds (Schoof et al., 1999).

In many countries, irrigation water is contaminated with arsenic that leads to oxidative decomposition of pyrite to form Fe^{+2} and Fe^{+3} sulfate and sulfuric acid, responsible for arsenic mobilization. In Southeast Asian countries, rice is grown mostly under waterlogged lowland conditions in which the physical and chemical condition of the soil affects the mobilization of arsenic and its uptake by rice (Das and Mandal, 1988). High arsenic concentrations in paddy rice has been considered an important source of arsenic intake (Bae et al., 2002; Ciminelli et al., 2017). Rice and beans contribute between 67% and 90% of the total arsenic intake from food (46%–79% from rice and 11%–23% from beans) (Ciminelli et al., 2017). People living in arsenic-contaminated areas in Bangladesh and India largely depend on rice for their calorific intake (Meharg, 2004a). The high level of arsenic concentrations in rice and beans potentially contribute to arsenic food intake in places where they are dietary staples.

1.5 MECHANISM OF ARSENIC IN GROUNDWATER

How does arsenic get into groundwater? When is arsenic dissolved in water? Has arsenic got into groundwater recently? Has water chemistry been changed by recent rapid pumping in Bangladesh to allow arsenic to enter into groundwater? These questions highlight a debate concerning the source and release mechanism of arsenic in groundwater. Arsenic is of natural and geological origin. Arsenic is thought to be closely associated with iron-oxides, and it is released from the geological strata underlying many arsenic-contaminated areas. Hydrological, geological, and geochemical studies provide a framework for understanding concentrations of arsenic in aquatic systems, which depend largely upon the pH and oxidation potential of water (Mariner et al., 1996). The most common oxidation states of arsenic in the environment are arsenite and arsenate. Arsenate is a thermodynamically stable and dominating form of the inorganic arsenic species in oxic water, whereas arsenite is the stable and dominating form of the inorganic arsenic species under reducing conditions (Ernest and Christoper, 1995; O'Neill, 1990). However, arsenate and arsenite may occur in oxidizing and reducing conditions, respectively, depending on environmental circumstances (Biswas, 2000). There are two main hypotheses concerning the release of arsenic in groundwater: the pyrite oxidation (aerobic) hypothesis and the oxyhydroxide reduction (anaerobic) hypothesis. Groundwater arsenic is found under oxidizing and reducing conditions. In oxidation condition, arsenic contamination is attributable to oxidation of arseniferous minerals, and, in reducing conditions, groundwater arsenic contamination occurs through exchange reactions with extractable arsenic associated with ferric oxyhydroxides (Wang and Mulligan, 2006).

Arsenic concentration is especially high in groundwater from pyrite-rich sedimentary aquifers. The oxidation hypothesis depends upon heavy groundwater withdrawal allowing oxygen to enter deeper water-bearing strata, thus inducing the oxidation that leaches out arsenic from arsenopyrite ores (Acharyya et al., 1999; Appelo and Postma, 1996; Das et al., 1995). The oxidation hypothesis also demonstrates that arsenic in groundwater is derived as a result of desorption and reductive dissolution of the surface reactive mineral phases such as hydrous ferric, aluminum, and manganese oxides present as coatings (disperse phase) in aquifer sediments (Nickson et al., 2000; Von Brumssen, 1999). Das et al. (1996) observed that arsenic concentrations in groundwater are from pyrite minerals containing arsenic and that bore-hole analyses show the presence of arsenic-rich iron-pyrite in sediment layers. Since iron-pyrite is not soluble in water, the question therefore arises of how arsenic from pyrites enters the water. They cited the oxidation of pyrites as the process to release arsenic into groundwater (Figure 1.3). The change of geochemical environment due to high withdrawal of groundwater for irrigation might have resulted in the decomposition of pyrites to ferrous sulfate, ferric sulfate, and sulfuric acid, and thus the arsenic in pyrites becomes available.

With excessive extraction of groundwater for irrigation, nonrecharge, fluctuating water table, and millions of boreholes caused by tubewell sinking, the aquifers have become aerated, transforming an anaerobic environment into an aerobic one. Air penetrates from the surface, oxidizes arsenopyrite, and releases arsenic into groundwater. During the wet season, the aquifers are saturated with water, so there is very little or no oxygen; therefore, there is no oxidation of arsenic. Elevated arsenic concentrations in groundwater are associated with the compaction caused by groundwater withdrawal (Welch et al., 1988). In the case of Bangladesh, changes in geochemical environment due to the high withdrawal of groundwater after 1975 at the outset of the "green revolution" resulted in the decomposition of pyrites, which led to the release of arsenic into groundwater.

Some scientists have disputed the pyrite oxidation hypothesis (Lalor et al., 1999; Mok and Wai, 1994; Nickson et al., 2000; Routh and Hjelmquist, 2011; Rowland et al., 2006), theorizing that the

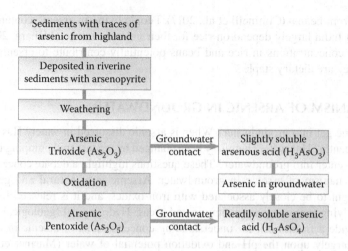

FIGURE 1.3 Arsenic transformation in groundwater through oxidation process. (After Hassan, M. M. 2003. *Arsenic Toxicity in Bangladesh: Health and Social Hazards.* Unpublished PhD Thesis, Durham: Durham University, United Kingdom.)

lowering of the water table has no role in arsenic poisoning. Such a mechanism is incompatible with the redox chemistry of water. Arsenic produced this way would be adsorbed to iron-oxyhydroxide (FeOOH), the product of oxidation (Kim et al., 2011; Mok and Wai, 1994; Thornton, 1996) rather than being released to groundwater. Nickson et al. (1998) proposed the oxyhydroxide reduction hypothesis for the mobilization of arsenic in Bangladesh groundwater. Arsenic is released when arseniferous iron-oxyhydroxides are reduced in anoxic groundwater (Bhattacharya et al., 1997; Islam et al., 2004a; Smedley and Kinniburgh, 2002; Stüben et al., 2003), a process that solubilizes iron and increases bicarbonate concentrations (Nickson et al., 2000). Sedimentary iron oxyhydroxides are known to scavenge arsenic (Mok and Wai, 1994), and, in aquifer sediments, arsenic correlates poorly with concentrations of iron. These relations strongly suggest that arsenic in groundwater beneath the Ganges plain is derived by reductive dissolution of iron oxyhydroxides in the sediment, which is known as the "oxyhydroxide reduction hypothesis" (Nickson et al., 2000). High arsenic concentrations in groundwater result from evaporative concentration and dissolution relations (or redox reaction) and are affected by adsorption (Welch and Lico, 1998). Amini et al. (2008) showed the probability of the occurrence of groundwater arsenic concentrations across the world resulting from "reducing" and "high pH/oxidizing" conditions (Figure 1.4). Using some proxy surface information, the models explained 77% of arsenic variation in reducing regions and 68% of arsenic variation in high-pH regions (Amini et al., 2008).

Arsenic is naturally present in groundwater and is freed by bacteria that break down the mineral sediments from volcanic-lithic fragments (Figure 1.5). The 16S rRNA gene sequences identification in the top soil indicated that microbes and organic matters from different sources are potential mechanism in arsenic mobilization in the aquifers (Li et al., 2014a; Mumford et al., 2012; Ye et al., 2017). Kondo et al. (1999) pointed out that mechanisms of arsenite and arsenate elution from soil involved anion exchange with hydroxide ion (OH−) and reductive labialization of arsenic through the conversion of arsenate to arsenite.

There is a line of literature that argues high arsenic contents have natural causes including processes like reductive dissolution of ferric oxide and hydroxides in Bangladesh and West Bengal in India (Nickson et al., 2000), dissolution of volcanic glass and high mobility of arsenic under high pH conditions at the Pampean region of Argentina (Smedley et al., 2005; Bhattachary et al., 2006), and recent geothermal activity in the Andes in Chile (Romero et al., 2003). Reducing groundwater in Quaternary aquifers with elevated levels of arsenic has been noted in Taiwan (Tseng et al., 1968), Vietnam (Berg et al., 2001), West Bengal of India (Bhattacharya et al., 1997), and Bangladesh

Arsenic Poisoning through Ages

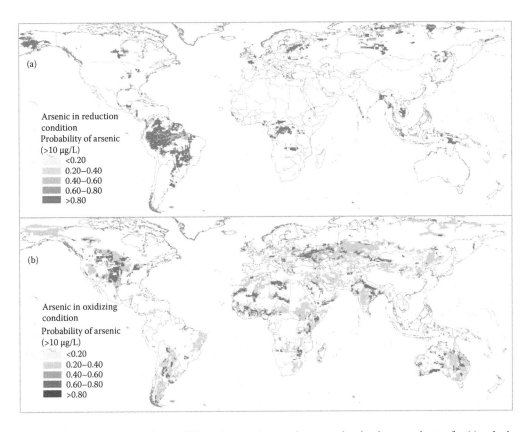

FIGURE 1.4 Modeled global probability of geogenic arsenic contamination in groundwater for (a) reducing groundwater conditions and (b) high-pH/oxidizing conditions where arsenic is soluble in its oxidized state. (Modified from Amini, M. et al., 2008. *Environmental Science and Technology* 42(10):3669–3675.)

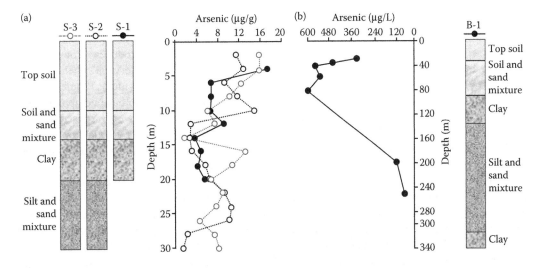

FIGURE 1.5 Lithology and arsenic concentrations with sediment depths in different boreholes: (a) three boreholes in Jianghan Plain of China and (b) Ghona union of Satkhira district of Bangladesh. (After Ye, H. et al., 2017. *Scientific Report* 7:42037 for China for (a) and author's own data from Southwest Bangladesh for (b).)

(Nickson et al., 1998; Zheng et al., 2005). Hydro-geochemical conditions leading to high concentrations of arsenic in groundwater in the Quaternary aquifers in Inner Mongolia are similar to those found elsewhere (Smedley et al., 2003). The geochemistry of arsenic is complex and is affected by dissolution-precipitation, oxidation-reduction, adsorption, and biologically mediated reactions (Varsányi et al., 1991; Welch et al., 1988). The mobility of arsenic in groundwater is closely related to the behavior of iron. During weathering of minerals containing arsenic, arsenic enters the surficial cycle mainly as soluble arsenate in which the element is present in a pentavalent state (Wilson and Hawkins, 1978). In reducing strata within the sediment, arsenic is released and diffuses upwards to the sediment-water interface where it is released into the overlying water.

1.6 HEALTH IMPACTS OF ARSENIC TOXICITY

Arsenic is found in the earth's crust at low levels (Kartinen and Martin, 1995) and is a contaminant in a wide variety of metal ores (Gochfeld, 1995). As a chemical substance, arsenic is of geological origin; it is a colorless, odorless and tasteless poison; and, to give an idea of its miniscule presence, it occurs at the equivalent of about one-third of a teaspoonful dissolved in the water of an Olympic-sized swimming pool (Meharg, 2005). It has also been known for centuries as a remedy and a murder weapon, but it releases into groundwater at unsafe levels for human consumption. Chronic exposure to arsenic is associated with a range of chronic diseases (Naujokas et al., 2013). Arsenic is also a known human carcinogen, being classified as such by the International Agency for Research on Cancer (IARC, 2012) and the US Environmental Protection Agency (EPA) (USEPA, 1988).

Arsenic is a potent carcinogen, leading to skin, bladder, liver, and lung cancers (IARC, 2012; Tapio and Grosche, 2006; Yoshida et al., 2004). Arsenic toxicity results in the formation of excess ROS, thus damaging organisms (Shi et al., 2004; Wang et al., 2001a). Arsenic is also known to cause cytotoxicity (Suzuki et al., 2007; Zhang et al., 2003) and genotoxicity (Benbrahim-Tallaa et al., 2005; Gentry et al., 2010). As little as 100 mg of arsenic trioxide can be lethal to humans (Jarup, 1992). Its toxicity to humans depends on the concentration and length of exposure. Arsenic toxicity starts in the human body when exposed to an excessive quantity of arsenic. The acute toxicity of arsenic is related to its chemical form and oxidation state. In human adults, the lethal range of inorganic arsenic is estimated at a dose of 1–3 mg/kg of arsenic. The symptoms of acute toxicity include severe vomiting, diarrhea, bloody urine, muscular cramps, gastrointestinal discomfort, convulsions, facial edema, and cardiac abnormalities (Benramdane et al., 1999; Hughes, 2002; Kamijo et al., 1998). Symptoms of acute toxicity may occur within a few minutes to hours of exposure. When arsenic is ingested in large amounts deliberately or inadvertently, it produces a constellation of severe and often fatal injuries to the cardiovascular, gastrointestinal, and nervous systems. In the acute form of effects, there is a considerable variation among different individuals. Some humans exposed to inorganic arsenic may ingest over 0.150 mg/kg/day, appearing to have no apparent ill-effects, while the characteristic signs of arsenic toxicity begin to appear to some exposed populations ingesting arsenic at oral doses of around 0.02 mg/kg/day (about 1.0 to 1.5 mg/day for an adult) (ATSDR, 2007). Doses of 0.600 to 0.700 mg/kg/day (around 50.0 mg/day in an adult or 3.0 mg/day in an infant) have caused death in some cases (ATSDR, 2007).

As a ubiquitous toxicant and carcinogenic element, groundwater arsenic is associated with a wide range of adverse human health effects (Clewell et al., 2016; Karagas et al., 2012; Kippler et al., 2016; Kumar et al., 2016a; Lin et al., 2013; Shankar et al., 2014). The noncancer toxic effects of arsenic include harm to the central and peripheral nervous systems, heart and blood vessel problems, and various skin lesions, such as hyperkeratosis as well as changes in pigmentation (Figure 1.6). It may cause birth defects and reproductive problems (NRC, 1999). The most common consequence caused by ingestion of arsenic-contaminated water is the development of various typical skin lesions like "raindrop" pigmentation on chest, back, and legs. Melanosis (known as hyperpigmentation), keratosis (a sensitive marker of arsenicosis), and hyperkeratosis (known as bilateral palmar-plantar

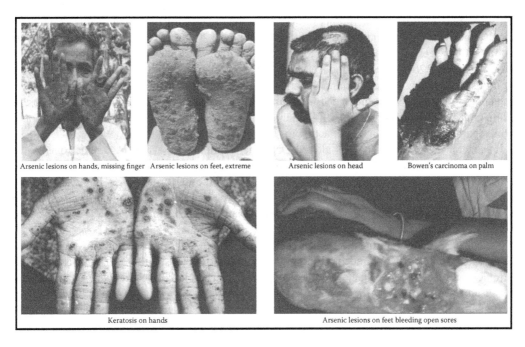

FIGURE 1.6 Chronic arsenic impacts on human health: keratosis to cancers. (From http://users.physics.harvard.edu/~wilson/arsenic/pictures/arsenic_project_pictures2.html)

thickening) are the common manifestations of chronic arsenic poisoning (Das and Sengupta, 2008; Hall, 2002; Haupert et al., 1996; Kile et al., 2011; Saha, 2003).

Skin cancer and various internal cancers, such as bladder, kidney, liver, and lung cancer, are possible with low level chronic exposure to arsenic (ATSDR, 2007). "Black Foot Disease" in the southwest coast of Taiwan, "Bell Ville Disease" in Córdoba Province of Argentina, and "Kai Dam" in Thailand are well-documented cases of health disorders due to groundwater arsenic poisoning. It is evident from Taiwan and Chile that skin cancers can appear after latency of about 10 years, while internal cancers, particularly bladder and lung, can appear after a latency of 30 years at a concentration of 50 μg/L of arsenic (Brown and Chen, 1995; Tsuda et al., 1995). Chronic exposure to elevated levels of arsenic is associated with substantial increased risk of a wide array of diseases including skin manifestations (Sarma, 2016); cancers of the lung (Sherwood and Lantz, 2016), bladder (Medeiros and Gandolfi, 2016), liver (Lin et al., 2013), skin (Fraser, 2012), and kidney (Hsu et al., 2013); neurological (Fee, 2016); diabetes (Kuo et al., 2015); and cardiovascular (Barchowsky and States, 2016) diseases. Epidemiological studies in children and adolescents have indicated a strong correlation between chronic arsenic exposure and cognitive dysfunction (Wasserman et al., 2007). Low exposures to inorganic arsenic (<50 μg/L) in drinking water can be the cause of arsenicosis symptoms for a lifetime. Morales et al. (2000) conclude that the lifetime risk of dying from cancer is 1 in 100 from consuming 50 μg/L and 1 in 50 from consuming 100 μg/L of arsenic in drinking water. Hassan et al. (2003) show that people who ingest arsenic between 10 and 50 μg/L daily are twice as likely to get arsenicosis symptoms as people who ingest at the relatively safe level (<10 μg/L).

Arsenic is genotoxic, inducing effects including deletion mutations, oxidative DNA damage, DNA strand breaks, sister chromatid exchanges, chromosomal aberrations, aneuploidy, and micronuclei (Basu et al., 2001; Hei et al., 1998; Hughes et al., 2011). Other effects of arsenic related to genotoxicity include gene amplification, transforming activity, and genomic instability (Rossman, 2003). Trivalent arsenicals, both inorganic and organic, are more potent genotoxins than the pentavalent arsenicals (Kligerman et al., 2003; Mass et al., 2001). The mechanism of genotoxic action of arsenic may result from generation of ROS, inhibition of DNA repair, and altered DNA methylation that may lead to genomic instability (Sykora and Snow, 2008). Arsenic hinders a series of gene proliferation processes and distorts signal transduction pathways (Shankar et al., 2014; Sinha et al., 2013; Wang et al., 2012). Arsenic-induced ROS has been correlated with alteration in cell signaling, apoptosis, and increase in cytokine production, leading to inflammation, which in turn results in formation of more ROS and mutagenesis, contributing to pathogenesis of arsenic-induced diseases (Eblin et al., 2006; Shankar et al., 2014). Gentry et al. (2010) highlighted the role of inhibition of DNA repair by arsenic as a mode of action for its carcinogenic effect.

Epigenetic mechanisms such as altered DNA methylation have a role in arsenic toxicity and carcinogenicity. Arsenic induces cytotoxicity by generating ROS (Selvaraj et al., 2013) and arsenic results in cytotoxicity by affecting the status of tumor-suppressor protein 53 (Huang et al., 1999; Yih and Lee, 2000). The protein 53 has a role in cell cycle regulation. Inhibition of its expression by hypermethylation of its promoter region could potentially lead to the development of cancer (Hughes et al., 2011). Genotoxicity occurs since ROS reacts with both deoxyribose and bases in DNA, causing base lesions and strand breaks. ROS are involved in oxidation of DNA, alteration of DNA repair, gene regulation mechanism, and threatening of gene stability (Shankar et al., 2014). Chronic exposure of cells to an elevated level of arsenic can result in induction of SAM depletion in cells, leading to loss of DNA methylation, and subsequently DNA hypomethylation in turn affects the genomic instability (Sciandrello et al., 2004). Enzymes involved in nucleotide excision repair (NER) and base excision repair (BER) are affected by arsenic (Hartwig et al., 2003; Schoen et al., 2004; Sykora and Snow, 2008). In their research, Andrew et al. (2003) found toenail arsenic levels to be inversely correlated with the expression of three NER genes in a small group of individuals exposed to arsenic in drinking water. This result suggests that carcinogenicity of arsenic includes inhibition of DNA repair under conditions of oxidative stress, inflammation, and proliferative signaling (Druwe and Vaillancourt, 2010; Hughes et al., 2011).

1.7 REGULATION FOR ARSENIC EXPOSURE

Arsenic is necessary to human beings, but an excess can cause harmful effects. Limits on arsenic exposure were set to avoid acute and chronic toxic effects. The first version of international standards for drinking water included arsenic in the category of toxic substances and established 200 µg/L as the allowable concentration in drinking water (WHO, 1958). In updated standards of 1963, the World Health Organization (WHO) lowered the allowable concentration to 50 µg/L (WHO, 1963). The WHO continued its review work to lower the guideline value for arsenic in drinking water by establishing a provisional guideline value of 10 µg/L in 1993 (WHO, 1993). This provisional guideline value of 10 µg/L has been adopted as the national standard for drinking water by a number of countries. However, many developing countries have retained the previous WHO guideline value of 50 µg/L as their national standard. The maximum contamination level (MCL) of 50 µg/L for some countries is grossly inadequate for protecting public health.

The arsenic limit set by Bangladesh is 50 µg/L (DoE, 1994). The WHO in 1984 issued Guidelines for Drinking Water Quality recommending a maximum value of 50 µg/L of arsenic in drinking water. However, the discovery of adverse health effects of continuous chronic exposure led the WHO to lower their recommendation to 10 µg/L in 1993 on a provisional basis (WHO, 1993). This provisional value now supersedes the "guideline value" of 1984 and is widely recommended as the permissible limit of ingesting arsenic from drinking water (Table 1.4). Tseng et al. (1996) pointed out skin pigmentation and keratosis among people who drank from arsenic-contaminated wells in Taiwan, while a very high incidence of lung, bladder, and other cancers were found in Taiwan (Chen et al., 1992) and in Chile (Smith et al., 1992). These convinced the WHO to recommend lowering the regulatory level for arsenic in water.

In the USA, the first drinking water standard was set at 50 µg/L in 1942 by the US Public Health Service (USPHS, 1943) by applying a safety factor to human toxicity data, and it did not consider the

TABLE 1.4
Current National Standards for Arsenic in Drinking Water in Some Countries

Arsenic Standard	Countries	Standard (μg/L)	Established Year
<10 μg/L	Australia	7	1996
10 μg/L	EU	10	1998
	Japan	10	1993
	Jordan	10	1991
	Laos	10	1999
	Mongolia	10	1998
	Syria	10	1994
	UK	10	2000
	USA	10	2001
	WHO	10	1993
25 μg/L	Canada	25	1999
	Mexico	25	1994
50 μg/L	Bangladesh	50	1994
	Bolivia	50	1997
	China	50	Unknown
	Egypt	50	1995
	India	50	Unknown
	Indonesia	50	1990
	Nepal	50	2003
	Philippines	50	1978
	Sri Lanka	50	1983
	Viet Nam	50	1989

Source: Ahmed, M. F. 2003. *Arsenic Contamination: Bangladesh Perspective*, ed. F.M. Ahmed. Dhaka: ITN, BUET, Bangladesh; Smedley, P. L. and D. G. Kinniburgh. 2002. *Applied Geochemistry* 17(5):517–568.

potential carcinogenicity of arsenic. Under the authority of the Safe Drinking Water Act of 1974, the USEPA issued a National Interim Primary Drinking Water Regulation for arsenic of 50 μg/L. This MCL of 50 μg/L could be a total fatal cancer risk of 1 in 100 (NRC, 1999), which does not protect public health. Therefore, the USEPA reduced the MCL to 10 μg/L in 2001. In preparing to develop an updated standard for arsenic in drinking water, the USEPA collected and compiled over 100,000 arsenic test results taken from 1980 to 1998 from more than 24,000 public water systems in 25 US states. These data reveal that arsenic in drinking water poses a significant public health risk, and over 56 million people in the 25 states consumed arsenic in water above the level of highest acceptable cancer risk (1 in 10,000). Therefore, on 22nd January 2001, the EPA issued the new "arsenic in drinking water standard" at 10 μg/L, reduced from 50 μg/L, which came into effect in 2006 to reduce the adverse health effects of arsenic poisoning (USEPA, 2001a).

The EU standard for arsenic permissible limit in drinking water is 10 μg/L, which was set in 1998. In reducing the content of arsenic in drinking water to a risk level of one in a million of lifetime risk calculated with a linear dose-response relationship, it is pointed out that the regulatory limit must be 1.5 parts per trillion (ppt), which is not attainable. The EU, therefore, enforced the MCL standard of 10 μg/L by 2004 to maintain a lifetime cancer risk level of 1 in 10,000. In the UK, the first regulatory limit of arsenic ingestion was set at 150 μg/L in 1900, and the figure was reduced threefold (50 μg/L) over the next century; by 2000 the limit was set at 10 μg/L to allow for a large safety factor (Table 1.4). An MCL of 7 µg/L is enforced in Australia, and some countries like Japan, Jordan, Laos, and Mongolia are following an MCL of 10 µg/L for minimizing health risk from arsenic poisoning (Smedley and Kinniburgh, 2002). Canada remains on 25 µg/L, while Bangladesh, China, India, Russia, and Sri Lanka are at 50 µg/L (Smith et al., 2002). The Department of Environment (DoE), Government of Bangladesh adopted the provisional value for the MCL of arsenic in drinking water (DoE, 1994).

The lack of a good animal model has forced scientists to rely on human evidence, particularly in studies of cancer in Taiwanese villagers exposed to arsenic from wells from the 1920s to the 1960s (Sun et al., 2006). Chiou et al. (2001) examined cases of urinary tract cancer in villagers exposed to arsenic levels as low as 10 µg/L to 50 µg/L. The guideline value for arsenic in potable water is based on the lifetime risk for cancer and was in 2003 lowered in Sweden from 50 µg/L to 10 µg/L (Svensson, 2007). The state of New Jersey, USA, in January 2006 lowered its guideline value from 50 µg/L to 10 µg/L in order to protect the public health (NJGS, 2006). Countries with a widespread arsenic problem have a guideline value still set at 50 µg/L, and the WHO has set the recommended Provisional Tolerably Weekly Intake to 15 µg/kg body mass, which translates to about 150 µg/day for an adult (Svensson, 2007). Establishing the drinking water limit for arsenic is not only a scientific problem but also a social problem in arsenic-prone countries.

1.8 CONCLUDING REMARKS

This chapter has mainly focused on arsenic poisoning and environmental health risk, chemistry of arsenic and its mechanism to release in groundwater, and drinking water arsenic regulatory principles. There has been an increasing interest in arsenic research over the last several years. Many empirical studies have been undertaken to explore the hydrological, geological, geochemical, and medical studies, and these provide a framework for discussing concentrations of arsenic, source of arsenic, its toxic nature, and its impact on human health. The case-control studies from Taiwan, Chile, Mexico, Argentina, India, and Bangladesh show the pattern of health problems caused by chronic arsenic ingestion. Arsenic affects multiple biological systems, sometimes years or decades after exposure reductions. The chronic ingestion of inorganic arsenic causes diseases from melanosis to cancer in terms of tracheal and bronchogenic carcinomas, hepatic angiosarcomas, and various skin cancers. It is noted that arsenic concentration is especially high in groundwater from pyrite-rich sedimentary aquifers. The aerobic hypothesis and the anaerobic hypothesis are the two established theories to explain the release of arsenic in groundwater. This chapter has also privileged the literature on arsenic speciation and different analytical methods of measuring arsenic concentrations in groundwater.

2 Groundwater Arsenic Catastrophe
The Global Scenario

> They put arsenic in his meat and stared aghast to watch him eat; they poured strychnine in his cup and shook to see him drink it up.
>
> **Alfred E. Housman**
> *English Classical Scholar and Poet, 1859–1936*

Groundwater arsenic poisoning is of increasing concern due to its high toxicity and widespread occurrence in the environment. High arsenic concentrations in groundwater have been detected in many countries around the world. In the last few decades a few major incidents of groundwater arsenic contamination have been reported in Argentina (Concha et al., 2010; Eichstaedt et al., 2015), Bangladesh (Kile et al., 2016a), Chile (Díaz et al., 2015), Xinjiang and Inner Mongolia in China (Wei et al., 2016), West Bengal in India (Chakraborti et al., 2016a), Mexico (Mendez et al., 2016), and Taiwan (Hsu et al., 2015; Tseng et al., 1968). In addition, about half of the countries around the world have arsenic contamination in their groundwater (Chakraborti et al., 2017a), and, where remediation systems are not in place, there is a hazard of poisoning. It is thought that about 300 million people worldwide are at risk of arsenicism, of whom 155 million live in the Bengal Delta, which covers Bangladesh and West Bengal of India.

Arsenic can occur in the environment in several oxidation states (−3, 0, +3, and +5), but, in natural water, it is mostly found in inorganic forms as oxyanions of trivalent arsenite, pentavalent arsenate, monomethylarsonic acid, and dimethylarsinic acid (Grund et al., 2005; Swami et al., 2014). The arsenate forms predominate in oxygenated deep-well water, but the trivalent species may be present where reducing conditions exist (Yalçin and Le, 1998). Inorganic arsenic is dissolved in groundwater and is more harmful than the organic arsenic present in food, and the trivalent forms of arsenic are more toxic than the pentavalent forms (Yamauchi and Fowler, 1994). Higher concentrations of arsenic can be found in some igneous and sedimentary rocks, particularly in iron and manganese ores. Arsenic may be released from these ores to the soil and groundwater.

High arsenic concentrations are also found in natural environments with geothermal systems (Bundschuh and Maity, 2015), sedimentary basins (Smedley and Kinniburgh, 2002), and metallic mineral deposits (Smedley et al., 2007). Arsenic is relatively soluble in hot or warm hydrothermal fluids (White, 1981) and commonly causes environmental problems downstream of hot spring systems (Axtmann, 1975). Most arsenic compounds are odorless and tasteless and readily dissolve in water. High arsenic concentrations are found in groundwater in a variety of environments, and most of the episodes are the result of natural occurrence. Anthropogenic sources of high concentration of arsenic have occasionally been identified, but these are beyond the scope of this book. There are several review works covering the arsenic-contamination scenario around the world (Herath et al., 2016; Hughes et al., 2011; Mandal and Suzuki, 2002; Murcott, 2012; Ravenscroft et al., 2009; Smedley and Kinniburgh, 2002). However, this chapter focuses on the pattern of inorganic arsenic concentrations in different countries of the globe with their sources and the health situation of the arsenic-exposed population.

2.1 WORLDWIDE ARSENIC CATASTROPHE

Arsenic is widely present around the world, particularly in soils, groundwater, and dietary staples. Occurrence of high concentrations of arsenic in drinking water has been recognized as a major public health concern in several parts of the world over the last couple of decades. A large number of aquifers are reported to be concentrated with arsenic of more than 10 µg/L. The natural biogeochemical processes and anthropogenic activities lead to the contamination of groundwater with elevated levels of arsenic. The primary source of arsenic in groundwater is predominantly natural (geogenic) and mobilized through complex biogeochemical interactions within various aquifer sediments and water (Herath et al., 2016).

Countries of particular concern in the South and Southeast Asian regions include Bangladesh, Burma, Cambodia, China, India, Nepal, Pakistan, Taiwan, Thailand, and Vietnam. Arsenic concentration in the Ganges-Meghna-Brahmaputra basin in South Asia is disastrous. Groundwater that is being used for drinking and cooking purposes in rural areas of this region has been found to be contaminated with arsenic exceeding 10 µg/L (Chakraborty et al., 2015; Maity et al., 2011). It is estimated that about 180 million people here are at risk from groundwater arsenic poisoning. Moreover, about 700,000 people have been affected with arsenic-induced chronic diseases in this region (Kim et al., 2011). Elevated levels of arsenic in groundwater of Latin American countries, mainly Argentina, Chile, and Brazil, are also documented, and cancerous health impacts of arsenic exposure are significant in these countries. There is arsenic contamination also documented in some European countries, for instance, France, Greece, Hungary, and Spain. Figure 2.1 and Table 2.1 provide an impression of the global spread of the many incidents of groundwater arsenic poisoning.

Arsenic concentrations have been reported as high as 5280 µg/L in La Pampa of Argentina (Smedley et al., 2005), 4730 µg/L in Bangladesh (Chakraborti et al., 2015), 21,000 µg/L in Antofagasta of Chile (Borgoño et al., 1977), 3700 µg/L in the South 24 Parganas of India (Chakraborti et al., 2009), 1100 µg/L in Mexico (Armienta and Segovia, 2008), and 3590 µg/L in Taiwan (Chiou et al., 2001) and higher than 100,000 µg/L in Brazil (Figueiredo et al., 2007). Arsenic exposure has been well known for many years in the area of endemic "black-foot disease" on the southwest coast of Taiwan (Blackwell et al., 1961). In Cordoba Province of Argentina, "the illness of Bell Ville" was reported as endemic arsenical skin disease, and cancer was first recognized as long ago as 1917 (Goyenechea, 1913). In Chile, almost 90% of the inhabitants of Antofagasta City were exposed to drinking water with a high arsenic content in the 1960s and 1970s (Byrne et al., 2010). Chronic arsenic poisoning was reported in Mexico between the 1960s and 1980s, and 21.61% of the exposed population showed at least one of the cutaneous signs of chronic arsenic poisoning (Espinosa et al., 2009). Although humans can be exposed to arsenic via various routes, the major source for exposure to inorganic arsenic is contaminated drinking water, especially in Bangladesh, China, India, and some regions in Central and South America.

2.2 ARSENIC IN ASIAN COUNTRIES

2.2.1 ALLUVIAL AND DELTAIC AQUIFERS, BANGLADESH

Groundwater arsenic contamination is one of the biggest environmental health disasters in Bangladesh. Groundwater in Bangladesh is the main source of public and private drinking water supplies, and presently much of that groundwater is contaminated with arsenic. Since the discovery of arsenic in 1993, by the Department of Public Health Engineering (DPHE) in Chapai Nawabganj district along the western border of Bangladesh with India, the areas identified as contaminated have been increasing at an alarming rate and the risk has spread over the whole country. Between the 1970s and 1980s, the Government of Bangladesh, along with the international aid agencies, undertook an ambitious plan to install tubewells as a means of providing an alternative for pathogen-free

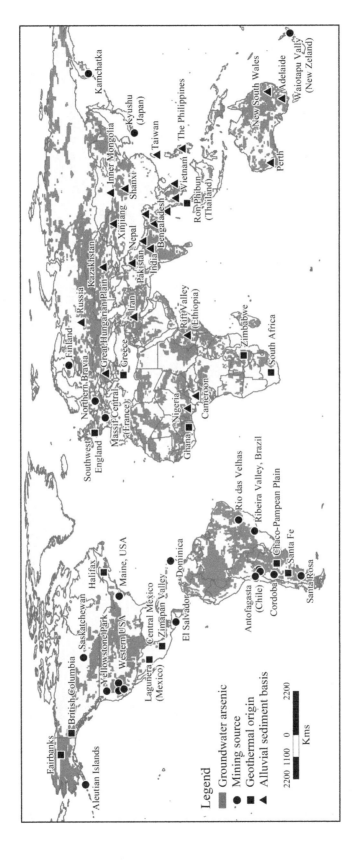

FIGURE 2.1 Geographical location of worldwide arsenic catastrophe. More than one-third of the countries in this globe are facing arsenic poisoning from their groundwater.

TABLE 2.1
Groundwater Arsenic Distribution and Problems in Some Countries Around the World

Country	Location	Maximum Concentration ($\mu g/L$)	Population Exposed[a]	Discovery Year	Environmental Conditions	References
Asian Countries						
Bangladesh	61 districts out of 64	4730	85 million	1993	Holocene alluvial sediments (floodplain and deltaic sediments), organic matter, reducing, high alkalinity, arsenic-rich sediments	Chakraborti et al. (2015); Hassan (2015); Edmunds et al. (2015); Rosenboom (2004)
Burma	Irrawaddy delta	630	3.4 million	2001	Holocene alluvial sediments, strong reducing	UNDP/UNCHS (2001); Van Geen et al. (2014)
Cambodia	Mekong delta	1610	300,000	1999	Holocene alluvial sediments, strong reducing, abundant hydrated ferric oxides, increased pH, iron phases in sediment aquifers	Berg et al. (2007); Kim et al., 2011; Phan et al. (2010, 2016)
China	Inner Mongolia, Shanxi, Xinjiang, etc.	2000	19.6 million	1960	Holocene alluvial and lacustrine aquifers, high reducing conditions, high alkalinity, high organic matter, arsenopyrite mining	Chen et al. (2017a); Guo et al. (2008); He et al. (2009); Rodríguez-Lado et al. (2013)
India	West Bengal, Bihar, Uttar Pradesh, Jharkhand, Punjab, Assam, Chattisgarh, Manipur	3880	70.5 million	1976	Holocene alluvial and deltaic sediments, oxidation of arsenic rich pyrite or anoxic reduction of ferric iron hydroxides in the sediments to ferrous iron	Chakraborti et al. (2003, 2004, 2009, 2016a); Mukherjee et al. (2006a); Nickson et al. (2007)
Iran	Kerman and Kurdistan province	1480	—	1981	Geogenic, restricted to rocks and sediments that originated from parent rocks	Chitsazan et al. (2009); Mosaferi et al. (2008)
Japan	Fukuoka; Kumamoto; Fukui; Takatsuki	25700	—	1990	Volcanic sediments, Holocene coastal sands, quaternary alluvium aquifer	Even et al., 2017; Mitsunobu et al. (2013); Tsuda et al. (1995)
Nepal	Terai region and Kathmandu Valley	2620	11 million	1999	Geogenic and the dissolution of arsenic-bearing rocks, sediments and minerals; changes in redox conditions, iron oxyhydroxides	Brikowski et al. (2014); Neku and Tendulkar (2003); NRCS (2000); Pokhrel et al. (2009)

(*Continued*)

TABLE 2.1 (Continued)
Groundwater Arsenic Distribution and Problems in Some Countries Around the World

Country	Location	Maximum Concentration (μg/L)	Population Exposed[a]	Discovery Year	Environmental Conditions	References
Pakistan	Parts of the Indus alluvial basin	2400	–	2000	Quaternary sediments, mainly of alluvial and deltaic origins; high percentage of fine to very fine sand and silt	Bibi et al. (2015); Fatmi et al. (2009); Rasheed et al. (2017); Rehman et al. (2016)
Taiwan	Southwest coast and Lanyang basin of the Northeast Taiwan	3590	200,000	1961	Arsenic-dissolved organic matter, groundwater-associated sediments, strongly reducing, artesian conditions, pyretic material or black shale	Blackwell et al. (1961); Cheng et al. (2017); Lin et al. (2006); Tseng et al. (1968)
Thailand	Ron Phibun District, Nakon Si Thammarat Province	5114	150,000	1988	Dredge quaternary alluvium, tin mining from bed-rock and alluvium, oxidation of disseminated arsenopyrite due to mining	Foy et al. (1992); Kim et al. (2011); Pechrada et al. (2010)
Turkey	Western Anatolia (Simav Plain), Sandıklı basin	7700	–	2003	Quaternary alluvium, tertiary limestone, volcanic sediments, sulfide-oxidation	Çolak et al. (2003); Davraz (2015); Gunduz et al. (2010)
Vietnam	Red River Delta and Mekong Delta	3050	10 million	1999	Pleistocene and Holocene sediments; strongly reducing, high alkalinity, arsenic-rich sediments	Berg et al. (2007); Nguyen and Itoi (2009); Stanger et al. (2005)
North America						
Canada	Nova Scotia, Saskatchewan, Ontario, British Columbia	3900	–	1937	Sulfide mineralization in volcanic rocks, high pH in groundwater, sorption to iron oxides, weathering and erosion of arsenic-containing rocks and soil	Bondu et al. (2017); Boyle et al. (1998); Dummer et al. (2015); Kwong et al. (2007); Wyllie (1937)
Mexico	Región Lagunera, Valle del Guadiana, Valle de Zimapán	24000	2 million	1958	Volcanic sediments, oxidation of sulfide, high pH, limestone aquifer, oxidation of arsenopyrite, dissolution of scorodite	Boochs et al. (2014); Cebrián et al. (1994); Del Razo et al. (1990); Rodríguez et al. (2004)
USA	Alaska, Arizona, California, Hawaii, Idaho, Nevada, Oregon, Utah, and Washington	6000	–	1937	Holocene and older basin-fill sediments, high pH, up-flow of geothermal water; dissolution of or desorption from iron oxide; dissolution of sulfide minerals	Luczaj and Masarik (2015); Flanagan et al. (2014); Southwick et al. (1983); Welch et al. (1999)

(Continued)

TABLE 2.1 (Continued)
Groundwater Arsenic Distribution and Problems in Some Countries Around the World

Country	Location	Maximum Concentration (μg/L)	Population Exposed[a]	Discovery Year	Environmental Conditions	References
South America						
Argentina	Chaco-Pampean Plain, Córdoba, Salta, Jujuy, La Pampa, Santa Fe	15000	2 million	1955	Tertiary-quaternary volcanic deposits, post-volcanic geysers and thermal springs, excessive irrigation strongly affects local geochemical and hydrochemical conditions	Auge, (2014); Nicolli et al. (1989, 2012); Panigatti et al., 2014; Robles et al. (2016)
Brazil	Ribeira Valley, Amapá State, Rio das Velhas, Minas Gerais, Rondonia State	100,000	–	–	Sulfide-rich gold-bearing rocks that constitute the aquifers	Bidone et al. (2016); Ciminelli et al. (2017); Figueiredo et al. (2007); Matschullat et al. (2000)
Chile	Antofagasta, Atacama, Calama and Tocopillae	27000	500,000	1959	Quaternary volcanogenic sediment, oxidizing, arid conditions, high salinity	Borgoño et al. (1977); Corradini et al. (2017); Herrera et al. (2014); Romero et al. (2003)
Europe						
France	Tinée and the Vésubie valleys, Vosges and the Pyrenees	263	17,000	1996	Sedimentary basin, oxidation, ore deposits containing arsenopyrite	Barats et al. (2014); Drouhot et al. (2014); Saoudi et al. (2012)
Finland	Pirkanmaa, Satakunta, Lapland	2230	–	1967	Alkaline volcanic rocks, mining	Jarva et al. (2008); Kurttio et al. (1999); Sorvari et al. (2007)
Germany	Triassic Keuper sandstones of Bavaria, Saxony, Wiesbaden	550	–	1992	Alluvium sediments, mineralized sandstone	Heinrichs and Udluft (1999); Schwenzer et al. (2001)
Greece	Chalkidiki and Thessaloniki of Northern Greece, Thessaly	1843	150,000	–	Thermal springs, mineralization, mining	Golia et al. (2015); Kelepertsis et al. (2006); Kouras et al. (2007)
Hungary	Great Hungarian Plain, Pannonian Basin	4000	400,000	1941	Quaternary alluvial sediments with the River Danube in the Great Hungarian Plain, reducing, humic acid	Nagy and Korom, (1983); Sugár et al. (2013); Varsányi (1989)

(Continued)

TABLE 2.1 (*Continued*)
Groundwater Arsenic Distribution and Problems in Some Countries Around the World

Country	Location	Maximum Concentration (μg/L)	Population Exposed[a]	Discovery Year	Environmental Conditions	References
Italy	Lombardia, Emilia-Romagna, and Veneto	1300	–	–	Shallow groundwater, hydrothermal, geothermal arsenic is common around the volcanic centers	Giuliano et al. (2005); Tamasi and Cini (2004); Vivona et al. (2005)
Romania	Danube basin, Transylvania region	176	36,000	1992	Quaternary alluvial sediments, strong reducing	Gurzau and Gurzau (2001); Lindberg et al. (2006)
Spain	Madrid Basin, Duero Basin	613	50,000	1998	Alluvial sediments; organic-rich; mineralization; associated with particular stratigraphic horizons, such as the Middle Miocene organic-rich Zaratan facies	García-Sánchez et al. (2005); Gómez et al. (2006); Sanz et al. (2001)
Sweden	Skellefteå	300	–	–	Alkaline volcanic rocks, glacial till	Svensson (2007)
Switzerland	Valais, Ticino, and Graubünden	170	14,000 in Valais	1989	Ore deposits and sediments, arsenic-bearing ferruginous limestones and ferruginous red clays	Haldimann et al. (2005); Pfeifer et al. (2010); Zayre et al. (2006)
UK	Midlands, Cornwall, Liverpool, Northwest England	355	–	–	Limestone, sandstone, estuarine alluvium, mining, alluvial or glacial aquifers	Edmunds et al. (1989); Middleton et al. (2016); Millward et al. (1997)
African Countries						
Barkina Faso	Mogtédo, Ouahigouya	1630	560,000	1979	Gold mineralization in Birimian (Lower Proterozoic) volcano-sedimentary rocks	Bretzler et al. (2017); Ravenscroft et al. (2009); Smedley et al. (2007)
Cameroon	Ekondo Titi	2000	4000	1986	Strongly reducing groundwater, reductive dissolution	Mbotake (2006); Ravenscroft et al. (2009)
Ethiopia	East African Rift Valley	96	–	–	Arsenic is enriched in hot springs, hydrothermal	Reimann et al. (2003)
Ghana	Wassa West, Obuasi, Accra, Bolgatanga, Brong-Ahafo	4500	–	–	Gold mining, ambient pH, redox conditions, sulfide oxidation and sorption	Bowell (1994); Buamah et al. (2008); Smedley (1996)

(*Continued*)

TABLE 2.1 (Continued)
Groundwater Arsenic Distribution and Problems in Some Countries Around the World

Country	Location	Maximum Concentration (µg/L)	Population Exposed[a]	Discovery Year	Environmental Conditions	References
Nigeria	Warri-Port Harcourt, Ogun State, Kaduna	750	—	—	Alluvial sediments, strongly reducing, slightly acidic	Edet and Offiong (2003); Gbadebo and Mohammed (2004); Oke (2003)
Oceania						
Australia	Victoria region, New South Wales	300,000	—	1976	Pyritic sediments, hydroxides and Fe oxyhydroxides, gold mining	Appleyard et al. (2006); Hinwood et al. (1999); Smith et al. (2003)
New Zealand	Waiotapu Valley, Rarangi Aquifer near Marlborough	8500	—	1939	Alluvial aquifers, reduced groundwater, geothermal water	Grimmett and McIntosh (1939); Webster and Nordstrom (2003)

[a] Exposed refers to population drinking water with country-wide permissible limit for inorganic arsenic, e.g., Bangladesh >50 µg/L, China >10 µg/L.

safe drinking water to the rural population to prevent diarrheal and other water-borne diseases (Atkins et al., 2007a). Approximately 97% of rural people of Bangladesh now have access to bacteriologically clean water through the installation of about 12 million tubewells. It is ironic that so many tubewells were installed in recent times for pathogen-free drinking water but that water is now contaminated with toxic levels of arsenic.

In a country with regular calamities (floods, cyclones, tidal surges, drought, famine, and diseases), groundwater arsenic poisoning presents a new dimension of hazard. It was estimated in 2002 that Bangladesh had about 11 million tubewells serving a population of around 133 million (World Bank, 2005). The likelihood of extensive contamination was confirmed in 1995 when chronic arsenicosis was first diagnosed by health professionals. In 1996, arsenic contamination was found in 7 districts, but this increased to 48 districts by mid-1997, and knowledge of contamination with elevated levels ($\geq$50 µg/L) has now extended to 61 districts out of 64. This high level of groundwater arsenic across the country is a hazard to human health and millions of people may die or suffer from the very serious consequences of consuming arsenic.

Arsenic contamination is especially frequent in the shallow aquifers. Until January 1997, over 3000 water samples from 27 districts in Bangladesh had been analyzed, and this revealed that 38% contained arsenic of more than 50 µg/L (Dhar et al., 1997). Furthermore, Ahmad et al. (1997) analyzed 294 tubewell water samples and found 29% samples to be contaminated at levels above 50 µg/L. Between September 1996 and June 1997, Biswas et al. (1998) looked at all 265 tubewells in the village of Samta in Jessore district and found that 91% contained arsenic at levels of more than 50 µg/L. In March 1998, it was reported that 46.50% of 9024 wells were above 50 µg/L and 9.79% wells were higher than 500 µg/L (Mandal et al., 1999). In continuing surveys of 42 districts, Chowdhury et al. (2000) reported the analysis of 10,991 water samples, of which 59% contained arsenic levels above 50 µg/L (Figure 2.2).

More than a decade ago, some 50,515 hand-pump tubewells from all 64 districts were analyzed, and 43% of the samples had arsenic of more than 10 µg/L, 27.5% of more than 50 µg/L; some 274 samples contained arsenic of more than 1000 µg/L, and the highest concentration was an astonishing 4730 µg/L (Chakraborti et al., 2004). In the Chiladi village of Noakhali district, 100% of tubewell water samples contained arsenic concentrations of more than 50 µg/L, 94% had more than 300 µg/L and 28% had more than 1000 µg/L (Chakraborti et al., 2002). Van Geen et al. (2007) found an elevated level (>50 µg/L) of arsenic in one-quarter and 10 µg/L in half of the 6000 tubewells looked at in 2000–2001 in Araihazar upazila near Dhaka. Hassan et al. (2003) analyzed 375 tubewell water samples in southwest Bangladesh (Satkhira district) and found 95% of the samples contaminated with an elevated level of arsenic. Hassan (2015) analyzed a total of 5594 tubewell water samples from Bagerhat, Khulna, and Satkhira districts and found that almost 87.7% tubewells fail to retain the World Health Organization (WHO) standard and 67.5% failed to maintain the Bangladesh standard permissible limit of 50 µg/L. Halim et al. (2010) found concentrations of dissolved arsenic in 38 shallow well water samples from eastern Bangladesh, ranging from 8.05 to 341.5 µg/L, with an average concentration of 95.14 µg/L, and 97% of the samples exceeded 10 µg/L. Islam et al. (2017) analyzed 47 groundwater samples from a shallow aquifer from Rangpur district and found arsenic concentration in the range of 0.5–42.8 µg/L, with the mean concentration of 8.80 ± 10.01 µg/L. Analyzing a total of 70 preselected groundwater sample points from shallow wells, deep wells, and dug wells at the Sadar upazila of Lakshimpur district, Bhuiyan et al. (2016) found arsenic concentration in the range between below detection level (BDL) and 404 µg/L, with the mean concentration of 85.54 ± 98.64 µg/L, and found that 10 samples retain the WHO limit and 50 samples retain the Bangladesh standard limit of 50 µg/L (Table 2.2).

The DPHE/United Nations Children's Fund (UNICEF) analyzed groundwater samples from 15 upazilas in southern Bangladesh more than a decade ago and found 66% of the analyzed 316,951 tubewell samples had an arsenic concentration of more than 50 µg/L, and all the tubewells in 574 villages had readings exceeding 50 µg/L (Rosenboom, 2004). In a random national survey of arsenic in groundwater using laboratory data for 3208 groundwater samples from the shallow Holocene

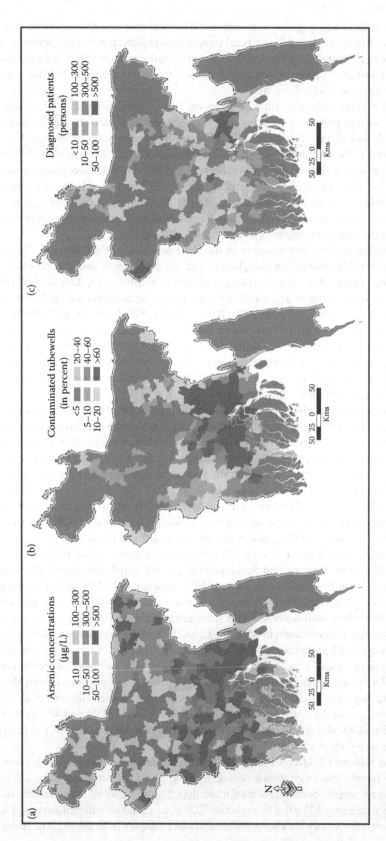

FIGURE 2.2 Groundwater arsenic concentrations and arsenic-affected patients in Bangladesh: (a) spatial pattern of arsenic concentrations—most of the areas in southern part of Bangladesh and some areas in north-eastern Bangladesh are found to be severely contaminated with arsenic, while the northern part of Bangladesh and hilly districts in south-eastern Bangladesh are found to be safe; (b) percent of contaminated tubewells—a significant number of contaminated tubewells are located in the central part of Bangladesh; and (c) arsenic-affected patient diagnosed—the maximum number of patients diagnosed with arsenic poisoning are in the areas of lower-right corner of central Bangladesh and areas in western part of Bangladesh.

TABLE 2.2
Arsenic Concentrations in Bangladesh in Different Research as well as Impacts of Arsenic on Human Health and Its Mitigation

References	Area	Sampling	Samples	Arsenic concentrations (µg/L)				Health Impacts	Mitigation
				Maximum	≤10	≤50	≥50		
Argos et al. (2014)	Araihazar, Narayangonj	Random	6000	–	50.0%	25.0%	25.0%	Malignant and nonmalignant lung disease	–
BGS/DPHE (1999)	Bangladesh	Random	2022	4727	–	65.1%	34.9%	Malignant and nonmalignant diseases	Policy and safe water options
BGS/DPHE (2001)	Bangladesh	Random	3208	2300	46.0%	27.0%	27.0%	Malignant and nonmalignant diseases	Policy and safe water options
Bhuiyan et al. (2016)	Lakshimpur Sadar upazila	Random	70	404	14.3%	71.4%	14.3%	–	–
Biswas et al. (1998)	Samta, Jessore	Census	265	–	–	9.0%	91.0%	Skin lesions, malignant and nonmalignant diseases	–
BRAC (1998)	Hajiganj	Census	11,954	–	–	7.2%	92.8%	Skin lesions	–
BUET/NEMIP (1997)	Northeast Bangladesh	Random	1210	268	–	66.8%	33.2%	–	Technological options
Chakraborti et al. (2004)	Bangladesh	Random	50,515	4730	43.0%	29.5%	27.5%	Skin lesions, malignant and nonmalignant diseases	Safe water options
Chakraborti et al. (2015)	Bangladesh	Random	54,000	4730	56.6%	15.9%	27.5%	Skin lesions, malignant and nonmalignant diseases	Policy
Chowdhury et al. (2000)	42 Districts	Random	10,991	–	–	41.0%	59.0%	Skin lesions	–
Das et al. (2009)	Bangladesh	Random	8000 villages	–	–	20.0%	80.0%	Skin lesions	–
DCH/SOES (1997)	Bangladesh	Patient affected	3673	4727	–	62.1%	37.9%	Malignant and nonmalignant diseases	Safe water options
Dhar et al. (1997)	27 districts	Random	3000	–	–	62.0%	38.0%	Malignant and nonmalignant diseases	–
Didar-Ul-Islam et al. (2017)	Patuakhali district	Random	18	19.24	77.8%	22.2%	–	–	–
DPHE (1996)	Bangladesh	Selected	2464	2500	–	73.7%	26.3%	–	Safe water options

(*Continued*)

TABLE 2.2 (Continued)
Arsenic Concentrations in Bangladesh in Different Research as well as Impacts of Arsenic on Human Health and Its Mitigation

References	Area	Sampling	Samples	Arsenic concentrations (μg/L)				Health Impacts	Mitigation
				Maximum	≤10	≤50	≥50		
Halim et al. (2010)	Eastern Bangladesh	Random	38	341.5	97.0%	–	3.0%	Skin lesions (melanosis and keratosis, pigmentation)	–
Hassan et al. (2003)	Ghona, Satkhira	Census	375	600	–	3.5%	96.5%	Skin lesions (melanosis and keratosis, pigmentation, gangrene)	Policy
Hassan (2015)	Jamira (Khulna); Magura (Satkhira); Dhopakhali and Gazalia (Bagerhat)	Census	5594	1500	12.3%	20.2%	67.5%		Arsenic-safe water table identification and policy
Islam et al. (2017)	Rangpur district	Random	47	42.8	21.3%	78.7%	–	–	–
Nickson et al. (1998)	Bangladesh	Random	46	335	–	69.6%	30.4%	–	–
Pesola et al. (2015)	Araihazar, Narayangonj	Random	6000	–	50%	25%	25%	CVD mortality	–
Rahman et al. (2015)	Matlab, Chandpur	Census	13,286	3644	39.2%	10.7%	50.1%	Prenatal arsenic exposure and drowning in children under 5	–
Rosenboom (2004)	15 Upazila in Southern Bangladesh	Census	316,951	–	–	34.0%	66.0%	Malignant and nonmalignant diseases	–
Van Geen et al. (2007)	Araihazar, Narayangonj	Random	6000	–	50.0%	25.0%	25.0%	–	Mitigation technologies
Van Geen et al. (2016)	Araihazar, Narayangonj	Census	48,790	–	44.0%	10.0%	46.0%	Intellectual function in 6-year-old children	Policy
Wasserman et al. (2007)	Araihazar, Narayangonj	Random	6000	–	50.0%	25.0%	25.0%		–
World Bank (2007)	Bangladesh	Bangladesh	3.04 million	–	–	71%	29%	Skin lesions (melanosis and keratosis, pigmentation)	Safe water options
Yunus et al. (2014)	Laksam, Comilla	Census	–	–	–	–	–	Skin lesions in women	–

aquifer (<150 m depth), it was reported that some 27% of shallow wells in Bangladesh contained more than 50 µg/L of arsenic, 46% exceeded 10 µg/L, and in some districts of southwest Bangladesh more than 90% of the tubewells were found to be affected (BGS/DPHE, 2001). It is noted that the Bangladesh Water Supply Arsenic Mitigation Project (BAMWSP) analyzed 3.04 million tubewell water samples in 190 upazila nationally using field testing kit (FTK) methods of which about 20% had concentration of arsenic above 50 µg/L (World Bank, 2007) and about 80% of the tubewells were found to be contaminated in some 8000 villages nationwide (Das et al., 2009). Analyzing a total of 54,000 tubewell water samples from all over the country, Chakraborti et al. (2015) found that about 56.6% of the water samples retain the WHO standard, 15.9% samples maintain arsenic within the range of 10–50 µg/L, and 27.5% water samples fail the Bangladesh standard limit as well as 7.5% tubewells containing arsenic at more than 50 µg/L.

Arsenic concentrations also vary with aquifer depths, and evidence shows that only 1% of tubewells deeper than 200 m have arsenic above 50 µg/L and only 5% exceed 10 µg/L of arsenic (BGS/DPHE, 2001). At depths of 12–24 m, the situation is bad (van Geen et al., 2003a); Harvey et al. (2006) have shown a distinct peak at approximately 30 m depth. Jakariya et al. (2009) also found a robust association between arsenic concentrations and aquifer depth between 30 m and 76 m. However, Nath et al. (2009) demonstrated a decreasing trend of arsenic concentration with depth, except for a slight increase observed between 70 and 80 m depth. Dug wells generally, but not always, have low arsenic concentrations, typically of less than 10 µg/L, even in areas of severe contaminated groundwater. In western Bangladesh, a tubewell having 30 m depth with around 2300 µg/L of arsenic is located just a few meters from a dug well of 8 m with a concentration of <4 µg/L (World Bank, 2005). Wells in the older Plio-Pleistocene sediments of the Barind and Madhupur tracts have low arsenic concentrations, and this is true also of groundwater in the north of the country.

Arsenic contamination of dietary staples is of recent concern in Bangladesh (Ahmed et al., 2016; Dittmar et al., 2010; Islam et al., 2016; McClintock et al., 2016). Rice, the staple food, shows high levels of arsenic above 1700 µg/kg in Bangladesh (Meharg and Rahman, 2003). Rice accumulates 10-fold higher inorganic arsenic than other grains (Davis et al., 2017). Smith et al. (2006a) reported that the mean total arsenic content in 46 rice samples was 358 µg/kg. In analyzing the total arsenic content of 150 paddy rice samples collected from different districts of Bangladesh, Duxbury et al. (2003) found the mean concentrations for Boro and Aman rice to be 183 µg/kg and 117 µg/kg, respectively. Das et al. (2004) found a range between 40 and 270 µg/kg. Williams et al. (2006) detected inorganic arsenic with a small percentage of dimethylarsenic acid in Bangladeshi rice, and they stated the mean level to be 130 µg/kg. Meharg et al. (2008) reported that 46% of brown rice in Bangladesh contains inorganic arsenic and 28% has some organic arsenic. Alam et al. (2003) found a range of arsenic in vegetables sampled in Samta village in Jessore of between 19 and 489 µg/kg. Smith et al. (2006a) measured a mean of 333 µg/kg in 39 vegetable samples, and the results of Das et al. (2004) were between 70 and 3990 µg/kg.

Arsenic-related health problems were first diagnosed in Bangladesh in 1987. Between 1993 and 1998, about 25 million were at risk of arsenic poisoning, and 20.6% out of 17,896 people were diagnosed with arsenicosis (Tondel et al., 1999). Over a decade ago, it was estimated by the BGS/DPHE (2001) that up to 35 million people were drinking groundwater with arsenic concentrations above 50 µg/L and up to 57 million were drinking water with more than 10 µg/L. Now the figure of risk of arsenic poisoning has increased to 85 million at levels over 50 µg/L, and, within the BAMWSP Project, some 40,000 patients were diagnosed across the country with arsenicosis symptoms (UNICEF, 2008). Skin disorders including hyper/hypopigmentation changes and keratosis are the most common external manifestations, although skin cancer has also been identified. Chakraborti et al. (2004) found that 83% of the analyzed 4536 hair samples, 93% of 4471 nail samples, and 95% of 1586 urine samples were contaminated with dangerous levels of arsenic. Arsenic-induced carcinogenic and noncarcinogenic malignancies in Bangladesh have been recognized in recent literature (Argos et al., 2014; Farzan et al., 2015; Jiang et al., 2015; Yunus et al., 2014, 2016; Pesola et al., 2015; Rahman et al., 2015). Apart from the health hazard, arsenic poisoning is causing widespread social problems. These include disrupted family life, the difficulty of arranging daughters' marriages, a lack of job offers for arsenic-affected candidates, and so on (Brinkel et al., 2009; Hassan et al., 2005). There is a tendency for unaffected people to maintain a safe distance from arsenic-affected people since they think that arsenicosis is like leprosy or another contagious disease. In rural Bangladesh, the people or communities affected by arsenicosis become isolated.

Arsenic distribution in Bangladesh is tremendously patchy because of the variation in sedimentary characteristics (Smedley and Kinniburgh, 2002). Arsenic in groundwater of Bangladesh comes from arsenic-rich Holocene sediments in the Ganges-Brahmaputra-Meghna river system, deposited over thousands of years along with the sands and gravels that comprise much of the surface layers. Groundwater from the Holocene aquifers contains arsenic at concentrations of up to 2300 μg/L (BGS/DPHE, 2001). A large range in the relative proportions of dissolved arsenate and arsenite exists in the groundwater of Bangladesh, but Zheng et al. (2005) found a high arsenic concentration with a strong ($\geq 70\%$) dominance of arsenite. Moreover, the distribution of arsenic in shallow aquifers in Bangladesh is thought to be controlled by the local hydrogeology (Edmunds et al., 2015; Harvey et al., 2006; Polizzotto et al., 2008). Preventing arsenic exposure by supplying safe water is the most vital mitigation approach (Mosler et al., 2010). Installation of deep tubewells is providing arsenic-safe water to about 12% of the affected population all over the country. Other approaches—arsenic-iron removal (AIRP), dug wells, pond sand filtration (PSF), rainwater harvesting, and piped water system—are so far reaching less than 1% of the population (Ahmed et al., 2006).

2.2.2 IRRAWADDY DELTA, BURMA

There are traces of arsenic in the Quaternary aquifer of the Irrawaddy delta in Burma, the Ayeyarwaddy Division, Nyaungshwe in the Shan State in southern Burma, Sittway township in the western coastal area, and Hinthada and Kyaungkone townships close to the south coast (Bacquart et al., 2015; Cha et al., 2016; UNDP/UNCHS, 2001; Van Geen et al., 2014). The traditional sources of water for domestic supply are dug wells, ponds, springs, and rivers, but many of these water sources have been superseded since 1990 by the development of shallow tubewells. It is estimated that more than 400,000 wells exist in the country as a whole and more than 70% of them are privately owned tubewells.

Little testing for arsenic in groundwater has been carried out so far in wells using the alluvial aquifer, but in an analysis of 1912 shallow tubewells in four townships in Ayeyarwaddy, Save the Children found that 22% of samples exceeded 50 μg/L of arsenic (World Bank, 2005). The United Nations Development Programme (UNDP) and the United Nations Centre for Human Settlements (UNCHS) detected arsenic at concentrations over 50 μg/L in 4% of 125 samples analyzed from Nyaungshwe in Shan State (UNDP/UNCHS, 2001). The Department of Medical Research, in collaboration with the Department of Health in Burma, conducted a small-scale health survey in February-March 2002 covering 15 villages in Thabaung and 10 villages from Kyonpyaw townships (Tun, 2002). A total of 99 well water samples from Thabaung and 74 household-water samples from Kyonpyaw were collected, and 66.6% and 36.5% from wells were found to have arsenic above 50 μg/L in Thabaung and Kyonpyaw, respectively (Tun, 2002). Van Geen et al. (2014) analyzed arsenic concentrations in 55 wells across seven villages in the lower Ayeyarwaddy basin and found elevated concentrations of arsenic (50–630 μg/L) in wells up to 60 m depth as well as low arsenic concentrations (<10 μg/L) in some shallow (<30 m) grey sands and in both shallow and deep orange sands.

The Water Resources Utilization Department (WRUD) of Burma carried out a survey of groundwater in Sittway, Hinthada, and Kyaungkone townships. In Sittway, WRUD (2001) used E-Merck field testing kits to check 58 shallow tubewells (STW) and 23 deep tubewells; they found that more than 70% of the STW contained arsenic of more than 10 μg/L, while only 4% of deep tubewells were found to be contaminated at the WHO guideline value. Salinity problems occur in some groundwater and surface waters in Sittway, and most tubewells are less than 15 m deep as a result. In Hinthada, some 31.6% of the sample tubewells were contaminated at the WHO standard and 13.3%

were at the Burma standard value of 50 µg/L. In Kyaunkone, the figure was 29.4% for 50 µg/L (WRUD, 2001). Well depths are typically around 30 and 50 m in Hinthada, although some deeper tubewells are between 55 and 70 m, and only one of these was found to be seriously contaminated.

2.2.3 Mekong Delta, Cambodia

Groundwater arsenic contamination in Cambodia was first investigated by JICA (Japanese International Cooperation Agency) in 1999 (Phan et al., 2010). High concentrations were found in the lower alluvial delta of the Mekong River and in the Kandal and Prey Veng Provinces (Kim et al., 2011; Kocar et al., 2008; Luu et al., 2009; Phan et al., 2016; Rowland et al., 2008). Groundwater from private tubewells remains popular in rural areas and in some small urban centers. More than 81% of the population in rural areas and about 60% in urban areas use groundwater; in contrast, only 15% of the population of Phnom Penh consumes well water (Sthiannopkao et al., 2008). The Government of Cambodia, with support from the WHO, conducted a survey of drinking water quality throughout the country in 2000. Analyzing a total of 88 groundwater samples, 9% were found to contain arsenic at more than 10 µg/L (Berg et al., 2007). In 2002 and 2003, some 5000 tubewells were analyzed in the northern part of the Kandal province (where several readings exceeded 500 µg/L) by 25 NGOs with field-testing kits provided by UNICEF (Halperin, 2003), and their study shows that 20% of the wells had arsenic levels above 50 µg/L and 50% were above 10 µg/L.

Buschmann et al. (2008) found the mean arsenic content of groundwater in Kandal province to be 233 µg/L, with the highest concentration of 855 µg/L. Kocar et al. (2008) reported that dissolved arsenic concentrations varied spatially, ranging up to 1300 µg/L in the groundwater aquifer and up to 600 µg/L in surficial clay pore water. Sthiannopkao et al. (2008) analyzed total arsenic concentrations in Prey Veng and Kandal Provinces and found 21 to 907 µg/L tubewell water in Kandal and from ND (nondetectable) to 691 µg/L in Prey Veng; some 54% of samples contained arsenic concentrations exceeding 10 µg/L. Analyzing 30 groundwater samples in Kandal province, Sthiannopkao et al. (2010) further found arsenic concentrations ranging between 5 and 1543 µg/L, with a mean of 454 µg/L. Taking samples from six villages in Kandal, they found the highest arsenic concentration in Prek Thom village (1068 µg/L) followed by Doun Sor (607 µg/L), Chounlork (377 µg/L), Tuol Tnort (271 µg/L), Phoum Thom (213 µg/L), and Poul Peal Ker (164 µg/L). The release of arsenic into the groundwater is seen here to be the result of microbial reductive dissolution of hydrated ferric oxides (Polizzotto et al., 2008).

Analyzing some 131 tubewell water samples collected from a rural floodplain in the south of the capital Phnom Penh, Buschmann et al. (2007) found arsenic concentrations ranging between 1.0 and 1340 µg/L, with a mean concentration of 163 µg/L, and 48% of the tubewells exceeded 10 µg/L. Berg et al. (2007) discovered groundwater arsenic concentrations between 1 and 1610 µg/L, with an average of 217 µg/L. Buschmann et al. (2007) looked at the variable concentrations in the Mekong delta and found mean values of 150 µg/L, with a standard deviation of 276 µg/L. Some 40% of their samples were found to contain over 50 µg/L, and 49% had over 10 µg/L. Elevated arsenic is found both in shallow water tables and in aquifers between 100 and 120 m depth in lower Mekong in Cambodia. In the lower Mekong delta, only 5.7% of all groundwater samples exceeded 50 µg/L, while 12.9% exceeded 10 µg/L (Stanger et al., 2005). The occurrence of elevated arsenic concentrations is sharply restricted to the close vicinity of the Bassac and Mekong River and the alluvium braided by these rivers. Analyzing some 90 samples, Buschmann et al. (2007) found high arsenic levels on the banks of the rivers Bassac and Mekong that averaged 232 µg/L, with a median value of 100 µg/L, while concentrations to the west and east of the rivers were below 10 µg/L (9 µg/L west of Bassac and 3 µg/L east of Mekong).

The population at risk of chronic arsenic poisoning is estimated to be 0.5–1.0 million in the Mekong delta. Gault et al. (2008) claim that arsenic in groundwater for drinking is positively correlated with both nail and hair samples in Cambodia. The mean was found to be 2.58 µg/g by Agusa et al. (2007) and 2.0 µg/g by Berg et al. (2007). Sthiannopkao et al. (2010) found high concentrations in 58 hair samples from the people of six villages in Kandal province, ranging between 0.06 and 30 μg/g, with a mean of 3.2 μg/g, while the mean concentration for 10 hair samples in Phnom Penh city was 0.13 μg/g. Analyzing urine samples by inductively coupled plasma mass spectrometry (ICP-MS), Phan et al. (2014) investigated 127 arsenicosis patients that had high levels of total arsenic in urine ranging from 3.76 to 373 μg/L, with the mean concentration of 78.7±69.8 μg/L, who showed skin signs of arsenicosis in the exposed population.

High arsenic concentrations (>100 μg/L) were found in areas of young alluvial sediments along the Mekong River, and this indicates the strong geological control of groundwater arsenic in Cambodia (Lawson et al., 2016). Arsenic concentrations in Cambodia are highest in reducing waters from Holocene sediments and are correlated with increasing pH. The young alluvium sediments in the Mekong floodplain have abundant hydrated ferric oxides in which arsenic has been sorbed (Buschmann et al., 2007). There is a relation between surface permeability, local recharge, and arsenic concentrations in shallow aquifers in the Mekong delta (Richards et al., 2017; Robinson et al., 2009).

2.2.4 Alluvial Plains, China

Elevated levels of arsenic concentrations in groundwater and associated health problems have been identified in different parts of China, especially in the provinces of Xinjiang, Inner Mongolia, Shanxi, Jilin, Ningxia, Anhui, Gansu, Heilongjiang, Henan, Hunan, Jiangsu, Liaoning, Sichuan, and Yunnan, as well as in certain suburbs of Beijing (Gao et al., 2017; Pi et al., 2015; Ren et al., 2016; Zhang et al., 2017b; Zhou et al., 2017). Arsenic-contaminated drinking water from groundwater sources was first recognized in the 1960s, and about 19.6 million people are at risk of being affected by the consumption of arsenic-contaminated groundwater (Rodríguez-Lado et al., 2013). The first area of endemic arsenicosis was discovered in Kuitun of Xinjiang province in 1983. In Kuitun, villagers used to drink river water containing a very low level of arsenic until the late 1960s, but they then switched to arsenic-contaminated groundwater at the beginning of the 1970s (Yu et al., 2007). Arsenic was found in the Kuitun and Usum County of Xinjiang province at a depth of 2–30 meters with a concentration of <10–68 μg/L, with a mean of 18 μg/L (Wang and Huang, 1994). In the Datong Basin of Shanxi Province, arsenic concentration ranged between 2.0 and 1300 μg/L (Wang et al., 1998), but later Wang et al. (2003) showed that 54.4% of 3083 tested wells exceeded 50 μg/L.

Eight provinces (Inner Mongolia, Shanxi, Xinjiang, Jilin, Ningxia, Qinghai, Anhui, and Beijing) are affected with a high level of groundwater arsenic poisoning, including 40 counties, 1047 villages, and a population of more than two million (Xia and Liu, 2004). Arsenic contamination in drinking water was found in these eight provinces to be between 50 μg/L and 2000 μg/L (Jin et al., 2003). Based on water quality data from 445,638 well water samples in 20,517 villages in 292 counties in 16 provinces between 2001 and 2005, Yu et al. (2007) found 21,155 samples (4.75%) to be contaminated (>50 μg/L). The highest concentration was recorded at 1000 μg/L in Anhui province of eastern China. Arsenic concentrations have been identified between 70 μg/L and 750 μg/L in deep artesian groundwater to the north of the Tianshan Mountains (Luo et al., 2006). Arsenic concentrations of more than 10 μg/L were found in 12% of the analyzed groundwater samples in the Xinjiang province, but concentrations in Chepaizi areas of the Xinjiang province ranged from 25 to 185 μg/L with a short distance spatial variability and increased arsenic with increasing aquifer depth (Zhang et al., 2017c).

In Inner Mongolia, excess arsenic concentrations (>50 μg/L) have been identified in groundwater from aquifers in the Huhhot Basin, with the maximum concentration of 1500 μg/L (He et al., 2009). Arsenic contaminations in Inner Mongolia are frequent in 655 villages of 11 counties; the concentrations range between <50 and 1800 μg/L, and some 1774 patients have been diagnosed there (Ma et al., 1999). Xia and Liu (2004) discovered the highest concentration of arsenic in Inner Mongolia to be 1860 μg/L. Bayingnormen (543 villages with arsenic concentrations over 50 μg/L) in the west and Tumet (with 81 villages at over 50 μg/L) in central Inner Mongolia are the two major endemic areas with the greatest challenge. In addition, arsenic concentration in well water of Gangfangying village of Inner Mongolia ranged from 1 to 1790 µg/L, with a mean of 130 µg/L having a standard deviation of 200 µg/L, and some 47.2% of the total of 303 wells had arsenic over 50 µg/L (Sun et al., 2006). Some shallow dug wells are also contaminated at concentrations of 556 µg/L in the Huhhot Basin (Smedley et al., 2003). Arsenic concentrations vary from 76 to 1093 µg/L in the Hetao plain (He et al., 2009), and in the same area Guo et al. (2008) found levels ranging between 0.58 and 572 µg/L, with 71% of the samples exceeding 10 µg/L. Moreover, the maximum groundwater arsenic concentration in the Datong Basin was 947.6 µg/L (Zhang et al., 2017b).

Chronic arsenicosis is a newly-emerging public health issue in China, and China lowered its drinking water standard for arsenic to 10 µg/L in 2007. Examining a total of 135,492 people from eight provinces (i.e., Inner Mongolia, Xinjiang, Shanxi, Jilin, Qinghai, Ningxia, Anhui, and Henan) between 2001 and 2003, some 10,096 cases of arsenicosis were diagnosed, 7.5% of whom had various degrees of skin lesions and serious symptoms such as cancers (Yu et al., 2007). Li et al. (2017) found an arsenic-induced cancer cluster of the HRB area. Following an ecological study of approximately 700 eight- to twelve-year-old children in the Datong and Jinzhong basins of Shanxi province, Wang et al. (2007a) claimed that exposure to arsenic lowered their IQ. In the Gangfangying village of Inner Mongolia, about 20% of 2080 individuals showed signs of arsenic-induced skin lesions (Sun et al., 2006). Wade et al. (2015) and Wang et al. (2007b) found an association between arsenic exposure and cardiovascular disease. Rodríguez-Lado et al. (2013) estimated that about 20 million people are at risk of being affected by the consumption of arsenic-contaminated groundwater. Ma et al. (1999) also confirmed a total of 1447 cases (81% of all cases) of arsenicism (mainly skin hyperkeratosis, depigmentation, hyperpigmentation, and skin cancer).

Arsenic contamination in China is considered to be geological (Gao et al., 2017), and the affected groundwater is found in alluvial-lacustrine aquifers containing relatively high organic matter (Chen et al., 2017b; Guo et al., 2008). Groundwater arsenic in Keshenketeng County was introduced by arsenopyrite mining, while in Hetao and Tumote groundwater, arsenic occurred naturally from Holocene alluvial-lacustrine aquifers under highly reducing conditions (Zhang et al., 2017b). Contrastingly, Zhang et al. (2002) suggested that the groundwater arsenic of the Hetao Basin was released from higher elevations, where mining had been carried out for a long time, and was then transported down the gradient.

2.2.5 Ganges-Brahmaputra Plain, India

Groundwater arsenic poisoning is documented in six states of the Ganga-Brahmaputra Plain (GB-Plain) of India, with 70.4 million people potentially at risk from arsenic toxicity (Chakraborti et al., 2016a). The Ganges alluvial plain in India has been identified with groundwater arsenic problems, and West Bengal is reported as the worst-affected area; other areas, such as Punjab, Haryana, and Himachal Pradesh, Bihar, Uttar Pradesh, Jharkhand, Assam, Chattisgarh, and Manipur, have been identified recently (Chakraborti et al., 2003, 2017a; Chatterjee et al., 2017; Kulkarni et al., 2017; Kumar et al., 2016b; Mukherjee et al., 2006a; Nickson et al., 2007). Arsenic concentration in groundwater was first discovered in 1976 in Chandigarh of north India (Datta, 1976), and in West Bengal, arsenic was first discovered in 1984 (Garai et al., 1984). Arsenic surveys in the villages of West Bengal by the School of Environmental Studies (SOES) at Jadavpur University, Kolkata, started in early 1988, and at that time 22 affected villages in five districts were found to be contaminated with arsenic (Figure 2.3).

Since 1988, Chakraborti et al. (2017a) have checked 140,150 hand-pump tubewell water samples from 19 districts of West Bengal, and 48.1% of the samples had arsenic over 10 µg/L, 23.8% above 50 µg/L (the Indian Standard), and 3.3% above 300 µg/L. The maximum concentration was recorded at 3700 µg/L. Analyzing water samples from 132,262 hand-pump tubewells with the silver diethyldithiocarbamate (Ag-DDTC) method in eight districts (Bardhaman, Haora, Hoogly, Malda, Murshidabad, Nadia, North 24-Parganas, and South 24-Parganas) of West Bengal, Nickson et al. (2007) found more than one-quarter (25.5%) of the tubewell water to be contaminated with above

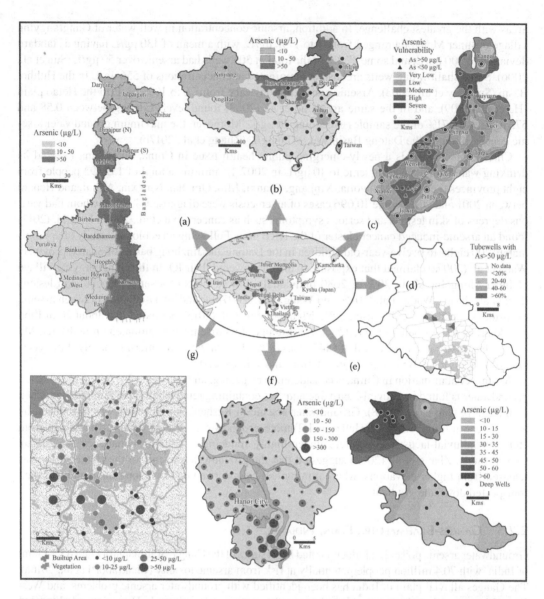

FIGURE 2.3 Groundwater arsenic concentrations in some regions of Asian countries: (a) West Bengal, India; (Das B., M. M. Rahman, B. Nayak et al., 2009. *Water Quality, Exposure and Health* 1:5–21) (b) China (Sun, G. 2004. *Toxicology and Applied Pharmacology* 198:268–271; Xia Y., and J. Liu. 2004. *Toxicology* 198(1–3):25–29); (c) Taiyuan Basin, China (Guo Q., Y. Wang, X. Gao et al. 2007a. *Environmental Geology* 52: 923–932); (d) Part of Inner Mongolia, China (Sun, G. 2004. *Toxicology and Applied Pharmacology* 198:268–271); (e) Maydavood area of Southeastern Iran (Chitsazan, M., M. S. Dorraninejad, A. Zarasvandi et al., 2009. *Environmental Geology* 58:727–737); (f) Hanoi, Vietnam (Berg, M., C. Stengel, P. T. K. Trang et al., 2007. *Science of the Total Environment* 372(2–3):413–425); and (g) Katmandu, Nepal (Chapagain, S. K., S. Shrestha, T. Nakamura et al., 2009. *Desalination and Water Treatment* 4:248–254.)

50 µg/L of arsenic and more than half (57.9%) had concentrations over 10 µg/L. Analyzing shallow groundwater in Murshidabad district of West Bengal, Kulkarni et al. (2017) investigated high arsenic concentrations (in Holocene sediments) in Beldanga (10–4622 µg/L at 35–45 m depth) and Hariharpara (5–695 µg/L at 6–37 m depth) sites, as well as low arsenic concentrations (Pleistocene sediments) in Nabagram (0–16 µg/L at 20–45 m depth) and Kandi (5–50 µg/L at 20–55 m depth).

In addition, in analyzing 4210 hand-pump tubewell water for arsenic in all the 141 administrative wards in Kolkata Municipal Corporation (KMC), Chakraborti et al. (2017b) found some 14.2% and 5.2% tubewells had over 10 µg/L and 50 µg/L of arsenic, respectively, but no tubewell deeper than 300 m had more than the 50 µg/L threshold and the maximum concentration was 800 µg/L at 20 m depth. Arsenic concentrations were found above 10 µg/L in 77 wards and over 50 µg/L in 37 wards, and the southern part of Kolkata city is more arsenic-prone than the northern and central areas (Chakraborti et al., 2017b).

Arsenic contamination in groundwater was first reported by Bihar in 2003 (Chakraborti et al., 2003). Conducting a three-year intensive research, it was calculated that some 39.02% of the analyzed 9500 samples contained arsenic above 10 µg/L and 23% contained over 50 µg/L, and some 525 of patients (5.5%) were diagnosed with various types of arsenic-related skin manifestations in the screening of 4513 persons (Mukherjee et al., 2006a). In Bihar, about 10.8% of tubewell water out of 66,623 samples from 11 districts were found to be contaminated with arsenic at 50 µg/L or more, while 28.9% had arsenic of more than 10 µg/L (Nickson et al., 2007). Analyzing groundwater arsenic concentrations in the middle Gangetic plain of Bihar, Kumar et al. (2016b) found arsenite as the dominant species in 73% of water samples and arsenate in 27% samples. In analyzing 19,961 tubewell water samples from Bihar, Chakraborti et al. (2017a) found 32.70% and 17.75% of the tubewells had arsenic above 10 µg/L and 50 µg/L, respectively, with the maximum concentration of 2182 µg/L. Moreover, arsenic concentrations in agricultural soil in Bihar varied between 3528 and 14,690 µg/kg, while in the subsurface sediments, concentrations ranged between 9119 and 20,056 µg/kg in Methrapur and 4788–19,681 µg/kg in Harail Chapar (Kumar et al., 2016b).

Groundwater arsenic contamination in Ballia district, Uttar Pradesh, was first discovered in September-October 2003 (Chakraborti et al., 2004). A total of 3901 water samples from Uttar Pradesh have been analyzed and 46.57% contained arsenic at over 10 µg/L and 30.47% above 50 µg/L; a total of 68 villages out of 91 selected from 10 blocks were affected, and surveys found 153 patients (15.47%) with skin lesions out of 989 people screened (Mukherjee et al., 2006a). In analyzing 4780 tubewell water samples from Uttar Pradesh, Chakraborti et al. (2017a) found 45.48% and 26.51% of the tubewells had arsenic above 10 µg/L and 50 µg/L, respectively, with the maximum concentration of 3191 µg/L. In Uttar Pradesh, Nickson et al. (2007) reported that some 21.5% samples out of 20,126 tested government-installed hand-pump sources contained arsenic within the range of 10–50 µg/L and 2.4% at concentrations of more than 50 µg/L.

Groundwater arsenic contamination was identified in some 22 villages of Rajnandgaon district of Chhattisgarh in 1999, and at that time it ranged between <10 µg/L and 880 µg/L (Chakraborti et al., 1999). Arsenical skin lesions were diagnosed in 42 of 150 adults and 9 of 58 children examined, and about 75% of the 150 people selected had arsenic in their hair above the toxic threshold level of 1.0 µg/g (Mukherjee et al., 2006a). Groundwater arsenic contamination was reported in 1976 from Chandigarh and in various villages of Punjab, Haryana, and Himachal Pradesh in northern India. Arsenic contamination was detected in groundwater of the Upper Brahmaputra plain and some 43% of the 137 water samples from hand-pump tubewells were found to be above 10 µg/L, 26% were over 50 µg/L, and the highest concentration was 490 µg/L (Chakraborti et al., 2004).

Jharkhand joined the states with groundwater arsenic contamination in December 2003 (Bhattacharjee et al., 2005). In Shahibganj district of Jharkhand, among the analyzed 9007 water sources, it was found that 3.7% were arsenic contaminated at 50 µg/L or more and 7.5% were at concentrations within the range of 10–50 µg/L (Nickson et al., 2007). In Jharkhand, Chakraborti et al. (2017a) analyzed some 146 tubewell water samples and found 25.34% and 15.4% of the tubewells had arsenic above 10 µg/L and 50 µg/L, respectively, with a maximum concentration of 880 µg/L.

In Assam, the highest levels were detected in Jorhat, Dhemaji, Golaghat, and Lakhimpur districts and ranged between 100 and 200 µg/L (Mukherjee et al., 2006a). On the banks of the Brahmaputra in Assam, samples from 5729 government hand-pump sources in 22 districts were analyzed for arsenic, and 6.3% samples were found to be concentrated with arsenic of more than 50 µg/L and 26.1% at concentrations between 10 and 50 µg/L (Nickson et al., 2007). In Assam, Chakraborti et al. (2017a) analyzed a total of 1590 tubewell water samples and found 47.61% and 15.47% of the tubewells had arsenic above 10 μg/L and 50 μg/L, respectively, with the maximum concentration at 383 μg/L. In Arunachal Pradesh, arsenic was detected in six districts, and the maximum concentration was 618 μg/L in Dibang district. In Tripura, arsenic was detected in parts of West Tripura, north Tripura, and Dhalai districts in a range between 65 and 444 μg/L. Chakraborti et al. (2017a) analyzed 628 water samples; 63% contained arsenic at over 10 μg/L, and 40% from Thoubal district of Manipur, one of the seven North Eastern Hill states of India exceeded 50 μg/L. In Nagaland, arsenic was detected in Mokokchung and Mon districts, and a remarkable number of groundwater samples contained arsenic beyond 10 μg/L (Mukherjee et al., 2006b).

Arsenic contamination in rice and vegetables is a recent concern in India (Bhattacharya et al., 2010; Islam et al., 2016; Mukherjee et al., 2017; Norra et al., 2005; Shrivastava et al., 2017). The total arsenic concentrations in rice, vegetables, and pulses cultivated in arsenic-affected West Bengal vary between <0.0003 and 1.02 mg/kg (dry weight). The highest mean arsenic concentration is found in potato (0.654 mg/kg), followed by Boro rice grain (0.451 mg/kg), Aman rice grain (0.334 mg/kg), cauliflower (0.293), and brinjal (0.279), while lower arsenic concentrations have been found in wheat (0.129 mg/kg) followed by garlic (0.126 mg/kg), lentil (0.096 mg/kg), beans (0.091 mg/kg), green chili (0.085 mg/kg), tomato (0.084 mg/kg), and lemon (0.012 mg/kg) as well as a low reading in turmeric (0.003 mg/kg) (Bhattacharya et al., 2010).

Resultant health problems due to chronic arsenic exposure were first identified in West Bengal in 1983 (Garai et al., 1984). All nine members of a family in Ramnagar village of South 24 Parganas, West Bengal, who were taking 3700 μg/L of arsenic from their own private tubewell were diagnosed with skin lesions—seven of them had severe arsenical skin lesions and four had cancer (Das et al., 2009). Some 96,000 people in West Bengal including children were randomly examined from highly arsenic-affected districts, and 9356 had skin symptoms (Chakraborti et al., 2009) —at least 100 had cancer and in a few hundred suspected Bowen's disease was detected (Mukherjee et al., 2006a). Arsenical neuropathy was observed in Patna district in 40.5% of 37 arsenicosis patients with 73.3% prevalence for predominant sensory neuropathy and 26.7% for sensor-motor (Chakraborti et al., 2016b). Analyzing water samples and examining 422 patients in Yadgir District of Karnataka, Tanga et al. (2016) found a prevalence of 2.38% for epidermal neoplasm and 10.9% for arsenic keratosis. In Domkal of Murshidabad, melanosis and leucomelanosis were diagnosed in 410 patients who consumed arsenic from their drinking water ranging from 10 μg/L/day to 600 μg/L/day (Sarkar, 2010). Examining 712 villagers from Patna district, 69 (9.7%) villagers had arsenical skin lesions (Chakraborti et al., 2016b).

The sources of arsenic in West Bengal are geogenic, and the affected areas of this region are mainly Holocene alluvial and deltaic sediments similar to those of large parts of Bangladesh. The source is the oxidation of arsenic rich pyrite or anoxic reduction of ferric iron hydroxides in the sediments to ferrous iron, thereby releasing the adsorbed arsenic to groundwater (Mandal and Suzuki, 2002). Arsenic is present in the alluvial sediments of the Gangetic Delta, and the sediments contain a large amount of clay and organic matters, which may retain and release arsenic in groundwater aquifers (Mukherjee et al., 2006b). Kumar et al. (2016b) also investigated that decomposition of organic matter present in dark and grey sections promote the redox conditions and trigger mobilization of arsenic into groundwater.

2.2.6 Kurdistan Province, Iran

The first case of chronic arsenic poisoning due to drinking water was recognized in Kurdistan province of the Islamic Republic of Iran in 1981. This case was of a woman with intense skin lesions who lived in a village in Bijar County. She had lost her legs as a result of gangrene (Mosaferi et al., 2008; Vanaei et al., 2006). The water supply in this area is drawn from deep wells, dug wells, and springs. In a survey of 18 villages in Kurdistan province, Mosaferi et al. (2008) found a mean arsenic concentration of 290 μg/L, with a maximum of 1480 μg/L, and there were readings over

1000 µg/L in some water samples. Water samples from deep wells tapping from the Maydavood aquifer of southwest Iran in June 2007 showed results ranging from 8 to 77 µg/L, and nearly 81% of the samples were above 10 µg/L (Chitsazan et al., 2009). Arsenic is highly concentrated in the hydrothermal warm springs (<38 °C), ranging from 15,900 to 30,500 µg/L in Kerman province of Central Iran (Khorasanipour and Esmaeilzadeh, 2015). Analyzing arsenic levels in groundwater, human hair, and urine with ICP-MS and AAS methods in Chelpu in northeastern Iran, Taheri et al. (2017) investigated arsenic levels in water between 12 and 606 µg/L; the figures in human hair ranged between 0.37 and 1.37 µg/g, and in urine the figures ranged between 9.0 and 271.4 µg/L. The source of arsenic in deep groundwater wells is geogenic and may be restricted to rocks and sediments that originated from parent rocks (Chitsazan et al., 2009; Khorasanipour and Esmaeilzadeh, 2015).

2.2.7 Japan

The volcanic rocks of the Japanese islands are rich in arsenic, at concentrations of up to 25,700 µg/L, and it is also present in geothermal waters and Quaternary alluvium. The detection of arsenic in drinking groundwater has received considerable attention in areas of Fukuoka, Kumamoto, Fukui, and Takatsuki (Even et al., 2017; Mitsunobu et al., 2013; Ogawa et al., 2012; Sawada et al., 2013). Arsenic concentrations in water from 34 wells in the Niigata Prefecture were measured between 1955 and 1959 as part of an historical cohort study using the Gutzeit method and ranged from ND to 3000 µg/L—6 wells had a ND concentration, 17 wells contained <1000 µg/L, and 11 wells contained ≥1000 µg/L (Tsuda et al., 1995). Groundwater arsenic was first detected in February 1994 in Ohkawa, and later 22.6% of 11,673 well water samples were found to be contaminated with over 10 µg/L of arsenic, with a maximum of 370 µg/L (Shimada, 1996). In analyzing 67 well water samples from the South Chikugo alluvial plain in Fukuoka Prefecture, some 43% of the samples were found to be arsenic contaminated with over 10 µg/L, with the maximum concentration of 293 µg/L (Kondo et al., 1999). During 1989–1991, some 22 well water samples from the southern part of Kumamoto City were tested at 5 to 66 µg/L of arsenic (Shimada, 1996). Groundwater samples from 14 wells in the western area of Fukui Prefecture had concentrations ranging from 11 to 50 µg/L. Substantial amounts of arsenic (11–60 µg/L) were detected in 10 wells in the eastern part of the Takatsuki City in 1990, and it was obvious that high arsenic concentrations were detected only in wells at depths of 49 to 68 m (Shimada, 1996). Oono et al. (2002) investigated arsenic concentration in the south of Osaka Prefecture for up to 10.6 µg/L in a Quaternary alluvial aquifer. Torres and Ishiga (2003) describe concentrations of up to 42 µg/L in Holocene coastal sands containing subneutral (pH: 6.1–7.1) and slightly oxic (DO: 1.5–4.0 ppm) groundwater at Yumigahma in western Japan. Sawada et al. (2013) investigated a significant dose-response trend with lung cancer risk.

2.2.8 Terai Alluvial Plain, Nepal

Groundwater arsenic contamination is a public health concern in the Terai region of Nepal. More than 90% of the population in the Terai region usually depend on groundwater for their drinking and domestic purposes (Neku and Tendulkar, 2003; Shrestha et al., 2003a). Some 200,000 shallow tubewells have been installed by different agencies in 20 Terai districts, serving about 11 million people (NASC/ENPHO, 2004). The Department of Water Supply and Sanitation, with financial support from the WHO, conducted the first study on arsenic contamination in the eastern Terai region (Jhapa, Morong, and Sunsari) in 1999, and the results of 268 samples showed that the highest concentration was 75 µg/L (NRCS, 2000). The Nepal Red Cross Society (NRCS), employing financial support from the Japanese Red Cross Society, analyzed 2000 Terai water samples and found that 3% exceeded 50 µg/L and 21% were above the WHO limit (JICA/ENPHO, 2005). Groundwater arsenic concentrations in six southern Terai districts of Nepal ranged from 0 to 770 µg/L, having the majority of the tested wells containing arsenic level below 10 µg/L and the high contamination levels found at depths of <50 m (Yadav et al., 2012). In January 2000, Patricia M. Halsey showed that 18% of 172 tested tubewells of the Terai region were contaminated at concentrations above 10 µg/L, with a highest detection of 111 µg/L (Halsey, 2000). In 2001, the Rural Water Supply and Sanitation Support Program (RWSSSP) found that 9.8% of 1508 samples in the districts of Rupandehi, Nawalparasi, and Palpa had over 10 µg/L of arsenic, with the maximum concentration of 2000 µg/L in Devdaha of Rupandehi district (Ngai et al., 2007). The testing of 287 tubewell water samples from Bagahi village of Rautahat district by Pradhan (2006) found that 6% were above 50 µg/L of arsenic.

Analyzing 14,932 hand-pump tubewell water samples in 20 districts of the Terai in 2002, it was found that 23% samples exceeded 10 µg/L, 5% were above the "Nepal Interim Arsenic Guideline" value of 50 µg/L, and 77% were safe with the WHO guideline value (Shrestha and Maskey, 2005). NASC/ENPHO (2004) reported that, out of 18,635 tubewell samples taken in 20 districts of the Terai, 7.4% had over 50 µg/L and the large majority (76.3%) had below the WHO guideline value. The highest prevalence (18.6%) of arsenicosis cases was found in Patkhouli village, where most tubewells (95.8%) were affected (Maharjan et al., 2006). The highest arsenic concentrations were recorded at 2620 µg/L in Rupandehi district, followed by 589 µg/L in Kapilbastu, 571 µg/L in Nawalparasi, 456 µg/L in Parsa, 270 µg/L in Banke, and 254 µg/L in Bara. The NRCS has so far analyzed 10,000 water samples in their program areas in Nepal and found a mean concentration of 11.9 µg/L, with a maximum of 456 µg/L (Shrestha et al., 2003b). Panthi et al. (2006) analyzed a total of 25,058 tubewell water samples from Nawalparasi district, of which 8% were found to be contaminated at $\geq$50 µg/L and 23% at $\geq$10µg/L. A wide range of arsenic occurrence (1–73 µg/L) was found in deep groundwater, and some 52% of deep groundwater samples exceeded the WHO standard for arsenic in drinking water (Chapagain et al., 2009). Ahmad et al. (2004) surveyed Goini village in Nawalparasi district and found readings from 104 to 1702 µg/L of arsenic, and in Thulo Kunwar, the figure ranged from 4 to 972 µg/L. In the lowlands of Nepal, Maharjan et al. (2005) found arsenic in a range of 3–1072 µg/L, with a mean concentration of 403 $\pm$ 229 µg/L. A total of 97.9% of tubewells had arsenic levels over 10 µg/L and 87.6% were above 50 µg/L. Arsenite is the dominant species in Nepal (>79%) (Pokhrel et al., 2009).

Around half a million people in the Terai are living at risk of arsenic poisoning and the rates of prevalence of arsenicosis in six districts (Nawalparasi, Parsa, Bara, Rautahat, Rupandehi, and Kapilavastu) of the Terai were calculated at 25.7%, 2.3%, 2.4%, 9.7%, 2.1%, and 3.9%, respectively (Maharjan et al., 2006). It is noted that the first arsenic patient was diagnosed in Nawalparasi district of Nepal and some 400 patients were identified as arsenicosis cases (Maharjan et al., 2006). Fewtrell et al. (2005) estimated 12.5 disability adjusted life years per 1000 population in Nepal due to excess arsenic in the drinking water (>50 µg/L). Analyzing 497 hair samples and 116 nail samples, Fewtrell et al. (2005) found that 95% of hair samples and 71% of nail samples had concentrations of more than the permissible limit. The sources of arsenic in Terai groundwater are geogenic, and the dissolution of arsenic-bearing rocks, sediments, and minerals contributes arsenic to the groundwater (Bhattacharya et al., 2003; Diwakar et al., 2015). The mobilization of arsenic in the Kathmandu Valley and Terai Basin is probably mainly related to changes in redox conditions. Iron oxides are abundant in sediments in these areas and show close association with arsenic, suggesting that iron oxyhydroxides could be the potential source (Brikowski et al., 2014; Gurung et al., 2007).

2.2.9 Indus Alluvial Plain, Pakistan

Pakistan has low levels of arsenic in groundwater, but there are hot spots in parts of the Indus alluvial basin that are causing concern (Fatmi et al., 2009). Groundwater pollution in Kalalanwala in Kasur district was first officially noted in July 2000, when a newspaper reported that residents of Kalalanwala suffered from a mysterious bone deformity disease. Arsenic contamination is also frequent in the Muzaffargarh district of Punjab Province; Jamshoro district of Sindh Province;

Faisalabad, Gujarat, and Jaranwala (Arshad and Imran, 2017; Bhowmik et al., 2015; Bibi et al., 2015; Farooqi et al., 2007; Rasheed et al., 2017; Rasool et al., 2015; Rehman et al., 2016). Rehman et al. (2009) analyzed a total of 63 groundwater samples from Faisalabad, of which more than half were found to be contaminated; the contamination figure was 34.5% out of 58 samples for Gujarat, 9.3% out of 43 samples for Peshawar, and more than half of the samples (52%) out of 58 from Jaranwala that were found to be contaminated with arsenic. Rasool et al. (2015) investigated groundwater arsenic concentrations ranging between 5.9 and 507 µg/L in Tehsil Mailsi of Punjab. Almost 97% of analyzed 35 sample groundwater sources in Badarpur and Ibrahimabad villages in Kasur district of Punjab were found to be unsafe following the Pakistan Standard guideline value of 50 µg/L, and the maximum concentration was recorded to be 3800 µg/L (Arshad and Imran, 2017).

In analyzing 49 groundwater samples from Muzaffargarh district, it was found that 43% of samples exceeded 50 µg/L of arsenic, 57% were over 10 µg/L, and 43% samples were safe within the WHO standard value. The highest reading was 906 µg/L (Nickson et al., 2005). The Pakistan Council of Research in Water Resources and UNICEF have reported that arsenic-contamination of groundwater is frequent in many areas of Punjab, ranging from 10 to 200 µg/L. Analyzing some 147 groundwater samples from Kalalanwala village, in Punjab, Farooqi et al. (2007) found arsenic ranging from 1 to 2400 µg/L, and 91% of the samples exceeded 10 µg/L. Elevated readings were found in the shallow-well water of four villages: 2400 µg/L in Kalalanwala and Kot Asad Ullah, 883 µg/L in Shamkey Bhatian, 672 µg/L in Manga Mandi, and 681 µg/L in Waran Piran Wala (Farooqi et al., 2007). Analyzing 153 samples from groundwater (>15 m depth) in Jamshoro district of the Sindh Province, Baig et al. (2009) found a range of arsenic concentrations at 13 to 106 µg/L for groundwater. In 2005, in collaboration with UNICEF, some 25,000 water samples were looked at for Rahim Yar Khan district and 53.76% were found to have high concentrations ranging from 20 to 500 µg/L (Ul-Haque et al., 2007). Assessing 228 groundwater sources in six villages, the maximum arsenic concentration was 3090 µg/L, with a median of 57.55 µg/L, and 89% of water sources exceeded 10 µg/L (Rasheed et al., 2017). Rasool et al. (2016) analyzed some 44 groundwater samples from Jala Jeem and Duniapur and found that arsenic exceeded 10 µg/L. Arsenic has also been found in dietary staples, and Arain et al. (2009a) observed that the leafy vegetables (spinach, coriander, and peppermint) contain higher levels (0.90–1.20 mg/kg) as compared to root vegetables (0.048 – 0.25) and grain crops (0.248 – 0.367 mg/kg) on a dried weight basis. Brahman et al. (2014) also investigated high arsenic content in grain crop, e.g., kidney bean compared to pearl millet and green gram.

More than 20% of the population in Punjab and around 36% of the population in Sindh are exposed to arsenic contamination above the WHO permissible limit (Ul-Haque et al., 2007). No confirmed cases of arsenic-related disease have been found in Pakistan, but epidemiologists are expecting severe health consequences in the future, and Brahman et al. (2016) discovered that children consuming elevated levels of groundwater arsenic from their drinking water are at a potential risk of chronic arsenic toxicity. Rehman et al. (2016) found that the hazard quotient (HQ) values for arsenic were <1, indicating a minimal noncancer risk in people living in southern districts of Khyber Pakhtunkhwa Province. Quaternary sediments, mainly of alluvial and deltaic origins, occur over large parts of the Indus Plain of Pakistan. The sediments are comprised mostly of coarse sand, containing a high percentage of fine to very fine sand and silt, and aquifers in these sediments are potentially susceptible to high groundwater arsenic concentrations.

2.2.10 QUATERNARY AQUIFER, TAIWAN

A high concentration of groundwater arsenic has been recognized in two areas in Taiwan: one on the southwest coast and another in the Lanyang Basin of the northeast. The southwest cluster was first identified in the 1960s as a major problem area because of its endemic "Blackfoot Disease" (BFD), a unique peripheral vascular disease associated with long-term ingestion of arsenic from artesian well water (Tseng et al., 1968). The Lanyang Basin on the northeast coast is an area of endemic toxicity, as is the Chainan flood plain (Lin et al., 2006). Blackwell et al. (1961) found levels of arsenic between

240 μg/L and 960 μg/L in 13 artesian wells in the southwest coast, and Kuo (1968) observed concentrations from 10 to 1800 μg/L, with a mean of 500 μg/L and almost half of the samples between 400 μg/L and 700 μg/L. Arsenic concentrations, mainly in the form of arsenate species in this region, are higher in water from deep artesian wells than in shallow wells. Lo (1975) reported a nationwide survey of arsenic content in drinking water from 83,656 wells. Some 18.7% of these had over 50 μg/L, and 2.7% were above 350 μg/L. Analyzing 3901 tubewell samples by the HG-FAAS method in 18 villages of four townships (i.e., Chiaohsi, Chuangwei, Tungshan, and Wuchieh) in the northeast, Chiou et al. (2001) yielded results varying from ND (<0.15 μg/L) to 3590 μg/L of arsenic.

Taiwan is the classic area for the identification of BFD and other peripheral vascular disorders. The Chianan plain is known for its unique cases of endemic BFD, a peripheral vascular disease (i.e., gangrene). This was first reported between 1910 and 1920 (Kao and Kao, 1954). This disease was found to be strongly correlated with the direct ingestion of high groundwater arsenic from deep artesian wells (Lin et al., 2006; Tseng, 1977). Apart from BFD, Taiwan is well known for increased incidence of lung, skins and bladder cancer caused by exposures to arsenic via drinking water for residents from the BFD endemic areas (Lamm et al., 2013; Lin et al., 2013; Wang et al., 2002). In a 12-year follow-up study of 8086 residents in the northeastern region, Chen et al. (2010a) found evidence linking arsenic in drinking water with increased urinary cancer risk in high exposure areas (>100 μg/L). Exposure to high levels of arsenic in drinking water (e.g., arsenic above 640 μg/L) is associated with the occurrence of liver cancer (Lin et al., 2013), and patients with arsenic-related bladder cancer may have poorer tumor characteristics and decreased cancer-specific survival rates (Yeh et al., 2015). Cheng et al. (2016) illustrated the relationship between arsenic intoxication and nonmelanoma skin cancers (cutaneous squamous cell carcinoma [SCC] and basal cell carcinoma [BCC]) through comparing the data of people living in the black foot disease endemic areas (BFDEA) and non-BFDEA in Taiwan. Examining 8854 participants from a nationwide health screening program, Cheng et al. (2017) showed that a high arsenic level in drinking water was a risk factor for rapid progression of chronic kidney disease (OR: 1.22 with 95% CI of 1.05–1.42, $p<0.01$).

On the southwest coast, high arsenic concentrations are found in groundwater from deep artesian wells (100–280 meters) abstracted from sediments that include fine sands, muds, pyretic material, or black shale (Thornton and Farago, 1997), while groundwater abstracted in northeast Taiwan is reported to be artesian, with a shallow depth ranging between 16 and 40 meters (Hsu et al., 1997). The groundwater is likely to be strongly reducing as the arsenic is found to be present largely as arsenite (Al Lawati et al., 2013; Das et al., 2016). Groundwater arsenic was mainly derived from groundwater-associated sediments (Yang et al., 2016), and arsenic-dissolved organic matter is important in driving the release of arsenic in shallow aquifer (Chen et al., 2016).

Along with the groundwater, humans may be exposed to arsenic through dietary intake. Fish in Taiwan, especially tilapia, are severely contaminated with arsenic in the BFD endemic areas (Chen and Liao, 2008; Liao et al., 2008). There is no effective therapy for arsenicosis, but potential treatments include reducing arsenic exposure and providing specific drugs for recovery and/ or preventing disease progression (Liao et al., 2008). In the early 1960s, a tap water supply system was implemented in BFD areas. After the mid-1970s, artesian well water was no longer used for drinking or cooking in the region. At present, groundwater is no longer the primary source for direct ingestion or domestic use in Taiwan (Lin et al., 2006).

2.2.11 RON PHIBUN, THAILAND

Chronic arsenic poisoning is documented in Ron Phibun District, Nakon Si Thammarat Province of southern Thailand. The problem occurs in an area where tin is mined from both bedrock and alluvium. Groundwater arsenic concentration was first identified in Thailand in 1988 (Kim et al., 2011). In Ron Phibun district, the span of arsenic concentration recorded was between 1.25 and 5114 μg/L in 23 shallow groundwater, and in 13 deep borehole samples, the concentrations varied from 1.25 to 1032 μg/L (Kohnhorst et al., 2005). Foy et al. (1992) reported arsenic levels in shallow wells from

20 to 2700 µg/L with piped water also having 70 µg/L. Zhang et al. (2001b) showed that 72.31% of the area of their study sites had arsenic levels exceeding 50 µg/L in the shallow groundwater. Intamo et al. (2016) investigated arsenic concentrations in soil in a limestone area of western Thailand and found high median concentrations of arsenic of 76 mg/kg, which are higher than the limits for agricultural and residential uses. In addition, a severe arsenic contamination (>1000 µg/L) with strongly reducing conditions was investigated in the city of Hat Yai of southern Thailand (Lawrence et al., 2000). Hat Yai obtains about half of its water supply from alluvial aquifers. Arsenic concentrations from 100 to 5100 µg/L have been found in shallow groundwater from Quaternary alluvial sediment in Ron Phibun District (Williams et al., 1998a). Deeper groundwater from an older limestone aquifer has been found to be less contaminated, although a few high arsenic concentrations do occur. Jindal and Ratanamalaya (2006: 228) reported that arsenic concentrations in shallow groundwater were found to exceed the WHO guideline for arsenic in drinking water. A total of 37 samples were collected from shallow aquifers around the lower Chao Phraya River basin area of Nakhonpathom Province, and arsenic concentrations in most of the shallow wells were found to be about 5 µg/L or less, with an average of 11 µg/L (Kim et al., 2011).

Skin problems such as keratosis and hyperpigmentation were first diagnosed in 1987 among the residents of Ronphibun district by the Thai Ministry of Public Health (Choprapawon and Porapakkham, 2001; Kim et al., 2011). A clinical survey conducted by the Ministry during 1987–1988 documented over 1000 cases of arsenic-induced skin disorders in people between the ages of 4 months and 85 years in Ron Phibun town. Most cases were of relatively mild disease, but 21.6% did have very significant lesions. Siripitaykunkit (2000) investigated chronic arsenism in Ron Phibun district and indicated that 24% of the population studied exhibited symptoms. In a follow-up study conducted by Pavittranon et al. (2003) a prevalence of melanosis and hyperkeratosis was found, of 5.99% and 8.67%, respectively. In a study of 7785 people in Ron Pibun township, three villages with a total of 19.9% of the population accounted for 60.9% of the cases (Pavittranon et al., 2003).

In 1992, a joint Thai-Japanese clinical survey was performed by examining blood samples from students in this area; it showed that over 89% of them had high levels of arsenic in their blood and 22% had skin lesions and hyperkeratosis (Paijitprapapon and Ramnarong, 1994). Elevated levels of arsenic in hair and fingernails were also found in 80% of the school age population (Choprapawon and Ajjimangkul, 1999). A follow-up study of 2400 school pupils in 1992 indicated that 89% had excess blood arsenic concentrations, with a 22% incidence of arsenical skin manifestations (Williams et al., 1998a). These health problems in the residents were related to the consumption of surface water and groundwater. The concentration of arsenic in these waters is up to 100 times higher than that of the WHO guideline value (Milintawisamai et al., 1997).

Contamination of shallow groundwater can be attributed to the oxidation of mining waste, which is rich in arsenopyrite and its alteration products, and finely disseminated arsenopyrite in the alluvium. The mobilization of arsenic is believed to be caused by oxidation of arsenopyrite, exacerbated by the former tin mining activities. In Ron Phibun district, tin and associated minerals, such as arsenopyrite and pyrite, have been extracted from granites (Pechrada et al., 2010; Wattanasen et al., 2006). Hazard mitigation programs initiated by the Thai Department of Mineral Resources focused on the provision of an alternative "safe" water supply. Due to this serious problem, the government prohibited further mining in the area.

2.2.12 Western Anatolia, Turkey

Groundwater arsenic poisoning is mainly documented in Western Anatolia and some regions of Turkey. Arsenic is one of the major pollutants in the Bigadiç district of western Turkey, and Gemici et al. (2008) found high spatial variations, from 33 to 911 µg/L, in groundwater samples. Analysis of 27 well water samples from the surficial aquifer in the Simav Plain of Western Anatolia of Turkey yielded average results of 99.1 µg/L, with a maximum of 561.5 µg/L and 7 samples out of

27 contained below 10 µg/L of arsenic (Gunduz et al., 2010). Gunduz et al. (2015) further examined groundwater arsenic concentrations ranging between 7.1 and 833.9 µg/L in the Simav Plain. Arsenic released from geological formations is considered to be the most important source in the Simav Plain. Arsenic concentrations in the Çeltikçi basin ranged from 3.0 to 7.3 µg/L in the wet season and <0.1–1.6 µg/L in the dry season (Özdemir, 2013); the figures were recorded between 1.1 and 19.7 µg/L in the wet season and 1.0–21.8 µg/L in the dry season in the Çöl basin (Balın, 2015). Analyzing 62 water samples from Aksaray Province of Turkey, 980 m above sea level, it was found that some 8% of the samples contain arsenic of more than 50 µg/L and 46.77% retained the safe level within the WHO and Turkish standards (Altaş et al., 2011). In the Sandıklı basin, arsenic concentrations in cold groundwater were found to range from <0.5 to 102.1 µg/L in the wet season and <0.5–126.6 µg/L in the dry season (Aksever, 2011), while arsenic in thermal water samples were analyzed to be between 514.4 and 866.9 µg/L (Davraz, 2015).

Arsenic-rich groundwater is associated with boron-rich sediments in the Kütahya area. Çolak et al. (2003) identified two distinct aquifers in the Kütahya area: a lower, karstified marble containing geothermal water with concentrations of 10–20 µg/L and a more important Upper Limestone aquifer containing up to 7700 µg/L. The sedimentary boron and arsenic were probably both derived from volcanic geothermal waters. Çolak et al. (2003) attributed arsenic mobilization to dissolution (oxidation) of realgar and orpiment by groundwater flowing along the base of the limestone. Kavaf and Nalbantcilar (2007) described more extensive contamination of river water with arsenic (up to 37 µg/L) and groundwater from springs and wells (up to 136 µg/L) drawn both from recent alluvium and the underlying Kütahya Formation, which consists of lavas and interbedded sediments. Gemici and Tarcan (2004) described arsenic-rich groundwater from around the Heybeli Spa at Afyon in western Turkey, which contains up to 1240 µg/L. In addition, arsenic is mobilized by sulfide-oxidation in concentrations of up to 463 µg/L in a karstic limestone aquifer in the Nif Mountain area southeast of İzmir (Simsek et al., 2008). Transported sediments from surrounding slopes have been deposited within the Simav plain to form the alluvial surface aquifer, which contains variable levels of iron oxide and arsenic along the vertical and horizontal cross-sections (Gunduz et al., 2010).

Dogan et al. (2005) conducted clinical investigations in two villages and identified widespread skin lesions, with an apparent dose-response relationship. At Dulkadir village, arsenic concentration in drinking water ranged from 300 to 500 µg/L, and 3 of the 56 people examined had skin lesions. At Igdekoy village, the water contained around 9000 µg/L of arsenic, and 33 of the 99 persons examined had skin disorders including keratosis, hyperkeratosis, hyperpigmentation, BCC, and Bowen's disease. Investigating a total of 402 death cases, Gunduz et al. (2015) concluded that cancers of lung (44.3%), prostate (9.8%), colon (9.8%), and stomach (8.2%) were comparably higher in villages with elevated arsenic levels in drinking water supplies. Furthermore, the majority of cases of liver, bladder, and stomach cancers were observed in villages with high arsenic levels (Gunduz et al., 2015).

2.2.13 RED RIVER BASIN AND MEKONG DELTA, VIETNAM

Groundwater is the main source of drinking water for communities in Vietnam, and approximately 13 million people (16.5% of the population) are drinking water from contaminated tubewells (Nguyen et al., 2009). Naturally occurring high arsenic in groundwater was first reported in 2001 in the Red River Delta of orthern Vietnam (Berg et al., 2001) and in 2005 in the Mekong Delta of southern Vietnam (Stanger et al., 2005). In the Red River Delta, high concentrations of up to 3050 µg/L were observed in the groundwater of the Gia Lam district, with a mean of 127 µg/L; in Thanh Tri district, the highest concentration was recorded to be 3010 µg/L, with a mean of 432 µg/L (Agusa et al., 2006). Arsenic in Ha Nam Province ranged from 3.0 to 486 µg/L, with a median of 256 µg/L, and in the Ha Tay Province, the figures ranged from 132 to 344 µg/L, and almost all of the groundwater samples (>95%) exceeded arsenic concentrations of 10 µg/L (Agusa et al., 2006). Nguyen et al. (2009) found elevated levels of arsenic in the groundwater of three villages (Vinh Tru,

Bo De, and Hoa Hau) of Ha Nam province in the Red River Delta with the average total arsenic concentrations of 348 µg/L, 211 µg/L, and 325 µg/L, respectively, while the arsenic level in groundwater of Nhan Dao village on the bank of the Red River was much lower than in other villages and half of the samples exceeded 10 µg/L (Nguyen et al., 2009). Arsenic in the groundwater of rural Hanoi in the Red River Delta area ranges between 1 and 3050 µg/L, with an average concentration of 159 µg/L and concentrations found in 72% tubewells above 10 µg/L, 48% above 50 µg/L, and 20% above 150 µg/L) (Berg et al., 2001).

Berg et al. (2007) identified concentrations from 1 and 845 µg/L, with a mean of 39 µg/L in the Mekong delta in Vietnam; arsenic was present in the 27% of tubewells over 10 µg/L, and 12% of the tubewells were above 50 µg/L. Hug et al. (2008) observed 20% of tubewells exceeding 10 µg/L and about 17 million people at risk of arsenic poisoning. Analyzing some 47 samples from the Mekong Delta, Nguyen and Itoi (2009) found total dissolved arsenic ranging between 1 and 741 µg/L. They also showed a variation with depth. Concentrations over 100 µg/L were identified in wells shallower than 100 m, and wells deeper than 150 m generally had low arsenic concentrations. Shinkai et al. (2007) show arsenic contaminations in Tien Giang Province and Dong Thap Province in the Mekong delta from 0.9–8.8 µg/L and 1.6–321 µg/L, respectively, with 27% of the shallow tubewells exceeding 10 µg/L in both regions. In the analysis of 68 groundwater samples from Dong Thap Province, adjacent to the Mekong River, Merola et al. (2015) found that some 53% of the wells had an arsenic content above 10 µg/L and wells close to the Mekong River had a maximum concentration of arsenic (329 µg/L); arsenic content in nail clippings collected from local residents was significantly correlated to arsenic in drinking water.

Health problems caused by arsenic ingestion through drinking water have been reported in Vietnam. Analyzing 68 water and 213 biomarker samples (human hair and urine), Agusa et al. (2014) investigated arsenic concentrations in groundwater in four districts (Tu Liem, Dan Phuong, Ly Nhan, and Hoai Duc) in the Red River Delta, with results in the range of <1 to 632 µg/L, with severe contamination found in the communities of Ly Nhan, Hoai Duc, and Dan Phuong districts. Human hair samples had arsenic levels in the range of 0.07–7.51 µg/g, and among residents exposed to arsenic levels ≥50 µg/L, 64% of them had hair arsenic concentrations higher than 1.0 µg/g, which is a level that can cause skin lesions, and urinary arsenic concentrations were 4.0–435 µg/g creatinine (Agusa et al., 2014). Approximately 0.5–1 million people in the rural Hanoi area are at a risk of chronic arsenic poisoning (Berg et al., 2007). Stanger et al. (2005) found serious localized health hazards, and some risk of low-level arsenic ingestion through contaminated rice and aquaculture. Phung et al. (2017) recently discovered that exposure to an elevated level of arsenic is a link to cardiovascular disease in Vietnam. Groundwater arsenic concentrations in the Red River Delta are related to the microbially induced reduction of iron oxyhydroxides (Berg et al., 2007; Eiche et al., 2017; Postma et al., 2012; Sø et al., 2017). In southwest Hanoi, the Holocene sediments contain high amounts of natural organic matter (Berg et al., 2008; Kuroda et al., 2017), and Larsen et al. (2008) observe that arsenic is mobilized in the Holocene aquifer in the Pleistocene aquifer on the Red River flood plain.

2.3 ARSENIC IN NORTH AMERICA

2.3.1 CANADA

Elevated levels of arsenic concentrations in groundwater are frequent in Halifax County, Nova Scotia, and Saskatchewan, Ontario, and British Columbia. Arsenic concentration in well water in Canada was first investigated by Wyllie (1937) who reported that water from some deep wells in Rocky Mountain areas of Ontario was known to contain large amounts of arsenic, ranging between 100 and 410 µg/L, as arsenic trioxide. High arsenic concentrations occur in groundwater are frequent in western Quebec, and 59 private well water samples reveal that more than half of the bedrock wells exceed the Canadian guideline value of 10 µg/L for arsenic, whereas shallow wells in unconsolidated surficial deposits are not contaminated (Bondu et al., 2017). In British Columbia, 60% of groundwater samples taken from an area of sulfide mineralization in Bowen Island exceeded 10 µg/L and 44% exceeded 50 µg/L, with a maximum of 580 µg/L (Boyle et al., 1998). Very high readings, up to 11,000 µg/L, have been reported in the vicinity of an abandoned arsenical wood preservative facility near Vancouver (Henning and Konasewich, 1984). White et al. (1963) reported a maximum dissolved arsenic concentration of 230 µg/L in Ellis Pool, Alberta.

Arsenic is widespread in private drinking water wells throughout the province of Nova Scotia in Canada, beyond the Canadian drinking water guideline of 10 µg/L (Goodwin et al., 2010) and approximately 46% of the population receive their drinking water from private wells (Dummer et al., 2015). Meranger et al. (1999) reported that levels of total soluble inorganic arsenic exceeding 50 µg/L were observed in 33%–93% of the wells in each of seven communities in Nova Scotia and the total measured levels varied from 1.5 to 738.8 µg/L, with concentrations of more than 500 µg/L in 10% of sample wells. A published groundwater chemistry atlas of Nova Scotia indicated that about 9.0% of 898 private well water samples across the province exceeded the current arsenic drinking water guideline of 10 µg/L, and several well water samples in Halifax County were over 3000 µg/L (Kennedy and Finlayson-Bourque, 2011). In analyzing 10,498 well water samples, it was reported that about 17% of the samples were found to have arsenic of more than 10 µg/L and the maximum concentration was recorded at 3900 µg/L (Dummer et al., 2015).

There have been chronic health problems in some areas. For instance, one person died of arsenic dermatosis in Ontario, with the other family members also afflicted by poisoning (Mandal and Suzuki, 2002), and Dummer et al. (2015) have explored the spatial variation in toenail arsenic concentrations in Nova Scotia. Examining 2000 pregnant women and their infants with biomarker samples between 2008 and 2011 from 10 Canadian cities, Ettinger et al. (2017) observed higher arsenic levels of total arsenic in blood in more than four-fifths of selected pregnant women and newborns.

The most common source of arsenic in Canada is sulfide minerals. Historical gold mining operations have led to elevated concentrations of arsenic in surficial soils throughout Canada, and the dissolved arsenic concentrations are higher than 25 µg/L in some places (Kwong et al., 2007; Saunders et al., 2010). The main natural sources of arsenic are weathering and erosion of arsenic-containing rocks and soil, while tailings from historic and recent gold mine operations and wood preservative facilities are the principal anthropogenic sources (Wang and Mulligan, 2006). Biswas et al. (2017) proved that sorption to iron oxides is a geochemical control in the release of arsenic to groundwater during pyrite oxidation.

2.3.2 Mexico

High levels of arsenic in drinking water were first identified as the cause of adverse health effects at Comarca Lagunera in Mexico in 1958 (Cebrián et al., 1994). A number of sites in Mexico were reported for the most arsenic contamination zones (Figure 2.4): (a) Región Lagunera, 8 and 624 µg/L (Del Razo et al., 1990); (b) Valle del Guadiana (located in the state of Durango, in north-central Mexico), 5 and 167 µg/L (Alarcón-Herrera et al., 2001); (c) Valle de Zimapán (in the state of Hidalgo, in central Mexico), 14 to 1097 µg/L (Armienta et al., 1997a); and (d) the city of Hermosillo (in the state of Sonora, in northern Mexico), up to 305 µg/L (Wyatt et al., 1998a).

The groundwater of the Región Lagunera, which represents the main source of drinking water for more than 2 million people, shows high arsenic concentrations (Boochs et al., 2014). Arsenic concentrations range from 5 to 750 µg/L in the Región Lagunera (Billib et al., 2012). In 128 well water samples in 11 counties in the Región Lagunera of the states of Durango and Coahuila, Del Razo et al. (1990) noted over 50 µg/L of arsenic in more than half, with a variance from 8 µg/L to 624 µg/L and an average of 100 ± 120 µg/L. Del Razo et al. (1994) also analyzed the drinking water of Santa Ana town in this region, finding 404 µg/L of arsenic; and the results of Hernández-Zavala et al. (1998) were 14.0 ± 3.1 µg/L in Nazareno, 116.0 ± 37 µg/L in Santa Ana, and 239.0 ± 88 µg/L in Benito Juárez. High arsenic levels (390 µg/L) in drinking water have also been reported in

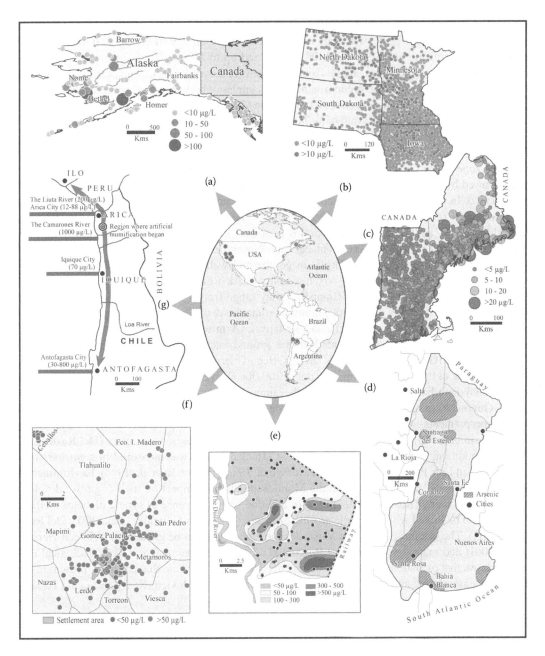

FIGURE 2.4 Groundwater arsenic concentrations in some regions of North American and South American countries: (a) Alaska, USA (Smith, D. P. 2013. Relationships between the health of Alaska Native communities and our environment - Phase 1, Exploring and communicating. US Geological Survey Fact Sheet: 2013-3066 (https://pubs.usgs.gov/fs/2013/3066/).); (b) Upper Midwest, USA (Erickson, M. L., R. J. Barnes. 2006. *Applied Geochemistry* 21:305–317); (c) New England, USA (Robinson, G. R. Jr., and J. D. Ayotte. 2006. *Applied Geochemistry* 21: 1482–1497); (d) Chaco-Pampa Plain, Argentina (Bundschuh, J., B. Farias, R. Martin et al., 2004. *Applied Geochemistry* 19(2):231–243); (e) Robles area, Argentina (Bundschuh, J., M. I. Litter, P. Bhattacharya. 2010. *Environmental Geochemistry and Health* 32(4):307–315); (f) Region Lagunera, Mexico (Del Razo, L. M., M. A. Arellano, M. E. Cebrián. 1990. *Environmental Pollution* 64:143–153); and (g) Chinchorro, Chile (Arriaza, B., D. Amarasiriwardena, L. Cornejo et al., 2010. *Journal of Archaeological Science* 37:1274–1278; Byrne, S., D. Amarasiriwardena, B. Bandak et al., 2010. *Microchemical Journal* 94:28–35.)

Coahuila (García-Vargas et al., 1994) and different towns of Zimapán valley in central Mexico, varying from 21 to 1070 µg/L, with only one safe sample at Dehdo well (21 µg/L) (Gómez-Arroyo et al., 1997). Armienta and Segovia (2008) discovered up to 1100 µg/L of arsenic in deep wells, and about 34% of samples from shallow and deep wells had over 50 µg/L. The pentavalent form of arsenic [As(V)] was found to be predominant in the samples of both Región Lagunera and Zimapán.

Wyatt et al. (1998b) showed high levels of arsenic (2 µg/L to 305 µg/L in 173 samples) in water from wells in the city of Sonora. Analyzing 74 well water samples from Durango City in the Guadiana valley in northern Mexico, Alarcón-Herrera et al. (2001) found more than 50 µg/L of arsenic in 59% of wells in the city and 48% in the rural area, with the highest concentration of 167 µg/L in the volcanic zone of the Guadiana valley. Gutiérrez-Pizano et al. (1996) found 32 times more arsenic than allowed by the national drinking water standard (25 µg/L) in well water in Acámbaro of central México. In the Los Altos de Jalisco region, amounts above the Mexican guideline value were isolated in 34% of 129 sample wells, and the highest values were found in Mexticacán (262.9 µg/L), Teocaltiche (157.7 µg/L), and San Juan de los Lagos (113.8 µg/L) (Hurtado-Jiménez and Gardea-Torresdey, 2006). In a study carried out between November 1994 and May 1996, Armienta and Segovia (2008) found up to 24,000 µg/L in geothermal wells and 8,000 µg/L in surface waters in Los Azufres in Michoacán state. Arsenic contamination in soil is also frequent. Some 73 soil samples collected in the semi-arid Zimapán Valley range from 4 to 14,700 mg/kg with a median value of 52 mg/kg (Ongley et al., 2007), and in some areas of the region, arsenic concentrations were found to range between <14 µg/L and 1090 µg/L (Armienta et al., 1997b).

The Región Lagunera is documented with a groundwater arsenic problem linked to chronic health problems (González-Horta et al., 2015; Ortega-Guerrero, 2016; Robles-Osorio et al., 2012). More than 21% of the exposed population out of 200,000 showed peripheral vascular disease or gastrointestinal disturbances and at least one of the cutaneous manifestations (Hernández-Zavala et al., 1998). Out of 120 sampled inhabitants of Zimapán, some 81% showed some degree of skin affectation (hyperpigmentation, hypopigmentation, and hyperkeratosis); hyperkeratosis was diagnosed in 26.5% (Armienta et al., 1997b) and the figure was 16.3% in the Región Lagunera (Del Razo et al., 1990). In addition, arsenic content in the hair of affected people was measured in a smelter community in San Luis Potosi of Mexico at an average of 9.9 µg/g (Cebrián et al., 1994). Environmental exposure to arsenic among children in the Morales area has an impact on their cognitive development in Torreón city (Rosado et al., 2007). In addition, the occurrence of type 2 diabetes mellitus was found to be linked with inorganic arsenic exposure (Coronado-González et al., 2007; Del Razo et al., 2011). The origin of arsenic in Mexico is mostly assumed to be geological (Alarcón-Herrera et al., 2001) and mainly volcanic (Del Razo et al., 1990). Geochemical and hydrogeological evidence shows that oxidation of arsenopyrite and dissolution of scorodite that naturally present in the aquifer release arsenic and contaminate the limestone aquifer in Mexico (Rodríguez et al., 2004).

2.3.3 THE USA

The occurrence of groundwater arsenic concentrations with more than 10 µg/L has been identified in Alaska, Arizona, California, Hawaii, Idaho, Nevada, Oregon, Utah, and Washington in the USA (Figure 2.4). About 40% from regulated water supplies out of analyzed 17,000 groundwater samples in the USA have arsenic concentrations over 1 µg/L, and about 5% of regulated water systems are estimated to have more than 20 µg/L of arsenic (Welch et al., 1999). Using a 25-state database of compliance monitoring from community systems, the USEPA (2001b) found that 5.3% of groundwater systems and 0.8% of surface water systems had over 10 µg/L. In a national retrospective groundwater study of 18,850 drinking water samples (2262 from community wells and 16,602 from private wells), the USGS found the 90th percentiles for community wells and private wells to be 8 µg/L and 13 µg/L, respectively (Focazio et al., 2000). The natural occurrences of arsenic in groundwater are most commonly the result of (a) the up-flow of geothermal water, (b) dissolution of or desorption from iron oxide, and (c) dissolution of sulfide minerals (López et al., 2012; Muñoz et al., 2016; Verplanck et al., 2008).

In analyzing 243 water samples, some 28% samples contained arsenic <50 µg/L, 40% had <100 µg/L, and about 20% were above 100 µg/L in well water and springs in Fairbanks, Alaska (Wilson and Hawkins, 1978). In Ester Dome near Fairbanks, readings are from <1 to 1160 µg/L, with a median value of 146 µg/L (Verplanck et al., 2008). Mueller (2002) indicated that some 38% of well water samples exceeded 10 µg/L of arsenic in Fairbanks. Arsenic concentrations as high as 15,000 µg/L have been encountered in wells of the Paleozoic aquifer of eastern Wisconsin, and in some townships, 20%–40% of the wells tested were above 10 µg/L (Luczaj and Masarik, 2015). Millard County, Utah, is reported to have groundwater arsenic concentrations of 1.8–210 µg/L, and the highest average concentration of arsenic in hair was 1210 µg/kg among the Hinckley residents and arsenate is the predominant (86%) arsenic species (Southwick et al., 1983: 210). Lewis et al. (1999) reported hypertensive heart disease, nephritis, neprosis, and prostate cancer among the people of the arsenic-affected areas in Utah.

In New Hampshire, about 12% of 794 randomly selected private wells exceed 10 µg/L of arsenic, with the maximum concentration of 180 µg/L (Peters et al., 2006). In southeast New Hampshire, Flanagan et al. (2014) found that 17% of 232 wells exceeded 10 µg/L, with the maximum concentration at 140 µg/L. Moreover, in the bedrock aquifer with the presence of arsenic-bearing arsenopyrite minerals, about 62% of 127 water samples exceeded 10 µg/L of arsenic, with the maximum concentration of 398 µg/L (Peters et al., 2006). Arsenic in well water in certain areas of Oregon ranged from ND amounts to 2000 µg/L (Whanger et al., 1977). The average arsenic concentration in blood levels was 32 µg/L in people living in areas of Oregon where water is known to be highly contaminated. In Lessen County, California, concentrations vary between 50 and 1400 µg/L (Goldsmith et al., 1972) and in the San Joaquin Valley, the figure ranges from <1 to 2600 µg/L (Fujii and Swain, 1995).

In the Piedmont regions of the Newark Basin, 23% of 53 selected wells in Pennsylvania (Peters and Burkert, 2008) and 15% of 94 sampled wells in New Jersey (Serfes et al., 2005) contained high (>10 µg/L) concentrations of arsenic. A dataset of 12,263 private wells showed that 12% of tested wells exceed the New Jersey Maximum Contamination Level (MCL) of 5 µg/L, and over 40% of the wells exceeding 5 µg/L of arsenic are in certain Piedmont regions (Spayd et al., 2015). Water chemistry data from 5023 groundwater samples collected between 1997 and 2007 in Pennsylvania show that the glacial aquifer in the central lowland province in northwest Pennsylvania also has high rates (20%) of arsenic with >10 µg/L, with the maximum concentration of 293 µg/L (Gross and Low, 2013). Welch and Lico (1998) reported high concentrations exceeding 100 µg/L, but with extremes of up to 2600 µg/L, in shallow groundwater from the southern Carson Desert.

In New Brunswick, some 6% of 10,555 samples of well water contained high (>10 µg/L) arsenic concentrations, with a maximum of 850 µg/L, and for a level of >10 µg/L, it was found in 18.4% of analyzed 11,111 well water samples in Maine, with the maximum of 3100 µg/L (Klassen et al., 2009). Foust et al. (2004) found 10–210 µg/L in 41 water samples from Verde Valley watershed in central Arizona, predominantly present as arsenate. In Vermont, 22% of 236 samples of private well water exceeded 10 µg/L of arsenic, with the maximum concentration of 170 µg/L of arsenic (Ryan et al., 2015). Analyzing 217 water samples from Central Appalachia (eastern Kentucky, western West Virginia, southeastern Ohio, and northeastern Tennessee), Shiber (2005) found that only 6% had arsenic exceeding 10 µg/L, but the highest was recorded at 84.3 µg/L in Ohio.

2.4 ARSENIC IN SOUTH AMERICA

2.4.1 Chaco-Pampean Plain, Argentina

Natural occurrences of arsenic have been documented in shallow groundwater in Argentina, particularly in the Chaco-Pampean Plain of northern and central Argentina, Córdoba, Salta, Jujuy, La Pampa, Santa Fe, and Tucumán Provinces, where drinking water wells contain 50 to 15,000 µg/L

(Alcaine et al., 2012; Bundschuh et al., 2012; Nicolli et al., 2012; Panigatti et al., 2014; Smedley et al., 2008). Arsenic intoxication in Argentina was first mentioned by Dr. Mario Goyenechea, specifying the origin from the ingestion of arsenical water by two patients from Bell Ville city in the province of Cordoba (Goyenechea, 1913). Therefore, the clinical manifestations were called "Bell Ville disease" (Auge, 2014). Figure 2.4 shows the spatial distribution of arsenic in parts of Argentina and some North and South American countries.

Groundwater arsenic concentrations in the Chaco-Pampean Plain are reported to be as high as 600 µg/L (Farías et al., 2003). The Chaco-Pampean Plain is densely populated and consumers rely heavily upon shallow aquifers for their drinking purposes. In Córdoba Province, sampling indicates arsenic concentrations between 6.0 and 11500 µg/L, with a median of 255 µg/L (Nicolli et al., 1989). In Tucumán Province, Nicolli et al. (2001) found the figures between 12 µg/L and 1660 µg/L, with a median of 46 µg/L. Groundwater arsenic concentrations were found to be between 0.7 and 1990 µg/L, and 91% of 45 sample well water exceeded 10 µg/L in the central region of the Chaco Province (Giménez et al., 2014). Analyzing 687 water samples in 192 locations, representing 86.4% of the total population from the province of Tucumán, Ojeda et al. (2014) observed that 67.7% of surveyed sites (130) had a concentration of arsenic within <10 µg/L, representing 90.8% of the population of the province. In La Pampa, the spectrum is <4.0 µg/L and 5280 µg/L, and the median is 145 µg/L (Smedley et al., 2008). The concentrations of arsenic in the drinking water of northern Argentina (San Antonio de los Cobres and Taco Pozo) are about 20 times higher than the WHO permissible limit (Concha et al., 1998).

About 87% of the provincial territory of Buenos Aires has groundwater that exceeds 50 µg/L of arsenic (Auge, 2014). Analyzing 33 well water samples from Trenque Lauquen County in Buenos Aires province, Moschione et al. (2014) found that two-thirds of the wells contained arsenic at more than 50 µg/L, and one-third contained arsenic of less than 50 µg/L. Campaña et al. (2014) found concentrations in shallow wells (up to 40 meters) in the southwestern part of Buenos Aires province from 15 to 267 µg/L and an increasing trend during the dry season. Ferral et al. (2014) observed arsenic concentrations in the Puelche aquifer in Mataderos of Buenos Aires that ranged from 10 to 25 µg/L and 40% of the samples exceeded the Argentine Food Code of 10 µg/L. In Robles County in the Río Dulce alluvial cone in the Santiago del Estero Province, Bundschuh et al. (2004) identified more than 48% of the 63 studied wells in shallow aquifers as being contaminated with arsenic, frequently above 1000 µg/L and sometimes up to 4800 µg/L. The deep groundwater is also contaminated but at the lower rate of <50 µg/L.

Concern about groundwater arsenic contamination in Argentina and its influence on human health was first raised by Ayerza (1918). The disease ascribed to arsenic contamination was later called "chronic endemic regional hydroarsenism" (HACRE, Hidroarsenicismo Crónico Regional Endémico). In addition, in Monte Quemado of Córdoba Province, there are skin manifestations caused by chronic arsenic ingestion, known collectively as the "Illness of Bell Ville" (Goyenechea, 1913). The occurrence of endemic arsenical skin disease and cancer was first recognized in 1942 in Argentina and epidemiological studies have highlighted the increased risk of skin, bladder, and lung cancers, as well as other cancers and noncancer diseases (Borgoño et al., 1977; Hopenhayn-Rich et al., 1998). A regular intake of drinking water containing more than 100 µg/L of arsenic leads to clearly recognizable signs of toxicity, and the ultimate symptom is skin cancer (Astolfi et al., 1981; Hopenhayn-Rich et al., 1998). Biagini (1972) followed 116 patients with clear signs of chronic arsenic disease over a number of years; after 15 years of follow-up, 78 people have died, 24 of them from cancer. Bates et al. (2004) found increased bladder cancer mortality in Córdoba Province with a 50-year latency among those ingesting contaminated drinking water.

Arsenic is dominantly present as arsenate, and metal oxides in the sediments are thought to be the main source of dissolved arsenic. The natural contamination of arsenic in groundwater is related to tertiary-quaternary volcanic deposits, together with post-volcanic geysers and thermal springs (Jha and Mishra, 2016: 351; Nicolli et al., 2012; Puntoriero et al., 2014; Smedley et al., 2005). The aquifers in the Pampean Plain comprise tertiary loess deposits, and, in the Chaco Plain, tertiary and quaternary alluvial and aeolian sediments are predominant (Bundschuh et al., 2012; Farías et al., 2003; Nicolli et al., 2012; Panigatti et al., 2014; Robles et al., 2016). In Monte Quemado of Córdoba Province, the problem seems to have been solved by building a canal that supplies the town with arsenic-safe water from Salta province (Mandal and Suzuki, 2002). Excessive irrigation in some areas of Santiago del Estero strongly affects the local geochemical and hydrochemical conditions and might be an important factor for arsenic mobilization in the areas (Claesson and Fagerberg, 2003).

2.4.2 BRAZIL

Arsenic contamination is prominent in Brazil, for instance, in the Ribeira Valley (Piririca belt), Amapá State (around potential gold deposits in granite-greenstone terrains), Rio das Velhas greenstone belt (Iron Quadrangle), northeastern region, and Rondonia State (Bidone et al., 2016; Ciminelli et al., 2017; Matschullat et al., 2000; Pontes et al., 2014; Rezende et al., 2015). A total of 18,670 arsenic analyses of stream sediments and soils are available in the geochemical data bank of the Geological Survey of Brazil, about 20% of which have arsenic levels higher than 100,000 µg/L (Figueiredo et al., 2007).

The Iron Quadrangle in the state of Minas Gerais in southeast Brazil is one of the richest mining areas in the world. Borba et al. (2003) estimated that over 300 years, nearly 400,000 tons of arsenic have been discharged into the rivers. The runoff water from some old gold mines in the region had very high arsenic contents, up to 2980 µg/L (Borba et al., 2003). Some 18 rivers samples have been analyzed at mean concentrations of 31 µg/L, with a maximum of 350 µg/L (Figueiredo et al., 2007). Campos (2002) identified 130–170 µg/L in shallow (12 m) domestic wells in São Paulo, which he attributed to excessive use of phosphatic fertilizers containing traces of arsenic. In 1998, the urinary arsenic of 126 schoolchildren in Minas Gerais had a mean concentration of 25.7 µg/L (range, 2.2–106 µg/L). There is 620 µg/g in summer and 1268 µg/g in winter in Rio de Carmo's stream sediments (Gonçalves et al., 2007). Borba (2002) detected 4000 µg/g in sediments in gold districts in the Carmo, Conceição, and das Velhas river basins, and in the groundwater of Ouro Preto and Mariana mines, arsenic was as high as 2800 µg/L.

The Ribeira Valley is contaminated with arsenic, and in drinking water, the maximum was 2000 µg/L in Santana, Amapá State (Figueiredo et al., 2007). Surface water is contaminated with 5 to 231 µg/L, but most samples contain <50 µg/L or even 10 µg/L, which is close to the 0.5 µg/L in residential tap water (Figueiredo et al., 2007). The maximum concentration of arsenic in sediment in the Ribeira Valley was recorded as 355 mg/kg (Toujague, 1999). The maximum arsenic concentrations were recorded as 27 µg/L in Mina do Chiquinho, 71 µg/L in Chafariz—Rua do Barão, 29 µg/L in Piedade-Tassara, and 224 µg/L in Biquinha da Santa Rita (Gonçalves et al., 2007). In the Nova Lima and Santa Bárbara municipalities of the Quadrilátero Ferrífero, Matschullat et al. (2000), studying the arsenic contaminations of 7- to 12-year-old children, found concentrations between 2 and 106 µg/g in their urine; 22% of these values were higher than 40 µg/g. Geographically, arsenic is distributed in the Quadrilátero Ferrífero rocks in close association with sulfide-rich, gold-bearing rocks (Borba et al., 2003; Deschamps et al., 2002). In the Ouro Preto municipality, arsenic contamination can be due to the rock types that constitute the aquifers (Gonçalves et al., 2007).

2.4.3 CHILE

High arsenic concentrations with severe poisoning from drinking water have been documented in the cities of Antofagasta, Atacama, Calama, and Tocopillae in Chile (Borgoño et al., 1977; Corradini et al., 2017; Herrera et al., 2014). Archaeologies of ancient Andean societies in northern Chile find them to have been highly contaminated with arsenic. The Chinchorro (a fishing community) and pre-Hispanic Azapa (an agrarian society), the earliest inhabitants of this region, were both already affected with arseniasis 7000 years ago (Arriaza et al., 2010). High concentrations of arsenic in hair samples from the ancient Chinchorro population concluded that they were significantly affected with arsenic poisoning throughout the generations (Byrne et al., 2010). These ancient populations were unknowingly poisoning themselves with the silent killer, arsenic.

Groundwater arsenic concentrations in Antofagasta were recorded between <100 and 21,000 µg/L (Borgoño et al., 1977; Santolaya et al., 1995) and in Calama City between <100 µg/L and >800 µg/L (Karcher et al., 1999). Between 1958 and 1970 almost 90% of the inhabitants (about 130,000) of Antofagasta were exposed to drinking water with a high arsenic content (800 µg/L) (Borgoño and Greiber, 1971). Drinking water arsenic contamination in the Atacama region of northern Chile is well known and has affected its population for many generations. Concentrations in Atacama Desert of Chile are variable, from <5 to 278 µg/L, and high concentrations are associated with high salinity (Leybourne and Cameron, 2008). In some parts of Tarapaca region, groundwater arsenic concentrations were found to be between 50 and 500 µg/L (Amaro et al., 2014).

The dermatological consequences of arsenicism were first noted in the children of Antofagasta in 1962 (Borgoño and Greiber, 1971). In a survey of 27,088 school children in northern Chile, some 12% were found to have the cutaneous changes of arsenicism; some 30% of these had suggestive systematic symptoms, and 11% had acrocyanosis (Mandal and Suzuki, 2002; Mukherjee et al., 2006a). Evidence of effects on the respiratory and cardiovascular systems, together with skin lesions, was also reported. Arsenic ingestion has been linked to increased bladder, kidney, and lung cancer mortality in Chile (Flynn et al., 2002; Fraser, 2012; Steinmaus et al., 2016). It has been estimated that around 7% of all deaths occurring in Antofagasta between 1989 and 1993 were due to past exposure to arsenic between 1955 and 1970 at concentrations of about 500 µg/L (Smith et al., 1998b). There is ample epidemiological evidence regarding the harmful effects of chronic arsenic exposure in the Chilean population (Hall et al., 2017; Hopenhayn-Rich et al., 2000; Rossman et al., 2004). The installation of an effective water treatment plant carried out by a process of adsorption with aluminum sulfate or iron hydroxide particles is being used for arsenic removal from water in Antofagasta.

The aquifers of Chile are composed mainly of volcanogenic sediments, and the sources of arsenic have been reported as quaternary volcanogenic sediments, minerals, and soil (Smedley and Kinniburgh, 2002). The contamination originates from arsenic-rich ores, and it is liberated by volcanic activity, which is common in the Andes region (Flynn et al., 2002). A common source of arsenic has been identified in the El Tatio geothermal field, and water there contains up to 27,000 µg/L (Romero et al., 2003). Chile also has soils contaminated with irrigation waters, and these contain between 86 and 446 mg/kg of arsenic compared to 64 mg/kg in control samples (Mandal and Suzuki, 2002). Arsenic contamination in the Elqui watershed area of Coquimbo Region has been assumed to be related to the quaternary volcanogenic sediments (Dittmar, 2004; Oyarzun et al., 2004).

2.5 ARSENIC IN EUROPE

2.5.1 France

Arsenic concentrations over 10 µg/L have been found in 45 administrative areas, and 17,000 people were found drinking water exceeding 50 µg/L in France (Grossier and Ledrans, 1999). The most contaminated groundwater is in and around the Massif Central, the Vosges, and the Pyrenees, and water between 10 and 50 µg/L has been located in sedimentary basins in the Aquitaine and Centre regions (Ravenscroft et al., 2009: 442). Bonnemaison (2005) noted that some 20% of thermal waters in France contain over 50 µg/L of arsenic and surface water contains about 0.73 µg/L of arsenic. Arsenic concentrations in the soil in a former gold mine in southern France were found to be high heterogeneous ranging from 29 to 18,900 µg/g (Drouhot et al., 2014). Elevated levels of dissolved arsenic concentrations were found in the Tinée and the Vésubie valleys, with the highest concentration of 263 µg/L and high heterogeneous distribution of arsenic in water is related to ore deposits containing arsenopyrite (Barats et al., 2014). Saoudi et al. (2012) observed a high level of urinary arsenic concentrations in the French adult population, with the geometric mean of 13.42 µg/L and 72.75 µg/L for the 95th percentile of arsenic.

2.5.2 Germany

Groundwater arsenic contamination in Germany is documented. In Saxony, only 2% of 150 sample wells were found to exceed the 1986 standard of 40 µg/L, but 40% exceeded 10 µg/L (Kevekordes et al., 1998). Heinrichs and Udluft (1999) described natural arsenic contamination, mostly at the 10 µg/L level, in the Upper Triassic Keuper sandstones of northern Bavaria. A maximum of 550 µg/L of arsenic was found in deep saline groundwater. The municipal wells, mostly 100–150 m depth, have concentrations ranging from 10 to 150 µg/L. The famous spa at Wiesbaden, which comprises 40 hot springs, has been known since 1886 to contain over 100 µg/L of arsenic (Schwenzer et al., 2001). In addition, Rott and Friedle (1999) reported arsenic between 15 and 38 µg/L in Rhine alluvium, and it is present mainly in the form of arsenite.

2.5.3 Greece

Elevated arsenic concentrations have been reported for many areas in Greece, especially Chalkidiki in the north. In several areas in northern Greece, people consume drinking water with arsenic concentrations of up to 30 µg/L. Arsenic in this area ranges from 1 to 1843 µg/L, with a mean concentration of 311 µg/L, and almost 65% of the wells examined exhibited more than the limit of 10 µg/L for drinking water (Kouras et al., 2007). In Thessaloniki, northern Greece, total arsenic is in the range between 4 and 130 µg/L, with a mean concentration of 46 µg/L; 80% of samples contain more than 10 µg/L, and about 55% have between 10 and 50 µg/L (Katsoyiannis and Katsoyiannis, 2006). Fytianos and Christophoridis (2004) reported that 13.5% of drinking water samples from 52 villages exceeded 10 µg/L of arsenic. In eastern Thessaly, the mean arsenic concentration in 26 water samples was 12 µg/L, with a maximum of 125 µg/L (Kelepertsis et al., 2006). Golia et al. (2015) investigated temporal variation of arsenic and heavy metals content in surface soils of the Almyros area in central Greece. Urban areas in Aksios and Kalikratia rely on arsenic-contaminated groundwater for their municipal supplies. Analyzing 21 samples, Katsoyiannis et al. (2007) found typical concentrations of 10–70 µg/L. A pronounced carcinogenic risk from arsenic exposure was evident from samples in the Kirki mining region of northeastern Greece (Nikolaidis et al., 2013).

2.5.4 The Great Hungarian Plain

In Hungary, the amount of arsenic in groundwater ranged from 60 and 4000 µg/L in samples tested between 1941 and 1983 (Nagy and Korom, 1983). Later, arsenic exceeding 50 µg/L was identified in groundwater from alluvial sediments associated with the River Danube in the southern part of the Great Hungarian Plain, the average being 32 µg/L (Varsányi, 1989). About half a million people in this region drink water containing more than 10 µg/L (Varsányi and Kovács, 2006). Arsenic has been found in the Great Hungarian Plain; the total arsenic concentration of 23 public well water samples were found to be between 7.2 and 210.3 µg/L, and 22 samples contained more than 10 µg/L of arsenic (Sugár et al., 2013). Analyzing 85 groundwater samples collected from drinking water wells of 80–560 m depth in Csongrád County, Varsányi et al. (1991) found a maximum reading of 150 µg/L. High levels have been reported in Pannonian Basin, and the arsenic content of sediments varies from 0.04 to 3.9 mg/kg in Bácsalmás; 0.51 to 9.6 mg/kg in Mindszent; 0.24 to 1.08 mg/kg in Tótkomlós; and 0.09 to 1.80 mg/kg in Szeghalom (Varsányi and Kovács, 2006). Börzsönyi et al. (1992) found arsenic in hair, confirming long-term exposure, and many cases of hyperkeratosis and hyperpigmentation were found in both adults and children in the contaminated area of southeast Hungary. Arsenic exposure above 10 µg/L during pregnancy was associated with an increased risk of congenital heart anomalies in general (adjusted OR:1.41; 95% CI: 1.28–1.56) (Rudnai et al., 2014). Positive associations between long-term exposure to inorganic arsenic <100 µg/L in drinking water and BCC in the skin were found in areas of Hungary, Romania, and Slovakia (Leonardi et al., 2012).

2.5.5 Italy

Elevated levels of arsenic concentrations are found along the Po, Adda, Adige, and Reno rivers in Lombardia, Emilia-Romagna, and Veneto in Italy. In Veneto, 21% of 1303 wells surveyed contained >10 µg/L of arsenic, 3% had more than 50 µg/L, and 2% had higher than 100 µg/L. The maximum arsenic concentrations reported were >400 µg/L in Lombardia, 480 µg/L in Veneto, and 1300 µg/L in Emilia-Romagna (Ravenscroft et al., 2009: 446). Tamasi and Cini (2004) described concentrations of up to 14.4 µg/L in springs originating from volcanic rocks in the Mount Amiata region of Siena. Vivona et al. (2005) reported arsenic concentrations of 4–52 µg/L in the Tiber River valley in a volcanic-alluvial aquifer, in volcanic aquifers of up to >50 µg/L, and where 10% of the wells exceeded 10 µg/L. In addition, geothermal arsenic is common around the volcanic centers of southern Italy, including Vesuvius, Etna, and Vulcano.

2.5.6 The Danube Basin: Romania, Slovakia, and Croatia

Arsenic concentrations are reported in Romania, Slovakia, and Croatia in the Danube Basin. The geographical distribution of arsenic in drinking water is especially noticeable in Bihor and Arad counties of the Transylvania region in Romania, near the Hungarian border (Lindberg et al., 2006). Samples collected between 1992 and 1995 show a maximum concentration of 176 µg/L in the associated aquifers (Gurzau and Gurzau, 2001). Estimates indicate that about 36,000 people are exposed to arsenic at 11 to 48 µg/L and about 14,000 at levels exceeding 50 µg/L (Gurzau and Gurzau, 2001). Elevated levels of arsenic concentrations in Slovakia were reported, and Rapant and Krčmová (2007) showed that groundwater arsenic concentrations in 1.2% samples exceeded 10 µg/L and in 0.3% exceeded 50µg/L. In eastern Croatia, Habuda-Stanić et al. (2007) recorded high arsenic concentrations in drinking water in and around the towns of Osijek (38 µg/L), Cepin (172 µg/L), and Andrijasevci (612 µg/L). Ćavar et al. (2005) estimated that 3% of the population of Croatia may be exposed to a serious health risk from arsenic in drinking water. Habuda-Stanić et al. (2007) stated that a total of 200,000 are exposed to arsenic over 10 µg/L. In Osijek and Vinkovci, groundwater is treated by coagulation-filtration, which reduces arsenic concentrations from around 250 to 40 µg/L, and other municipalities use rapid sand filtration, but this is no more effective (Ravenscroft et al., 2009: 430).

2.5.7 Scandinavia

Arsenic concentrations are reported in some Scandinavian countries, especially Finland and Sweden. In both the Pirkanmaa and Satakunta regions of Finland, arsenic concentrations are high (Backman et al., 2006; Jarva et al., 2008). Kurttio et al. (1999) found levels from <0.05 to 64 µg/L in 275 well water samples, and around 1% exceeded 10 µg/L of arsenic. Analyzing a total of 1151 water samples, Sorvari et al. (2007) found an average of 29.7 µg/L, with a standard deviation of 137 µg/L and a maximum of 2230 µg/L. Only 3% of samples exceeded 10 µg/L of arsenic, but the proportion was much higher in southwest Finland and parts of Lapland. In the Haukipudas area of northern Finland, Roman and Peuraniemi (1999) described groundwater containing up to 43 µg/L. Arsenic varies between 1000 and 2200 mg/kg in soils from mine sites (Parviainen et al., 2006: 46), and in arable soils, the average was 4.06 mg/kg, with a standard deviation of 1.03 mg/kg and a maximum of 6.8 mg/kg (Sorvari et al., 2007). Despite the very low exposure levels, Kurttio et al. (1999) found an association with the incidence of bladder cancer, while Hakala and Hallikainen (2004) reported limited health effects in the Finnish population and did not identify dermatological effects. Arsenic concentrations are reported in some parts of Sweden, but Skellefteå was found to be the worst affected. Analyzing some 44 tubewells and dug wells and 96 domestic wells from municipalities within Skellefteå, Svensson (2007) found spatial variations of arsenic concentrations between <0.5 µg/L and 300 µg/L, and some 29% of the drilled wells and 11% of the dug well samples had a concentration exceeding 10 µg/L.

2.5.8 Madrid and Duero Basin, Spain

High arsenic concentrations (>50 µg/L) have been detected in Spain in drinking water supplies from underground sources. In 1998, water samples from 353 water supplies in Madrid were analyzed; 74% of samples had <10 µg/L, 22.6% had levels of 10–50 µg/L, and 3.7% had >50 µg/L (Sanz et al., 2001). In the Madrid Basin, well water samples from depths of 50–476 m contained an average of 25 µg/L, with a maximum of 91 µg/L (Hernández-García and Custodio, 2004). Analyzing 28 water samples from different aquifer depths in the Tertiary Duero Basin of the North Iberian Meseta, arsenic was found to range between 20 and 260 µg/L (García-Sánchez et al., 2005). In an intensive survey of 514 wells in the Duero Basin, Gómez et al. (2006) reported a mean concentration of 41 µg/L, with a maximum of 613 µg/L. In addition, García-Sánchez and Alvarez-Ayuso (2003) analyzed arsenic concentrations of up to 52 µg/L in groundwater in the middle reaches of the Duero River in Salamanca Province, an area of mining activities. In addition, Morell et al. (2006) investigated arsenic concentrations of up to 14 µg/L in Triassic sandstones in the Mediterranean provinces of Castellon and Valencia. In analysis of 133 soil samples from Aragón, Navas and Machín (2002) found a peak concentration of 58.9 mg/kg, with a mean of 11.8 mg/kg and standard deviation of 10.9 mg/kg. In the Duero Basin, high arsenic concentrations in groundwater are closely associated with particular stratigraphic horizons, such as the Middle Miocene organic-rich Zaratan facies.

2.5.9 Switzerland

In Switzerland, areas with high arsenic concentrations have been found primarily in the thermal spring area of northern Switzerland, the Jura Mountains, and the Alps in the areas of Valais, Ticino, and Graubünden (Pfeifer et al., 2007, 2010; Zayre et al., 2006). The discovery of natural arsenic in Swiss groundwater started in 1989, when a Canadian mining company started an exploration campaign on the site of the former gold mine of Costa-Astano in southern Switzerland (Pfeifer et al., 2010). Analysis of 336 drinking water samples from public water supplies in 1998 from the canton of Graubünden found low concentrations (<10 µg/L) in 93% of the samples, but some 6% had between 10 and 50 µg/L, and 0.9% exceeded 50 µg/L; the maximum was recorded for 170 µg/L (Pfeifer and Zobrist, 2002). Drinking water in Valais was not tested for arsenic until 1999; since then, it has been determined that level of arsenic in drinking water contains between 5 and 50 µg/L (Pfeifer and Zobrist, 2002) and in 2004, about 14,000 people in this canton were exposed to arsenic from drinking water at >10 µg/L (Haldimann et al., 2005). In the Jura mountains, arsenic concentrations in drinking water were recorded below 10 µg/L, and in the canton of Ticino, the maximum concentration was recorded at 80 µg/L (Haldimann et al., 2005). Moreover, Arsenic concentrations were also recorded in Schinzach (25 µg/L), Baden (38 µg/L), Zurzach (123 µg/L), Saxon Valroc (15 µg/L), Combioula (24 µg/L), and Leukerbad (27 µg/L) (Pfeifer et al., 2010). Weathering and erosion of rocks containing arsenic releases this element into soils, sediments, and natural waters. Ore deposits and sediments in the canton of Valais have been known for some time to contain arsenic, and in the Jura mountains, arsenic-bearing ferruginous limestones and ferruginous red clays are at the origin of elevated arsenic concentrations in drinking water.

2.5.10 UNITED KINGDOM

Arsenic contamination in groundwater is distinct in the United Kingdom (Burgess and Pinto, 2005; Kavanagh et al., 1998; Millward et al., 1997). A nationwide survey of trace elements found that 10–15% of waters from sandstone aquifers contain 10–50 µg/L of arsenic (Edmunds et al., 1989), and there are no reports of adverse health effects. The most affected bedrock aquifer is the Triassic Sherwood Sandstone, which is a major water resource in the Midlands and northwest England. The maximum value in the Midlands was found to be 26 µg/L. In northwest England, 11.5% of 672 drinking water samples exceeded 10 µg/L, and 1.6% exceeded 50 µg/L. In the Lower Greensand, 11% of samples exceeded 10 µg/L, with a maximum of 20 µg/L in reducing waters (Shand et al., 2003). Middleton et al. (2016) found increased levels of arsenic (>10 µg/L) in several private water supplies in Cornwall, and some people were confirmed as having higher levels of arsenic in their urine, meaning they were consuming the toxin. The highest concentrations were recorded in Liverpool (355 µg/L), the Carlisle Basin (233 µg/L), Manchester (215 µg/L), Cheshire (57 µg/L), and the Vale of York (Shand et al., 1997; Griffiths et al., 2003, 2005). In the confined Chalk of North Humberside, Smedley et al. (2004) found that 30% of reducing groundwater exceeded 10 µg/L of arsenic, with a maximum concentration of 63 µg/L. A local arsenic occurrence (25 µg/L) was noted in the Lincolnshire limestone (Griffiths et al., 2006), but in other UK limestone aquifers, all the samples were found to be below 10 µg/L (Ravenscroft et al., 2009: 441). Alluvial or glacial aquifers in the UK experience arsenic contamination (AGMI, 2004).

2.6 ARSENIC IN AFRICA

2.6.1 BURKINA FASO

Groundwater arsenic is a natural phenomenon in Burkina Faso. In 1979, IWACO carried out an investigation of groundwater from three boreholes in a village close to Mogtédo in the central region Faso, and this found concentrations in the range between 200 and 1600 µg/L (Smedley et al., 2007). A number of sample hand-pumped boreholes and dug wells from Ouahigouya in northern Burkina Faso are contaminated with arsenic ranging from <0.5 to 1630 µg/L, with a median of 15 µg/L, although most of the analyzed groundwater samples have arsenic lower than 10 µg/L (Ravenscroft et al., 2009: 477; Smedley et al., 2007). Analyzing some 1498 well water samples from rural Burkina Faso, Bretzler et al. (2017) found arsenic concentrations above 10 µg/L in 14.6% of analyzed water samples, and around 560,000 people are potentially exposed to arsenic-contaminated groundwater in Burkina Faso. In analyzing some 200 groundwater samples from boreholes with hand pumps in the region, a range of concentrations from <0.8 to 560 µg/L was observed (Smedley et al., 2007). Skin disorders (melanosis, keratosis, and fewer skin tumors) were identified among the populations in some villages in northern Burkina Faso. The villagers now use alternative low-arsenic boreholes in the village, and the affected patients have apparently recovered. The high arsenic in groundwater is thought to be mobilized from zones of gold mineralization in Birimian (Lower Proterozoic) volcano-sedimentary rocks (Ahoulé et al., 2015; Bretzler et al., 2017; Smedley et al., 2007).

2.6.2 CAMEROON

Groundwater arsenic contamination is a natural phenomenon in Cameroon. Until the 1970s, drinking water was drawn predominantly from surface-water sources, and waterborne diseases such as cholera and dysentery caused hundreds of deaths. However, during the last decades, over 80% of potable water needs in Ekondo Titi have been provided by about 300 private and public wells tapping shallow aquifers, which contain variable amounts of arsenic (Mbotake, 2006). Arsenic was first identified in the groundwater here in 1986, following geochemical studies, but this information was effectively unknown in Ekondo Titi until late 2000. The occurrence of arsenic concentration ranges from 100 µg/L in shallow aquifers to 2000 µg/L in the deeper aquifers in the coastal alluvium, and Mbotake (2006) estimated that about 4000 people are probably at risk of arsenic poisoning. High concentrations occur in strongly reducing groundwater, suggesting that arsenic is mobilized by reductive dissolution (Ravenscroft et al., 2009: 478).

2.6.3 Ethiopia

Groundwater arsenic concentrations are frequent in deep and shallow wells, springs, and rivers along the Ethiopian section of the East African Rift Valley. Reimann et al. (2003) found pervasive water quality problems in Ethiopia. Analyzing 138 drinking water samples throughout the Ethiopian part of the Rift Valley, separated into water drawn from deep wells (>60 m deep), shallow wells (<60 m deep), hot springs (T > 36°C), springs (T < 32°C), and rivers, they found that only 9 samples (6.5%) were contaminated with arsenic above 10 µg/L, with a maximum of 96 µg/L. Arsenic is enriched in hot springs, which may impact upon surface waters. Reimann et al. (2003) noted clusters of deep wells with elevated levels of arsenic in the center of the Rift Valley, which they attributed to hydrothermal sources.

2.6.4 Ghana

High arsenic concentrations have been noted in soils, tubewells, boreholes, and river water close to the mines around the town of Obuasi of the Upper East region in Ghana (Asante et al., 2008). The first published report of arsenic pollution was at the Ashanti Gold Mine in southwest Ghana, where arsenic concentrations in the soil were 189–1025 mg/kg and in the groundwater were 86–557 µg/L (Bowell, 1994). Norman et al. (2001) tested 127 drilled and 76 dug wells in southwest Ghana using the Arsenator™ field kit and found that water from 18% of drilled wells and 3% of dug wells exceeded 5 µg/L of arsenic, with the maximum reading of 2000 µg/L. Arsenic concentrations vary from <1 to 64 µg/L and <1 to 141 µg/L in Obuasi and Bolgatanga, respectively (Smedley, 1996). The concentrations are low in the shallow groundwater, but increase at the greater depths of 40–70 m below ground level in Obuasi and 20–40 m in Bolgatanga. In Ashanti region, some 7% and 1% of the well water samples had arsenic levels above 10 µg/L and 50 µg/L, respectively, and up to 29% of the well water samples in Atwima district are more than 10 µg/L (Buamah et al., 2008).

Smedley et al. (1996) reported concentrations of arsenic of up to 175 µg/L in surface water and up to 65 µg/L in groundwater in the Obuasi gold mining area, and deep mine exploration boreholes (70–100 m) have relatively low arsenic contents at 5–17 µg/L. Arsenic analysis of 150 boreholes in the southeastern part of Ghana (Accra, Eastern, and Volta regions) revealed concentrations in the range between 2 and 39 µg/L, with only 2% of boreholes tested having arsenic concentrations exceeding 10 µg/L in drinking water (Kortatsi et al., 2008). In the western region, some 12% of sampled wells were found to have levels over 10 µg/L (Buamah et al., 2008). Well water samples analyzed in Wassa West and Wassa Amenfi districts exhibited high levels of arsenic, with 13% and 24% of the wells, respectively, exceeding 10 µg/L. Remarkably, 6% of the wells in these two districts had above 50 µg/L. Obiri (2007) analyzed concentrations of arsenic of up to 4500 µg/L at Dumasi in Wassa West district. In addition, the Brong-Ahafo region had the least incidence of groundwater arsenic, with only 5% of its wells exceeding 10 µg/L (Buamah et al., 2008).

Analyzing the hair and nail samples from inhabitants of Wassa West district, a major gold mining area in Ghana, arsenic was present in hair (0.07–0.95 µg/g), while the levels in nail samples ranged from 0.08 to 3.90 µg/g (Serfor-Armah et al., 2009). Generally, levels of arsenic in the hair are less than the WHO recommended value of 1.00 µg/g; however, five nail samples were above the maximum WHO guideline value of 1.08 µg/g (Serfor-Armah et al., 2009). Sulfide minerals such as arsenopyrite and pyrite are present in the Birimian basement rocks of both the Obuasi and Bolgatanga areas, and these form the dominant arsenic sources (Smedley, 1996).

2.6.5 NIGERIA

Groundwater arsenic contamination is evident in Nigeria. Analysis of 1908 samples by UNICEF from boreholes, dug wells, and water vendors in eight hydrological regions in Nigeria found that all the samples complied with a 10 µg/L threshold; however, samples from dug wells and boreholes along the Warri-Port Harcourt axis are contaminated with arsenic (280–750 µg/L) (Ravenscroft et al., 2009: 480). To the east of the delta, Edet and Offiong (2003) measured arsenic concentrations between 10 and 35 µg/L in surface waters of the Lower Cross River basin. There is a maximum of 200 µg/L and average concentration of 76 µg/L in dug wells in the limestone area of Ogun State in the southwest of Nigeria (Gbadebo and Mohammed, 2004). In an environmental impact assessment at Kaduna in north-central Nigeria, Oke (2003) reported high arsenic concentrations in surface water and wells dug in laterite.

2.7 ARSENIC IN OCEANIA

2.7.1 AUSTRALIA

Groundwater arsenic contamination has been detected in some parts of Australia. In an investigation of the relationship between environmental exposure to arsenic from contaminated soil and drinking water and the incidence of cancer in the Victoria region, particularly around gold mining areas, arsenic concentrations in groundwater were found to be in the range of <1 to 300,000 µg/L, and in surface water were <1 to 28,300 µg/L (Hinwood et al., 1999). Arsenic concentrations of up to 7000 µg/L were measured in shallow groundwater in the Gwelup area in Perth, and in 1976 no arsenic was detected in any of the investigated shallow wells, whereas in 2004 after extensive dewatering and peat excavation, concentrations were in excess of 1000 µg/L (Appleyard et al., 2006). Smith et al. (2003) reported natural arsenic in Holocene coastal barrier sands in the Stuarts Point Coastal Sands Aquifer of New South Wales (maximum 70 µg/L) and the underlying Yarrahapinni Fractured Rock Aquifer (maximum 337 µg/L). Arsenic varied from 7.4 to 396 mg/kg in soils of Ballarat-Creswick area of Central Victoria, and the average in soil (39.0 mg/kg) was markedly higher than the Environmental Protection Agency recommended maximum for soils and the Australian and New Zealand Environment Conservation Council's environmental level of 20 mg/kg (ANZECC, 1992). The mean level (39 mg/kg) in soils is well above the normal global range (6.0 mg/kg), and extreme levels (>190.0 mg/kg) in soils were found in areas of mine tailings sites in Australia (Sultan, 2007). In addition to anthropogenic sources such as mining activities and pesticide use, different minerals present a significant source of natural arsenic contamination to the environment.

2.7.2 NEW ZEALAND

Arsenic contamination is evident in some parts of New Zealand. An early case of arsenic poisoning from well water was described by Grimmett and McIntosh (1939) in the Waiotapu Valley in the Central Volcanic Plateau (CVP). The maximum arsenic concentration in geothermal water in the CVP is 8500 µg/L (Webster and Nordstrom, 2003). Springs and drains in the area contained up to 2000 µg/L of arsenic, and some drinking water wells also contained arsenic of up to 50 µg/L, although no cases of arsenicosis were reported (Grimmett and McIntosh, 1939). Approximately 10% of 157 monitoring sites have concentrations of >10 µg/L, with a maximum concentration of 260 µg/L (Ravenscroft et al., 2009: 484). The Waikato River, the longest river in New Zealand, contains geothermal arsenic originating from the CVP; it is the source of water supply to the city of Hamilton and contains up to 150 µg/L of arsenic (McLaren and Kim, 1995). Similarly, Mroczek

(2005) reported geothermal arsenic contributing to a background of 21 µg/L in the Tarawera River. Arsenic concentration ranged between 2.0 and 21 µg/L in the Rarangi Shallow Aquifer (RSA) near Marlborough (Wilkinson, 2005). The shallow (5–8 m) wells in the RSA that are used for drinking, dairy farming, and irrigation contain <1.0–43 µg/L of arsenic.

2.8 OTHER AREAS WITH ARSENIC CONTAMINATION

Arsenic poisoning has been reported in other countries. In Asia, high arsenic concentrations have been detected in groundwater in Laos (Chanpiwat et al., 2011), the Philippines (Lomboy et al., 2017; Sy et al., 2017), the Eastern Caucasus foothills in Russia (Zakharova et al., 2002), and Jurassic rocks of the South Mangyshlak zone in Kazakhstan (Kortsenshteyn et al., 1973). High levels of arsenic concentrations have also been found in well water in Kampong Sekolah on the western coast of Malaysia (Chow, 1986), along the Upper Citarum River in the west of Java, Indonesia (Sriwana et al., 1998), in eastern Afghanistan (Saltori, 2004), and in Mongolia (MoH, 2004). In an analysis of 867 samples from 21 provinces and Ulan Bator, some 10.3% of the samples contained arsenic and clinical examinations found evidence of arsenicosis in 16.5% of the exposed population (MoH, 2004).

In Latin America, arsenic was discovered in the groundwater of Nicaragua, Bolivia, El Salvador, Ecuador, and Honduras and recently in Uruguay, Colombia, Guatemala, Costa Rica, Cuba, and Venezuela (Bundschuh et al., 2010). In the southern Poopó basin of the Bolivian highlands, arsenic was revealed in shallow drinking water wells, from BDL to 207 g/L (Muñoz et al., 2014). In the Rio Tambo watershed of the north-central Andean region of Ecuador, Cumbal et al. (2006) reported arsenic (970–5080 µg/L) in geothermal sources. In El Salvador, Sancha and Castro (2001) reported arsenic concentrations of 150–770 µg/L from Ilopango Lake, which is used as a source of water for 300,000 people. Sancha and Castro (2001) reported arsenic in various Andean rivers of Peru, especially the Rio Locumba, where concentrations are around 500 µg/L. Espinoza and Bundschuh (2008) analyzed arsenic concentrations ranging between <10 and 122 µg/L in the Sebaco-Matagalpa Valley of eastern Nicaragua, where 37% of well water exceeded 10 µg/L.

Arsenic concentrations are also noticeable in some parts of Europe. In the Czech Republic, arsenic in shallow water contains up to 1500 µg/L (Drahota et al., 2006), 10–30 µg/L in eastern Denmark (Jessen et al., 2005), up to 28 µg/L in Ireland (Toner et al., 2005), and 33 µg/L in Lithuania (Ravenscroft et al., 2009: 450). In Norway, Frengstad et al. (2000) identified only 1% of 476 groundwater samples as exceeding 10 µg/L, with a maximum of only 19 µg/L. Kralj (2004) reported maximum concentration of 589 µg/L at Ljubljiana in west-central Slovenia. In the Neogene alluvial aquifer in northern Belgium, Coetsiers and Walraevens (2006) reported up to 60 µg/L, and in southwest Poland, the maximum concentration was 40 µg/L (Dobrzynski, 2007). This contamination is attributed to oxidation of pyrite and arsenopyrite. Arsenic contamination is not widely distributed in Africa. There is some evidence in Botswana (Huntsman-Mapila et al., 2006), Zambia (Amini et al., 2008), Zimbabwe (Grinspan and Biagini, 1985), and Egypt (Saad and Hassanien, 2001), but South Africa (Okonkwo, 2007; Röllin et al., 2017) is the most prominent.

2.9 NATURAL GEOCHEMICAL PROCESSES OF ARSENIC MECHANISM

Groundwater is the most important source of pathogen-free drinking water because much untreated surface water is contaminated with fecal bacteria, causing cholera, dysentery, diarrhea and other water-borne diseases. Background concentrations of arsenic in groundwater are not a problem in many countries since they are either less than the detectable limit or less than the WHO guideline value of 10 µg/L. Concentrations are the result of the strong natural influence of water-rock interactions and the favorable physical and geochemical conditions in aquifers for the mobilization of arsenic. Arsenic is particularly mobile at pH values typically found in groundwater (pH 6.5–8.5) under both oxidizing and reducing conditions (IARC, 2004).

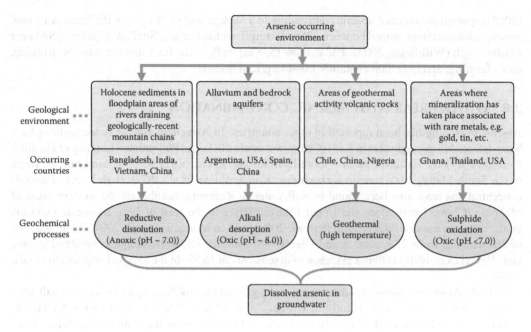

FIGURE 2.5 Natural geochemical processes that release arsenic into groundwater in different arsenic-contaminated countries of the world. (Modified from Herath, I., M. Vithanage, J. Bundschuh et al., 2016. *Current Pollution Reports* 2(1):68–89.)

Geochemical processes including reductive dissolution, alkali desorption, sulfide oxidation, and geothermal processes that can initiate the natural release of arsenic from aquifer materials into groundwater (Bundschuh et al., 2013; Herath et al., 2016) (Figure 2.5). The mechanism of releasing arsenic into groundwater may occur in a variety of geochemical environments; however, an important occurrence of arsenic is in the young alluvial basins adjacent to active mountain belts (Herath et al., 2016). The geochemical processes, except sulfide oxidation, occur independently in the subsurface environment. Nevertheless, tapping of groundwater with tubewells leads to widespread distribution of dissolved arsenic in the subsurface (Bhattacharya et al., 2009). The reductive dissolution of arsenics-bearing Fe(III) oxides and sulfide oxidation in aquifer sediments are thought to be the most significant geochemical triggers that release arsenic into groundwater to a large extent (Bauer and Blodau, 2006). Adsorption controls the mobility of arsenic into groundwater. Responses of adsorption between arsenic and mineral surfaces are important in controlling dissolved concentrations of arsenic in groundwater.

The amount of arsenic dissolved in water is attributed mostly to dissolution-precipitation, oxidation-reduction, and adsorption (Welch et al., 1988; Ferguson and Gavis, 1972); however, the importance of hydrogeology and bacterially mediated reactions is also emphasized (Saunders et al., 2005; Dowling et al., 2002; Nickson et al., 2000). There are three major types of natural geological conditions giving rise to high levels of arsenic in groundwater (IARC, 2004): (a) aquifers composed of rocks or sediments enriched with arsenic-containing minerals of geogenic origin; (b) aquifers containing sediments coated with iron oxyhydroxide (FeOOH) phases enriched in arsenic through hydrological action, where arsenic is mobilized into pore-water by reducing conditions; and (c) aquifers enriched in arsenic through high rates of evaporation in arid areas, leading to increased mineral concentration in groundwater. In addition, during the weathering of arsenic-containing minerals, arsenic enters the water mainly as soluble arsenate.

Geochemical conditions similar to the alluvial sediments in Bangladesh exist in the Red River alluvial tract in the city of Hanoi, Vietnam, where FeOOH reduction is thought to have led to the high arsenic levels recorded in groundwater (Berg et al., 2001). Smedley and Kinniburgh (2002) outline that the reducing conditions observed in the Bengal Delta and Vietnam aquifers are similar to those in the regions of Taiwan, Inner Mongolia, and Xinjiang of China, and Hungary that suffer from elevated levels of arsenic in groundwater. In Finland, the highest levels of arsenic were found in groundwater from wells drilled in Precambrian bedrock (Lahermo et al., 1998). Geothermal water and high evaporation rates are associated with arsenic concentrations ≥ 10 μg/L in groundwater of the USA (Welch et al., 2000).

2.10 CONCLUDING REMARKS

High concentrations of naturally occurring arsenic in groundwater are reported in countries of Southeast Asia, some countries in Europe, some countries in Latin America, and some parts of the USA. Southeast Asian countries, including Bangladesh, India, China, Vietnam, Taiwan, Thailand, and Nepal, are of particular concern because of widespread arsenic-related diseases. The western part of the USA, including Michigan, Minnesota, South Dakota, Oklahoma, and Wisconsin as well as Maine, is contaminated with groundwater arsenic poisoning. Arsenic contamination is also documented in some European countries, for instance, Greece, Hungary, Romania, Croatia, Serbia, Germany, Turkey, Spain, and the UK. Moreover, many Latin American countries, including Argentina, Chile, Bolivia, Mexico, and El Salvador, are exposed to elevated levels of arsenic concentrations in their groundwater. This worldwide distribution of arsenic in groundwater poses a significant risk to human health, and about 300 million people worldwide are at risk of arsenic poisoning. The primary source of groundwater arsenic is natural and derived predominantly from interactions between groundwater and aquifer sediments of minerals, including pyrite, arsenopyrite, and other sulfide minerals. The geochemistry of arsenic is a function of multiple oxidation states, speciation, and redox transformation. Mobilization of arsenic in groundwater is controlled by adsorption onto metal oxyhydroxides and clay minerals.

3 Groundwater Arsenic Discontinuity

Spatial Mapping, Spatial Planning and Public Participation

A map is the greatest of all epic poems. Its lines and colours show the realization of great dreams.

Gilbert H. Grosvenor
Editor of National Geographic, 1903–1954

Water resources are a prerequisite for human development and progress. Whether groundwater or surface water resources, they are currently under increasing pressure worldwide (DeNicola et al., 2015; Gain and Giupponi, 2015; Hoekstra, 2016; Liu et al., 2016; Shady, 2008; Vörösmarty et al., 2010). Urban expansion together with high population density and industrial development as well as agricultural activities has resulted in increased water demand. Groundwater contamination with inorganic arsenic is one of the most important environmental issues in the world, and this groundwater plays a foremost contributing role in the livelihood of people and the economy of countries as well. The natural geochemical processes are directly or indirectly affecting the chemical composition of groundwater gradually (Islam et al., 2017). Moreover, the excessive unplanned use of groundwater resources in particular has contributed significantly to altering water quality, leading to groundwater contamination, yet the availability of safe drinking water is a determining factor in public health (Montes et al., 2016). Elevated levels of arsenic concentrations in groundwater can create hazardous effects for the people who are exposed to this groundwater.

Groundwater is purportedly the main source of untreated pathogen-free safe drinking water for more than one-third (2.4 billion) of the total population on the globe (WHO, 2015). But Bangladesh has many water-related problems and groundwater systems in Bangladesh experience significant pressure on their quantity and quality due to an increased water abstraction for irrigation and household purposes and consequent lowering of water levels. It is ironic that the very tubewells installed to provide pathogen-free drinking water are found to be contaminated with toxic levels of arsenic that threaten the health of millions of people in Bangladesh (Hassan and Atkins, 2011).

Geostatistics and Geographical Information System (GIS) have been used as a management and decision tool in the spatial discontinuities of groundwater quality as well as groundwater arsenic concentration (Antunes et al., 2014; Cinti et al., 2015; Delbari et al., 2016; Flanagan et al., 2016; Golia et al., 2015; Hassan et al., 2006; Marko et al., 2014). Geostatistics relies on both statistical and mathematical methods to create surfaces for groundwater arsenic concentrations (Liu et al., 2006). GIS, at the same time, is considered an automated decision-making system with mapping capabilities for the geographically referenced information (Achour et al., 2005; Berke, 2004; Burrough and McDonnell, 2000; Maguire, 1991) in preparing spatial mapping for investigating the historical and currently existing arsenic situations in groundwater. Groundwater contamination is a multidimensional phenomenon that has a spatial component, and GIS is very well suited to groundwater applications. GIS, in recent times, has been widely used in groundwater vulnerability mapping with geochemical and isotopic analysis in the complex heterogeneous geological environments (Jasmin and Mallikarjuna, 2015; Martínez-Graña et al., 2014; Montes et al., 2016; Olea et al., 2017). Groundwater vulnerability mapping can be used to monitor the geographic extent of groundwater contamination and environmental risk as well as to formulate sustainable spatial policies for preventing aquifer pollution from groundwater contaminants (Kourgialas and Karatzas, 2015; Rebolledo et al., 2016; Sikdar and Chakraborty, 2017). Spatial mapping for arsenic concentrations involves two basic stages: (a) interpolation and extrapolation of values at locations for unsurveyed areas and (b) display of the results. These two stages can be performed with geostatistics and stochastic simulations. Geostatistics has the advantage of considering the spatial continuity of the data and the capability of displaying the uncertainty associated with the modeling, which can be accomplished through semivariogram (Olea et al., 2017; Webster and Oliver, 2007: 2).

There is a complex pattern of spatial variability of arsenic concentrations in groundwater with differences between neighboring wells at different scales and changes in aquifer depth (Bhuiyan et al., 2016; Hassan and Atkins, 2006; Islam et al., 2017; Peters and Burkert, 2008). In a study from Bangladesh, Yu et al. (2003) found a statistically insignificant positive correlation between observed arsenic concentrations and shallow tubewell density. Mukherjee and Fryar (2008) observed elevated dissolved arsenic (>10 μg/L) in a majority of the deep groundwater samples in West Bengal, India. Guo et al. (2012) show a large spatial variation of groundwater arsenic concentrations in the Hetao Basin of Inner Mongolia, and Pi et al. (2016) investigated the spatial variability of total arsenic concentrations in groundwater in Datong basin, northern China, ranging from 5.47 μg/L to 2690 μg/L with an average of 697 μg/L of arsenic. Analyzing with multivariate geostatistical modeling, Andrade and Stigter (2013) inspected the spatio-temporal arsenic occurrence in shallow alluvial groundwater under agricultural land in central Portugal.

Public participation GIS (PPGIS), in recent times, is an important issue in spatial planning for arsenic mitigation. A PPGIS has been developed to integrate local people's perceptions and analyze their knowledge as part of "participatory development" (Abbot et al., 1998) and for future spatial decision-making as the representation of "multiple realities for single uses" (Cinderby, 1999). A PPGIS attempts to design and adapt GIS that specifically addresses the needs of participant communities (Bryan, 2011; Harris and Weiner, 1998; Harvey and Chrisman, 1998; Panek, 2015; Saunders et al., 2016) and to promote a "bottom-up" policy of development by incorporating local concerns and knowledge within a spatial database (Cinderby, 1999). A PPGIS can be used as a new window to view GIS practices in a social setting within the domain of "information-democracy" (Dervin, 1994). A PPGIS for community empowerment requires the "demand-driven" rather than "technology-driven" conventional "top-down" applications. The materials presented in this chapter are aimed at stressing the importance of the spatial analysis of groundwater arsenic concentrations and how this spatial analysis with public participation contributes to spatial policy formulation. Mapping with geostatistical predictions is *important* to understanding the complex processes of groundwater arsenic concentrations. The coastal belt in southwest Bangladesh has been selected as a basis for developing this chapter.

3.1 SPATIAL CHARACTERISTICS: WHAT IS SPECIAL ABOUT SPATIAL ANALYSIS?

GIS can be defined with "decision by position" for spatial analysis and spatial decision-support system (SDSS). This spatial analysis in GIS is an automated system that can be applied for monitoring the groundwater contaminants and their spatial discontinuity. GIS offers display capabilities that may assist with the identification of areas with special properties. A number of frequency-based methods are well-established in processing geochemical data that can be collected from groundwater contaminants (El-Makky, 2011; Zuo et al., 2013). Traditional statistical techniques in terms of probability graphs, multivariate analysis, and trend surface analysis have been widely applied in processing geochemical data. The traditional methods ignore spatial structures of information for geochemical patterns (Wang et al., 2015). Spatial statistical techniques can be applied for the pattern of spatial features of geochemical attributes (Cinti et al., 2013; Luz et al., 2014). The spatial distribution and variation of geochemical information can be quantified and analyzed with geostatistics (Goovaerts, 1999; Hamedani et al., 2012), fractal and multifractal models (Zuo et al., 2015), and spatial autocorrelation (Delbari et al., 2016).

GIS is a powerful tool for analyzing the spatial distribution of groundwater arsenic concentrations at local and regional scales. It is a computer-based technology for integrating spatial and nonspatial information into a common environment for spatial analysis, mapping, and graphic display as well as spatial decision-making (Burrough and McDonnel, 2000). GIS with integrated statistical approaches can be used for modeling the spatial distribution of groundwater quality. GIS, with its geostatistical methods, can explore spatial structure of data for better representation. Several authors have attempted GIS-based groundwater models to understand the pattern of contamination (Bonsor et al., 2017; Ghosh and Parial, 2014; Gorgij et al., 2017; Sikdar and Chakraborty, 2017). Hossain and Piantanakulchai (2013) attempted to predict arsenic contamination risk in groundwater using a GIS and classification tree method. Hassan (2005) developed a PPGIS combining GIS and participatory rural appraisal (PRA) for assessing the mitigation needs of arsenic-affected communities.

Existing methods for groundwater contamination mapping can be classified into a number of distinguishing categories (Gogu and Dassargues, 2000; Gogu et al., 2003; Sinan and Razack, 2009; Yang and Wang, 2010): (a) deterministic models based on physical processes, (b) statistical models, and (c) overlay and index methods. Deterministic methods use mathematical models to simulate the complex phenomena of groundwater contaminants, mainly the inorganic arsenic for this chapter. Statistical models depend on extensive arsenic databases from groundwater and relevant attributes. Overlay and index methods combine factors controlling arsenic contamination from groundwater—their main advantages are that they can be applied over large areas, which makes them suitable for regional scale assessments (Thapinta and Hudak, 2003). A number of overlay and index methods are well known for groundwater contamination analysis (e.g., DRASTIC system, GOD system, AVI method, SINTACS method, etc.) (Aller et al., 1987; Draoui et al., 2008; Foster, 1987).

Map interpolation with overlay and buffer operations as well as index methods can be useful in analyzing the overall arsenic situation in groundwater. GIS has strong spatial overlay capabilities because of its mathematical and programming facilities (Kim et al., 2004; Morra et al., 2006), which allow different map data to be combined in determining different "problem regions" of arsenic pollution. Thus, overlaying settlement within the buffer zone facilitates the generation of information on arsenic "problem regions" (Hassan, 2005). Overlay and index methods evolved in three broad application domains in GIS: (a) relevant information, (b) logical and mathematical analytical approaches, and (c) decision-support systems (Eastman et al., 1995). In order to prepare a spatial decision-support policy, a multi-criteria analysis (MCA) can be applied to prioritize the relative importance of different criteria (Chenini et al., 2010).

In order to design successful remediation strategies and protect public health from any groundwater contamination or from toxic substances, it is vital to identify the areas that are most vulnerable to groundwater contamination. This analysis is possible with spatial operation for spatial and temporal distribution of toxic substances for groundwater contamination (Dokou et al., 2015). Geostatistics offer an effective way to estimate (a) contaminant concentrations in unsampled areas using variants of kriging (Varouchakis et al., 2012; Venteris et al., 2014) and (b) the probability that their concentrations will exceed a pre-specified threshold using indicator kriging (Lu et al., 2007).

Management of groundwater contamination is a difficult task due to the spatial heterogeneity of aquifers and their numerous complex physical processes and chemical reactions (Kourgialas and Karatzas, 2015). This complexity may cause a contaminant to change its physical state or chemical form and finally change the degree of contamination in an aquifer (Civita, 2010). Moreover, groundwater monitoring is time-consuming and excessively costly to adequately define the geographic extent of contamination at a regional scale. Therefore, the best possible way to manage groundwater contamination is to identify the spatial distribution of risk areas from contamination and areas vulnerable to contamination (Kourgialas and Karatzas, 2015; Mimi and Assi, 2009).

The risk and vulnerability maps are useful tools for minimizing groundwater arsenic contamination with monitoring systems in contaminated areas. Groundwater risk can be assessed by combining the vulnerability and hazard assessments. The highest groundwater contamination risk appears when hazards occur in a highly vulnerable zone (Ravbar and Goldscheider, 2007).

"Aquifer vulnerability" generally refers to the sensitivity of groundwater quality to a contaminant based on the aquifer's characteristics (Margat, 1968). These aquifer characteristics can be changed by human activities, mainly the excessive withdrawal of groundwater for irrigation, industrial, and domestic purposes.

3.2 SPATIAL VARIABILITY: WATER QUALITY INVESTIGATION

Groundwater is an essential resource that provides drinking water to billions of people around the world, and assurance of drinking water safety is important to a community. The assessment and protection of groundwater quality is a crucial issue for managing demands for using this water resource for drinking, agricultural, and industrial purposes. With the combined effects of population growth, intense groundwater abstraction has caused widespread depletion and degradation of groundwater resources (Chen et al., 2017a; Gorgij et al., 2017). Therefore, groundwater quality assessments have become increasingly important, and a number of criteria have been developed for assessing water quality, for instance, the WHO guidelines for a range of drinking water pollutants (Amiri et al., 2014). There are currently a number of techniques that can be used to measure the spatial variability of chemical and physical properties of groundwater.

There is an increasing interest in geostatistical techniques and multivariate analysis for exploring the spatio-temporal variations of chemical properties and contamination of groundwater (Dehghanzadeh et al., 2015; Mizan et al., 2017; Yazdanpanah, 2016). Geostatistical approaches can be utilized to achieve a sustainable exploitation of groundwater resources (Narany et al., 2014) as well as to assess surface water quality (Belkhiri and Mouni, 2014). Geostatistics with kriging methods can design water quality monitoring in different aquifers with hydrochemical evaluation by spatial mapping (Machiwal et al., 2012; Ou et al., 2012). Geostatistics can also emphasize spatial variation of major factors that influence groundwater quality (Belkhiri and Narany, 2015; Masoud, 2014).

Spatial discontinuity of arsenic concentration has been reported in Bangladesh (Hassan and Atkins, 2011; Radloff et al., 2017), West Bengal in India (Biswas et al., 2014), China (Cai et al., 2015; Guo et al., 2014; Ma et al., 2016), Chianan Plain of Taiwan (Sengupta et al., 2014), Mekong Delta of Vietnam (Wilbers et al., 2014), the southern Pampa of Argentina (Díaz et al., 2016), the Duero River Basin of Spain (Pardo-Igúzquiza et al., 2015), Nova Scotia in Canada (Dummer et al., 2015), New England in the USA (Ayotte et al., 2006; Yang et al., 2012), the Águeda watershed area in Portuguese district of Guarda and the Spanish provinces of Salamanca and Caceres (Antunes et al., 2014), and so on.

Point data are usually inadequate for identifying the safe and contaminated as well as vulnerable areas. Therefore, appropriate interpolation methods are required to map the spatial pattern of groundwater quality. Geostatistical algorithms can be used to assess the uncertainty of the predicted values instead of the traditional interpolation methods (Al-Omran et al., 2017; Delbari et al., 2016; Jang et al., 2010). Several studies have been done to investigate spatial variability of groundwater quality using geostatistics. Elumalai et al. (2017) deployed Ordinary Kriging (OK) and Inverse Distance Weighted (IDW) methods for spatial interpolation mapping for groundwater contamination in a coastal city of Richards Bay in South Africa. Javed et al. (2017) used OK for spatial assessment of groundwater quality in Jhelum city, Pakistan. Using the OK interpolation method, Al-Abadi et al. (2017) mapped the groundwater potential in the Altun Kupri Basin in the northeast Kirkuk Governorate of Iraq. Using two deterministic methods (e.g., IDW and Radial Basis Function) and two stochastic interpolation methods (e.g., OK and Universal Kriging), Adhikary and Dash (2017) predicted the spatial variation of groundwater up to a depth of 20 meters in Delhi, India.

Indicator Kriging (IK) has been used frequently for analyzing the risk of groundwater contamination; for instance, Liu et al. (2004) used IK to assess arsenic contamination in an aquifer in the Yun-Lin coastal area of Taiwan. Goovaerts et al. (2005) studied the spatial variability of arsenic in groundwater in southeast Michigan with an IK prediction method. Lee et al. (2007) used IK for estimating the probability distribution of arsenic concentrations and for an evaluation of potential health risk of arsenic-affected groundwater in northeastern Taiwan. Dash et al. (2010) used IK to evaluate the probability of exceedance of groundwater quality in Delhi, India. Using OK and IK, Adhikary et al. (2010) mapped groundwater pollution in India. Using the geostatistics with IK, Hassan and Atkins (2011) analyzed the spatial discontinuity of groundwater arsenic in analyzing 375 water samples from southwest Bangladesh. Piccini et al. (2012) illustrated an application of IK for mapping the probability of exceeding nitrate contamination thresholds in Central Italy. Antunes and Albuquerque (2013) used IK for the evaluation of potential arsenic contamination in abandoned mining areas in Portugal, and the IK probability maps show a high probability for arsenic contamination.

Hossain and Sivakumar (2005) proposed OK to predict safe and unsafe zones at unsampled locations using groundwater arsenic contamination. The OK has been used in prediction of arsenic in groundwater. Hill et al. (2009) used arsenic concentration of water samples from wells shallower than 150 m with a Monte Carlo framework for the assessment of kriging estimation. Nas (2009) studied the spatial structure of various water quality parameters in the groundwater of Konya, Turkey, using OK. Hu et al. (2005) studied the spatial variability of the concentration of nitrates and electric conductivity in the groundwater of north China plain by using the OK and inverse kriging methods. Diggle et al. (1998) introduced a model-based geostatistics (MBG) approach (i.e., model-based OK and model-based Bayesian OK) for computing uneven spatial data distribution with uncertainty associated with predicted spatial pattern. Banerjee et al. (2008) introduced a Gaussian spatial predictive model that is based on hierarchical models with a Bayesian framework.

With the advection dispersion model, Radloff et al. (2017) explored the distribution of inorganic arsenic concentration with depth and time in a shallow Holocene aquifer at Araihazar in Bangladesh. Based on dense groundwater contamination data for 2000–2008 and kriging interpolation methods in GIS, Dokou et al. (2015) developed an average groundwater contamination map for Greece. Using multiple statistical analysis and a geostatistical interpolation approach, Belkhiri and Narany (2015) explored the spatial variability of groundwater quality with determining factors and mechanisms controlling this variability on the Ain Azel Plain in Algeria. Using multivariate analyses and geostatistics, Cai et al. (2015) investigated the pattern of spatial distribution and source of arsenic and heavy metals in southeast China and Han et al. (2013) examined the spatial and temporal patterns of groundwater arsenic in the shallow and deep groundwater of Yinchuan Plain, China. Andrade and Stigter (2013) used multivariate geostatistical modeling to assess distribution of arsenic in shallow alluvial groundwater under agricultural land in central Portugal.

Díaz et al. (2016) inspected the spatial distribution of arsenic and other associated elements in loess soils and waters and their control factors of the southern Pampa of Argentina. Applying geostatistical models, Dummer et al. (2015) observed the regional distribution of arsenic concentrations in private well water and spatial variation of high toenail arsenic concentrations (arsenic levels $\geq$85th percentile) with high well water arsenic concentrations ($\geq$5.0 µg/L) in Nova Scotia, Canada, and evaluated the geological and environmental features associated with higher levels of arsenic in well water. Applying geostatistical logistic regression models, Ayotte et al. (2006) and Yang et al. (2012) noticed the prediction of spatial probability of arsenic concentrations in groundwater in New England of the USA. Mehrjardi et al. (2008) compared the performance of IDW and kriging in analyzing groundwater quality with spatial pattern. Ahmadian (2013) compared the relative performance of classical geostatistical kriging methods in analysis of groundwater quality in Tehran-Karaj Plain. Beg et al. (2011) used the IDW method for depicting the spatial prediction of fluoride concentration in unsampled locations in the groundwater of Chhattisgarh, India.

Biswas et al. (2014) explored the pattern of spatial, vertical, and temporal variation of arsenic in shallow aquifers (<50 m) of the Bengal Basin, mainly in the Nadia District of West Bengal, India, and its regulation by geochemical processes. Seasonal variation of arsenic concentrations is significant for the shallow aquifer. Arsenic concentrations at 2–4 week intervals in different tubewells in Bangladesh show temporal variations within the shallow aquifers (Cheng et al., 2005; Dhar et al., 2008; Planer-Friedrich et al., 2012). This substantial variation can be due to the heavy withdrawal of water from this aquifer. However, seasonal variation of groundwater arsenic concentrations can be caused with scavenging of arsenic by colloidal aggregates formed in the presence of Dissolved Organic Carbon at high concentrations of major ions during the dry season (Biswas et al., 2014). During the monsoon, arsenic can be remobilized because of the dispersion of colloidal aggregates by dilution with recharge water (Planer-Friedrich et al., 2012). Majumder et al. (2014) also found organic colloidal aggregates as a potential scavenger of arsenic in the aquifer of the Bengal Basin. Vertical shifting of groundwater layers can be associated with the seasonal variation of arsenic in shallow aquifers (Cheng et al., 2005; Dhar et al., 2008).

Temporal variations of arsenic concentrations at different sites show different trends. Temporal change of arsenic concentrations has been reported in Bangladesh (Cheng et al., 2005; Planer-Friedrich et al., 2012), West Bengal in India (Farooq et al., 2011), Nepal (Brikowski et al., 2014), Inner Mongolia in China (Duan et al., 2015; Guo et al., 2013; Han et al., 2013), the Duero River Basin in Spain (Mayorga et al., 2013), Nevada and Snohomish County, Washington, in the USA (Steinmaus et al., 2005; Thundiyil et al., 2007), Ouro Preto of Brazil (Gonçalves et al., 2007), and the Zimapán Valley in Mexico (Rodríguez et al., 2004). Several studies have documented that arsenic concentrations in some wells might vary dramatically by season (Brikowski et al., 2014; Farooq et al., 2011; Guo et al., 2012), while the variation at other sites was inconsistent from year to year (Dhar et al., 2008), and some studies suggest the absence of any significant change in arsenic concentrations (Cheng et al., 2005; Thundiyil et al., 2007). Anthropogenic or natural factors responsible for the observed groundwater arsenic concentration changes include dilution by recharge of water with low arsenic concentrations, seasonal changes in redox conditions, pumping rates, water table depths, or shifts in direction of hydraulic gradient (Brikowski et al., 2014; Dhar et al., 2008; Guo et al., 2012). Because the distribution of groundwater arsenic is extremely heterogeneous in both the vertical and horizontal dimensions, a proper policy formulation for arsenic mitigation is problematic. In understanding the nature of spatial and temporal variability of arsenic concentrations, geostatistics and GIS applications have been utilized in this chapter. Borehole lithology is considered to delineate the pattern of arsenic concentrations with depth and lithology.

3.3 SPATIAL CONTINUITY: THEORIES OF ESTIMATION

3.3.1 Geostatistical Estimation

Geostatistics is the practical application of the theory of regionalized variables which regards spatial properties as random variables (Jalali et al., 2016). There are two families of mapping methods in geostatistics (a) kriging is a generalization of least squares that provides single estimated values by minimizing the prediction error and (b) stochastic simulation instead provides multiple maps that honor the data and the style of spatial fluctuation (Olea et al., 2017). Geostatistics allows for preparing maps of continuous variations of groundwater arsenic attributes to an area where samples at few locations have been taken for analysis. Geostatistics specializes in analysis and interpretation of geographically referenced data (Ağca et al., 2014; Bhuiyan et al., 2016; Bolstad, 2008; Deutsch, 2002; Goovaerts, 1997; Kumari et al., 2014). It is a distribution-free procedure and is based on a theory of regionalized variables whose values vary from place to place (Journel and Huijbregts, 1978; Isaaks and Srivastava, 1989). Based on a stochastic model, the geostatistical approach relies on both statistical and mathematical methods to create surfaces and to assess the uncertainty of predictions for regionalized variables (Bastante et al., 2008; Ghosh and Parial, 2014; Liu et al., 2006; Uyan, 2016; Xu et al., 2005). Geostatistics represents an appropriate method of prediction and is used for spatial estimation taking spatial variability into account. It provides reliable estimations of phenomena at unsampled locations where no measurements are available.

The term geostatistics was first introduced by Matheron (1963) on the basis of the theory of Danie Gerhardus Krige, a South African mining engineer (Krige, 1951) for estimating reserves of ore. Since then, geostatistics has been expanding as a spatial method for analyzing spatial discontinuities of geographical and environmental problems. In the 1980s, Professor John Aitchison from Hong Kong University (Aitchison, 1982) developed a compositional data analysis procedure that is nowadays known as the *"log-ratio approach"* (Pawlowsky-Glahn and José Egozcue, 2016). Geostatistics is currently applied in diverse disciplines for spatial continuity. An advantage of geostatistics is its use of quantitative measures of spatial correlation, commonly expressed as variograms (Uyan and Cay, 2013). A prerequisite for geostatistical interpolation is the definition of a model describing the spatial autocorrelation between the observations.

The theory and technique of geostatistics were developed to find the "best linear unbiased estimator (BLUE)" of unknown parameters (Goovaerts, 1997). A commonly used measure is the variogram, which describes the spatial structure and mathematical relationship between the variance of pairs of observations (data points) and the distance separating these observations (h) (Mabit and Bernard, 2007). Until recently, maps of environmental variables have primarily been generated by using empirical knowledge. Unlike the traditional approaches to mapping, geostatistics can provide more accurate maps since geostatistics relies on the "actual measurements and semi-automated algorithms" (Hengl, 2009).

Geostatistics takes into account spatial correlation between neighboring observations and includes different approaches spanning from conditional estimator to simulation, either a parametric or indicator approach (Uyan, 2016). Kriging interpolation is regarded as the optimal interpolation and is the most widely used in geostatistics. It estimates an unknown point by making use of a known point. Geostatistical results, using kriging techniques, are efficient when data for variables are distributed normally (Wu et al., 2014; Uyan et al., 2015). Groundwater arsenic discontinuities can be analyzed with a spatial interpolation process that estimates spatial arsenic concentrations at unsampled points from a surrounding set of measurements. When the local variance of sample values is controlled by the relative spatial distribution of these samples, geostatistics can be used for spatial prediction or spatial interpolation.

"Spatial prediction" or "spatial interpolation" aims at predicting values of arsenic concentrations over an area of interest on maps. "Spatial prediction," in general, can be applied to both "interpolation" and "extrapolation" (Dubois and Galmarini, 2004). Therefore, the term "spatial prediction" will be used in this chapter to explore "interpolation" and "extrapolation" of groundwater arsenic concentration. In geostatistics, "interpolation" corresponds to cases where concentrations can be estimated with surrounded sampling events and "extrapolation" is the prediction at locations outside the practical range where there is insufficient or no information to make significant predictions (Hengl, 2009). The geostatistical approach is mainly based on two different operative steps: (a) semivariogram calculus and modeling and (b) the kriging interpolation technique.

3.3.2 Semivariogram Estimation

Semivariogram is the geostatistical method to find the correlation between the samples. Two functions are commonly used in geostatistics: (a) semivariogram (or simply called the variogram) and (b) covariance (Rouhani and Mayers, 1990). Semivariogram is the best method to characterize the structure of spatial continuity (Júnior et al., 2015). A semivariogram can be calculated for each catchment using the average squared difference between all pairs of values that are separated by the corresponding distance lag. Semivariogram measures the average dissimilarity among the observations at pairs of locations as a function of the separation distance vector h (lag). Semivariogram calculated from the sampled data is called "experimental semivariogram" (Yeşilkanat et al., 2015), and spatial continuity can be analyzed with spatial autocorrelation based on an experimental semivariogram. A semivariogram can be computed in different directions to detect any anisotropy of the spatial variability. An anisotropic model generally includes geometric anisotropy and zonal anisotropy. The geometric anisotropy yields variograms having the same structural shape and maximum variability (sill) but a direction-dependent range for the spatial correlation. The zonal anisotropy, on the other hand, can be defined by sills varying with direction (Goovaerts, 1997).

A variogram is a means of local estimation in which each estimate is a weighted average of the observed values in the neighborhood. The weights mainly depend on fitting the variogram to the measured points. A variogram is scale-dependent, and different lag distances have to be tested until a sufficient number of pairs to represent the model are found. The effective lag distance cannot be more than half of the maximum distance between data (Mabit and Bernard, 2007). Variogram quantifies the spatial variability of the random variables between two sites. It can be computed in different directions to detect any spatial anisotropy of the spatial variability (Lee et al., 2007). It is necessary to adopt a geometric anisotropic model that yields variograms with the same structural shape and variability (sill + nugget) but a direction-dependent range for the spatial correlation (Deutsch and Journel, 1998).

A kriging treatment quantifies the variability of arsenic in the form of a semivariogram, which graphically expresses the relationship between semivariance and the sampling distance (Srivastava, 1996). Semivariance is a measure of spatial dissimilarity of an event, while covariance and correlation are both measures of the similarity of the head and tail values. Semivariance is the moment of inertia or spread of the h-scattergram about the 45° (1 to 1) line that can be shown on a scatter plot. The correlation versus lag is referred to as the "correlogram," and the semivariance versus lag is the "semivariogram." The covariance versus lag is generally just referred to as the covariance function (Bohling, 2005). The semivariogram analysis consists of the experimental semivariogram (calculated from the data) and the semivariogram model (fitted to the data).

In the literature, the terms variogram and semivariogram are often used interchangeably. By definition, $\gamma(h)$ is semivariogram and the variogram is $2\gamma(h)$. The experimental semivariogram, $\gamma(h)$, is half the average squared difference between pairs of data $Z(x_i)$ and $Z(x_i + h)$ at locations x_i and $x_i + h$. An estimate of the semivariogram with $N(h)$ the number of sampling pairs separated by a distance of h(lag) is given by the following equation (Chiverton et al., 2015; Goovaerts, 1997; Isaaks and Srivastava, 1989; Matheron, 1963):

$$\gamma(h) = \frac{1}{2N(h)}\left[\sum_{i=1}^{N(h)}\{Z(x_i) - Z(x_i + h)\}^2\right]. \quad (3.1)$$

Referring to Equation 3.1, there are three variographic parameters: (a) "nugget effect" (C_0), (b) "sill" ($C_0 + C_1$), and (c) "range" (a). The "nugget effect" is a measurement of spatial discontinuity at small distances which represents the value of the initial variability, micro-scale variation, or measurement error. In theory, a semivariogram value at the origin (i.e., 0 lag) should be zero. If it is significantly different from zero or very close to zero for lags, then this semivariogram value is referred to as the nugget. The nugget effect should normally explain up to 30% of the variance. The nugget represents variability at distances smaller than the typical sample spacing, including measurement error. It is estimated from the empirical semivariogram $\gamma(h)$ at $h = 0$. The nugget effect is a constant of 1 except at h (lag) equal to zero, where the variance is zero (Pyrcz and Deutsch, 2014):

$$\gamma(h) = \begin{cases} 0 & \text{if } h = 0 \\ 1 & \text{if } h > 0 \end{cases}. \quad (3.2)$$

A "sill" is an estimate of sample variances under the assumption of spatial independence. It represents the maximum variability reached at a distance equal to the "range." It is the semivariance value at which the variogram reaches the asymptotical value. It can also be used to refer to the "amplitude" of a certain component of the semivariogram (Bohling, 2005). A "range" is the lag distance at which sample data is spatially independent and the semivariogram reaches the sill value, i.e., the distance where data are no longer correlated. The increasing trend of semivariogram is an indicator of the rate at which the "influence" of a sample decreases with increasing distances from the sample site. This critical distance is called the "range" of a semivariogram. It gives a precise definition to the notion of "zone of influence." At distances beyond the "range," the sample spatial data becomes totally independent (Mazzella and Mazzella, 2013). Apparently, autocorrelation is essentially zero beyond the "range" (Bohling, 2005).

The experimental semivariogram, $\gamma(h)$, is fitted with a theoretical model, which may be spherical, exponential, Gaussian, pentaspherical, or circular, to determine the nugget effect (C_0), the sill ($C_0 + C_1$), and the range (a). These models provide information about the structure and pattern of spatial variation (Olea, 1994; Wang et al., 2015). Figure 3.1 shows the pattern of theoretical models for experimental semivariogram for arsenic concentration in an area of southwest Bangladesh. The spherical model is one of the most commonly used semivariogram models and is known as the "Matheron model" represented by a simple polynomial expression. The spherical model can be defined as (Isaaks and Srivastava, 1989: 374; Oliver and Webster, 2015: 33):

$$\text{Spherical: } \gamma(h) = \begin{cases} 0 & \text{if } h = 0 \\ C_0 + C_1 \left[\frac{3}{2}\left(\frac{h}{a}\right) - \frac{1}{2}\left(\frac{h}{a}\right)^3 \right] & \text{if } h \leq a. \\ C_0 + C_1 & \text{if } h > a \end{cases} \quad (3.3)$$

The spherical model is said to be transitive because it reaches a finite sill at a finite range (Olea, 1994), and this model shows a progressive decrease of spatial autocorrelation (equivalently, an increase of semivariance) to a distance, beyond which autocorrelation is zero. The range for the spherical model can be computed by setting 95% of the "sill" value. The spherical model is a commonly encountered variogram shape, which increases in a linear fashion and then curves to a sill of 1 at a distance of 1. This spherical model is related to the volume of intersection of two spheres separated by some distance.

A graph of the cross-variogram for the arsenic data shows $\gamma(h)$ as a function of lag distance h, and the model illustrates features common to the arsenic semivariogram (Gerlach et al., 2001): (a) $\gamma(h)$ increases from smaller to larger lags but a limiting "sill" is always found, (b) $\gamma(h)$ approaches for small lags suggesting a large "nugget effect," and (c) the spherical semivariogram model gives good and acceptable fits to $\gamma(h)$. In our spherical semivariogram model, the nugget variance is not too large, and there is a distinct range and sill.

3.3.3 KRIGING ESTIMATION

Kriging is mainly designed to model spatial variability. It is a generic term for a range of least squares methods to provide the "best linear unbiased predictions" (BLUP), best in the sense of minimum variance (Oliver and Webster, 2014). Interpolation methods can be used in the estimation of spatial continuity of variables to infer the values to nonsampling locations. This estimation method in geostatistics is known as kriging, an interpolation approach that provides an optimal estimative of regionalized variables with minimum variance and without bias, using a theoretical variogram (Bressan et al., 2009; Ghosh and Parial, 2014; Isaaks and Srivastana, 1989). Kriging is useful for estimating spatial interaction of various data (Im and Park, 2013; Martin and Simpson, 2005). Kriging is a method for linear optimum appropriate interpolation with a minimum mean square error. It is known to be an exact estimator in the sense that observation points are correctly reestimated (Uyan and Cay, 2013).

Kriging provides a solution to groundwater arsenic problems of estimation based on a continuous model of stochastic spatial variation. Kriging provides an estimation of the value of spatially distributed variables and assesses the probable error associated with the estimates (D'Acqui et al.,

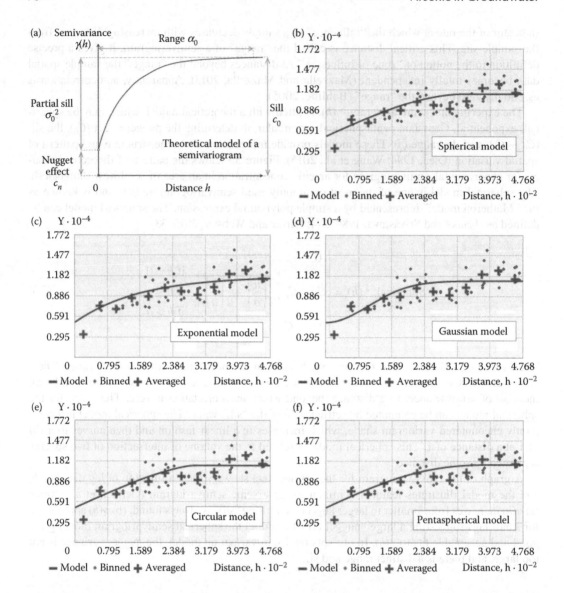

FIGURE 3.1 Different theoretical models of experimental semivariogram. The shapes of a number of models (e.g. spherical, exponential, Gaussian, circular, pentaspherical) were developed from arsenic attributes collected from Ghona union in Satkhira district of southwest Bangladesh. The shape of a spherical model shows a steady increase up to a distance equal to the "range" and then reaches a plateau, the exponential model reaches the threshold value with an asymptotic trend, a graph of a Gaussian model has a parabolic form near the origin, in a pentaspherical model, experimental semivariogram increases with distance until it reaches the sill value at the distance equal to the model range, and a circular model reaches the threshold value with a steady increase up to a distance equal to the model range and then reaches a plateau.

2007). In its original formulation, a kriged estimate at a place is simply a linear sum or weighted average of the data in its neighborhood (Webster and Oliver, 2007: 154). Two functions in kriging, in terms of semivariogram and the covariance, are commonly used for describing the spatial correlation structure of the variables of interest (Rouhani and Mayers, 1990). The semivariogram analysis consists of the experimental semivariogram (calculated from the data) and the semivariogram model (fitted to the data).

Kriging is a technique of making optimal, unbiased estimates of regionalized variables at unsampled locations using the structural properties of the semivariogram and the initial set of data values (David, 1977). Weights in the kriging approach are based not only on the distance between the measured points and the prediction location but also on the overall spatial arrangement among the measured points and their values (Teegavarapu and Chandramouli, 2005). The general equation for the kriging estimating prediction value $\hat{Z}(S_0)$ is

$$\hat{Z}(S_0) = \sum_{i=1}^{n} \lambda_i Z(S_i), \quad (3.4)$$

where $\hat{Z}(S_0)$ is the prediction value for location S_0, n is the number of measured sample points surrounding the prediction location, λ_i is the weight obtained from fitted variogram, and $Z(S_i)$ is the observed value at location S_i. The values of weight (λ_i) can be estimated by minimizing the kriging (error) variance (σ^2) given by (Marko et al., 2014):

$$\sigma^2 = \sum_{i=1}^{n} [\hat{Z}(x) - Z(x)]^2. \quad (3.5)$$

Kriging uses a weight model that ensures more response to nearby points as in the method of weighted mean in conventional statistics theorem (McGrath et al., 2004). This kriging is defined as BLUE mathematically, which means the identification of weight according to the condition that estimated errors are minimum. In recent years, researchers have studied a number of kriging estimations, but the OK, IDW, and IK are suitable for analyzing the spatial pattern of the groundwater arsenic continuity.

Ordinary Kriging is a geostatistical approach for estimation and linear interpolator that estimates a value at a point of a region for which the variogram is known, without prior knowledge about the mean of the distribution (Choudhury, 2015; Dokou et al., 2015). In OK, a random function model is used in which the bias and error variance can both be calculated and then weights are chosen for the nearby samples such that they ensure that the average error for the model is zero and the modeled variance is minimized. In order for the estimator to be unbiased in OK, the sum of these weights needs to equal one (Isaaks and Srivastava, 1989). The estimation equation is a linear weighted combination of the form (Journel and Huijbregts, 1978):

$$\hat{Z}(S_0) = \sum_{i=1}^{n} \lambda_i Z(S_i) \quad \text{with} \quad \sum_{i=1}^{n} \lambda_i = 1. \quad (3.6)$$

OK weights λ_i are allocated to the known values in such that they sum to unity (unbiasness constraint) and they minimize the kriging estimation variance (Delbari et al., 2016). The weights are determined by solving the following system of equations (Isaaks and Srivastava, 1989):

$$\begin{cases} \sum_{j=1}^{n} \lambda_j \gamma(x_i, x_j) + \mu = \gamma(x_i, x_0), \quad i = 1, \ldots, n \\ \sum_{j=1}^{n} \lambda_j = 1 \end{cases} \quad (3.7)$$

where $\gamma(x_i, x_j)$ is the average semivariance between pairs of data locations, μ is the Lagrange parameter for the minimization of kriging variance, and $\gamma(x_i, x_0)$ is the average semivariance between the location to be estimated (x_0) and the i_{th} sample point.

OK can be used for spatial pattern of groundwater arsenic concentrations because of its high uneven distribution. OK is used to estimate values when data point values vary or fluctuate around a constant mean value (Serón et al., 2001). It is applied for an unbiased estimate of spatial variation of a component. The estimation variance of OK is used to generate a confidence interval for the corresponding estimate assuming a normal distribution of errors (Goovaerts et al., 2005). The unknown local mean is filtered from the linear estimator by making the sum of kriging weights to one. OK also provides a measure of uncertainty attached to each estimated value through calculating the OK variance (Delbari et al., 2016):

$$\sigma^2(x_0) = \sum_{i=1}^{n} \lambda_i \gamma(x_i, x_0) + \mu. \tag{3.8}$$

This estimation variance may be used to generate a confidence interval for the corresponding estimate assuming a normal distribution of errors (Goovaerts, 1997).

Indicator Kriging proposed by Journel (1983) is one of the most efficient nonparametric methods in geostatistics (Bastante et al., 2008) due to its ability to take data uncertainty into account. The IK interpolation method can be used for estimating a conditional cumulative distribution function at an unsampled location (Braimoh and Onishi, 2007). It is the basis of some estimate algorithms and sequential indicator simulations. IK can model for estimating the probability distribution of spatial variables on the basis of surrounding observations. IK makes no assumption regarding the distributions of variables, and a 0–1 indicator transformation of data is adopted to ensure that the predictor is robust to outliers (Cressie, 1985).

IK is a nonlinear, nonparametric form of kriging in which continuous variables are converted to binary ones (indicators). It can handle distributions of almost any kind, and empirical cumulative distributions of estimates can be computed, thereby providing confidence limits for them. It can also accommodate "soft" qualitative information to improve prediction (Webster and Oliver, 2007: 155). In IK, no assumption is required about the variable distribution. It is very useful for the development of contaminant concentration maps in cases where the identification of areas of high concentrations exceeding the parametric values for drinking water or other uses is of interest. For the areas with lower concentrations, the knowledge of the exact concentration values is not as important, as long as they remain below the parametric value (Dokou et al., 2015).

In an unsampled location, the values estimated by IK represent the probability that does not exceed a particular threshold. Therefore, the expected value derived from indicator data is equivalent to the cumulative distribution of the variable (Smith et al., 1993) by calculating conditional probabilities. These probabilities can be estimated by transforming the indicator variables $I(x_i; Z_k)$ into either zero or one. For a continuous variable $Z(x_i)$, the indicator variable $I(x_i; Z_k)$ where Z_k is the desired threshold limit, can be defined as (Choudhury, 2015; Goovaerts, 1997):

$$I(x_i; Z_k) = \begin{cases} 1 & \text{if } Z(x) \leq Z_k \\ 0 & \text{if } Z(x) > Z_k \end{cases}, \quad k = 1, \ldots, K \tag{3.9}$$

where K is the number of cutoffs. The experimental indicator semivariogram, $\gamma_I^*(h)$ is then defined for every set of indicators at each cutoff Z_k as (Delbari et al., 2016):

$$\gamma_I^*(h) = \frac{1}{2N(h)} \sum_{i=1}^{N(h)} [I(x_i; Z_k) - I(x_i + h; Z_k)]^2 \tag{3.10}$$

where $N(h)$ is the number of pairs of indicator transforms $I(x_i; Z_k)$ and $I(x_i + h; Z_k)$ separated by vector h. The conditional cumulative distribution function (ccdf) at location x_0 is then obtained by the IK estimator as follows (Delbari et al., 2016):

$$F(x_0; Z_k \mid (n)) = I^*(x_i; Z_k) = \sum_{i=1}^{n} \lambda_i I(x_i; Z_k) \tag{3.11}$$

where $I^*(x_i; Z_k)$ is the estimated indicator value at unsampled location x_0 and λ_i is the weight assigned to the known indicator value $I(x_i; Z_k)$. These discrete probability functions must be interpolated within each class (between every two parts of ccdf) and extrapolated beyond the minimum and maximum values to provide a continuous ccdf covering all the possible ranges of the property of interest (Goovaerts, 1997).

IK determines the probability of arsenic indicator by using samples in the neighborhood. To conduct IK, arsenic concentrations can be transformed into an indicator variable and the variogram function can be evaluated in horizontal directions to identify the anisotropic variation present in groundwater arsenic concentrations. In this chapter, three threshold limits (i.e., the first quartile, median, and the third quartile) were selected for arsenic indicator analysis. An omnidirectional variogram is advantageous for analyzing spatial structures of arsenic concentrations. The arsenic interpolation map produced by the IK prediction method was constrained by spherical semivariogram fit. The experimental semivariogram was computed from arsenic data and a mathematical model was fitted to arsenic values by weighted least-squares approximation.

The *Inverse Distance Weighted* approach is the nongeostatistical method of interpolation which assumes that grades vary in a deposit according to the inverse of their separation (raised to some power). This method does not account for "nugget variance" or other aspects of the "variogram." Spatial interpolation with the IDW method can be suitable for analysis of spatial pattern of groundwater arsenic concentrations because of its exact interpolation capability. The IDW interpolator is a point estimation technique based on the weighting of a random function for a particular cell node of a grid (Serón et al., 2001). It assumes that each input point has a local influence that diminishes with distance. It weights the points closer to the processing cell greater than those further away, hence the name IDW interpolation or Inverse Squared Distance (ISD) interpolation (Ashraf et al., 1997). In the IDW interpolation method, the maximum and minimum values in the interpolated surface can only occur at sample points (Longley et al., 2001). A specified number of points, or all points within a specified radius, can be used to determine the output value for each location. The IDW interpolation method can be calculated using the following equation:

$$\lambda_i = d_{i0}^{-p} \Big/ \sum_{i=1}^{n} d_{i0}^{-p} \quad \sum_{i=1}^{n} \lambda_i = 1. \tag{3.12}$$

where n is the number of measured sample points surrounding the prediction location and λ_i is the weight obtained from fitted variogram. Since the distance becomes larger, the weight is reduced by a factor of P. The quantity d_{i0} is the distance between the prediction location, S_0, and each of the measured locations, S_i. In the IDW method, the surface is driven by local variation and calculation depends on the power parameter and neighborhood search strategy. The power parameter controls the influence of surrounding points upon the interpolated value. A higher power results in less influence from distant points, while the optimal power value is determined by minimizing the Root-Mean-Square Prediction Error (Tsanis and Gad, 2001). If the power value is 0, there is no decrease with distance.

In addition, "search neighborhood" in the IDW is used in prediction of an unmeasured location. The shape of the search neighborhood is based on an understanding of spatial locations and spatial autocorrelation of the dataset (Hassan and Atkins, 2011). Generally, the search area of the IDW method is a circular weighting, but the area could be elliptical or even directional in order to remove the strong influence of local anomalous values due to clustered data surrounding the estimation point (Isaaks and Srivastava, 1989; Serón et al., 2001). The spatial arsenic interpolation maps produced with the IDW method are based on the weighting of a random function for the sample study points.

3.3.4 Cross-Validation and Prediction Error

The validation and the sufficiency of the developed model semivariogram can be tested and evaluated via a technique called cross-validation (Isaaks and Srivastana, 1989: 352). Cross-validation can be used to compare the performance of different kriging techniques using different variograms. Cross-validation, which involves partitioning data into training and testing sets, is commonly used to obtain a reliable estimation for model performance (Chang et al., 2013). Cross validation can produce a low-bias estimator for the generalization abilities of a statistical model and therefore provides a sensible criterion for model selection and performance comparison.

In the prediction maps of arsenic concentrations, cross-validation can be used to compare the prediction performance of different interpolation algorithms. The cross-validation can be applied to improve and control the quality of the applied geostatistical model and thus the results of the spatial analysis. The differences between measured and predicted values provide a quality control for the model of computation for arsenic concentrations. Figure 3.2 shows the cross-validation of prediction error of the IK prediction for spatial arsenic discontinuity. It is noted that IK is one of the suitable prediction methods for analysis of spatial arsenic discontinuity.

Cross-validation can be performed by eliminating sequentially each data point from the dataset and then estimating it using the constructed model (Dokou et al., 2015; Yeşilkanat et al., 2015). The kriged values $\hat{Z}(x_i)$ are compared with the observed ones $Z(x_i)$, and various statistical measures are calculated in order to compare the goodness of fit of each model (Lark, 2000; Oliver and Webster, 2015: 68). These are mean error, root mean square error, average standard error, mean squared deviation ratio, mean standardized error, and root mean square standardized error. All these error propagations can be measured with different equations, and they are available in different literature (Gong et al., 2014; Hu et al., 2004; Isaaks and Srivastana, 1989: 352; Lark, 2000; Oliver and Webster, 2015: 68; Webster and Oliver, 2007: 191).

Cross-validation estimation is obtained by leaving one sample out and using the remaining data. This test allows the assessment of the goodness of fitting (avoid sufficiency) of the variogram model (type, parameter estimates), the appropriateness of neighborhood, and type of kriging used. The interpolation values are compared to the real values and then the least square error models are selected for regional estimation (Uyan and Cay, 2013). Based on the cross-validation results of the data exploration and the variogram analysis, the measured data are converted to surface maps. Cross-validation indicators and the model parameters (nugget, sill, and range) help us to choose a suitable model for the prediction maps of arsenic concentrations.

3.4 DESCRIPTIVE STATISTICS AND REGRESSION

Descriptive statistics can be used to calculate the normal statistical parameters (e.g., frequency, minimum, maximum, mean, standard deviation, etc.) for groundwater arsenic concentration. The correlation matrix measures how fit the variance of each variable can be explained by relationships with each other, and the terms, according to Liu et al. (2003), can be categorized as "strong," "moderate," and "insignificant" to correlation matrix analysis and refer to absolute values as >0.75 as strong, 0.75–0.50 as moderate, and 0.50–0.30 as insignificant. When more than two variables are

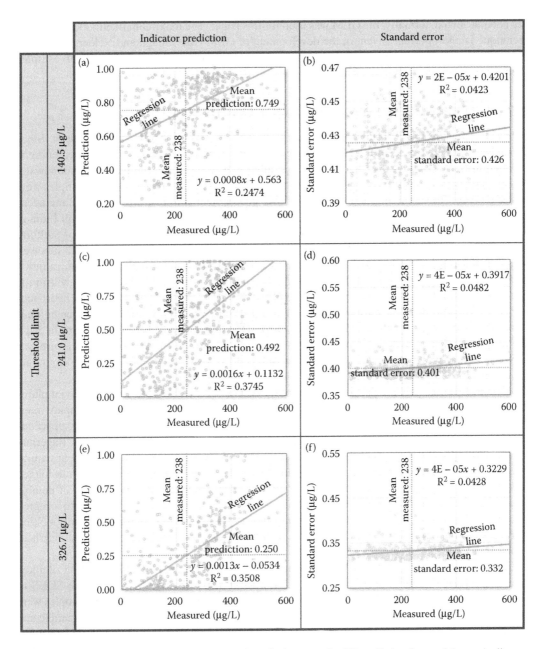

FIGURE 3.2 Properties of cross-validation and prediction-error for IK prediction for spatial arsenic discontinuity in Ghona union of Satkhira district of southwest Bangladesh.

considered simultaneously, a multiple linear regression model can be used to evaluate their interdependency (Adhikary et al., 2009). Generalized Linear Models (GLM) can be suitable for identifying the association between arsenic concentrations and aquifer depths as well as tubewell installation age. The GLM is a mathematical extension of linear models that do not force data into unnatural scales and thereby allow for nonlinearity and nonconstant variance structures in the data (Jin et al., 2005; Khuri, 2001; McCullagh and Nelder, 1989; Nelder and Wedderburn, 1972). They are based on an assumed relationship (link function) between the mean of the response variable and the linear combination of the explanatory variables. Hypothesis testing applied to the GLM does not require normality of the response variable, nor does it require homogeneity of variances. Since arsenic data are not distributed normally, the GLM was used for this paper. The maximum likelihood estimation technique is an important advent in the development of GLM (McCullagh and Nelder, 1989). The Newton-Raphson (maximum likelihood) optimization technique is used in this chapter to estimate the GLM, and the STATA statistical software was utilized to calculate the GLM.

3.5 ATTRIBUTES FOR SPATIAL ARSENIC CONCENTRATION

Tubewell screening is important priority work for arsenic data collection in Bangladesh. Since groundwater arsenic concentrations are uneven in spatio-temporal aspects, a census screening is a prerequisite. Accordingly, census information rather than any random sampling for groundwater arsenic data were collected for spatial arsenic analysis. A total of 375 tubewell water samples were collected from Ghona union (the fourth order local government administrative unit in Bangladesh after division, district, and sub-district) from southwest Bangladesh near the border with India in 2000 and in 2010. Moreover, all the 2650 tubewell water samples from Magura union of Satkhira district, 1082 tubewell water samples from Dhopakhali union of Bagerhat district, 1722 tubewell water samples from Gazalia union of Bagerhat district, and 140 water samples from Jamira union of Khulna district were collected for spatial arsenic analysis. All the selected areas are within the coastal belt since the coastal area has been identified as a problem area due to complex hydrogeological conditions and adverse water quality, which make safe water supply difficult. Ghona union is about 17.26 km^2 in area with 3782 households, having the population density of 1026/km^2 in 2011; the population density was measured by 1052/km^2 (area: 12.69 km^2) in Dhopakali; the figure was 1081/km^2 (area: 20.0 km^2) for Gazalia, 1030/km^2 (area: 27.57 km^2) for Magura, and 1071/km^2 (area: 20.11 km^2) for Jamira (BBS, 2011).

The area is geologically a part of the quaternary deltaic sediments of the Ganges alluvial plains and tidal plains (Dowling et al., 2002), and the subsurface geology of the area has complex interfingerings of coarse- and fine-grained sediments from numerous regressions and transgressions throughout geological time (Umitsu, 1993). Like other areas in the coastal belt, Khulna, Bagerhat, and Satkhira districts are associated with an acute drinking water crisis. Due to frequent natural disasters, permanent water and sanitation technologies have not developed in these water-logged areas. Therefore, people of this coastal area commonly face water scarcity. Availability of potable water itself is limited, and most of the traditional water technologies do not work in this region because of its different hydrogeophysical characteristics. Using pond and river water for cooking purposes is a regular practice in the area. This region is often considered one of the most diarrhea-prone areas of the country.

Arsenic concentrations from Ghona union were analyzed with a laboratory method of Flow Injection Hydride Generation Atomic Absorption Spectrometry (FI-HG-AAS) from the SOES of Jadavpur University, Bengal, India. In order to prevent adsorption losses, the collected samples were preserved by acidification with a drop of concentrated nitric acid in each 10 ml of water sample and placed in a refrigerator at a temperature below 4°C until the data were analyzed with FI-HG-AAS. The method is characterized by high efficiency, low sample volume, reagent consumption, improved tolerance of interference, and rapid determination (Le et al., 1992; Samanta and Chakraborti, 1997). With a 95% confidence interval, the minimum detection limit of the FI-HG-AAS method is 1.0 µg/L, and the quantification limit is 3.0 µg/L (Samanta et al., 1999), which is excellent for arsenic research. On the other hand, all the samples from other aforementioned areas were tested with the HACH field-testing kits in 2014 financed by the ICCO Cooperation, the Netherlands.

Apart from water quality data, a number of relevant information (e.g., geographical locations of tubewell position, tubewell depth, installation year, users, etc.) was collected with observation and face-to-face questionnaire surveys (Hassan, 2015). GIS information, for instance, all the points (mainly the tubewells), lines (e.g., boundary information, roads, etc.), and polygon (e.g., settlement area, water bodies, etc.) information, was collected through extensive field visits with Global Positioning Systems (GPS) (Model: Garmin GPSMAP 62STC), small-scale map data (RF 1:3960), and Google satellite imageries. This GPS has a high-sensitivity receiver with the facilities of a preloaded base map with topographic features. The geographical location of all the tubewells were also plotted on "mouza maps" (lowest level administrative territorial unit in Bangladesh) having a scale of 1:3960.

3.6 GROUNDWATER ARSENIC CONCENTRATIONS: PATTERN AND ARRAY

3.6.1 PROBABILISTIC SCALE: HOW TO QUANTIFY THE PATTERN?

There has been an uneven concentration of groundwater arsenic at the study site. Some tubewells are highly contaminated with arsenic and some are less so. Arsenic concentrations in groundwater can be classified into different categories based on different permissible limits (Table 3.1): (a) the WHO permissible level (<10 μg/L), (b) the Bangladesh standard maximum contaminant level (<50 μg/L), (c) a moderate contamination level (50–100 μg/L), (d) a high contamination level (100–300 μg/L), and (e) a severe contamination level (>300 μg/L). These figures can be framed into two different broad categories on the basis of the official Bangladesh standard daily maximum tolerable limit of 50 μg/L. These are (a) a safe level (<50 μg/L) and (b) a contamination level (>50 μg/L). The pattern of arsenic concentrations in five different areas in the coastal belt of Bangladesh shows a diversified shape that varies considerably and unpredictably over distances of a few meters within the settlement area (Table 3.1).

The term "contamination" in this chapter refers to the occurrence of arsenic concentrations in tubewells at levels above the Bangladesh standard permissible limit of 50 μg/L. The maximum arsenic concentration in Ghona was recorded as 600 μg/L, with the mean concentration of 238 ± 117 μg/L; the maximum figure was found to be 1500 μg/L in Magura with the mean concentration of 155.81 ± 173.75 μg/L; the maximum was 500 μg/L in Dhopakhali and Gazalia with the mean concentrations of 163.01 ± 135.17 μg/L and 92.90 ± 96.45 μg/L, respectively, and the maximum concentration was recorded as 50 μg/L in Jamira, with the mean concentration of 6.39 ± 9.51 μg/L (Table 3.1). It is interesting that Jamira union is completely safe from arsenic toxicity from drinking water—no arsenic was detected in more than half (52.86%) of the total sampled tubewell water.

3.6.2 SPATIAL SCALE: WHAT ARE THE AREAL DISCONTINUITY AND AT WHAT EXTENT?

The probability map developed with different geostatistics and stochastic methods shows the spatial discontinuity of arsenic concentrations. Figure 3.3 shows the spatial variation of arsenic concentrations in Bangladesh. Following the national arsenic data of 3534 water samples all over the country, it is observed that arsenic concentrations are heavily concentrated near the middle and southern parts of Bangladesh and low levels of arsenic and arsenic-safe zones can be identified in the northern, middle, and southeastern parts of Bangladesh (Figure 3.3). About 24.73% of the analyzed tubewells were found to be contaminated with arsenic above the Bangladesh Standard Limit of 50 μg/L.

The IK prediction method shows a highly uneven spatial pattern of arsenic concentrations in micro level (Figure 3.4). Different threshold limits show different patterns of spatial arsenic continuity in Ghona union. The measured arsenic concentrations are 140.5 μg/L, 241 μg/L, and 326.7 μg/L at the first, second (median), and the third quartile of the frequency distribution of arsenic concentrations. The variogram for IK with nugget effect, range, and sill for spherical model shows the local erratic variation of arsenic concentrations (Hassan and Atkins, 2011). The minor range of continuity in the variogram with large nugget effect results from the high and uneven spatial discontinuity of arsenic concentrations. The upward trend between nugget and partial sill indicates some regional spatial trend in the arsenic data (Figure 3.1). There are visible discrepancies in spatial continuity of arsenic concentrations in the central and eastern parts of Ghona union, and the pattern is very uneven. The spherical model shows a distinct character about semivariogram at different threshold limits, which explains an erratic variation of arsenic concentration over space.

TABLE 3.1
Descriptive Statistics of Arsenic Concentrations in Different Study Sites at the Coastal Belts of Bangladesh

Area	WHO Limit (≤10 µg/L)	DoE Limit (≤50 µg/L)	Moderate (51–100 µg/L)	High (101–300 µg/L)	Severe (>300 µg/L)	Total Tubewell	Maximum Concentration (µg/L)	Mean with Std. Deviation (µg/L)
Ghona union, Satkhira	4 (1.07)	13 (3.47)	32 (8.53)	200 (53.33)	126 (33.60)	375	600	238.0 ± 117.0
Magura union, Satkhira	366 (13.81)	368 (13.89)	1022 (38.57)	449 (16.94)	445 (16.79)	2650	1500	155.81 ± 173.75
Dhopakhali union, Bagerhat	64 (5.92)	155 (14.33)	355 (32.81)	360 (33.27)	148 (13.68)	1082	500	163.01 ± 135.17
Gazalia union, Bagerhat	146 (8.48)	580 (33.68)	632 (36.70)	293 (17.02)	71 (4.12)	1722	500	92.90 ± 96.45
Piprail in Jamira Union, Khulna	113 (80.71)	27 (19.29)	–	–	–	140	50	6.39 ± 9.51

Data Source: Field Survey, 2001, 2010, and 2014.

Note: Figures in parentheses indicate the net percent of the surveyed tubewells, and the arsenic data are calculated by descriptive statistics. In Piprail, arsenic was found to be absent in 74 (52.86%) tubewells.

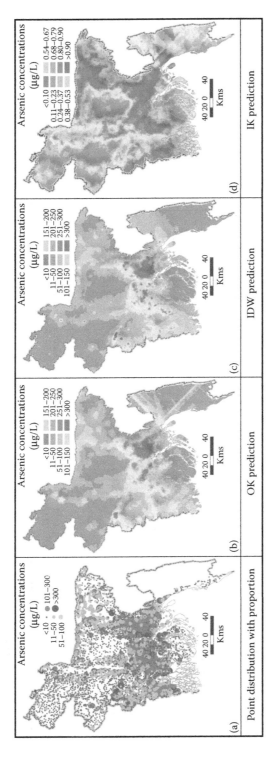

FIGURE 3.3 The pattern of groundwater arsenic concentrations and spatial discontinuity of groundwater arsenic in Bangladesh with different GIS and geostatistical methods: (a) arsenic concentration in each tubewell with point proportion following the low concentration with a small dot and high concentration with a bigger circle; (b) the pattern of spatial arsenic concentration with the ordinary kriging prediction method; (c) spatial variation of arsenic concentrations with the IDW prediction method; and (d) spatial arsenic variation with the IK prediction method. It is noted that this figure is completely different from Figure 2.2 of Chapter 2, which was prepared based on the choropleth mapping concept.

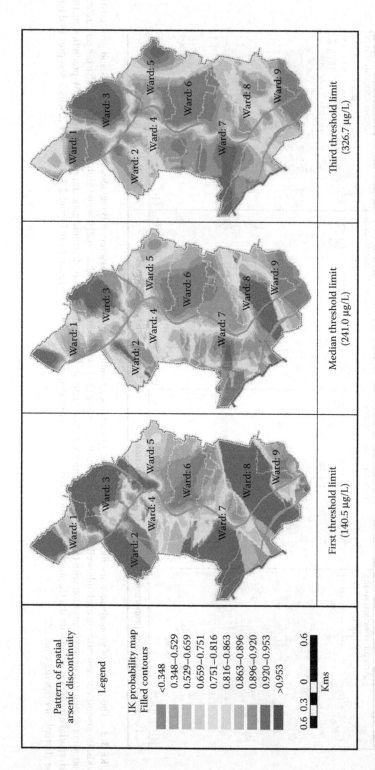

FIGURE 3.4 The pattern of spatial arsenic concentrations and its variations at three different threshold limits of IK prediction method. Arsenic-safe zones as well as arsenic-contaminated zones are found in different areas with different degrees in an area (Ghona union of Satkhira district) of Southwest Bangladesh.

TABLE 3.2
Arsenic Concentration Pattern with IK Prediction Method at Three Different Threshold Limits

Major Groups	Concentration (μg/L)	Detailed Classification	Area Covered (hectares)		
			First (140.5 μg/L)	Median (241.0 μg/L)	Third (326.7 μg/L)
Safe Level	<10	WHO Permissible Limit	49.2 (2.85)	52.6 (3.05)	103.6 (6.0)
	10–50	Bangladesh Standard Limit	67.4 (3.90)	83.5 (4.84)	87.4 (5.06)
Contamination Level	50–100	Moderate Contamination	257.5 (14.92)	321.9 (18.65)	281.2 (16.29)
	100–300	High Contamination	468.2 (27.13)	649.8 (37.65)	636.5 (36.88)
	>300	Severe Contamination	1023.7 (59.31)	618.2 (35.82)	617.3 (35.76)
	–	Total:	1726 (100.0)	1726 (100.0)	1726 (100.0)

Data Source: Field Survey, 2001.

Note: Figures in parentheses indicate the net percent of the surveyed tubewells. (The dataset is classified on the basis of different permissible limits of daily arsenic ingestion, and the interpolated area has been calculated by ArcGIS Software).

Arsenic concentration in groundwater is highly uneven over space. The pattern of arsenic concentrations varies considerably and unpredictably over distances of a few meters, which results in large nugget variances of the spherical variogram (Figure 3.1). Considering both the WHO and Bangladesh standard permissible limits, the safe zones identified in the IK estimation are especially in the central and southern parts of Ghona union in a scattered manner, while the contaminated zones are concentrated in the west, northeast, and eastern parts of Ghona. Table 3.2 shows the arsenic concentration pattern at three different threshold limits.

3.6.3 Depth Scale: Which Aquifer Is Contaminated at Which Level?

Arsenic concentrations in all the study sites are highly uneven with depth. Elevated levels of arsenic concentrations are found in the shallow aquifer, while low concentrations can be found in the deep aquifer. The product moment of correlation results ($r = 0.238$, $r = 0.131$ and $r = 0.293$) show an "insignificant" increasing tendency of arsenic concentrations with the increase of aquifer depth with erratic variations in Magura, Dhopakhali, and Gazalia, respectively. On the other hand, the correlation values ($r = -0.216$ and $r = -0.0624$) show that there is a weak decreasing tendency of arsenic concentrations with the increase of aquifer depth in Ghona and Jamira.

The adjusted mean smoothed lowess curve with a bandwidth of 0.8 shows an increasing trend of arsenic concentration up to a depth of 80 m in Magura, 20 m in Dhopakhali, and 22 m in Gazalia with some erratic fluctuations and regional variations. In Piprail, the pattern of lowess curve shows a different trend with a negative pattern up to a 60 m depth and then moves with a steady state toward a depth of 467 m (Hassan, 2015). The lowess curve shows an increasing trend of arsenic concentration up to a depth of 75 m in Ghona union, with some erratic fluctuations and regional variations at the depth between 30 m and 60 m (Figure 3.5). Moreover, the three-dimensional view of arsenic

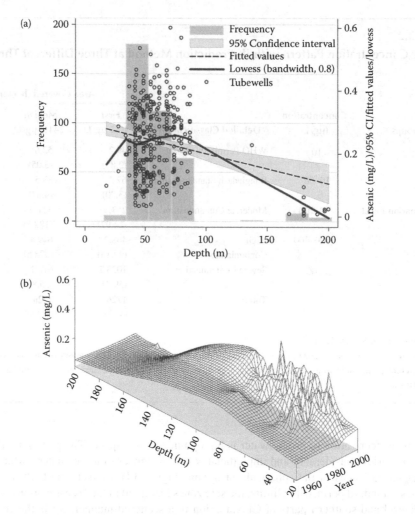

FIGURE 3.5 Relationships between arsenic concentrations and aquifer depth: (a) concentrations in different aquifer depths and (b) three-dimensional view of arsenic concentrations with depth and time.

concentration shows that arsenic concentrations are high, uneven, and erratic within 40–100 m depth and high between 100 and 140 m depth.

Arsenic concentrations are also found to be highly uneven at certain depths (Hassan, 2015). A sharp variation in arsenic concentrations was identified in 51 tubewells having a range between 34 μg/L and 428 μg/L at a depth of 42 m and a substantial variability for 28 tubewells was found between 37 μg/L and 515 μg/L at a depth of 55 m. Variation was also identified in seven tubewell water samples having a range of 3.0–93.0 μg/L at the depth of 180 m (Figure 3.6). In Magura, a strong arsenic variation was identified in 206 tubewells at the depth of 13 m having a range of 0–300 μg/L, with a mean concentration of 23.38 ± 35.46 μg/L, and a substantial arsenic variability for 318 tubewell water at a depth of 57 m was found between no detection limit and 1000 μg/L, with a mean concentration of 167.35 ± 152.46 μg/L (Hassan, 2015). In Dhopakhali and Gazalia, significant arsenic variations were identified at different depths; for instance, in 556 tubewell water samples at a depth of 20 m, a range of arsenic concentrations were recorded for 0–500 μg/L, with a mean concentration of 165.0 ± 133.32 μg/L, and in Gazalia, arsenic variability for 694 tubewell water samples at a depth of 20 m was found between no detection limit and 500 μg/L, with a mean concentration of 99.99 ± 101.62 μg/L (Hassan, 2015).

Groundwater Arsenic Discontinuity

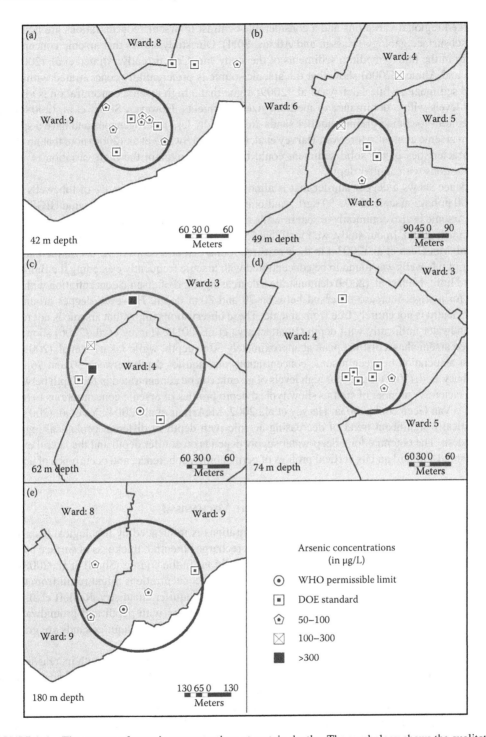

FIGURE 3.6 The pattern of arsenic concentrations at certain depths. The symbology shows the qualitative variation of arsenic concentrations at very close distances within a certain depth in an aquifer. In shallow aquifers (e.g. a depth of 42, 49, 62 and 74 m), a significant number of tubewells were found to be located within few meters from each other, but extensive variations of arsenic concentrations are noticeable in the tubewells. In deep aquifers (e.g. 180 m depth), arsenic concentrations in tubewells were almost within the safe level, mainly within 50 μg/L of arsenic. (Adapted from Hassan, M. M. and P. J. Atkins. 2011. *Journal of Environmental Science and Health, Part A* 46(11):1185–1196).

Marked regional variations and a considerable contrast in arsenic concentrations are noticeable in the subsurface geology (Hassan and Atkins, 2011). Our study shows that arsenic concentration is higher in the fine to medium sediments of the study site. Concurrently, Ahmed et al. (2004) and Burges and Ahmed (2006) show that the arsenic source is preferentially concentrated within fine-grained sediments, while Jakariya et al. (2009) argue that a high arsenic concentration is visible at aquifer levels with a dominance of medium-sized sediments. However, Sharif et al. (2008) claim that medium- to coarse-grained aquifer sands are generally less heterogeneous and have a spatially uniform arsenic content. Moreover, Harvey et al. (2002) and Swartz et al. (2004) note that no chemical characteristics of the solid sediment could be found to explain the high variation of arsenic concentrations with aquifer depth.

Evidence shows a deeper aquifer that is almost free from arsenic, only 1% of tubewells deeper than 200 m have arsenic above 50 μg/L, and only 5% exceeded 100 μg/L of arsenic (BGS/DPHE, 2001). Arsenic is also commonly absent in wells shallower than 5 m, especially in dug wells (Burges and Ahmed, 2006). In our study, we found that 75% of the deep tubewells failed to retain the WHO standard. Van Geen et al. (2003a) in their research found that at a depth of 12–24 m, the maximum number of tubewells was found to be concentrated with arsenic frequently exceeding the Bangladesh standard limit. Nath et al. (2009) demonstrate a decreasing trend of arsenic concentration with depth, except for a slight increase observed between 70 and 80 m depth. However, deeper groundwater (>80 m depth) is not entirely free from arsenic. These observations imply that arsenic is not released to groundwater uniformly with depth (Bhattacharya et al., 2011). Harvey et al. (2006) showed that dissolved arsenic has a distinct peak at approximately 30 m depth, while Jakariya et al. (2009) show a robust association between arsenic concentrations and aquifer depth between 30 and 76 meters. Choudhury et al. (2017) found that high levels of arsenic can be concentrated in the depth between 20 and 50 meters. A number of studies show typical depth profiles of arsenic concentrations to be "bell shaped" (Van Geen et al., 2003a; Harvey et al., 2002; McArthur et al., 2004). Yu et al. (2003) show a statistically significant trend of decreasing arsenic with depth in different geological regions of Bangladesh. The potential for a deep-water supply depends on aquifer depth and the lateral extent of a substantial aquitard and its vertical profiles of permeability, coherence, and occurrence of aquitard.

3.6.4 Spatial Variation: What Is the Possible Mechanism?

Spatial variability of groundwater arsenic concentrations is influenced by lithological heterogeneity, which is controlled by sediment geochemistry, recharge potential, thickness of surface aquitard, local flow dynamics, and the degree of reducing conditions in the aquifer (Sharif et al., 2008). This incongruity in the relation between grain size and arsenic concentrations might result from the textural properties of the aquifer sands at different depths and aquifer chemistry. Radloff et al. (2017) investigated that the increasing trend of arsenic concentrations with depth and groundwater age can be attributed to the evolution of the aquifer over 100–1000 years as aquifer sands are gradually flushed of their initial arsenic content.

A wide range of arsenic concentrations can be maintained in groundwater with increasing depth governed by the history of flushing and local recharge rates (Radloff et al., 2017). The causes of the heterogeneous arsenic distribution include geologic setting, hydro-geochemical conditions, sources of organic matter, water-rock interactions, groundwater flow, anthropogenic influences, and so on (Díaz et al., 2016; Freikowski et al., 2013; Harvey et al., 2002; Kocar et al., 2008; Neumann et al., 2010). Such spatial variability naturally leads to the concern that groundwater arsenic concentration may change over time as well (Brikowski et al., 2014; Duan et al., 2015; Dhar et al., 2008; Guo et al., 2013; Thundiyil et al., 2007). Based on 46 groundwater samples from the well-known BFD area of southwestern Taiwan, Sengupta et al. (2014) investigated the role of local hydrogeology in spatial variation of arsenic concentration in groundwater wells.

3.7 PUBLIC PARTICIPATION GIS: THEORETICAL FRAMEWORK

There has been a remarkable interest in integrating GIS into participatory planning for dealing with spatial information and decision-making. A PPGIS has been developed in combination with a PRA and GIS modeling. Traditionally, GIS has been considered to represent a "top-down" and technology driven approach in spatial decision-making processes (Hassan, 2005). Incorporating the spatial and cognitive knowledge of local communities is essential in designing and implementing sustainable spatial planning for arsenic concentration mapping and mitigation. Lack of community involvement and support are thought to be unsustainable in spatial arsenic mitigation planning.

Mapping is one of the first steps toward a spatial understanding of existing problems (Luansang et al., 2012). Literature on participatory mapping highlights the role of community mapping as a tool to empower communities. Participatory maps, also known as community maps or indigenous maps, are defined by Perkins (2007) as local maps that are produced collaboratively by local people and that incorporate local social, cultural, and economic information. Participatory community maps generally have two broad uses: the first is to act as countermaps that challenge existing spatial documents, while the second is to supplement formal planning through incorporation of local knowledge (Kienberger, 2014; Padawangi et al., 2016; Robbins, 2003).

Participatory mapping has rapidly been propagated in recent times, marking a distinction between grassroots' perspectives and official arrangements of functions of land (Panek, 2015). A PPGIS has been given different research vocabulary in different areas, such as "public forum GIS," "bottom-up GIS," "critical cartography," "collaborative mapping," "community integrated GIS," "counter mapping GIS," "participatory Remote Sensing," "participatory GPS," "ground mapping," "participatory stone mapping," and "participatory sketch mapping" (Baldwin and Oxenford, 2014; Blanke et al., 2014; Cadag and Gaillard, 2012; Dana, 2010; Hassan, 2005; Malve et al., 2016). This concept implies the involvement of a local community in the design and development of a GIS application.

Conventional GIS focuses on digital representations of social and environmental phenomena that best reflect their "expert viewpoint" (Weiner et al., 1995) rather than on "lay perceptions" or a "bottom-up approach." Moreover, a GIS is frequently used for digital map production and in some cases stands accused of transforming bad data into impressive looking maps (Weiner and Harris, 2003). GIS has not been viewed as a method in a participatory process, but, as a "top-down" approach, it encourages the separation of the planning process from the community people in the development paradigm (Jordan and Shrestha, 2000). A GIS can contribute to the social and spatial marginalization of communities via differential access to data through the digital representation, epistemologies, and the multiple realities of landscape represented in GIS (Harris and Weiner, 1998). A people-orientated GIS has the capability to promote a local development plan incorporated with the digital database.

A PPGIS is assumed to be cost-effective and more accurate over a full-blown GIS, and it provides a critical complement to grassroots efforts that are undertaken to empower communities (Bauer, 2009; Carver, 2003; Ghose, 2001; Panek, 2015). In recent times, the expert driven GIS technology is used for incorporating participatory methods to explore land reforms (Weiner et al., 1995), land cover change (Elliot and Campbell, 2001; Mapedza et al., 2003), disaster risk reduction (Cadag and Gaillard, 2012; Singh, 2014), land use planning (Puginier, 2001; Sandström et al., 2003), mapping of illegal settlements (Livengood and Kunte, 2012), rural development (Trang, 2004), natural resource management (Dangles et al., 2010; Kalibo and Medley, 2007; McCall and Minang, 2005; Tripathi and Bhattarya, 2004), and forest management (Kyem, 2002) to assist communities in redefining themselves and their territories (Smith, 1995) to incorporate into conventional decision-making processes (Bird, 1995).

3.8 GIS AND SPATIAL ARSENIC MITIGATION PLANNING

3.8.1 Suitable Location for Arsenic-Safe Tubewell Installation

Identification of a suitable arsenic-safe aquifer with GIS and geostatistics is an important objective for this section. Suitability analysis is a process of systematically identifying or rating potential locations with respect to a particular use. The geostatistical approach has identified the spatial determination for suitable areas for tubewell installation with aquifer depths and water tables. Which areas will get priority in getting access to safe drinking water? The answer to this question can detect the safe water "command areas" and safe water "demand areas" (Hassan and Ahamed, 2017). In doing this, the safe and contaminated areas in Dhopakhali union were identified following the concentration levels of inorganic arsenic with geostatistical approach, mainly ordinary kriging. The safe water command areas were identified in some parts of the southeastern and middle of Dhopakhali. The people who are living within the high and severe contamination zones need safe water options. About 87.47% (342 hectares) of the settlement area is within the unsafe zones, and tubewell technology is not suitable in the arsenic-contaminated areas. Apart from shallow tubewell, people are habituated to using untreated pond water for their drinking and cooking purposes, but this water source was excluded for safe water "command areas." Only the tubewell technology was considered for this Spatial Decision-Support System (Figure 3.7).

We have classified the aquifer based on the existing water table, and they were categorized as (a) lower than 20 m depth, (b) between 20 and 24 m depth, and (c) more than 24 m depth. At the depth of lower than 20 meters, there are suitable areas for arsenic-safe water option for installation of tubewells, and they are mainly located in the southern part of Dhopakhali. At the depth of 20–24 meters, there is an arsenic-safe water table, and they are distributed and scattered in the northeastern, central, and southern parts of the study site. Moreover, at the depth more than 24 meters, arsenic-safe water can be withdrawn mainly from the northern, central, and southern parts of the site. Fieldwork showed that the subsurface geology in Dhopakhali is not suitable for installing deep tubewells, and no deep tubewell exists in Dhopakhali. Moreover, the deep aquifer is heavily concentrated with sodium chloride. The arsenic-safe areas at different depths have been identified at the micro level, and people can easily locate which sites are best fitted for getting arsenic-safe water and at which depth (Table 3.3). Moreover, this planning can be helpful for future strategic design to provide alternative technological options in providing safe drinking water.

3.8.2 Spatial Deep Tubewell Planning with PPGIS

What are the suitable sites for installing additional deep tubewells for an arsenic mitigation option? How many deep tubewells will be needed and what is the basis of each? People's perceptions on this issue are mainly focused on the "threshold distance" (the distance people could travel maximum for collecting arsenic-free water from a deep tubewell), while others explained their comments related to the "population size," "number of households," and sometimes the "area of the neighborhood." Some participants considered different "school and madrassah" (religious school) as the suitable areas to install deep tubewells. Generally, local people thought that the "threshold distance" covers a buffer distance from 300 m to 500 m for a deep tubewell.

Deep tubewell planning for arsenic mitigation can be examined through the combination of a PRA and group interviews using GIS maps. Participatory mental mapping involves mapping the layers of different voices overlaid on GIS maps (Figure 3.8). The incorporation of different participatory mental maps into a digital spatial database allows the use of conventional GIS techniques to achieve a greater understanding of the safe tubewell planning. The combination of different datasets has enhanced the understanding of both the local community and the "expert" viewpoints.

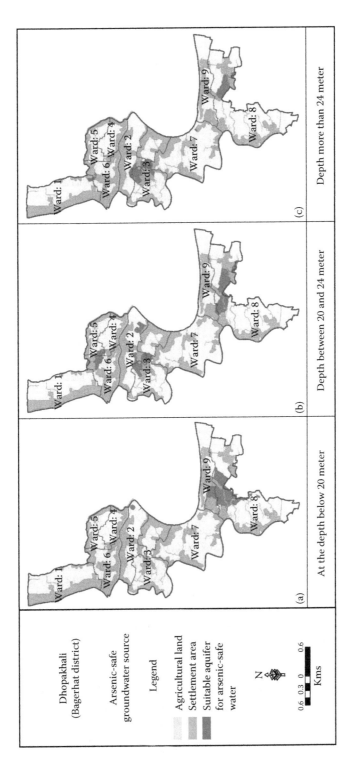

FIGURE 3.7 Suitable arsenic-safe aquifer at different depths: (a) at the depth below 20 meters and the southern part covers almost all of the safe water table; (b) at the depth between 20 and 24 meters and the northern, central, and southern parts cover the safe zones and (c) at the depth more than 24 meters and northern, central, and southern parts cover the safe zones. People can get their access to arsenic-safe water with shallow tubewells. (Adapted from Hassan, M. M. 2015. *Scanning and Mapping the WASH Situation in Coastal Bangladesh: Problems and Potential*. Dhaka: Geo-Ecological Research Team (GeRT).

TABLE 3.3
Suitable Area for Arsenic-Safe Tubewell Installation in Dhopakhali Union

Depth (meter)	Administrative Wards	Village	Settlement Cluster (Para)
<20	Ward: 2	Masokhali	Purbo Para
	Ward: 3	Barodaria	Purbo Para
	Ward: 7	Vasa	Uttar Para, Haldar Para
	Ward: 8	Shanpukuria	Paschim Para, Dakshin Para
		Sherokhali	Dakshin Para
	Ward: 9	Vasa	Purbo Para
		Boga	Purbo Para
20–24	Ward: 1	Kamarganti	Purbo Para
	Ward: 2	Masokhali	Dakshin Para
	Ward: 3	Barodaria	Purbo Para
		Boyarsinga	Purbo Para
	Ward: 4	Madhovkathi	Hawlader Bari
	Ward: 6	Jhalodanga	Moddho Para, Hazi Bari
	Ward: 7	Sitabari	Uttar Para
		Vasa	Uttar Para
	Ward: 8	Sherokhali	Dakshin Para
		Shanpukuria	Paschim Para
	Ward: 9	Vasa	Maddho Para, Purbo Para
		Boga	Paschim Para
>24	Ward: 1	Kamarganti	Purbo Para
	Ward: 2	Masokhali	Dakshin Para, East Para
	Ward: 3	Barodaria	Purbo Para
	Ward: 6	Jhalodanga	Hazi Bari

Note: The safe water table can be accessible in some settlement clusters, and people have shallow tubewells in the identified areas within the settlement clusters.

Participating farmers and marginal farmers pointed out that if deep tubewells are installed within 500 m walking distance, people could collect water conveniently, while some thought that every 60–70 households are suitable for a deep tubewell. Accordingly, they pointed out locations for six additional deep tubewells that could fulfil the demands of arsenic-safe water for the unserved area from Ghona union (Figure 3.8a). Participating school teachers outlined that deep tubewell planning should be based mainly on "school and madrassah." They thought that deep tubewells should be installed in the densely populated areas since it is expensive and people can easily get access to it. Participants from this group identified nine points for installing deep tubewells, which cover eight schools and one madrassah (Figure 3.8b). They thought that about 400–500 pupils attend a school, and it is not possible to stop them drinking contaminated tubewell water since there is no alternative. Participants from local political leaders considered mainly the "school and mosque" for deep tubewell planning. They argued that 13 additional deep tubewells would be required for arsenic-safe water in Ghona if considering the schools and mosques rather than the "threshold distance" and "population size" (Figure 3.8c). Some participants envisaged that the school and mosque committee can easily maintain tubewells provided to them. In overlaying and reviewing different mental maps with locations of proposed deep tubewells prepared by three focus groups, only 17 deep tubewells have been identified after filtering out the duplication and triplication of the same location. From the composite mental mapping overlaid, it can be seen that three deep tubewells will go for Wards 1, 2, 3, and 5 each; two tubewells will go for Ward 8; and one tubewell will go for Wards 6, 7, and 9 each, while no deep tubewell will be required for Ward 4 (Figure 3.8d–e).

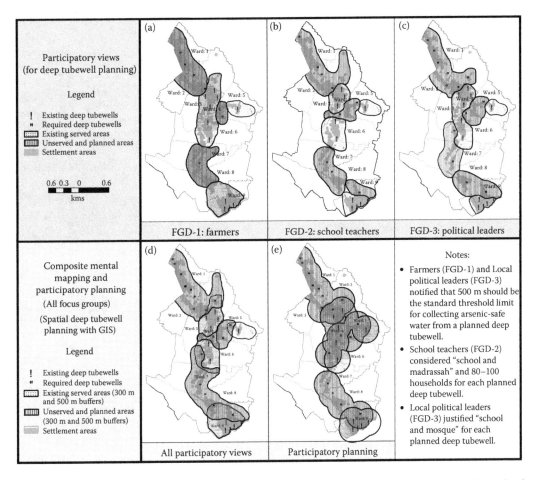

FIGURE 3.8 Participatory views of spatial deep tubewell planning: views (a) from farmers; (b) from school teachers; (c) from local political leaders; (d) combining all the participatory views and (e) transformation of participatory planning into GIS for installing deep tubewells for arsenic-safe water for the unserved areas of Ghona union. (Adapted from Hassan, M. M. 2005. *Health Policy* 74(3):247–260.)

3.9 CONCLUDING REMARKS

An attempt has been made to analyze the spatial arsenic discontinuity with geostatistics that shows the pattern of arsenic concentrations in different areas. The nugget effects, sills, and ranges have shown locally erratic variations of arsenic concentrations. The deviating relationships between aquifer depths and arsenic concentrations are also a central issue. Deep tubewells were also found to be contaminated, but at a low level of concentration. There is no uniformity in spatial arsenic concentrations corresponding to surface geology. This unevenness is due to the aquifer characteristics and surface geology. The Ganges-deltaic flood plain deposits contain very high and erratic concentrations of arsenic. The sharp micro-level variation of arsenic concentrations in the same aquifer raises a number of issues. Is geological variability the main cause of the differences? The variation of arsenic concentrations over time also raises the issue of the mechanism of arsenic in Bangladesh groundwater. What are the reasons for the variation of arsenic concentration with tubewell age: heavy withdrawal of groundwater for irrigation and domestic uses or geological origin and lithology? Therefore, more research is needed on depth-specific distribution in different geological settings to investigate the nature of arsenic mobilization as well as the role of organic matter in the mechanism of arsenic release in groundwater.

4 Chronic Arsenic Exposure to Drinking Water
An Environmental Health Concern

> There is still a controversy as to whether Napoleon was poisoned with arsenic. And the French say the British did it and the British say the French did it, but he died before the test for arsenic was available.
>
> **Michael M. Baden**
> *Physician and Forensic Pathologist, 1934*

Arsenic is a naturally occurring element and common metalloid chemical substance that ubiquitously exists in both organic and inorganic forms in the environment. Inorganic arsenic dissolved in groundwater is a potent environmental toxic element and has been recognized as a "human poison" (Hughes et al., 2011). It is more toxic than organic arsenic present in food and is frequently reported to be an environmental pollutant as well as present a serious health concern due to its potential notorious toxicity (Bloom et al., 2016). Ingestion of arsenic from drinking water can cause cancer, deformities, and mutations (Yoon et al., 2016; Zhang et al., 2017c). Inorganic arsenic was first classified as a carcinogen in 1980 by the International Agency for Research on Cancer (IARC, 1980). The repeated consumption of water containing arsenic is a major health risk, and about 300 million people globally are being poisoned by inorganic arsenic in their drinking water.

The long-term deleterious effects of arsenic have been well known for the last couple of centuries. Inorganic arsenic compounds are skin and lung carcinogens, and there is ample evidence of carcinogenicity in humans. As a ubiquitous toxicant and carcinogenic element, groundwater arsenic is associated with a wide range of adverse human health effects (Chakraborti et al., 2017a; Clewell et al., 2016; Chung et al., 2013a; Edmunds et al., 2015; Karagas et al., 2012; Kippler et al., 2016; Kumar et al., 2016a; Wei et al., 2017). Chronic exposure to elevated levels of arsenic is associated with substantial increased risk of a wide array of diseases including skin manifestations (Sarma, 2016; Wei et al., 2017); cancers of the lung (Sherwood and Lantz, 2016), bladder (Medeiros and Gandolfi, 2016), liver (Lin et al., 2013), skin (Karagas et al., 2015), and kidney (Cheng et al., 2017); neurological disorders (Fee, 2016); diabetes (Kuo et al., 2015); and cardiovascular effects (Barchowsky and States, 2016) diseases.

Arsenic contamination of drinking water is a global public health issue. The mode of arsenic poisoning to human health mainly depends on the chemical forms of arsenic, arsenic exposure, and duration of exposure. Early epidemiological studies concerning drinking water arsenic exposure to human health showed noncancerous skin lesions, such as hyperpigmentation and keratoses of the palms and soles of the feet. A number of epidemiological investigations in endemic areas with a high concentration of arsenic (i.e., 100–1000 µg/L) in Argentina, Bangladesh, Chile, India, and Taiwan have found significant cancerous health risks that are linked to chronic ingestion of arsenic-contaminated water (Clewell et al., 2016). Since the mid-1980s, epidemiological studies focused on cutaneous manifestations and cancers, but have more recently studied cardiovascular diseases and noninsulin-dependent diabetes (Marchiset-Ferlay et al., 2012).

Inorganic arsenic in the human body is largely dependent on its two most common valence states in aqueous oxic environments: arsenate and arsenite forms. Arsenite is considered to be more toxic when compared with arsenate (Oremland and Stolz, 2003). The most deceptive and dangerous aspect of arsenic toxicity is its very slow and insidious development. The prevalence of chronic arsenicism shows a dose-response relationship with concentration of arsenic in drinking water (Fu et al., 2014). Humans are exposed to different inorganic and organic arsenic species present in water, food, and other environmental media. The materials presented in this chapter are aimed at providing the impacts of groundwater arsenic poisoning on human health and its mechanism.

4.1 TOXIC EFFECTS OF INORGANIC ARSENIC

Epidemiological evidence of the health effects of arsenic exposure is well-established. Chronic exposure to low-level environmental arsenic compounds at ppb or μg/L to a few ppm or mg/L has been linked to morbidity and mortality associated with diseases affecting millions of people worldwide (Dangleben et al., 2013; Prasad and Sinha, 2017). Hutchinson (1887) first noted more than 130 years ago that an unusual number of skin tumors developed in patients treated with arsenicals. More recently the International Agency for Research on Cancer (IARC) has determined that inorganic arsenic compounds are skin and lung carcinogens in humans (IARC, 2012). The impact of arsenic poisoning on human health ranges from acute lethality to chronic effects. Arsenic toxicity starts in the human body when exposed to an excessive quantity of arsenic.

Acute Lethality: Acute poisoning results from an ingestion of large quantities of arsenic over a short exposure time. Acute toxicity of arsenic is related to its chemical form and oxidation state. Acute effects are caused by ingestion of inorganic arsenic compounds, mainly arsenite (IARC, 2012). The acute effects of arsenic toxicity represent only its most severe form, and the effects are well known from the incidence of suicidal, homicidal, and other nefarious activities. The characteristic signs of arsenic toxicity in acute form can appear in exposed populations ingesting arsenic at oral doses of around 0.02 mg/kg/day (about 1.0 to 1.5 mg/day for an adult) (ATSDR, 2007), while Opresko (1992) has estimated the acute lethal dose of inorganic arsenic to humans to be about 0.600 to 0.700 mg/kg/day (around 50.0 mg/day in an adult or 3.0 mg/day in an infant). Symptoms of acute toxicity may occur within a few minutes to hours of exposure.

1. *Dire symptoms*: The symptoms of acute toxicity include muscular cramps, facial edema, and cardiac abnormalities; shock can develop rapidly as a result of dehydration (Kamijo et al., 1998; WHO, 1981). Moreover, stomach pain and nausea may lead to shock, coma, and even death (Saha et al., 1999). When arsenic is ingested in large amounts deliberately or inadvertently, it produces a constellation of severe and often fatal injuries to the cardiovascular, gastrointestinal, and nervous systems (NRDC, 2000).
2. *Gastrointestinal effects*: The clinical features initially invariably relate to the gastrointestinal system and are nausea, vomiting, colicky abdominal pain, and profuse watery diarrhea (Ratnaike, 2003). Severe acute abdominal pain may be possible as the impact of acute arsenic poisoning. Possible symptoms following oral ingestion include, severe intestinal pain, severe vomiting, diarrhea, muscle cramps, severe thirst, coma, and death (Hughes et al., 2011). If the patient survives the acute symptoms, there is often peripheral nervous system damage (Gilbert, 2004). Respiratory failure and pulmonary edema are common features of acute poisoning (Lerman et al., 1980). Some clinical features are acute psychosis, a diffuse skin rash, toxic cardiomyopathy, and seizures (Ghariani et al., 1991). Metabolic changes with acute arsenic poisoning are also reported (Ratnaike, 2003).
3. *Death*: There are reports of death due to ingestion of high doses of arsenic. Ingestion of 70–180 mg/L (about 1–3 mg/kg) of arsenic trioxide can be fatal, but initial effects may be delayed for several hours (Gilbert, 2004). In addition, epidemiological studies reveal that a lethal dose of inorganic arsenic in acute poisoning ranges from 100 to 300 mg/L (i.e., 2–5 mg/kg) of arsenic (Ellenhorn, 1997). It is also evident that oral exposure to arsenic in water at 60 mg/L can kill promptly (ATSDR, 2007).

Chronic Arsenic Exposure to Drinking Water

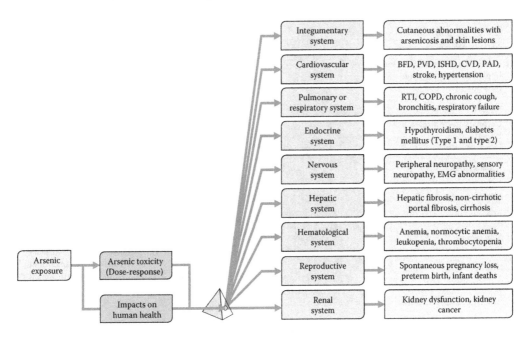

FIGURE 4.1 Significant effects of chronic arsenic poisoning on major human organ systems.

Chronic Effects: Chronic poisoning occurs due to consumption of arsenic-contaminated water for a long time. Repeated low levels of exposure to inorganic arsenic from drinking water over an extended period of time can produce health effects. It is estimated that it takes about 5–15 years to develop chronic arsenicosis symptoms; over time, the symptoms can become more pronounced, and, in some cases, internal organs including the liver, kidneys, and lungs can be affected (Figure 4.1). In the most severe cases, cancer can occur in the skin and internal organs and limbs can be affected by gangrene (Lin et al., 2013; Yu et al., 2006). The latency period differs from patient to patient depending on the amount of arsenic ingested, the nutritional status of the exposed population, the immunity level of the exposed individual, and the duration of arsenic ingestion. There are four known stages of chronic arsenicosis developed with drinking water arsenic ingestion (Chakraborti et al., 2017a; Romero-Schmidt et al., 2001; Saha et al., 1999). They are

1. *Preclinical*: The patient does not have any visible bodily symptoms. Arsenic can mainly be detected in their biomarkers, i.e., urine or body tissue samples. A line of literature shows that the early signs of chronic arsenic exposure are garlic odor on the breath, excessive perspiration, muscle tenderness and weakness, and changes in skin pigmentation (Ratnaike, 2003).
2. *Clinical*: A number of symptoms can be visible on the patient's skin at this stage. The skin is quite sensitive to arsenic and skin lesion (hyperkeratosis and dyspigmentation) has been observed even at the low exposure levels (e.g., 5–10 µg/L) of arsenic in drinking water (Yoshida et al., 2004). The first and earliest visible symptom of chronic arsenic exposure is skin lesions that are characterized by leukomelanosis and melanosis on the trunk and keratosis of the palms and soles (Mazumder et al., 2010; Nriagu et al., 2012). Hyperpigmentation and hyperkeratosis on the palms and soles can occur in a couple of months with repeated ingestion of 0.4 µg/kg/day (Borgoño et al., 1980). These arsenicosis symptoms are mainly dose and time dependent. Hyperkeratosis usually follows the initial pigmentation changes and may then proceed to skin cancer, notably squamous cell skin cancers (Yu et al., 2006).

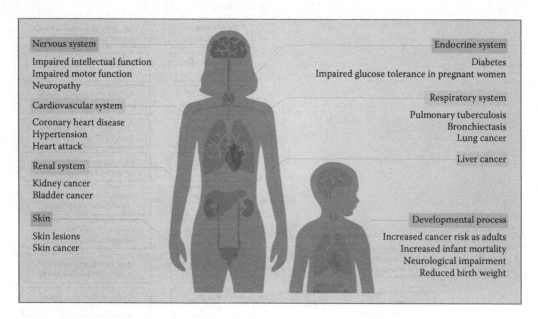

FIGURE 4.2 Impacts of chronic arsenic exposure from drinking water on human health. Arsenic can be responsible for both the malignant and nonmalignant effects on human health. (After HRW. 2016. *Nepotism and Neglect: The Failing Response to Arsenic in the Drinking Water of Bangladesh's Rural Poor.* New York: Human Rights Watch.)

3. *Complications*: Clinical symptoms become more pronounced and internal organs are affected. Enlargement of liver, kidneys, and spleen has been reported (Figure 4.2). At this complication stage, chronic inorganic arsenic exposure is associated with an increased risk of cardiovascular effects (Chen and Karagas, 2013), pulmonary disorders (Dauphiné et al., 2011), reproductive effects (Bjørklund et al., 2017), neurological effects (Parvez et al., 2011a), and diabetes mellitus (Jovanovic et al., 2013). Moreover, peripheral neuropathy is common in persons chronically exposed to arsenic-contaminated drinking water (ATSDR, 2007).
4. *Malignancy*: At this stage, gangrene, tumors, or cancers can affect skin or internal organs (e.g., lung, liver, bladder, and kidney). Chronic ingestion of arsenite through contaminated water results in accumulation of arsenite and methylation of arsenite (MA^{3+}) in vital organs (i.e., liver, kidney, heart, and lungs), muscles, and neuronal tissues leading to the incidence of atherosclerosis, hypertension, ischemic heart disease, diabetes, hepatotoxicity, nephrotoxicity, neurotoxicity, and cancer of the skin, bladder, and lungs (Medeiros and Gandolfi, 2016; Mandal, 2017; Sherwood and Lantz, 2016).

4.2 NONCARCINOGENIC EFFECTS OF CHRONIC ARSENIC EXPOSURE

4.2.1 Integumentary System (Dermal Effects)

The integumentary system is an organ system comprising skin, hair, nails, exocrine glands, and nerve receptors that protect the body from infections. Cutaneous abnormalities are well-known early signs of chronic inorganic arsenic poisoning (Chen et al., 2009; Yamaguchi et al., 2016). In humans, skin is the most sensitive target organ for chronic arsenic exposure (Pei et al., 2013; Yoshida et al., 2004), and skin lesions like raindrop pigmentation, palmar and plantar hyperkeratosis, and hypo- and hyper-pigmentation are regarded as hallmarks of arsenic toxicity (Bhattacharjee et al., 2013a; Samal et al., 2013). Among these skin lesions, palmar and plantar hyperkeratoses are known as pre-malignant skin lesions, which later develop into skin cancer like basal cell carcinoma (BCC), squamous cell carcinoma (SCC), and Bowen's disease (BD) (Banerjee et al., 2011; Tseng et al., 1968). Malignancy from these diseases may occur after about 15–20 years of clinical onset of the disease as monocentric or multicentric SCC. Malignancy in other organs, e.g., lungs, bladder, liver, uterus, etc., may also develop (Chakraborti et al., 2017a). A line of literature shows the association between arsenic exposure to drinking water and the occurrence of skin lesions (Nriagu et al., 2012; Sarma, 2016). Skin lesions are generally referred to as the early manifestation of chronic arsenic intoxication, and hypopigmented skin spots have been associated with arsenic cytoxicity and melanocytes, being more sensitive to arsenic than other cell types of the skin (Graham-Evans et al., 2004) (Table 4.1).

The most common consequence caused by ingestion of arsenic-contaminated water is the development of various typical skin lesions like "raindrop" pigmentation on the chest, back, and legs. Melanosis (known as hyperpigmentation) and hyperkeratosis (known as bilateral palmar-plantar thickening) are the two common manifestations of chronic arsenic poisoning (Çöl et al., 1999; Kile et al., 2011; Saha, 2003). Melanosis is considered an early-stage skin lesion and the most common manifestation, whereas keratosis is considered a sensitive marker of more advanced stages of arsenicosis (Sengupta et al., 2008; Tseng et al., 1968). Melanosis is marked by raindrop-shaped discolored spots, diffuse dark brown spots, or diffuse darkening and hardening (roughness) of the skin on the limbs and trunk (Nriagu et al., 2012). Keratosis may appear as a uniform thickening or as discrete nodules. Simple keratosis usually appears as bilateral thickening of the palms and soles, while in nodular keratosis, small protrusions appear on the palms and soles, with or without nodules on the dorsum of hands, feet, or legs (Mazumder et al., 1998a; Naujokas et al., 2013; Ratnaike, 2003). Keratosis is the most frequent manifestation preceding transformation into arsenic-related skin cancer. Hyperkeratosis is defined as extensively thickened keratosis easily visible from a distance (Breton et al., 2007). Leukomelanosis (hypopigmentation) is the alternate darkened light spots present anywhere on the body, while pigmentation occurs as diffuse dark brown spots, or less discrete diffuse darkening of the skin, or has a characteristic rain drop appearance (Mazumder et al., 2010; Naujokas et al., 2013; Smith et al., 2000b). Pigmentation and keratotic lesions are the most common health effects found in populations exposed to arsenic-contaminated drinking water in Taiwan, Chile, and Argentina (Figure 4.3).

Chronic exposure of low doses of arsenicals causes a "milk and white" complexion, and together with these symptoms various skin eruptions occur (Pimparkar and Bhave, 2010). The skin is quite sensitive to arsenic and, skin lesion has been observed even at the exposure levels in the range of 5–10 μg/L in drinking water (Yoshida et al., 2004). Arsenic-associated skin lesions are possible with low levels of arsenic contamination (<50 μg/L) in drinking water (Argos et al., 2011; Das et al., 1996; Smith and Steinmaus, 2009). In studies of BFD endemic areas in Taiwan, the prevalence rates of hyperpigmentation, hyperkeratosis, and skin cancers have been found to be 183.5, 71.0, and 10.6 per 1000 population, respectively (Tseng et al., 1968). Analyzing arsenic concentrations in 14 villages of West Bengal, Chakraborty and Saha (1987) reported that the lowest arsenic concentration in drinking water that produced arsenic dermatosis was 200 μg/L. Analyzing arsenic-induced skin lesions from keratosis and hyperpigmentation in 7683 subjects exposed to arsenic-contaminated drinking water from West Bengal, India, Mazumder et al. (1998a) claimed that keratosis was common in people consuming arsenic-contaminated water in the low exposure category (<50 μg/L) and hyperpigmentation in the highest exposure category of >800 μg/L of arsenic. According to their investigation, skin lesions usually develop after 5–10 years of exposure to inorganic arsenicals.

After diagnosing 12,195 patients with different types of skin lesions, Chakraborti et al (2002) concluded that exposure to ≥300 μg/L of arsenic in drinking water for up to two years may cause arsenical skin lesions. Examining 96,000 individuals from nine arsenic-affected districts of West Bengal and 18,841 individuals from 31 districts of Bangladesh, Chakraborti et al (2013a) identified 9356 and 3762 patients, respectively, having arsenical skin lesions and 58.6% of individuals with

TABLE 4.1
Epidemiology of Arsenic Exposure to Drinking Water and Prevalence of Non-carcinogenic Effects on Human Health

Reference	Location	Study Design	Time	Comparison (Exposed vs. Reference)	Disease Pattern	Outcome	Results (95% CI)
Dermal Effects: Melanosis, Keratosis, Pigmentation, and Hyperkeratosis of Palms and Soles							
Argos et al. (2011)	Bangladesh (Araihazar)	Longitudinal	2000–2009	Considered 10,182 participants ($\leq$10.0 μg/L vs. >10.0 μg/L).	Chronic arsenic exposure from drinking water was associated with increasing incidence of skin lesions with increasing arsenic exposure, even at low levels of arsenic exposure.	HR	1.17 (0.92–1.49) in 10.1–50.0 μg/L class; 1.69 (1.33–2.14) in 50.1–100 μg/L class; 1.97 (1.58–2.46) in 100.1–200 μg/L class; and 2.98 (2.40–3.71) in >200 μg/L class
Breton et al. (2007)	Bangladesh (Pabna)	Case control (792/792)	2001–2003	About 80% of controls from "low-exposure" arsenic (<50 μg/L) communities and 20% controls from "high-exposure" (>50 μg/L) areas.	Associated with increased risk of skin lesions including melanosis and keratosis.	OR	1.93 (1.15–3.19)
Chakraborti et al. (2013b)	India (Karnataka)	Case-referent	2009	171 individuals for skin lesions, 59 water and 170 biological (human hair and nail) samples.		Spearman two-tailed correlation	Hair: r = 0.682; and Nail: r = 0.685
De Chaudhuri et al. (2008)	India (North 24 Parganas, Nadia, and Murshidabad districts of West Bengal)	Case control (229/199)	—	Analyzed arsenic content in drinking water, nail, hair, and urine samples.	58.6% out of 171 screened individuals were diagnosed with skin lesions (melanosis and keratosis). Three PNP polymorphisms (His20His, Gly51Ser, and Pro57Pro) are associated with arsenic-induced skin lesions.	OR	His20His: 1.69 (1.08–2.66) Gly51Ser: 1.66 (1.04–2.64) Pro57Pro: 1.66 (1.05–2.66)
Fatmi et al. (2013)	Pakistan (Khairpur district of Sindh province)	Cross sectional	2008–2009	Considered chemical analyses of 707 water samples and 534 individuals for screening skin lesions.	Screening a total of 534 individuals, it diagnosed some 72 cases of prevalence of arsenical skin lesions.	Descriptive statistics	—

(Continued)

TABLE 4.1 (*Continued*)
Epidemiology of Arsenic Exposure to Drinking Water and Prevalence of Non-carcinogenic Effects on Human Health

Reference	Location	Study Design	Time	Comparison (Exposed vs. Reference)	Disease Pattern	Outcome	Results (95% CI)
Kile et al. (2011)	Bangladesh (Pabna)	Case control (859/868)	2001–2003	Median arsenic exposure: 23 µg/L (range: <1–1480 µg/L) and median total urinary arsenic: 53.9 (range: 26.1–114.1 µg/L for controls).	Common skin lesions diagnosed were leukomelanosis (46.4%), melanosis (44.0%), keratosis (15.6%), hyperkeratosis (4.6%), and skin cancer (0.09%).	OR	1.56 (1.15–2.12)
Liu et al. (2013)	China (Xinjiang)	Case control (178/179)	—	Considered arsenic level ranging 47–427 µg/L (averaging 190 µg/L) in endemic sites and 10–17 µg/L in the control sites.	Skin lesion related to chronic arsenic exposure was found to be 31.1% in 175 surveyed villagers from the endemic area.	*T*-tests, one-way ANOVA and Fisher Exact tests	—
Mazumder et al. (1998a)	India (West Bengal)	Cross-sectional	1995–1996	Considered 7683 participants (<50 µg/L vs. >800 µg/L).	A clear relationship was apparent between water levels of arsenic and the prevalence of keratosis and hyperpigmentation.	MSR	Keratosis: 1.6 (1.0–2.4); and hyperpigmentation: 1.2 (0.8–1.8)
Nriagu et al. (2012)	India (northern Nadia, West Bengal)	Case control (100/100)	2006–2007	Considered 200 participants including 100 cases with skin lesions and 100 controls with no lesions.	Skin lesions in areas where people are exposed to high levels of arsenic in their drinking water. An SNP (rs9989407) strongly contributes to development of arsenic-induced skin lesions.	OR	0.587 (0.247–1.394)
Pei et al. (2013)	China	Cohort	2010	Considered 75 subjects with different exposure levels and length of exposure with age.	Hypopigmentation and keratosis are recognizable at early stage when other adverse effects are not yet observable.	Multiple regression	$r = 0.857$ (with length of exposure and age)
Pierce et al. (2012)	Bangladesh	Case control (1085/1794)	2000–2009	Urinary arsenic metabolite concentrations and 259,597 genome-wide single nucleotide polymorphisms (SNPs) for exposed individuals.	Risk of skin lesions through interaction with arsenic exposure. One of five genetic variants (rs9527) is associated with skin lesions.	PE for OR	—

(*Continued*)

TABLE 4.1 (Continued)
Epidemiology of Arsenic Exposure to Drinking Water and Prevalence of Non-carcinogenic Effects on Human Health

Reference	Location	Study Design	Time	Comparison (Exposed vs. Reference)	Disease Pattern	Outcome	Results (95% CI)
Rahman et al. (2006)	Bangladesh (Matlab)	Case-referent	2002–2003	Considered 1830 referents and 504 diagnosed patients.	Individuals exposed to a time-weighted mean drinking water arsenic concentration of 10–49 µg/L since 1970 had an increased risk of skin lesions compared with individuals exposed to <10 µg/L.	OR	Male: 3.25 (1.43–7.38); Female: 1.66 (0.65–4.24)
Wei et al. (2017)	China (Bameng region, western Inner Mongolia)	Cross-sectional	–	Considered a total of 207 adult women aged ≥20 years.	The nail content of arsenic and urinary concentrations of DMA, MMA, and inorganic arsenic were significantly higher in women with skin lesions than in those without skin lesions.	Descriptive statistics	–
Vascular Effects: Ischemic Heart Disease, Peripheral Vascular Disease, Cerebral Infarction, and Hypertension							
Abhyankar et al. (2012)	World wide	Systematic Review	1966–2011	Selected 11 cross-sectional studies for final review.	Environmental exposure to arsenic is linked to hypertension in persons living in arsenic-endemic areas.	OR	1.27 (1.09–1.47)
Afridi et al. (2011)	Pakistan (Hyderabad)	Cohort	2007–2008	Biological samples of 130 MI (myocardial infarction) patients (77 male and 53 female) with ages 45–60 years vs. 61 healthy persons (33 male and 28 female) of the same age group were collected.	Levels of arsenic concentrations in blood samples show an association between inorganic arsenic ingestion and MI incidence.	RR	3.41 (1.67–6.99)
Chen et al. (2011a)	Bangladesh (Araihazar)	Prospective Cohort	2000	Considered 2000 urine samples and 198 deaths from diseases of circulatory system as well as mean arsenic concentrations in well water (266 µg/L [>80th] vs. 3.7 µg/L [<20th percentile]).	Exposure to arsenic in drinking water is adversely associated with mortality from diseases of circulatory system (i.e., CVD, Cerebrovascular disease, and ISHD).	RR	1.46 (0.96–2.20)

(Continued)

TABLE 4.1 (Continued)
Epidemiology of Arsenic Exposure to Drinking Water and Prevalence of Non-carcinogenic Effects on Human Health

Reference	Location	Study Design	Time	Comparison (Exposed vs. Reference)	Disease Pattern	Outcome	Results (95% CI)
Chen et al. (2013b)	Bangladesh	Case Cohort	–	Groundwater arsenic exposure levels were 0.1–864.0 µg/L in both men and women.	Different levels of exposure to arsenic in drinking water are adversely associated with CVD, heart diseases, and stroke.	RR	CVD: 1.49 (1.06–2.11) Heart diseases: 1.54 (1.02–2.31) Stroke: 1.38 (0.84–2.27)
Cheng et al. (2010)	Taiwan (southwest and northeast Taiwan)	Cohort	1971–2005	Mortality data in BFD endemic area vs. Lan-Yang Basin in northeast Taiwan and arsenic concentrations in well water.	A significant dose-response relationship established between chronic arsenic exposure to drinking water and prevalence of CVD.	SMR	1.05 (1.01–1.10)
Gong and O'Bryant (2012)	USA (3 rural counties, Texas)	Cross sectional	–	Considered 509 subjects and estimated residential drinking water levels 8.1 µg/L (>75th) vs. 4.1 µg/L (<25th percentile).	Exposure to low-level (around 10 µg/L) arsenic in drinking water is associated with increased mortality of coronary heart disease.	OR	1.10 (1.00–1.21)
Hall et al. (2017)	Northern Chile	Case control (612/654)	2007–2010	Compared to those in the lowest category for lifetime highest 5-year average arsenic exposure (<60 µg/L), those in the middle (60–623 µg/L) and upper (>623 µg/L) exposure categories.	Both the middle and upper exposure categories are positively associated with hypertension.	OR	1.49 (1.09–2.05) in the middle category, and 1.65 (1.18, 2.32) in the upper category
Huang et al. (2007)	Taiwan (south western)	Cross sectional	–	Considered 871 subjects with 372 diagnosed with hypertension based on a positive history and with urinary arsenic concentrations.	Cumulative arsenic exposure (CAE) from drinking water has been shown to be associated with hypertension in a dose-response pattern.	OR	1.47 (1.04–2.07)
James et al. (2015)	USA	Case Cohort	–	Groundwater arsenic exposure levels were 1–88 µg/L in both men and women aged 20–74 years.	Significant associations were found between inorganic arsenic exposure and coronary heart disease.	RR	3.10 (1.10–9.11)
Khan et al. (2010a)	Bangladesh (Samta village, Jessore)	Cross sectional	–	Considered 120 exposed vs. 120 nonexposed populations.	PAD prevalence is associated with high arsenic concentrations having abnormal Ankle-brachial Blood Pressure Index (ABSPI).	RR	5.8 (1.26–26.85)

(Continued)

TABLE 4.1 (Continued)
Epidemiology of Arsenic Exposure to Drinking Water and Prevalence of Non-carcinogenic Effects on Human Health

Reference	Location	Study Design	Time	Comparison (Exposed vs. Reference)	Disease Pattern	Outcome	Results (95% CI)
Liao et al. (2012)	Taiwan (3 villages in southwest Taiwan)	Cohort	2002–2009	Some 380 vs. 296 individuals from endemic and nonendemic areas, respectively, and CEI from well water levels (>14.7 vs. <17.4 ppm/year).	A clear relationship was apparent between cumulative arsenic exposure and CVD mortality.	HR	1.89 (0.50–7.10)
Medrano et al. (2010)	Spain	Ecological	1998–2003	Municipal drinking water levels (<1 µg/L vs. >10 µg/L).	Low to moderate levels of arsenic concentrations in drinking water are associated with mortality of CVD, CHD and Cerebrovascular disease.	SMR	CVD: 1.10 (1.08–1.12) CHD: 1.18 (1.15–1.22) Cerebrovascular Disease: 1.04 (1.01–1.08)
Wu et al. (2010)	Taiwan (Lanyang Basin, northeast Taiwan)	Cohort	1972–2008	Selected 504 participants having inorganic arsenic exposure (≤50 µg/L vs. 50–300 µg/L).	Elevated levels of arsenic concentrations in drinking water are associated with CVD mortality (22 deaths with CMR of 4.45).	RMR	2.07 (0.74–5.80)
Zhang et al. (2013)	China (Wuyuan County of Inner Mongolia)	Cross sectional	2010	Blood pressure of 405 individuals and drinking water with an arsenic content of <50 µg/L.	Prevalence of hypertension in groups with the duration of arsenic exposure ≤30 years, >30–50 years, and >50 years are 25.37%, 51.98%, and 58.56%, respectively.	OR	1.45 (0.63–3.35) with 30–50 years exposure; and 2.95 (1.31–6.67) with >50 years exposure
Respiratory System: Increased Mortality from Pulmonary Tuberculosis, Bronchiectasis, and Lung Cancer							
Ahmed et al. (2017)	Bangladesh (Matlab Upazila)	Cohort	2001–2004	Assessed urinary arsenic that was collected from mothers during early pregnancy and their children aged 4.5 and 9 years. Assessed lung function of 540 children aged 9.	Prenatal arsenic exposure is related to impaired lung function, while childhood exposure may increase airway inflammation, particularly in boys.	β-value	Boys (4.5 years): 0.89 (0.13–1.66); and Boys (9 years): 0.88 (0.16–1.61)

(Continued)

TABLE 4.1 (Continued)
Epidemiology of Arsenic Exposure to Drinking Water and Prevalence of Non-carcinogenic Effects on Human Health

Reference	Location	Study Design	Time	Comparison (Exposed vs. Reference)	Disease Pattern	Outcome	Results (95% CI)
Mazumder et al. (2000)	India (West Bengal)	Case control	—	Considered 7683 participants of all ages who were highly exposed to drinking water arsenic (≥ 50 μg/L vs. <50 μg/L).	Associated with increased risk of cough, chest sounds and shortness of breath.	POR	Cough (male: 5.0 and female: 7.8); Chest sounds (male: 6.9 and female: 9.6; Breath shortness (male: 3.7 and female: 23.2
Mazumder et al. (2007a)	India (West Bengal)	Case control	—	Considered 258 individuals (108 with arsenical skin lesions vs. 150 without arsenic lesions).	Associated with increased risk of bronchiectasis in population with arsenical skin lesions.	OR	10.1 (2.7–37.1)
Milton et al. (2001)	Bangladesh (Barisal, Laxmipur, and Madaripur districts)	Case control	—	Considered 94 arsenical skin lesions vs. 124 unexposed cases and 136–1000 μg/L of drinking water arsenic concentration.	An association between arsenic exposure and chronic bronchitis investigated.	CPR	Male: 1.6 (0.8, 3.1); and Female: 10.3 (2.4, 43.1)
Parvez et al. (2010)	Bangladesh (Araihazar)	Cross sectional	2000–2002, 2002–2004, and 2004–2006	A total of 7.31%, 9.95%, and 2.03% of 11,746 participants having a chronic cough, breathing problem, or blood in their sputum, respectively.	This prospective cohort study found a dose-response relationship between arsenic exposure and clinical symptoms of respiratory symptoms.	HR	1.11 (0.95, 1.29) and 1.35 (1.16, 1.56) for urinary arsenic for the third and fifth quintiles, respectively
Recio-Vega et al. (2014)	Mexico (Comarca Lagunera)	Cross sectional	2000–2013	Some 358 children aged 6–12 years having individual exposure were assessed based on urinary concentration of inorganic arsenic, and lung function was assessed by spirometer.	More than 57% of the subjects had a restrictive spirometric pattern. High urinary arsenic level was in children with restrictive lung patterns.	β-value	DMA: 0.002 (0.006, 0.005) Inorganic Arsenic: −0.003 (−0.005, −0.008)
Smith et al. (2011)	Chile (Regions II and V)	Ecological	1958–2000	Compared mortality data in different time frame of arsenic exposure, and estimation with linear interpolation.	Arsenic is an immunosuppressant and also a cause of chronic lung disease.	MRR	2.1 (1.7, 2.6)

(Continued)

TABLE 4.1 (*Continued*)
Epidemiology of Arsenic Exposure to Drinking Water and Prevalence of Non-carcinogenic Effects on Human Health

Reference	Location	Study Design	Time	Comparison (Exposed vs. Reference)	Disease Pattern	Outcome	Results (95% CI)
Tsai et al. (1999)	Taiwan (BFD-endemic area)	Ecological	1971–1994	HAA vs. No HAA (High Arsenic Area).	Mortality from "bronchitis" among residents of the BFD areas in Taiwan was significantly higher than nearby reference population and the total population.	SMR	1.53 and 1.95
von Ehrenstein et al. (2005)	India (West Bengal)	Cross sectional	1998–2000	Considered a cohort of 287 participants who were exposed to drinking water arsenic (up to 500 µg/L).	Higher risk of common respiratory symptoms with lower FEV_1 (forced expiratory volume measured in 1 sec) and FVC (forced vital capacity) among people with arsenical lesions.	OR	FEV_1: 256.2 mL (113.9, 398.4); and FVC: 287.8 mL (134.9, 440.8)
Metabolic Effects: Diabetes Mellitus							
Chen et al. (2011b)	Taiwan (Changhua County)	Cross sectional	2002–2005	1043 subjects having urinary arsenic concentrations of >200 µg/g vs. ≤35 µg/g creatinine.	A higher risk for DM in subjects with a total urinary arsenic >75 µg/g creatinine as compared with low urinary arsenic concentrations.	OR	2.22 (1.21, 4.09)
Islam et al. (2012b)	Bangladesh	Cross sectional	2009	Selected 1004 individuals having 9% with prevalence of T2D, and considered elevated levels of arsenic in drinking water (>50 µg/L).	An association between chronic arsenic exposure through drinking water and Type 2 DM. Risks are generally higher with longer duration of arsenic exposure.	OR	1.9 (1.1, 3.5
James et al. (2013)	USA (Colorado)	Case-cohort	1984–1998	Considered 141 cases of DM diagnosed vs. a sub-cohort of 488 participants who were disease free at baseline.	Exposure to low-level inorganic arsenic in drinking water is associated with increased risk for type II DM in population based on a comprehensive lifetime exposure assessment.	HR	1.27 (1.01, 1.59)

(*Continued*)

TABLE 4.1 (Continued)
Epidemiology of Arsenic Exposure to Drinking Water and Prevalence of Non-carcinogenic Effects on Human Health

Reference	Location	Study Design	Time	Comparison (Exposed vs. Reference)	Disease Pattern	Outcome	Results (95% CI)
Jovanovic et al. (2013)	Serbia (middle Banat region)	Cross sectional	2006–2009	Exposed population with mean arsenic concentration of 56 µg/L vs. unexposed population with below detection limit (2 µg/L).	Low levels of arsenic exposed people are at high risk for the occurrence of type 2 diabetes in comparison to unexposed population.	OR	1.22 (1.09, 1.36)
Kim and Lee (2011)	South Korea	Cross sectional	2007–2009	Measuring urinary arsenic metabolites in 200 participants with total urinary arsenic concentrations.	An association between total urinary arsenic concentration and the prevalence of DM established in a representative sample of women with environmental arsenic exposure.	OR	Female: 1.502 (1.038, 2.171); All participants: 1.312 (1.04, 1.655)
Kuo et al. (2015)	USA (North and South Dakota)	Prospective Cohort	1989–1999	A total of 1694 diabetes-free participants aged 45–75 years were recruited in 1989–1991 and followed through 1998–1999. Used proportions of urinary inorganic arsenic, MMA, and DMA over their sum as the biomarkers of arsenic metabolism.	Over 11,260 person-years of follow-up, 396 participants developed diabetes. Arsenic metabolism, particularly lower MMA%, was prospectively associated with increased incidence of diabetes.	HR	MMA%: 1.07 (1.00, 1.15)
Lai et al. (1994)	Taiwan (BFD-endemic area)	Cross sectional	—	Examined 891 adults residing in villages of BFD-endemic area (100–15,000 µg/L/yr vs. >15,000 µg/L/yr).	A positive dose-response association between cumulative exposure to arsenic from drinking water and the prevalence of DM.	OR	6.61 (0.86, 51.0) for cumulative arsenic exposure of 100–15,000 µg/L/yr; and 10.05 (1.30, 77.9) for >15,000 µg/L/yr
Rahman et al. (1998)	Bangladesh	Cross sectional	1996	Considered 163 exposed with arsenical keratosis vs. 854 unexposed population.	Arsenic exposure is a risk factor for DM, and an association between arsenic exposure and DM was examined.	CPR	5.2 (2.5, 10.5)

(Continued)

TABLE 4.1 (Continued)
Epidemiology of Arsenic Exposure to Drinking Water and Prevalence of Non-carcinogenic Effects on Human Health

Reference	Location	Study Design	Time	Comparison (Exposed vs. Reference)	Disease Pattern	Outcome	Results (95% CI)
Tsai et al. (1999)	Taiwan (BFD-endemic area)	RCO	1971–1994	HAA vs. No HAA (High Arsenic Area).	An increased prevalence of diabetes in women compared with that in men occurred after 40 years of age in areas with high levels of inorganic arsenic in drinking water.	OR	1.46 (1.28, 1.67)

Neurological Effects: Peripheral Neuropathy Involving Sensory and Motor Systems, Cranial Nerve Involvement, Hearing Loss, Mental Retardation, Encephalopathy

Reference	Location	Study Design	Time	Comparison (Exposed vs. Reference)	Disease Pattern	Outcome	Results (95% CI)
Gardner et al., 2013	Bangladesh	Cohort	2001–2009	Based on 1505 mother-infant pairs and early-life exposure to arsenic concentrated in maternal and child urine.	Inverse association was investigated between concurrent exposure to arsenic and children's weight and height, age-adjusted Z scores, and growth velocity at age 5 years.	Z-score	−0.33 kg (−0.60, −0.06)
Hamadani et al. (2011)	Bangladesh (Matlab)	Cohort	1994	Considered children born between May 2002 and December 2003 with 5 years of age and urinary arsenic concentrations.	Adverse effects of inorganic arsenic exposure through drinking water on IQ (Verbal and Full-Scale) in pre-school girls at 5 years of age.	β-value	VIQ: −2.4 (−3.8, −1.1); FSIQ: −1.4 (−2.7, −0.1)
O'Bryant et al. (2011)	USA (Cochran County and Parmer County, Texas)	Cross sectional	—	Considered 434 adult participants for epidemiological study of cognitive aging among rural people. Less than 50 µg/L urinary arsenic or water arsenic was undertaken.	Long-term low-level exposure to arsenic is significantly associated with decreased capacity in executive function, mental acuity, and verbal skills.	GIS analysis and Correlation Coefficient	Executive functioning (CLOX 1 B(SE): −0.225 (0.080), p: 0.005)
Parvez et al. (2011a)	Bangladesh	Cross sectional	2008	Considered 304 children aged between 8 and 11 years and arsenic and manganese concentrations in drinking water, blood, urine, and toenails.	Arsenic exposure (water arsenic, urinary arsenic, and toenail arsenic) are inversely associated with children's intellectual function with total motor function scores.	β-value for TMC	−3.63 (−6.72, −0.54)

(Continued)

TABLE 4.1 (Continued)
Epidemiology of Arsenic Exposure to Drinking Water and Prevalence of Non-carcinogenic Effects on Human Health

Reference	Location	Study Design	Time	Comparison (Exposed vs. Reference)	Disease Pattern	Outcome	Results (95% CI)
Roy et al. (2011)	Mexico (Torreón)	Cohort	—	Considered 526 children aged 6–7 years. Total urinary arsenic (50–100 µg/L urinary or water arsenic), MMA, and DMA concentrations were measured and behavior of selected children assessed.	Examined poorer scores in arsenic exposed children on measures of language and vocabulary and a modest association with hyperactive behavior using the attention deficit hyperactive disorder (ADHD) index. Higher urinary DMA was associated with higher ratings on the Oppositional, Cognitive Problems and ADHD Index.	β-value	DMA: Oppositional (2.5: −0.7, 5.6); Cognitive (2.5: −0.2, 5.4); Hyperactive (1.9: −0.9, 4.7); and ADHD index (2.1: −0.4, 4.7)
Sen and Biswas (2012)	India (West Bengal)	Cross sectional	More than 13 months	Considered a total of 1169 adult arsenicosis patients aged 18–65 years who used to drink arsenic-contaminated tubewell water with 25–900 µg/L for 20 years.	Increased risk of psychiatric disorder, depression, anxiety. Of the 1169 participants, some 18.99% of the arsenicosis patients had psychiatric ailments and 16.34% had clinically significant psychopathology.	GHQ-12 and 18-item BPRS	GHQ Score >2 and BPRS Score >30
Wang et al. (2007a)	China (Shanyin county, Shanxi province)	Cross sectional	2003	Selecting 524 children between 8 and 12 years of age who had been exposed to either arsenic or fluoride.	It is significant that children's intelligence and growth can be affected by high concentrations of arsenic.	Correlation coefficient	−0.201 ($p < 0.01$)
Wasserman et al. (2004)	Bangladesh (Araihazar)	Cross sectional	2002	Investigation of intellectual function in 201 children 10 years of age and arsenic concentrations in 196 well samples. Analysis of blood lead (BPb) and hemoglobin (Hgb) concentrations from 107 agreed children.	Exposure to inorganic arsenic is associated with reduced intellectual function in a dose-response manner.	β-value	Full-scale: −11.3 (4th quartile) Performance score: −9.7 (4th quartile)

(Continued)

TABLE 4.1 (Continued)
Epidemiology of Arsenic Exposure to Drinking Water and Prevalence of Non-carcinogenic Effects on Human Health

Reference	Location	Study Design	Time	Comparison (Exposed vs. Reference)	Disease Pattern	Outcome	Results (95% CI)
Hepatic Manifestation: Fatty Degeneration, Noncirrhotic Portal Fibrosis, Cirrhosis, Hepatomegaly, Cholestasis							
Das et al. (2012)	India (Murshidabad district, West Bengal)	Case control	2009	Considered 103 arsenic exposed individuals having at least 10 years of exposure (both with and without skin lesions) vs. 107 unexposed individuals. Liver function tests were conducted to determine the hepatic dysfunction.	Significant increasing trend of serum levels of bilirubin, ALT, AST, and ALP in the arsenic exposed population when compared to the unexposed group with similar socioeconomic status.	Mann Whitney test	ALT: 36.60 ± 11.07 ($p < 0.0012$) AST: 51.72 ± 11.34 ($p < 0.0001$) ALP: 274.03 ± 84.62 ($p < 0.0001$)
Ghosh (2013)	India (West Bengal)	Controlled, cross sectional, observational	2008–2009	Some 73 arsenic affected patients with raindrop pigmentation who consumed water containing arsenic ≥ 50 μg/L and having hair and nail arsenic level >0.6 μg/L vs. 25 nonsmoker healthy controls.	A significant number of patients (31.5%) with hepatomegaly due to elevated levels of arsenic ingestion from drinking water.	Sample t-test and χ^2-test	—
Hernández-Zavala et al. (1998)	Mexico (Lagunera Region)	Cross sectional	1995	Some 51 individuals in three groups, one in each village, e.g., Nazareno, Santa Ana, and Benito Juárez, who consumed water containing arsenic 14–300 μg/L in these three study villages.	Statistically significant increases in the serum concentrations of conjugated (direct) and unconjugated bilirubin and the increased serum ALP activity in the highly exposed individuals in Santa Ana and Benito Juárez.	Correlation	Relationship between ALP activity ($r = 0.56$), direct bilirubin ($r = 0.38$), indirect bilirubin $r = 0.33$, total bilirubin, ($r = 0.45$) and the urinary concentration of total arsenic.
Islam et al. (2011)	Bangladesh (Bheramara, Chuadanga and Jessore districts)	Cross sectional	2009	Selected a total of 200 residents living in arsenic-endemic areas in Bangladesh. Arsenic in drinking water, hair and nails were measured by ICP-MS.	A novel exposure and dose-response relationship between arsenic exposure metrics and serum hepatic enzyme activity was found.	Correlation	$r = 0.55$ between water and hair arsenic concentrations and $r = 0.49$ between water and nail arsenic concentrations

(Continued)

TABLE 4.1 (Continued)
Epidemiology of Arsenic Exposure to Drinking Water and Prevalence of Non-carcinogenic Effects on Human Health

Reference	Location	Study Design	Time	Comparison (Exposed vs. Reference)	Disease Pattern	Outcome	Results (95% CI)
Santra et al. (1999)	India (West Bengal)	Cohort	—	248 patients with evidence of chronic arsenic toxicity vs. 23 control subjects underwent clinical and laboratory examinations including liver function tests and HBsAg status.	Large number of patients with liver disease (190 out of 248, 76.6%) due to chronic arsenicosis from arsenic-contaminated drinking water and NCPF is the predominant lesion in this population.	—	Arsenic content in liver: 1.46 ± 0.42 for cases; and 0.16 ± 0.04 ($p < 0.001$) for control. Maximum was recorded for 6 mg/Kg
Hematopoietic Effects: Bone Marrow Hypoplasia, Aplastic Anemia, Anemia, Leukopenia, Thrombocytopenia, Impaired Folate Metabolism, Karyorrhexis							
Heck et al. (2008)	Bangladesh (Araihazar)	Cross sectional	2000–2002	Hemoglobin measures, skin lesions, arsenic exposure, and nutritional and demographic information from 1954 participants in the health effects of arsenic longitudinal study.	High arsenic exposure and anemia are correlated.	Correlation	—
Islam et al. (2004b)	Bangladesh (Chapai Nawanganj)	Cross sectional	—	Some 115 exposed individuals diagnosed as arsenicosis patients and age-matched 120 unexposed subjects were enrolled as control group.	Higher prevalence of leukopenia and lymphocytosis in arsenicosis patients were associated.	Correlation	—
Reproductive Effects: High Perinatal Mortality, Low Birth Weight, Spontaneous Abortions, Stillbirths, Pre-Eclampsia, Congenital Malformation							
Ahamed et al. (2006)	Bangladesh (Eruani Village of Comilla district)	Cross sectional	1997–2005	201–500 μg/L vs. with no arsenic.	LBW is associated with high exposed group of inorganic arsenic ingestion	PPR	2.35 (0.24, 23.47)
				501–1200 μg/L vs. with no arsenic.	LBW and PB are associated with high arsenic-contaminated drinking water.	PPR	LBW: 3.36 (0.70, 15.05) and PB: 4.20 (0.51, 34.67)
Bloom et al. (2016)	Romania (Timisoara County)	Exploratory Study	December 11 to January 2013	Prospectively followed 122 women with singleton deliveries with water consumption weighted by arsenic.	Investigated an association between exposure of higher average arsenic (10 μg/L) ingestion and lower birth weight.	OR	1.23 (0.78, 2.01)

(Continued)

TABLE 4.1 (Continued)
Epidemiology of Arsenic Exposure to Drinking Water and Prevalence of Non-carcinogenic Effects on Human Health

Reference	Location	Study Design	Time	Comparison (Exposed vs. Reference)	Disease Pattern	Outcome	Results (95% CI)
Chakraborti et al. (2003)	India (Bihar)	Cross sectional	—	463–1025 µg/L vs. 7–459 µg/L arsenic.	Low birth weight (LBW) and preterm birth (PB) are associated with the use of arsenic-contaminated water by 463–1025 µg/L.	PPR	LBW: 8.33 (1.03, 67.14) and PB: 3.33 (0.92, 12.11)
Chakraborti et al. (2004)	India (West Bengal)	Cross sectional	—	401–1474 µg/L vs. 200–400 µg/L of arsenic.	A high rate of spontaneous abortion was examined for the consumption of 401–1474 µg/L against 200–400 µg/L.	—	182 losses/1000 pregnancies against 95 losses/1000 pregnancies
Cherry et al. (2010)	Bangladesh	Ecological	2001–2003	934 infant mortality occurring in designated area and arsenic levels in tubewell water containing ≥50 µg/L vs. <10 µg/L.	Increased infant mortality.	OR	1.20 (0.90, 1.59)
Chou et al. (2012)	Taiwan	Prospective cohort	—	1 unit vs. 0.2–3.9 µg/g (creatinine).	Low birth weight.	OR	1.02 (0.78, 1.34)
Farzan et al. (2017)	USA (New Hampshire)	Cohort	2009	Maternal urinary arsenic levels at gestational weeks 24–28 and levels of inflammatory biomarkers in plasma from 563 pregnant women and 500 infants' cord blood.	Arsenic exposure during pregnancy may affect markers of vascular health and endothelial function in both pregnant women and children.	β	145.2 ng/ml (4.1, 286.3)
Fei et al. (2013)	USA (New Hampshire)	Epidemiologic	—	Some 133 pregnant women with arsenic level in urine and placental tissue sample.	Infant birth weight.	Coefficient estimation	−0.009 (0.032, −0.001)
Gelmann et al. (2013)	Romania (Arad County)	Cross sectional	—	>9.0 µg/L vs. ≤9.0 µg/L.	Low birth weight.	67% (different between groups)	P: 0.019

(Continued)

TABLE 4.1 (Continued)
Epidemiology of Arsenic Exposure to Drinking Water and Prevalence of Non-carcinogenic Effects on Human Health

Reference	Location	Study Design	Time	Comparison (Exposed vs. Reference)	Disease Pattern	Outcome	Results (95% CI)
Guan et al. (2012)	China (Dalian City)	Cross sectional	—	125 mother–infant pairs and arsenic-affected area (590 μg/L) vs. arsenic-free area (1 unit vs. <25 μg/L maternal blood).	Birth Weight (a 1 μg/L increase in maternal blood arsenic is associated with a 20 g decrease in birth weight).	β	−20.0 (−39.60, −0.40)
Huyck et al. (2007)	Bangladesh (42 villages in Sirajdikhan Upazila of Munshigani district)	Prospective	—	49 women ≥18 years of age and arsenic levels in maternal hair at first visit (1 unit vs. 0.14–3.28 μg/g (maternal hair at 1st prenatal visit, μg/g)).	Low birth weight (<2750 g) is associated with the use of arsenic-contaminated water.	OR	2.50 (0.74, 8.33)
Kwok et al. (2006)	Bangladesh (261 villages from Faridpur Sadar, Matlab and Shahrasti Upazila)	Cross sectional	2002–2003	Examined 2006 pregnant women chronically exposed to arsenic in drinking water to find out relationships between arsenic exposure and selected reproductive health outcomes.	Investigated statistically a slight association between arsenic exposure and birth-defects.	OR	1.005 (1.001, 1.010)
Mukherjee et al. (2005)	India (West Bengal)	Cross sectional	—	284–400 μg/L vs. <3 μg/L of arsenic.	An increased LBW and PB relative to 18 pregnancies (284–400 μg/L) in 7 reference women (<3 μg/L).	PPR	LBW: 2.45 (0.61, 9.88)
				401–1474 μg/L vs. <3 μg/L of arsenic.	An increased LBW and PB relative to 18 pregnancies (401–1474 μg/L) in 7 reference women (<3 μg/L).	PPR	LBW: 2.57 (0.59, 11.20) and PB: 3.43 (0.83, 14.13)
Myers et al. (2010)	China (Inner Mongolia)	Cross sectional	—	>100 μg/L vs. <20 μg/L arsenic.	Infants born to mothers consuming high arsenic-contaminated water were 50 g heavier, on average, than those born to mothers consuming water with <20 μg/L of arsenic.	β	50 (20, 80)
				>50 μg/L vs. <50 μg/L of arsenic.	Preterm birth.	OR	1.02 (0.72, 1.44)

(Continued)

TABLE 4.1 (Continued)
Epidemiology of Arsenic Exposure to Drinking Water and Prevalence of Non-carcinogenic Effects on Human Health

Reference	Location	Study Design	Time	Comparison (Exposed vs. Reference)	Disease Pattern	Outcome	Results (95% CI)
Neamtiu et al. (2015)	Romania (Timis County)	Case-control	—	Inorganic arsenic-contaminated drinking water used by pregnant women and urinary arsenic level, DDM, and MMA.	Inorganic arsenic exposure from drinking water was associated with urinary arsenic biomarkers among pregnant women and spontaneous loss.	Spearman rank correlation coefficient	Total urinary arsenic: 0.35 (−0.12, 0.68) and $P = 0.13$; MMA: 0.15 (−0.31, 0.56) and $P = 0.52$; DMA: 0.31 (−0.16, 0.66), $P = 0.18$
Rahman et al. (2009)	Bangladesh	Prospective	—	1 unit vs. 6–100 µg/L (an increase of average inorganic arsenic in urine, µg/L).	Birth weight is associated with the use of arsenic-contaminated water.	β	−1.68 (−2.90, −0.46)
Shirai et al. (2010)	Japan (Tokyo)	Prospective cohort	—	1 unit vs. 9.81–1603 µg/L (creatinine).	Birth weight.	No effect	No effect
Vall et al. (2012)	Spain (Tenerife)	Cross sectional	—	0.1–31.4 µg/L vs. <0.10 µg/L (meconium arsenic concentration).	Birth weight. Preterm birth.	β β	223.8 (p: 0.043) 0.1 (P: 0.813)
Xu et al. (2011)	China (Shanghai City)	Cross sectional	—	1 log unit vs. 0.63–30.45 maternal blood.	Among male newborns, a ten-fold increase in the maternal arsenic level is associated with a 354.41 g decrease in birth weight.	β	−354.41 (−677.53, −31.28)
Renal Failure: Proximal Tubule Degeneration, Clinical Nephritis and Proteinuria							
Chen et al. (2011b)	Taiwan (Changhua County)	Cross-sectional	2002–2005	Analysis of urine and plasma specimen from 1043 subjects.	Renal dysfunction rates significantly increased when the urinary arsenic rose above 75 µg/g creatinine for both tubular (β_2MG >0.154 mg/L) and glomerular function (eGFR <90 mL/min/1.73 m^2).	OR	—
Chiu and Yang (2005)	Taiwan (four townships in BFD endemic area, Putai)	Ecological	1971–2000	273 male deaths and 248 female deaths. Groundwater arsenic data. Historical well water median of 780 µg/L and current supply <10 µg/L Chen et al. (1962).	Nephritis, nephrosis, nephrotic syndrome mortality validated by death certificate using ICD-9 code.	SMR	1.05 (0.96–1.14)

(Continued)

TABLE 4.1 (Continued)
Epidemiology of Arsenic Exposure to Drinking Water and Prevalence of Non-carcinogenic Effects on Human Health

Reference	Location	Study Design	Time	Comparison (Exposed vs. Reference)	Disease Pattern	Outcome	Results (95% CI)
Feng et al. (2013)	China (Ying County of Shanxi province)	Case-control	Prior to 2003 and to 2008	Considered 97 cases with >50 μg/L of arsenic and 970 nonpatients from low (<10 μg/L), medium (10–50), and high (>50 μg/L) exposure groups.	The median level of α1-MG in urine samples of patients group was much higher (15.39 μg/mg Cr) than in three arsenic exposure groups. The figure for urinary α1-MG shows statistically significant for renal dysfunctions upon arsenic exposure.	AUC (area under the curve) and OR	AUC: 0.70; and OR: 1.013 (1.0, 1.025)
Hawkesworth et al. (2013)	Bangladesh (Matlab)	Cohort	2000–2004	1334 children assessed at 4.5 years and arsenic concentrations in urine.	A marginal inverse association between infancy urinary arsenic and eGFR at 4.5 years.	Simple regression analysis	−33.4 ml/min/1.72 m^2 (−70.2, 3.34) and P trend 0.08
Hong et al. (2004)	China (Guizhou province)	Case-control	—	Considered 122 cases with high urinary arsenic level (>200 μg/g) in polluted area compared with 123 residents with nonpolluted area.	The calculated BMD/LBMD of urinary arsenic level for 354.40/261.50 μg/g creatinine in combination for Uβ$_2$-MG, UALB, and UNAG suggesting an excess level of 10% risk of renal dysfunction above the background from LBMD.	Correlation Coefficient and Regression Analysis	As with Uβ$_2$MG: 0.47 As with ULAB: 0.63 As with UNAG: 0.51
Hsueh et al. (2009)	Taiwan (Taipei)	Case-control	—	354 (125 cases vs. 229 controls) and urine arsenic ≥20.7 μg/g.	MDRD-eGFR <60 ml/min/1.73 m^2.	OR	4.34 (1.94, 9.69) and p trend <0.01
Kong et al. (2012)	China (Hong Kong)	Cross-sectional	—	Considered 60 cases against 60 controls and arsenic assessment in urine.	Urine albumin/creatinine ratio of >3.5 mg/mmol.	Median IQR of arsenic	In cases: 4.06 (3.28, 7.70) In controls: 4.45 (2.70, 13.26)
Palaneeswari et al. (2013)	India (Chennai)	Case-control	—	Considered 50 cases against 50 controls and arsenic assessment in blood. Mean (SD) blood arsenic (μg/L) ESRD: 3.20 (0.42); and Control: 2.30 (0.00).	End stage renal disease (ESRD).	P value	P value <0.01

(Continued)

TABLE 4.1 (Continued)
Epidemiology of Arsenic Exposure to Drinking Water and Prevalence of Non-carcinogenic Effects on Human Health

Reference	Location	Study Design	Time	Comparison (Exposed vs. Reference)	Disease Pattern	Outcome	Results (95% CI)
Smith et al. (2012)	Chile (Antofagasta region)	Ecological	—	Historical mean water arsenic in Antofagasta 870 µg/L and rest of Chile <10 µg/L (mean of 14 µg/L in 1984). SMR of Antofagasta region of Chile. Some 14 male and 14 female deaths (born 1940–1957); 6 male and 6 female deaths (born 1958–1970).	Chronic renal disease mortality validated by the Ministry of Health using ICD-9 codes.	SMR	2.03 (1.48, 2.76)
Tsai et al. (1999)	Taiwan (four townships in BFD endemic area, Putai)	Ecological	—	Considered 206 male deaths and 196 female deaths. Historical content of wells ranged from 250 to 1140 µg/L (median 780 µg/L).	Nephritis, nephrotic syndrome, nephrosis mortality validated by death certificate using ICD-9 code. Groundwater arsenic data from Chen et al. (1962).	SMR	1.10 (1.00, 1.22)
Zheng et al. (2013)	USA	Cross-sectional	—	Considered 3821 persons with urinary arsenic and arsenic concentration in drinking water is <100 µg/L and albumin/creatinine ratio of ≥30 mg/g.	Found a positive dose-response association between urine arsenic and albuminuria.	PR	1.24 (1.07, 1.43) with 9.7–15.6 µg/g urinary arsenic; and 1.55 (1.35, 1.78) with ≥15.6 µg/g urinary arsenic with p trend <0.01

ABI: Ankle-brachial Blood Pressure Index; ALP: Alkaline Phosphatase; ALT: Alanine Transaminase; AST: Aspartate Transaminase; AUC: Area under the Curve; β: Average change per unit increase of exposure; BPRS: Brief Psychiatric Rating Scale; CEI: Cumulative Exposure Index; CHD: Coronary Heart Disease; CKD: Chronic Kidney Disease; CMR: Crude Mortality Rate (MR: deaths/1000 person-years); CPR: Crude Prevalence Ratios; eGFR: estimated glomerular filtration rate; CVD: Cardiovascular Disease; ESRD: End Stage Renal Disease; FSIQ: Full Scale IQ; GHQ: General Health Questionnaire; HAA: High Arsenic area; HBsAg: Hepatitis B Surface Antigen; HR: Hazard Ratio (discrete time); ICP-MS: Inductively Coupled Plasma Mass Spectroscopy; LBW: Low Birth Weight; MI: Myocardial Infarction; MRR: Mortality Rate Ratio; NCPF: Noncirrhotic Portal Fibrosis; PAD: Peripheral Arterial Disease; PB: Preterm Birth or Delivery; PE: Proportion Explained; POR: Prevalence Odds Ratio; PPR: Prevalence Proportion Ratio; PR: Prevalence Ratio; OR: Odds Ratios; RCO: Retrospective Cohort; RERI: Relative Excess Risk for Interaction; RMR: Relative Mortality Risk; RR: Relative Risk; TMC: Total Motor Composite Score; VIQ: Verbal IQ.

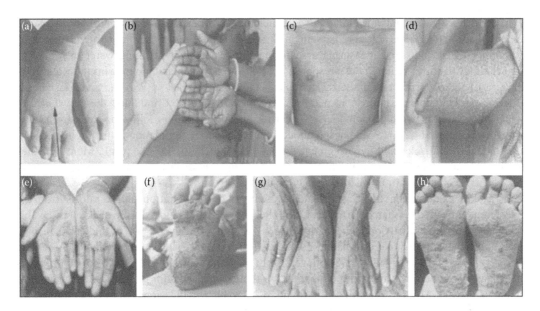

FIGURE 4.3 Arsenical skin lesions: (a) nonpitting edema; (b) diffuse melanosis; (c) spotted melanosis; (d) leuco-melanosis; (e) diffused and nodular keratosis on palm; (f) spotted keratosis on sole; (g) dorsal keratosis; (h) an arsenicosis patient with severe keratosis who died of lung cancer. (After Chakraborti, D. et al. 2017a. *Hydrogeology Journal* (doi:10.1007/s10040-017-1556-6).)

at least one of the related symptoms of melanosis and keratosis. They calculated a positive and strong correlation between age and arsenic concentrations with more than 80% of the dermatological symptoms linked to patients aged 25 years or older. Moreover, analyzing arsenic concentrations in 171 hair and nail samples (0.456–6.976 μg/g for hair and 0.696–9.815 μg/g for nail) in India, Chakraborti et al (2013b) found that few people have arsenical skin lesions; they considered the rest of the individuals as "sub-clinically affected." One study from West Bengal showed that the lowest level of arsenic ingested by a confirmed case was 115 μg/L and the average latency for skin lesions was estimated to be 23 years from first exposure (Haque et al., 2003).

Urinary arsenic levels are higher among those with severe skin lesions compared with those without. Analyzing arsenic concentrations in drinking water and urinary 8-OHdG (8-hydroxy-2′-deoxyguanosine) from China, Pei et al (2013) examined the staining of PMN (polymorphonuclear leukocytes) nuclei that is frequently found along with the elevated amounts of cell debris in individuals with skin lesions. Screening a total of 207 adult women in the Bameng region of Inner Mongolia, it was found that urinary dimethyl arsenic (DMA), monomethyl arsenic (MMA), and higher nail arsenic concentration are positively associated with risk of skin lesions. The nail content of arsenic and urinary concentrations of DMA, MMA, and inorganic arsenic were significantly higher in women with skin lesions (271.82, 61.72 and 60.00 mg/kg, respectively) than in those without skin lesions (137.83, 29.35 and 31.04 mg/kg, respectively) (Wei et al., 2017). Screening a total of 534 individuals with a diagnosis of arsenical skin lesions and examining the prevalence rate with reference to arsenic concentrations in different categories, Fatmi et al. (2013) found the highest prevalence rate at the concentration level of 100–199 μg/L (15.2 cases) followed by ≥400 μg/L (13.5 cases) and 300–399 (12.8 cases). In a study from Sri Lanka, Jayasumana et al. (2013) investigated abnormal skin manifestations including hyperpigmentation of 54.4% of palms and 39% of soles with relatively low keratosis in 23.2% of palms and 17.6% of soles with high urinary arsenic levels in 125 patients affected by chronic kidney disease. Smith et al. (2000b) report that arsenic-induced skin lesions occur among Atacameno people in northern Chile, despite a good nutritional status.

There is inter-individual variation in arsenic metabolism efficiency and susceptibility to arsenic toxicity. Jaafar et al. (1993) observed arsenical skin lesions of keratosis and hyperpigmentation by examining the patients and measuring arsenic magnitudes in their drinking water in Malaysia. Guo et al. (2001) investigated a dose-response relationship between arsenic in drinking water (50–1860 µg/L) and occurrence of arsenical dermatosis (44.8% and 37.1% in Wuyuan and Alashan out of 1176 and 433 subjects studied, respectively) in Inner Mongolia, China, and the prevalence of arsenical dermatosis was highest in the >40-year-old age group. In their follow-up study from Inner Mongolia, Guo et al. (2007b) diagnosed about 22% of the study population as having hyperkeratosis on palms and soles and hyperpigmentation and depigmentation of the trunk. Moreover, they reported 35 subjects with keratosis and 5 subjects with pigment disorder who drank water containing <50 µg/L of arsenic in Inner Mongolia (Guo et al., 2007b).

In a skin examination of 75 male subjects with hypopigmentation and keratosis from an arsenic-endemic area of Shanxi Province, China, Pei et al. (2013) found a degree of skin lesions that exhibited a positive association with length of exposure (LOE) and age ($r = 0.765$ and 0.750, respectively) and a strong positive correlation coefficient for age and LOE (i.e., $r = 0.857$). Assessing the prevalence of skin lesions of 178 villagers from endemic and 179 villagers from control sites in Xinjiang Province in China with biomarkers in their urine, including arsenic, porphyrins, and malondialdehyde (MDA), Liu et al. (2013) found 31.1% to have skin lesions related to chronic arsenic exposure from the endemic area against none in the control group. The average urinary arsenic (117 ± 8.3 µg/g of creatinine) from the endemic villagers was significantly higher than that of the controls (73.6 ± 3.2 µg/g of creatinine) with $p < 0.001$. When the urinary arsenic was higher than 150 µg/g of creatinine, MDA and porphyrins were higher in the endemic villagers compared to the controls (Liu et al., 2013).

A study conducted in Bangladesh reported a strong association between arsenic exposure and skin lesions. Melanosis and keratosis occurred in residents (36 out of 167 patients) exposed to low arsenic levels (≤ 10 µg/L). Analyzing a case-control study in Bangladesh with 859 cases and 868 controls, Kile et al. (2011) showed that persons with skin lesions were more likely to drink from tubewells containing higher concentrations of arsenic (39.0 µg/L vs. 11.4 µg/L), to have higher concentrations of arsenic in their urine (72.5 µg/L vs. 54.0 µg/L), to have a higher proportion of MMA in their urine (13.3% vs. 12.0%), and to have a lower proportion of DMA in their urine (74.6% vs. 75.9%) compared with persons without skin lesions. In a follow-up analysis of 1085 individuals with arsenic-induced premalignant skin lesions and 1794 controls, Pierce et al. (2012) showed an association with a risk of skin lesions. In a case-control study of 176 cases and age- and village-matched referents from Bangladesh, McDonald et al. (2007) found that the prevalence of skin lesions was 0.37% in people exposed to arsenic concentrations below 5 µg/L, 0.63% at 6–50 µg/L, and 6.84% at 81 µg/L and the relative risk (RR) of skin lesions was found to increase threefold at concentrations above 50 µg/L ($p < 0.05$). Considering 10,182 participants in a longitudinal study with different arsenic categories (10.1–50 µg/L, 50.1–100 µg/L, 100.1–200 µg/L, and ≥ 200.1 µg/L against ≤ 10 µg/L) from Bangladesh, Argos et al. (2011) showed an association between increasing arsenic exposure from drinking water with increased incidence of skin lesions, even at low levels of arsenic exposure. Arsenic-induced skin lesions are always regarded as nonmalignant, but in rare cases, skin cancer may occur with hyperkeratosis (Çöl et al., 1999).

Using data from arsenic-affected Bangladeshi individuals, the first genome-wide association study (GWAS) of arsenic-related metabolism and toxicity phenotypes conducted by Pierce et al. (2012) showed poisoning mechanisms by which arsenic affects human health. In analyzing more than 2000 individuals, it was identified that genetic variants that influence arsenic metabolism can influence risk for arsenical skin lesions through interaction with arsenic exposure (Pierce et al., 2013). Moreover, Nriagu et al. (2012) examined the strongest association of genotype and skin lesions for an SNP (single nucleotide polymorphism) of rs9989407 (p-value: 0.058; χ^2: 5.69). Apart from skin, arsenic deposition in other keratin-rich areas of the body such as nails may lead to manifestations such as the formation of distinct white lines (Mee's lines) in the nails of fingers and toes (Ratnaike, 2003).

4.2.2 VASCULAR SYSTEM

Exposure to inorganic arsenic has been linked to various vascular alterations affecting both the small and large blood vessels (Table 4.1). Much of the early work on arsenic and vascular disease focused on the effects in small vessels (i.e., Blackfoot Disease—BFD and peripheral vascular diseases—PVD), while later research has been directed at effects in larger vessels (cardiovascular disease—CVD and cerebrovascular diseases) (Afridi et al., 2011; Barchowsky and States, 2016; Gong and O'Bryant, 2012; James et al., 2015; Moon et al., 2012; Phung et al., 2017; States et al., 2009; Tsuji et al., 2014). A number of epidemiological studies have shown that long-term exposure to high levels of arsenic exposure (>300 μg/L) from groundwater is associated with elevated risks for an array of blood pressure effects including pulse pressure (PP) and mean arterial blood pressure (MAP), cardiac ischemic disease, ischemic heart disease (ISHD), peripheral arterial disease (PAD), stroke, and hypertension (Abhyankar et al., 2012; Farzan et al., 2013a; Kwok et al., 2007; Li et al., 2008; Tseng, 2008; Wu et al., 2008; Zhang et al., 2013). Moreover, elevated risk of CVD was associated with the exposure category with arsenic concentrations of more than 50 μg/L (Sohel et al., 2009; Chen et al., 2011a; James et al., 2015).

Arsenic is associated with an increased rate of various vascular ailments through metabolic disorder (Wang et al., 2007c). The vascular damage in BFD seems to be the result of an early destruction of vascular endothelial cells (Pimparkar and Bhave, 2010). A number of epidemiological studies from Bangladesh (Chen et al., 2011a), Taiwan (Tseng et al., 2003), and the United States (Chen and Karagas, 2013) show an increased morbidity and mortality with CVD including heart and blood vessels with elevated levels of inorganic arsenic exposure (Barchowsky and States, 2016; Liao et al., 2012; Osorio-Yáñez et al., 2013). CVD mortality with high levels of arsenic ingestion (from >100 μg/L to >300 μg/L) has been proven from a cohort study in rural Bangladesh (Chen et al., 2011a). Sohel et al. (2009) found a clear dose-response relationship, with an increased risk of mortality from CVD even at relatively low levels of arsenic exposure (10–49 μg/L). However, in an ecological study of municipal drinking water levels of arsenic in Spain, Medrano et al. (2010) found an increased risk of CVD, coronary heart disease, and cerebrovascular disease mortality in municipalities with arsenic concentration levels of <1.0 μg/L compared to >10.0 μg/L.

In an ecological study with mortality due to CVD in 30 US counties between 1968 and 1984, Engel and Smith (1994) found excess mortality rates for diseases of the arteries and anomalies of the circulatory system in areas where arsenic concentrations were more than 2 μg/L. Lewis et al. (1999) investigated a statistically significant increased mortality for hypertensive CVD and other heart diseases among cohort males (SMR: 2.20) and females (SMR: 1.73) from the residents of Millard County, USA. Analyzing chronic diseases among 1185 residents in Wisconsin, USA, Zierold et al. (2004) investigated the odds ratio (OR) for heart attack (95% CI) of 2.1 for the group exposed to arsenic above 10 μg/L compared with the group exposed to <2 μg/L of arsenic. Navas-Acien et al. (2005) performed a systematic review of the epidemiologic evidence from 1966 to 2005 on the association between arsenic exposure and cardiovascular outcomes and in a follow-up of 15 years; they found a significantly higher mortality from CVD. Wu et al. (2012) analyzed the association between an elevated level of inorganic arsenic ingestion (50–300 μg/L) and CVD mortality with RMR 2.07 (95% CI: 0.74, 5.80) having CMR (Crude Mortality Ratio) of 4.45. Moreover, a low level of arsenic exposure is associated with increased mortality of CVD (Gong and O'Bryant, 2012).

Analyzing age-adjusted mortality rates, Wu et al. (1989) examined the dose-response relation between levels of arsenic exposure from artisan well water and risks of vascular diseases among patients in a BFD area. Chen et al. (2011b) found that exposure to arsenic in drinking water is adversely associated with mortality from diseases of the circulatory system (HR: 1.46; 0.96–2.20) in terms of CVD (HR: 1.92; 1.07–3.43), cerebrovascular disease (HR: 1.07; 0.54–2.12), and ISHD (HR: 1.94; 0.99–3.84). Various biological mechanisms for cardiovascular toxicity define a causal relationship with a high level of arsenic exposure that is an established risk factor for CVD (Saposnik, 2010; Stea et al., 2014). Considering a total of 380 subjects from an arseniasis-endemic area and 303 from a nonendemic area of southwest Taiwan as well as analyzing various urinary arsenic species, Liao et al. (2012) observed a significant dose-response relationship between arsenic exposure and lactate dehydrogenase (LDH). LDH is a marker of arsenic toxicity associated with CVD mortality (HR: 3.98) in a dose-dependent manner.

Chronic arsenic exposure has been associated with an increased prevalence of PVD (Chen et al., 2007; Yang, 2006). Tsai et al. (1999) investigated mortality from PVD in the arsenic-endemic areas of Taiwan during 1971–1994. In a cohort of 789 BFD patients followed for 15 years, Chen et al. (1988a) reported that there was a significant increase in the number of deaths from PVD among residents of the BFD endemic area of Taiwan. Examining 582 adults from BFD endemic villages and using multiple logistic regression analysis, Tseng et al. (1996) showed a close relationship between long-term arsenic exposure and the prevalence of PVD. This PVD has been named "BFD" in Taiwan after its clinical appearance (Tseng et al., 2005). Tseng (2002) found prevalence rates of PVD for those with a cumulative arsenic exposure (CAE) of 0, 0.1–19.9, and >20 mg/L/years were 4.4, 11.6, and 19.8%, respectively, and the respective OR was 1.00, 2.77, and 4.28 after adjustment for potential confounders. Yang (2006) examined the SMR of PVD among residents in the BFD endemic areas in Taiwan from 1971 to 2003 with the gradual decline of mortality due to PVD for about 25–27 years after cessation of arsenic-contaminated drinking water. This disease was first reported in the early 20th century and was confined to the southwestern coast of Taiwan, where people used artesian well water from as deep as 100–300 meters underground (Tseng, 2007).

Arsenic has been associated in a dose-dependent manner with hypertension (Abhyankar et al., 2012; Chen et al., 1995; Dastgiri et al., 2010; Hall et al., 2017; Mazumder et al., 2012; Li et al., 2013a; Stea et al, 2014; Wang et al., 2011). Hypertension is a major risk factor for other vascular diseases, which account for more than 30% of global mortality (Smith and Steinmaus, 2009). An individual is considered to have hypertension when a systolic blood pressure of 160 mmHg or more in combination with a diastolic blood pressure of 95 mmHg or more (Rahman et al., 1999). There is an impact of long-term exposure to arsenic in drinking water on blood pressure, pulse pressure (PP), and mean arterial blood pressure (MAP) (Jiang et al., 2015; Zhang et al., 2013). By the large, MAP reflects the heart and hemal function and PP reflects blood vessel elasticity. Both the MAP and PP are considered exceptionally important biomarkers for assessing cardiovascular disease (Baguet et al., 2000; Benetos et al., 1997).

Epidemiological studies show relationships between arsenic exposure and PP (Huda et al., 2014; Islam et al., 2012a). Examining a total of 1481 subjects exposed to arsenic-contaminated drinking water and 114 unexposed subjects aged 30 years or more from Bangladesh, Rahman et al. (1999) showed a significant dose-response relationship between arsenic exposure and increased blood pressure. Analyzing a total of 483 subjects (322 from arsenic-endemic and 161 from nonendemic areas) in Bangladesh, Huda et al. (2014) found associations of arsenic exposure with plasma uric acid (PUA) levels and its relationship with hypertension. They are significantly ($p < 0.001$) higher PUA levels in subjects living in arsenic-endemic areas than those in nonendemic area. Using population-weighted average levels of arsenic in drinking water, Lisabeth et al. (2010) found a significant increased risk of ischemic stroke in Michigan, USA, comparing with average levels of arsenic of 22.3 µg/L versus 4.5 µg/L.

Conducting a cross-sectional study of 405 villagers of Inner Mongolia with blood pressure who had been drinking water with an inorganic arsenic content of <50 µg/L, Zhang et al. (2013) show that the OR for prevalence of abnormal PP and MAP was 1.06 (0.24–4.66) and 0.87 (0.36–2.14) in the group with 30–50 years of exposure and was 2.46 (0.87–6.97) and 3.75 (1.61–8.71) for the group with >50 years exposure, compared to the group with arsenic exposure ≤30 years, respectively, (Zhang et al., 2013). Examining a total of 382 men and 516 women residing in villages from BFD-endemic areas in Taiwan, Chen et al. (1995) proved the prevalence of hypertension as the long-term effect of arseniasis, declaring that "the higher the cumulative long-term arsenic exposure, the higher the prevalence of hypertension" with a 1.5-fold increase in arsenic-exposed residents with age-sex adjusted prevalence of hypertension compared to residents in nonendemic areas. In addition,

Mazumder and Dasgupta (2011) suggest a relationship between arsenic exposure and noncirrhotic portal hypertension. Lewis et al. (1999) identified hypertensive heart disease as related to chronic arseniasism.

Arsenic exposure has also been associated with an increased incidence of cerebrovascular disease with the complications of myocardial injury, cardiac arrhythmias, and cardiomyopathy (Benowitz, 1992; Chiou et al., 1997; Manna et al., 2008). Analyzing some 8102 subjects exposed to arsenic-contaminated drinking water from Taiwan, Chiou et al. (1997) concluded that long-term exposure to inorganic arsenic is associated with an increased prevalence of cerebrovascular disease, especially cerebral infarction, with the multivariate-adjusted OR for arsenic content in well water of <0.1, 0.1–50, 50.1–299.9, and >300 µg/L being 1.00, 3.38, 4.47, and 6.90, respectively. There is also an association between chronic arsenic exposure and ISHD mortality. Tseng et al. (2003) evaluated the association between chronic arsenic exposure and ISHD in 462 subjects living in the BFD-hyperendemic villages and prevalence of ISHD was highest in the age group >60 years (30.7%) and lowest in the age group 30–39 years (4.9%).

Consistent with the association with risk factors and CVD, an association has been found between arsenic exposure and atherosclerosis (Wang et al., 2010; Hsieh et al., 2011; Simeonova and Luster, 2004). Exposure to inorganic arsenic from drinking water leads to development of carotid atherosclerosis, and Wang et al. (2002) with their study from Taiwan found a significant biological gradient between exposure to inorganic arsenic and carotid atherosclerosis. Moreover, Wang et al. (2007c) discovered that genetic polymorphism of glutathione S-transferase P1 (GSTP1) and tumor protein p53 have joint effects on susceptibility of carotid atherosclerosis with arsenic exposure of >50 µg/L. Exposure to elevated levels of chronic inorganic arsenic from drinking water leads to a prolonged QT-interval (Chen et al., 2013a; Mumford et al., 2007; Wang et al., 2009). Chronic arsenic exposure can increase QT/QTc depression linking to atherosclerotic diseases. Mordukhovich et al. (2009) and Wang et al. (2010) show that low arsenic levels have a positive tendency to alter QT interval duration. Subsequent QT prolongation was observed in women at long-term exposure to inorganic arsenic ranging from 0.1 to 790 µg/L (Chen et al., 2013a).

The impact of chronic arsenic ingestion on vascular effects may be due to functional polymorphism in genes related to arsenic metabolism (Pimparkar and Bhave, 2010). The mechanisms whereby arsenic causes cardiovascular disease, hypertension, and hyperlipidemia are currently believed to be related to arsenic-induced oxidative stress (Ma et al., 2017; Stea et al., 2014). Oxidative stress has been reported to cause endothelial dysfunction and even cell apoptosis (Li et al., 2014b). Once entering the body, arsenic and its metabolites generate free radicals, which damage proteins, fatty acids, DNA, and RNA, and cause oxidative stress or death to cells (Gong et al., 2011). Genetic predisposition plays an important role for coronary heart disease (Kullo and Ding, 2007), and certain genotypes of the heme oxygenase-1 gene have been shown to be a risk factor for coronary heart disease when exposed to arsenic (Wu et al., 2010).

4.2.3 Pulmonary System (Respiratory Effects)

The obstructive and restrictive disorders of the pulmonary system are a major concern regarding chronic arsenic poisoning (Parvez et al., 2013; Smith and Steinmaus, 2009). Elevated levels of chronic arsenic exposure are associated with risk of reduced lung function and nonmalignant respiratory effects having lung disease as well as respiratory tract infections (RTI) in populations living in arsenic-endemic areas (Ahmed et al., 2017; Arain et al., 2009b; De et al., 2004; Mazumder, 2007a; Smith et al., 2011; Von Ehrenstein et al., 2005). Epidemiological studies reported chronic cough, chronic obstructive pulmonary disease (COPD), and interstitial lung disease as common respiratory complications among the affected population (Smith et al., 2006b; Parvez et al., 2011b). Moreover, exposure to arsenic in drinking water increases the incidence of bronchiectasis, a pulmonary disease characterized by chronic infection, inflammation, irreversible bronchial damage, and respiratory failure (Ilowite et al., 2008; Parvez et al., 2010).

Based on a cohort study conducted from Antofagasta in Chile where 180 residents were exposed to drinking water containing arsenic at 800 μg/L, Borgoño et al (1977) reported that 38.8% of 144 subjects with arsenical skin lesions complained of chronic cough, compared with 3.1% of 36 subjects without any skin lesion. From a study among 398 school children in Antofagasta, Zaldívar and Ghai (1980) found an excessive cough (38%) among school children with skin lesions. High mortality and incidence of COPD and bronchiectasis among adults and children were also recognizable in arsenic-endemic areas of Chile. Smith et al. (2011) revealed an increased mortality from arsenic associated pulmonary tuberculosis in the Chilean cohort. In the same Chilean cohort, increased mortality from bronchiectasis was significant for those exposed to arsenic during early life with a SMR of 50.1, and there were 6- to 7-fold increases in lung cancer mortality rates resulting from early life exposures (Smith et al., 2006b). The arsenic-exposed Chilean population indicate a long latency pattern of increased lung, kidney, and bladder cancer mortality continuing for >25 years after exposures ended (Marshall et al., 2007; Yuan et al., 2010). An ecological study revealed that the mortality from "bronchitis" among residents of the BFD areas in Taiwan was significantly higher than that of the nearby reference population and the total population, with respective SMR of 1.53 and 1.95 (Tsai et al., 1999). Conducting an ecological study with retrospective designs, including small sample sizes or measured arsenic exposure from Inner Mongolia, Guo et al. (2007b) reported a 13-fold increased risk of cough and a high prevalence of bronchitis among people living in arsenic-exposed villages.

A number of epidemiological studies from Bangladesh and India show that individuals with skin lesions or drinking water contaminated with high arsenic concentrations (>500 μg/L) had a risk of high prevalence of respiratory symptoms (i.e., cough and breathing problems) of between 2 and 15 times greater than individuals without skin lesions (De et al., 2004; Von Ehrenstein et al., 2005). Islam et al. (2007) found a high prevalence of respiratory complications including breathing problems, chest sound, asthma, bronchitis, and cough in diagnosed 125 arsenicosis patients exposed to arsenic-contaminated drinking water (mean concentration of 216 μg/L) in Bangladesh. Recent prospective cohort studies found that low to moderate levels of arsenic exposure in the Bangladesh population resulted in impaired lung function and tuberculosis (Parvez et al., 2013). In a cross-sectional study in 241 nonsmoking individuals from a large cohort with a wide range of arsenic exposure (0.1–761 μg/L) from drinking water in Bangladesh, Parvez et al. (2008) found a positive association between urinary arsenic and serum level of Clara Cell Protein (CC16) that provides a novel biomarker for adverse effects of arsenic exposure on respiratory lung function. Exposure to elevated levels of drinking water arsenic *in utero* and during childhood is linked with lung function and nonmalignant lung diseases (Dauphiné et al., 2011; Recio-Vega et al., 2014). Using Health Effects of Arsenic Longitudinal Study (HEALS) information from Ariahazar Upazila of Bangladesh, Parvez et al. (2010) discovered that a total of 15.95% of 11,746 adult participants had experienced at least one respiratory symptom and some 7.31%, 9.95%, and 2.03% participants completing four years of active follow-up reported having a chronic cough, breathing problems, or blood in their sputum, respectively. Ahmed et al. (2017) found that prenatal arsenic exposure is related to impaired lung function, while childhood exposure may increase airway inflammation, particularly in boys.

Carrying out lung function tests on 17 patients from arsenic-affected areas of West Bengal, India, Mazumder et al. (1998b) identified features of restrictive lung disease in 9 (53%) and combined obstructive and restrictive lung disease in 7 (41%) cases. Examining some 7683 participants of all ages who had arsenic-associated skin lesions and who were highly exposed to arsenic ($\geq$50 μg/L) from the South 24-Parganas district of West Bengal, Mazumder et al. (2000) found the prevalence odds ratio (POR) for both males and females to be 5.0 and 7.8, respectively, for cough; 6.9 and 9.6, respectively, for chest sounds; and 3.7 and 23.2, respectively, for shortness of breath.

Studying 156 arsenic exposed (>600 μg/L) individuals with skin lesions from West Bengal, De et al. (2004) reported that 57% of subjects had respiratory symptoms and 53% had restrictive lung disease as well as 41% of the selected participants having both obstructive and restrictive lung diseases. A cohort of 287 participants who were exposed to low doses of arsenic (up to 500 μg/L) from West Bengal, India, reported a higher risk of common respiratory symptoms with lower FEV_1 (forced expiratory volume measured in 1 sec) and FVC (forced vital capacity) among people with arsenical lesions (Von Ehrenstein et al., 2005). Conducting an epidemiological study on 258 individuals (108 with arsenical skin lesions and 150 without skin lesions) from an arsenic-endemic population in West Bengal, Mazumder et al. (2005) reported the incidence of bronchiectasis in a population with arsenical skin lesions. The mean bronchiectasis severity score was 3.4 ± 3.6 for 27 participants with skin lesions and 0.9 ± 1.6 for 11 participants without these lesions and the adjusted OR was found to be 10.1 (95% CI: 2.7, 37.1). They also investigated a 10-fold increased risk of chronic obstructive pulmonary disease (COPD) among 108 people with arsenical skin lesions compared with 150 subjects without lesions (Mazumder et al., 2005).

The association between arsenic exposure and nonmalignant respiratory effects remains strong and significant, but the mechanism of arsenic-induced nonmalignant respiratory effects is unclear. However, tissue inflammation by deposition of arsenic on the epithelium and damage to lung function is possible (Parvez et al., 2008, 2010). De et al. (2004) suggest that arsenic may induce lung toxicity by inflammation mediated through the immune response. It is also suggested that deposition of arsenic in the lungs may increase inflammation and cause fibrosis that culminates in pulmonary cellular dysfunction (Gerhardsson et al., 1988; Saady et al., 1989). It is also suggested that there is a role for antimicrobial peptide HBD1 (human β-defensin-1) against pulmonary pathogens relevant to bronchiectasis (Yuan et al., 2010).

4.2.4 ENDOCRINE SYSTEM (METABOLIC EFFECTS)

Arsenic is a well-known disruptor of the endocrine system (Davey et al., 2008) including thyroid, thyroid hormone (Naujokas et al., 2013), pancreas (Lu et al., 2011), and gonads (Dávila-Esqueda et al., 2012; Shen et al., 2013). Thyroid plays a critical role in regulating the metabolism in the human body and is responsible for the production of tri-iodothyronine (T3) and thyroxine (T4) hormones (Ciarrocca et al., 2012). Exposure to low levels of arsenic through groundwater (2–22 μg/L) is associated with hypothyroidism (Gong et al., 2015). It is evident that arsenic accumulates in the pancreas, and insufficient insulin production by pancreatic β-cells is considered to be one of the distinct complications of the heterogeneous disorder causing diabetes. Two types of diabetes can be recognizable: Type 1 Diabetes Mellitus (T1DM) develops upon destruction of pancreatic β-cells by the host immune system and Type 2 Diabetes Mellitus (T2DM) is characterized by the development of insulin resistance in the body. Based on experimental and epidemiological evidence, high inorganic arsenic exposure and T2DM reportedly share a strong association with each other (Beck et al., 2017). Apart from high concentrations, long-term exposure to inorganic arsenic may also increase the risk of T2DM (Bräuner et al., 2014; Hsu et al., 2013; Lawrence et al., 2008; Mendez et al., 2016; Wang et al., 2014a).

The endocrine system includes all of the glands of the body that produce hormones. Metabolic syndrome has been recognized as an important risk factor for Diabetes Mellitus (DM). A line of epidemiological studies from Taiwan, Bangladesh, and Mexico indicates a dose-response relationship between high chronic exposure to inorganic arsenic in drinking water (>100 μg/L) and the prevalence of noninsulin-dependent DM (i.e., NIDDM or T2DM) (Andra et al., 2013; Del Razo et al., 2011; Gribble et al., 2012; Islam et al., 2012b; Jiang et al., 2012; Jovanovic et al., 2013; Kim et al., 2013). T2DM accounts for 90–95% of all cases of diabetes and is a major public health problem worldwide, whereas less than 10% of the cases are insulin-dependent DM (i.e., IDDM or T1DM) (Wild et al., 2004). DM is a group of metabolic diseases characterized by hyperglycemia resulting from defects in insulin secretion by pancreatic β-cells and/or insulin action on peripheral tissues (Huang et al., 2011).

Determining the history of symptoms, a positive association having a statistical significance between DM and exposure to arsenic has been found in Bangladesh. Rahman et al. (1998) reported a significant increase of the crude prevalence ratio of DM as a result of arsenic-contaminated drinking water, evaluated to be 5.2 (95% CI: 2.5, 10.5) among individuals with arsenical skin lesions compared with those who did not have such lesions after adjusted for age, sex, and body mass index (BMI). Nabi et al. (2005) found that the prevalence of DM among chronic arsenic-exposed individuals in Bangladesh was approximately 2.8 times higher than that in the unexposed individuals. In a cross-sectional study with 1004 participants from Bangladesh, Islam et al. (2012b) examined an increased risk of T2DM with high levels of arsenic exposure (>50 µg/L), and the OR (OR: 1.9, 95% CI: 1.1, 3.5) shows a dose-response pattern with a significant increasing trend in relation with the increasing exposure category. Moreover, research in Bangladesh has also confirmed the increased risk of T2DM with modest levels of arsenic exposure (Pan et al., 2013).

Analyzing oral glucose tolerance and examining 891 adults residing in villages of BFD-endemic areas, Lai et al. (1994) first reported a positive association between cumulative exposure to arsenic from drinking water and the prevalence of DM with a multivariate-adjusted OR of 6.61 and 10.05 for those who had cumulative arsenic exposures of 100–15,000 µg/L/yr and >15,000 µg/L/yr, respectively, compared with those who were unexposed. In a cohort study conducted from arsenic-endemic areas of Taiwan, Tseng et al. (2000) reported an incidence of DM in residents exposed to arsenic in well water with the concentration ranging between 700 and 930 µg/L. Analyzing urine and plasma specimens from 1043 subjects between 2002 and 2005 in Taiwan, Chen et al. (2011b) found a significantly higher risk for DM (OR: 2.22; 95% CI: 1.21, 4.09) in subjects with a total urinary arsenic >200 µg/g creatinine as compared with subjects whose urinary arsenic was ≤35 µg/g creatinine after the adjustment for potential confounders. Chiu et al. (2006) demonstrated that women are more vulnerable than men with the prevalence OR of diabetes in the arseniasis-endemic areas in Taiwan.

In a case-control study in Coahuila, Mexico, Coronado-González et al. (2007) evaluated the effect of exposure to elevated levels of inorganic arsenic in drinking water (20–400 µg/L) and the occurrence of T2DM. The study established that subjects with intermediate total urinary arsenic levels (63.5–104.0 mg/g creatinine) are at a twice higher risk of diabetes (OR: 2.16; 95% CI: 1.23, 3.79), but the risk was almost three times higher in individuals with elevated levels (>104 µg/L) of total urinary arsenic (OR: 2.84; 95% CI: 1.64, 4.92). In a cross-sectional study from South Korea measuring urinary arsenic metabolites in 200 participants with total urinary arsenic concentrations, Kim and Lee (2011) reported an association between total urinary arsenic concentration and the prevalence of DM with OR in female participants and all participants of 1.502 (95% CI: 1.038, 2.171) and 1.312 (95% CI: 1.040, 1.655), respectively, for the level of urinary total arsenic concentration.

Steinmaus et al. (2009) hypothesized a threefold increase of T2DM over low concentrations of urinary arsenic in their study from the USA. In addition, Gribble et al. (2012) and Kim et al. (2013) found evidence of arsenic associated diabetes in the USA. In a case-cohort study including 141 cases of DM diagnosed between 1984 and 1998 in Colorado, USA, James et al. (2013) found that exposure to low-level inorganic arsenic (<100 µg/L) in drinking water is associated with increased risk (HR: 1.27; 95% CI: 1.02, 1.64 per 15 µg/L) for T2DM; the study also found that, for every 15 µg/L increase in arsenic concentration in the drinking water, the risk for DM increased by 27% (95% CI: 1%, 59%) after adjusting for ethnicity and time varying measures of BMI and physical activity. In a cross-sectional study, Navas-Acien et al. (2008) investigated a positive association between total urine arsenic and the prevalence of T2DM in people with low to moderate arsenic exposure (Huang et al., 2011). In the United States, Meliker et al. (2007) found an increased risk of mortality due to DM from inorganic arsenic exposure in drinking water (Male SMR: 1.28; CI: 1.18–1.37; and Female SMR: 1.27; CI: 1.19–1.35). Jovanovic et al. (2013) in their cross-sectional study in Siberia found an association between inorganic arsenic exposure and the occurrence of T2DM with a significantly higher OR for the exposed population (mean arsenic: 56 µg/L) in comparison with the unexposed population (mean arsenic: 2 µg/L).

A number of epidemiological studies show unclear and controversial effects of lower levels of exposure to inorganic arsenic on diabetes risk. Lewis et al. (1999) in their epidemiological study from the USA found no significant increase in mortality (SMR) from diabetes DM among subjects exposed to high concentration of arsenic (<200 µg/L) in drinking water. For Steinmaus et al. (2009), there was no evidence of T2DM occurrence at low arsenic exposure in the US population. In a cross-sectional study of 1185 residents in Wisconsin, USA, Zierold et al. (2004) calculated the OR for DM of 1.4 (95% CI: 0.8, 2.3) for arsenic exposure levels of 2–10 µg/L and 1.1 (95% CI: 0.5, 2.2) for >10 µg/L level with the referent group <2 µg/L. In a case-control study in southern Spain, Ruiz-Navarro et al. (1998) found that 38 participants with DM had similar total urinary arsenic concentrations (average: 3.44 µg/L) compared with 49 control participants (average: 3.68 µg/L). Moreover, Chen et al. (2010c) from a cross-sectional study in Bangladesh found no association between arsenic exposure from drinking water and diabetes mellitus. Furthermore, a cross-sectional study conducted in Inner Mongolia has also confirmed no profound association between arsenic exposure and increased risk of T2DM (Li et al., 2013a).

However, the biological mechanisms for an association between chronic arsenic exposure and increased risk of DM are unknown, but one reasonable mechanism for inorganic arsenic diabetotoxicity is the modification of the expression of genes involving both peripheral insulin resistance in adipose, liver, and muscle cells, and insufficient insulin production because of pancreatic β-cell dysfunction (Campbell, 2009; Lu et al., 2011; Paul et al., 2007). Common mechanisms involved in arsenic-induced diabetes are insulin resistance and pancreatic β-cell dysfunction. These complex mechanisms are initiated by activating Reactive Oxygen Species (ROS), nuclear factor-κB (NF-κB), cytokines (tumor necrosis factor α [TNFα] and interleukin-6 [IL-6]), and inhibition of peroxisome proliferator activated receptor γ (PPARγ) (Kaneto et al., 2007; Kim and Lee, 2011; Rezaei et al., 2017; Tseng, 2004). Additionally, apoptotic death/damage of pancreatic β-cell due to increased levels of ROS and oxidative stress is also considered to be a common mechanism in increasing the risks of diabetes due to inorganic arsenic exposure (Lu et al., 2011).

4.2.5 Nervous System (Neurological Effects)

The nervous system refers to the complex network of neurons or nerve cells that transmits nerve impulses or signals between different parts of the body. Ingestion of chronic inorganic arsenic from drinking water can cause various neurological disorders (Abdul et al., 2015; Mukherjee et al., 2003; Mundey et al., 2013; Wasserman et al., 2011). Peripheral neuropathy and painful muscle spasms are predominant and common neurological complications of arsenic toxicity. This neuropathy usually begins as numbness in the hands and feet, but later may develop into a painful "pins and needles" sensation (Rodríguez et al., 2003; Sengupta et al., 2008). Repeated exposures to lower levels of arsenic (0.03–0.1 mg/kg/day) are typically characterized by a symmetrical peripheral neuropathy (Franzblau and Lilis, 1989; Szuler et al., 1979). Tyler and Allan (2014) investigated high neurotoxicity manifestations, such as a decline in hearing and ability to taste, learning and memory deficits, impacted vision, and numbness in arsenic-affected villagers of Inner Mongolia.

Arsenic exposure in drinking water is linked to significant neurological impairments in children and adults with impaired cognitive abilities and motor functions (Chen et al., 2009; Dong and Su, 2009; Gong et al., 2011; Parvez et al., 2011a). A typical feature of arsenic-induced neurotoxicity is peripheral neuropathy occurring due to damage to peripheral nerves (Mathew et al., 2010). Tseng (2003) studied the existence of subclinical sensory nerve defects in residents of BFD-endemic villages in Taiwan characterized by long-term arsenic exposure from drinking water. Analyzing an electrophysiological study of 88 arsenicosis patients in West Bengal, sensory neuropathy was found in 24 (27.3%), motor neuropathy in 13 (14.7%), and abnormal electromyography (EMG) in 5 (5.7%) cases (Mukherjee et al., 2003). Mazumder et al. (1998b) found abnormal electromyography in 10 (34.5%) and altered nerve conduction velocity and EMG in 11 (37.9%) cases in analyzing objective evaluation of neuronal involvement in 29 patients. Abnormal EMG findings, suggestive mostly of sensory neuropathy, were reported in 10 out of 32 participants exposed to drinking arsenic-contaminated well water (60–1400 µg/L) in Canada (Hindmarsh et al., 1977). Mukherjee et al. (2003) found peripheral neuropathy in 45%, sensory neuropathy in 37.1%, and sensory motor effects in 7.7% of patients among 700 patients in West Bengal having arsenical skin lesions. Having examined 255 arsenicosis patients with arsenical skin lesions who had been taking arsenic-contaminated drinking water for a long time having high mean concentrations of arsenic in their biomarkers (hair: 2280 µg/kg and nail: 3970 µg/kg), Rahman et al. (2014) identified symptoms and signs of neurotoxicity and 32.9% were clinically detected to be suffering from peripheral neuropathy.

Children's intelligence was found to be inversely related to the arsenic levels after adjusting for confounders. There is documentation of poorer cognitive function, reflected in poorer memory skills, slower processing speed, deficits in verbal skill, and lower IQ among children exposed to arsenic (Calderon et al., 2001; Von Ehrenstein et al., 2007; Wang et al., 2007a). Cognitive impairments were noticed in children at 6 and 10 years of age (Wasserman et al., 2007), and Hamadani et al. (2011) reported impairments in verbal and full-scale IQ in girls but not in boys. In a cross-sectional study with neurobehavioral tests in 49 adolescents drinking elevated levels of arsenic-containing well water (ranging from undetectable levels to 3,590 µg/L), with no measure of individual exposure and 60 controls in Taiwan, Tsai et al. (2003) found significant impact on the pattern of memory with low Performance Raw Scores and switching attention by long-term cumulative exposure to arsenic after adjusting for education and gender. Considering 524 children between 8 and 12 years of age who had been exposed to arsenic at concentrations of 142 ± 106 µg/L (medium arsenic group) and 190 ± 183 µg/L (high arsenic group) in drinking water compared with the control group in Shanxi province, Wang et al. (2007a) examined that the mean IQ scores decreased from 105 ± 15 for the control group to 101 ± 16 for the medium arsenic group and to 95 ± 17 for the high arsenic group, which suggests that children's intelligence and growth can be affected with high concentrations of arsenic. Neuropsychological development was also found to be poor in Mexican children exposed to arsenic (Calderon et al., 2001).

Analyzing elevated levels of arsenic concentrations (300–1584 µg/L) in drinking water, Ahamed et al. (2006) discovered a number of neurological symptoms of sensory neuropathy in children under the age of 16 with arsenical skin lesions in Eurani village from Comilla District of Bangladesh. Neuropathies from arsenic toxicity were recorded in 100 (60%) subjects out of 166 neuropathy patients (Ahamed et al., 2006), and the majority of the examined subjects presented with sensory features of distal paresthesias (57%), limb pains (18.7%), and distal hypesthesias (47%) that outnumbered motor features of distal limb weakness and atrophy. Wasserman et al. (2004) found that an increased level of drinking water arsenic is associated with reduced intellectual function in a dose-response manner, where children exposed to >50 µg/L of arsenic achieved significantly lower performance IQ scores than did children with <5.5 µg/L of arsenic in drinking water. Compared with the lowest quartile of arsenic exposure with covariate adjustment, the fourth quartile had significantly lower scores on Full-Scale ($\beta = -11.3$) and Performance Raw Score ($\beta = -9.7$). The relationship between and the loss of Full-Scale Raw Score (−3.8 and −6.4 points) is associated with arsenic concentrations in drinking water of 10 µg/L and 50 µg/L, respectively (Wasserman et al., 2004).

Conducting a cross-sectional study of 137 subjects from a larger cohort in Bangladesh, Hafeman et al. (2005) concluded that increased arsenic exposure having cumulative and urinary measures is associated with evidence of subclinical sensory neuropathy. Considering 304 children aged between 8 and 11 years and arsenic and manganese concentrations in drinking water, blood, urine, and toenails, Parvez et al. (2011a) argued that exposure to arsenic in drinking water at relatively low concentrations is inversely associated with children's intellectual function with total motor function scores. Analyzing the urinary arsenic concentration (20–238 µg/L) of children aged 5 years born between May 2002 and December 2003, Hamadani et al. (2011) found an adverse impact of arsenic exposure on Verbal and Full-Scale IQ in pre-school girls at 5 years of age. The β-values for VIQ ($\beta = -2.4$; 95% CI: −3.8, −1.1) and FSIQ ($\beta = -1.4$; 95% CI: −2.7, −0.1) show a consistent negative association with urinary arsenic in girls. Moreover, based on 1505 mother-infant pairs in rural Bangladesh, Gardner et al. (2013) evaluated the association between early-life exposure to arsenic, assessed via concentrations in maternal and child urine, and children's weights and heights up to age

5 years, during the period 2001–2009, and they investigated an inverse association between concurrent exposure to arsenic and children's weight and height, age-adjusted Z-scores, and growth velocity at age 5 years. Moreover, Alzheimer disease and its associated disorders (decreased capacity in executive function, mental acuity, and verbal skills) have been reported due to long-term low-level arsenic exposure (O'Bryant et al., 2011). Oxidative stress is thought to be one of the major causes of arsenic-induced neurotoxicity. An important mechanism involved in arsenic-induced neurotoxicity is the disorganization of cytoskeletal framework either by altering protein composition of cytoskeleton or hyper-phosphorylation of proteins (Abdul et al., 2015; Mundey et al., 2013).

4.2.6 Hepatic System

The accumulation of arsenic in the liver following repeated exposures is relatively high and prone to increased hepatic toxicity (Ratnaike, 2003). The metabolism of inorganic arsenic takes place in the liver where it is partially detoxified and later excreted through the urine (Watanabe and Hirano, 2013). A number of epidemiological studies in humans exposed to inorganic arsenic have noted symptoms of hepatic injury (Boyer, 1999; Ghosh, 2013; Mazumder, 2007b; Islam et al., 2011; Jomova et al., 2011; Morris et al., 1974). The liver swells and becomes tender, and the analysis of blood sometimes shows elevated levels of hepatic enzymes (Armstrong et al., 1984; Franzblau and Lilis, 1989). These effects are observed after repeated exposure to doses of 0.01–0.1 mg/kg/day of arsenic, although doses as low as 0.006 mg/kg/day have been reported to be causative with chronic exposure (Hernández-Zavala et al., 1998). The early clinical symptoms of arsenic-induced liver disease include bleeding from esophageal varices, ascites, jaundice, or enlargement of the liver (Jomova et al., 2011), while in later stages of severe toxicity, hepatic lesions may appear along with other complications such as hepatic fibrosis, Noncirrhotic Portal Fibrosis (NCPF), and cirrhosis with the likely chance of liver failure (Kapaj et al., 2006). Increased levels of bilirubin, alanine transaminase, aspartate transaminase, and alkaline phosphatase in serum could influence liver function upon chronic arsenic exposure (Das et al., 2012).

Franzblau and Lilis (1989) reported increases in alanine aminotransferase (ALT) and aspartate aminotransferase (SAT) serum activities of individuals acutely exposed to elevated levels of arsenic via drinking water (9,000–11,000 µg/L). Hernández-Zavala et al. (1998) found hepatomegaly, splenomegaly, and liver diseases in people living in the Lagunera Region in Mexico. Individuals chronically exposed to arsenic via drinking water showed predominantly conjugated hyperbilirubinemia and increased serum alkaline phosphatase (S-ALP) activity, suggesting the presence of cholestasis in Santa Ana and Benito Juárez villages where the mean arsenic concentrations were 116 ± 37 µg/L and 239 ± 88 µg/L, respectively (Hernández-Zavala et al., 1998). Considering 200 arsenic-exposed individuals and analyzing arsenic concentrations in drinking water, nails, and hairs, Islam et al. (2011) found respective activities of S-ALP, AST, and ALT to be significantly increased in the high-exposure groups compared to the lowest-exposure groups before and after adjustments were made for different covariates.

The incidence of hepatomegaly was found to have a linear relationship proportional to increasing exposure to arsenic in drinking water. NCPF has been identified as the predominant lesion in liver histology as the result of chronic arsenic toxicity. Mazumder (2007b) examined the relationships between NCPF and arsenic exposure to drinking water and found increased levels of arsenic in serum, hair, nails, and liver tissue in NCPF patients. Hepatomegaly was found in 62 out of 67 members of families who drank arsenic-contaminated water (200–2000 µg/L) in West Bengal, whereas it was found in only 6 out of 96 people who drank safe water in the same area; some 13 of those arsenic-exposed hepatomegaly patients were further investigated in a hospital, and all showed various degrees of portal zone expansion and fibrosis on liver histology (Mazumder and Dasgupta, 2011). Liver enlargement can be possible after drinking arsenic-contaminated water. Considering 103 arsenic-exposed individuals having at least 10 years of exposure and 107 unexposed individuals from Murshidabad district of West Bengal in India and the results of liver function tests, Das et al.

(2012) observed a significant increasing trend of serum levels of bilirubin, ALT, AST, and S-ALP in arsenic-exposed population when compared to the unexposed group with similar socio-economic status. The results show that liver injury or hepatic dysfunction can be possible with chronic arsenic exposure.

Examining hepatic abnormalities in 248 patients with clinical evidence of chronic arsenic toxicity who had been exposed to arsenic-contaminated drinking water for up to 15 years and some 23 control subjects from the same area, it was found that 77% of these 248 patients had hepatomegaly and 20% had splenomegaly, while portal fibrosis on liver histology was found in 63 out of 69 cases of hepatomegaly who were biopsied (Santra et al., 1999). Liver profiles in 93 patients with hepatomegaly revealed elevated ALT in 25.8%, high aspartate aminotransferase (SAT) in 6.3%, elevated S-ALP in 29%, and high serum globulin (>3.5 g/dL) in 20.7% of patients (Santra et al., 1999). Analyzing some 73 arsenic affected patients with raindrop pigmentation who consumed water containing arsenic $\geq$50 µg/L and having hair and nail arsenic level >0.6 µg/L and 25 nonsmoker healthy controls, Ghosh (2013) found 31.5% of the patients with hepatomegaly due to elevated levels of arsenic ingestion from drinking water. In a study with 7683 people from arsenic-affected districts of West Bengal, Mazumder (2005) identified 10.2% ($p < 0.001$) of 4216 subjects with hepatomegaly having high arsenic exposure.

Increased oxidative stress and ROS activity following chronic arsenic exposure leads to the activation of key kinase signaling molecules such as C-Jun N-Terminal Kinases (JNK), p38 Mitogen Activated Protein Kinase (p38 MAPK), and Cytochrome-P450, which are responsible for inducing hepatocyte apoptosis/cell injury (Suzuki and Tsukamoto, 2006). On the other hand, increased activity of ROS following arsenic exposure may also induce lipid peroxidation and further cause hepatic cell damage and hepatic toxicity (Bashir et al., 2006).

4.2.7 HEMATOLOGICAL SYSTEM

Arsenic exposure affects the hematopoietic system which includes the bone marrow, spleen, and erythrocytes. Hematological abnormalities have been reported in chronic arsenic poisoning, and a number of epidemiological investigations have noted associations between arsenic exposure and anemia (Heck et al., 2008; Lerman et al., 1980; Selzer and Ancel, 1983; Zhang et al., 2014a). Anemia, leukopenia, and thrombocytopenia are common effects of arsenic poisoning in humans at doses of 50 µg/kg/day or more (Mazumder and Dasgupta, 2011; Pakulska and Czerczak, 2006). Moreover, bone marrow depression is also a well-known manifestation while megaloblastic erythropoiesis occurs occasionally due to arsenic intoxication (Feussner et al., 1979). These effects may be due to both a direct, cytotoxic, or hemolytic effect on the blood cells (Armstrong et al., 1984; Lerman et al., 1980) and a suppression of erythropoiesis (Fincher and Koerker, 1987).

Normocytic anemia is common with arsenic exposure and a low level of hemoglobin may modify the health effects of arsenic exposure. Analyzing some 55 people exposed to arsenic-contaminated drinking water for approximately 5 years in Niigata Prefecture of Japan, Terada et al. (1962) reported a pattern of anemia, leucopenia, and thrombocytopenia among half of the patients having arsenical skin lesions. Anemia was reported in all 13 patients exposed to arsenic-contaminated groundwater ranging between 200 µg/L and 2000 µg/L in a study in West Bengal, India (Mazumder et al., 1988). Further research there has shown that some 156 people exposed to arsenic-contaminated water (50–14,200 µg/L) have an incidence of anemia in 47.4% of cases (Mazumder et al., 1998b).

Islam et al. (2004b) investigated relationships of clinical complications with nutritional status and the prevalence of leukopenia among arsenic exposed patients living in villages in Bangladesh. Analyzing a total of 115 exposed individuals diagnosed as arsenicosis patients and age-matched 120 unexposed subjects, they established a higher prevalence of leukopenia and lymphocytosis in arsenicosis patients. Analyzing hemoglobin, skin lesions, arsenic exposure, and nutritional and demographic information from 1954 Bangladeshi HEALS participants, Heck et al. (2008) investigated a negative association between arsenic exposure (urinary arsenic >200 µg/L) and hemoglobin

(<10 d/L) among all men and among women. They also analyzed an association between high arsenic exposure and anemia in Bangladesh. Conversely, some researchers found that hematological effects are not observed in all cases of arsenic exposure. Harrington et al. (1978) examined complete blood counts of 184 exposed people in Alaska, where the average concentration of well water was recorded at 224 µg/L and no association was observed between estimated daily arsenic ingestion and any hematological abnormality. Likewise, no association of anemia was found in people drinking arsenic-contaminated well water in two towns in Utah (180 µg/L and 270 µg/L) (Southwick et al., 1983).

4.2.8 Renal System

The renal system in humans contains organs that filter wastes from the blood stream and remove wastes from the body as urine. These organs include the kidneys that produce urine, ureters for the passage, urinary bladder for storage, and urethra for removing of urine. The kidney is the main organ involved in the excretion of arsenic and its metabolites (Chen et al., 2011b; Feng et al., 2013; Smith et al., 2012; Zheng et al., 2014), and a number of epidemiological studies have focused on N-acetyl-β_2-glucosaminidase (NAG), β_2-microglobulin (β_2-MG), micro albumin (mALB), and retinol binding protein (RBP) (Buchet et al., 2003; Chiu and Yang, 2005; Halatek et al., 2009; Huang et al., 2009; Kong et al., 2012). During the process of arsenic excretion through the renal system, accumulation of arsenic in kidneys leads to cytotoxicity in renal tissue (Madden and Fowler, 2000).

Analyzing the levels of urinary β_2-MG, mALB, and NAG in residents from polluted and nonpolluted areas of Guizhou province in China, Hong et al. (2004) confirmed the dose-response relationship between arsenic level and early kidney dysfunction indicators in urine. The calculated benchmark dose (BMD) and the lower confidence limit on the benchmark dose (LBMD) were used to estimate the critical concentration of arsenic in urine. The BMD/LBMD of urinary arsenic level was estimated for 163.61/136.98 µg/g creatinine for β_2-MG, 121.91/102.11 µg/g for mALB, and 171.88/144.44 µg/g for NAG, suggesting the lower confidence limit of urinary arsenic in residents for renal dysfunction for a 10% level of risk above the background obtained from LBMD (Hong et al., 2004). The results suggest that chronic arsenic exposure can cause chronic kidney damage and even kidney cancer. Hawkesworth et al. (2013) investigated marginal inverse association between infant urinary arsenic and estimated glomerular filtration rate (eGFR) in a study from Bangladesh. Considering 50 cases against 50 controls and arsenic assessment in blood from India, Palaneeswari et al. (2013) identified a link with end stage renal disease (ESRD).

Hsueh et al. (2009) reported that total urinary arsenic was associated significantly with chronic kidney disease (CKD) in a dose-response relationship. Buchet et al. (2003) noted that urinary arsenic was a significant predictor of renal tubular and glomerular dysfunction. Nordberg (2010) assessed increased levels of urinary β_2-MG and NAG as markers of renal dysfunction, which was increased in response to arsenic exposure. Feng et al. (2013) and Hong et al. (2004) showed that mALB can be used as a sensitive biomarker of early renal dysfunction caused by arsenic exposure. However, Feng et al. (2013) claimed that urinary arsenic only indirectly reflected renal damage because it varied rapidly with arsenic exposure and it could represent renal dysfunction only after a long period of metal exposure. Zhang et al. (2006) investigated that urinary RBP and $\alpha 1$-MG in medium and severe arsenic poisoning groups as well as transferrin (TRF) in severe poisoning group were significantly higher compared with the control group ($p < 0.05$), suggesting that renal tubular damage became more severe with increased arsenic poisoning. The CKD with an eGFR <60 mL/min/1.73 m^2) in Taiwan was calculated at 11.9% (Wen et al., 2008).

Analyzing urine and plasma specimens from 1043 subjects between 2002 and 2005, Chen et al. (2011b) observed that renal dysfunction rates significantly increased when the urinary arsenic rose above 75 µg/g creatinine for both tubular (β_2-MG >0.154 mg/L) and glomerular function (eGFR <90 mL/min/1.73 m^2). They found that the adjusted OR for an abnormal β_2-MG (>0.154 mg/L) was significantly higher in subjects with urinary arsenic >35 µg/g creatinine as compared with the reference group adjusted for age, sex, living area, cigarette smoking, diabetes, and hypertension. The risk for abnormal β_2-MG and estimated glomerular filtration rate (eGFR <90 mL/min/1.73 m^2) both increased around 2-fold ($p < 0.05$) in subjects with urinary arsenic >75 µg/g creatinine as compared with those with urinary arsenic ≤35 µg/g creatinine adjusted for all the risk factors plus lead, cadmium, and nickel (Chen et al., 2011b). Moreover, high urinary arsenic, tubular proteinuria, and high urine neutrophil gelatinase associated lipocalin (NGal) levels (>300 µg/mg creatinine/dL) were reported in CKD cases (Jayasumana et al., 2013). From a cross-sectional study, Zheng et al. (2013) investigated a positive dose-response association between urine arsenic and albuminuria.

Lipid peroxidation and renal cell damage due to increase ROS activity and oxidative stress are important mechanisms in arsenic-induced renal toxicity and kidney complications (El-Demerdash et al., 2009; Kokilavani et al., 2005). Increasing the expression of Hemeoxygenase-1, Mitogen Activated Protein Kinase (MAPK), and other signaling pathways that regulate transcription factors (e.g., activating transcription factor-2, activator protein-1, and ETS domain containing protein ELK-1) causes renal toxicity (Parrish et al., 1999; Sasaki et al., 2007; Singh et al., 2011).

4.2.9 Reproductive System

Arsenic is a well-known teratogen and affects pregnancy and fetal development. Chronic arsenic exposure to drinking water leads to adverse reproductive outcomes in terms of spontaneous pregnancy loss (SPL) (Bloom et al., 2016; Neamtiu et al., 2015; Quansah et al., 2015), increased preterm birth/delivery (PD) (i.e., live delivery prior to 37 weeks of completed gestation) (Challis et al., 2009; Kile et al., 2016b), low birth weight (LBW) (i.e., neonatal weight <2500 g at term) (Fei et al., 2013), congenital malformations (Kwok et al., 2006), neonatal deaths (i.e., deaths within the first 28 days following delivery) (Kile et al., 2016b), infant deaths (i.e., deaths within the first postnatal year) (Hopenhayn-Rich et al., 2003; Sohel et al., 2010), spontaneous abortion (i.e., loss of a clinically recognized pregnancy prior to 20 weeks completed gestation), and the rate of stillbirth (i.e., pregnancy losses following 20 weeks completed gestation) (Vahter, 2009). It is evident that exposure during pregnancy to an elevated level of arsenic-contaminated drinking water is associated with impaired fetal and infant development and survival (Hopenhayn-Rich et al., 2003; Von Ehrenstein et al., 2006). Arsenic affects sex organs and may cause fertility issues in both genders, and arsenic may induce gonad dysfunction through declined testosterone synthesis, apoptosis, and necrosis in males (Dávila-Esqueda et al., 2012; Shen et al., 2013). In addition, elevated level of arsenic exposure can alter immune response during pregnancy and early life of pregnant women (Björklund et al., 2017; Farzan et al., 2013a; Heaney et al., 2015; Kile et al., 2014) as well as female fecundity (Susko et al., 2017).

Following epidemiological evidence, Vahter (2009) observed a positive association between the consumption of arsenic-contaminated drinking water of at least 10 µg/L but often >50 µg/L, and the spontaneous loss of clinically recognized pregnancies. In a cross-sectional study conducted in rural Bangladesh, there was an increase in preterm delivery (PD) for 359 pregnancies in 96 women consuming arsenic-contaminated water of >50 µg/L for at least five years compared to 309 pregnancies in 96 referents exposed to arsenic less than <20 µg/L (Ahmad et al., 2001). There was also a 2.2-fold greater likelihood of spontaneous abortion in 96 women consuming arsenic-contaminated water of >50 µg/L and a 2.5-fold increase in stillbirth than among those consuming drinking water with arsenic less than 20 µg/L (Ahmad et al., 2001). In a prospective cohort study of 2924 pregnant women and urinary arsenic concentrations during 2002–2004 in Matlab, Bangladesh, Rahman et al. (2010) found an increased risk of infant mortality with increasing arsenic exposure during pregnancy (HR: 5.0, 95% CI: 1.4, 18) as well as spontaneous abortion (OR: 1.4, 95% CI: 0.96, 2.2). In an ecological study, Cherry et al. (2010) investigated increased infant mortality (OR: 1.20, 95% CI: 0.90, 1.59) with exposure to ≥50 µg/L of arsenic in drinking water compared to <10 µg/L.

From a cross-sectional study conducted in West Bengal, India, Mukherjee et al. (2005) found an increased incidence of Prevalence Proportion Ratio (PPR) for LBW (2.45, 95% CI: 0.61, 9.88) relative to 18 pregnancies (284.0–400.0 μg/L) in 7 reference women (<3 μg/L). Following a cross-sectional study, Ahamed et al. (2006) detected an increased rate of spontaneous abortion and stillbirth among 113 pregnancies in 40 women in the severely arsenic-affected village of Eruani in Bangladesh in association with arsenic-contaminated drinking water of 201–500 μg/L (200 losses/1000 pregnancies; 250 stillbirths/1000 live births) or arsenic concentration of 501–1200 μg/L (268 losses/1000 pregnancies; 171 stillbirths/1000 live births) compared to an urban population with no presumed exposure (170 losses/1000 pregnancies; 51 stillbirths/1000 live births). Ahamed et al. (2006) also examined increased LBW for 18 women exposed to 501–1200 μg/L of arsenic (PPR: 3.36, 95% CI: 0.70, 15.05) and 4 women exposed to 201–500 μg/L of arsenic (PPR: 2.35, 95% CI: 0.24, 23.47) compared to 18 unexposed referents. Women in the high exposure group also reported more PD (PPR: 4.20 and 95% CI: 0.51, 34.67). In an exploratory study with 122 women having singleton deliveries in Romania, Bloom et al. (2016) investigated an association of higher average arsenic (10 μg/L) ingestion with a lower birth weight (Z-score ($P = 0.021$). In a case-control study from Timis County of Romania, Neamtiu et al. (2015) investigated an association between arsenic exposure and spontaneous loss. Fei et al. (2013) also explored high-dose of arsenic exposure with LBW (Coefficient estimate: −0.009, 95% CI: −0.032, −0.001).

Based on a cross sectional study of 533 married women, arsenic exposure through drinking water (0–1710 μg/L) has shown a strong association with spontaneous abortion, still birth, and neonatal death in Bangladesh (Milton et al., 2005). In this study, there was a 2.5-fold increase in the odds for reported spontaneous abortion (OR: 2.5 and 95% CI: 1.5, 4.3), and stillbirth (OR: 2.5, 95% CI: 1.3, 4.9) was detected for exposures up to ≥50 μg/L of arsenic compared to <50 μg/L of arsenic in drinking water. Analyzing a prospective cohort study of 29,134 pregnant Bangladeshi women, Rahman et al. (2007) reported a 14% increased risk for recognized spontaneous pregnancy loss, inclusive of abortion and stillbirth (RR: 1.14), among women consuming water containing arsenic with >50 μg/L compared to those consuming water with <50 μg/L of arsenic. In addition, following an ecological study of 31,000 pregnancies in Bangladesh, Cherry et al. (2008) found an increased prevalence rate of stillbirth among women (OR: 1.80) exposed to ≥50 μg/L of arsenic in drinking water compared to <10 μg/L, adjusted for numerous covariates. From a prospective study among 52 women with singleton pregnancies of less than 28 weeks completed gestation in Bangladesh, Huyck et al. (2007) found higher LBW in the high exposed group with the increase of a 1 μg/g in the level of total maternal hair arsenic. Rahman et al. (2009), from their prospective study on 1578 Bangladeshi women with singleton pregnancies, found a decreasing pattern of birth weight with an increase of 1 μg/L of inorganic arsenic in urine.

Chronic arsenic exposure and birth defects are linked (Wu et al., 2011). In a cross-sectional study of 202 married women from West Bengal, India, Von Ehrenstein et al. (2006) reported a significant association between the rate of stillbirths and exposure to arsenic from drinking water exceeding 200 μg/L compared to <50 μg/L (OR: 6.07). Mazumder (2008) examined a six-fold increased rate of stillbirths in 207 pregnancies in West Bengal in women exposed to high concentrations of arsenic (>200 μg/L) in their drinking water. In a cross-sectional study conducted among 18 married women in West Bengal, India, Chakraborti et al. (2004) examined a higher rate of spontaneous abortion for consumption of 401–1474 μg/L of arsenic (182 losses/1000 pregnancies) in comparison with the consumption of 200–400 μg/L of arsenic from drinking water (95 losses/1000 pregnancies). In another cross-sectional study of 17 women from West Bengal exposed to inorganic arsenic with 284–400 μg/L and 401–1474 μg/L, Mukherjee et al. (2005) reported an increased LBW (PPR: 2.45 and 2.57, respectively) and PD (PPR: 2.45 and 3.43, respectively) relative to 18 pregnancies in 7 reference women (<3 μg/L). Furthermore, women exposed to contaminated wells for more than 10 years had higher LBW (PPR: 1.83, 95% CI: 0.81, 4.13) but lower PD (PPR: 1.83, 95% CI: 0.39, 1.73) than women with less than 10 years of exposure (Mukherjee et al., 2005).

In a cross-sectional study, Guo et al. (2003) reported increased odds for spontaneous abortion (OR: 2.7) among 224 Mongolian women residing in villages with drinking water arsenic concentrations >50 μg/L compared to 99 women residing in villages with arsenic concentrations <50 μg/L. In a cross-sectional study for 142 women with singleton pregnancies receiving prenatal care at Shanghai City hospital and analyzing total arsenic in maternal blood (mean: 4.13 μg/L, range: 0.63–30.45) at the time of delivery, Xu et al. (2011) observed that among 71 male newborns, a ten-fold increase in the maternal arsenic level was associated with a −354.41 g decrease in birth weight. In another study analyzing 125 women from a primary delivery center in Dalian City and total arsenic in maternal blood collected upon admission for delivery (median: 5.30 μg/L, range: 0–25), Guan et al. (2012) found that a 1 μg/L increase in maternal blood arsenic was associated with a −20 g decrease in birth weight (95% CI: −39.60, −0.40).

In a prospective study among 3872 women residing in arsenic-contaminated areas and 14,387 women residing in a control area free from arsenic contamination in Taiwan, Yang et al. (2003) noticed an association between drinking well water with arsenic contamination (range: 0.15–3585 μg/L) and preterm delivery (OR: 1.10 and 95% CI: 0.91, 1.33) along with low birth weight (β: −29.05 and 95% CI: −44.55, −13.55). In a cross-sectional study among 38 women residing in four villages with arsenic-contaminated drinking water >10 μg/L and four villages accessing uncontaminated water sources (<1 μg/L) in Romania with uncomplicated, full-term singleton deliveries in 10 years, Gelmann et al. (2013) found no association for arsenic-contaminated drinking water and birth weight, but women with a LBW delivery were significantly more likely to have had the sum of inorganic arsenic in urine and its metabolites >9 μg/L (67%) compared to women with a normal birth weight delivery (10%). In a cross-sectional study conducted in Spain among 96 women and their newborns with maternal completion, Vall et al. (2012) found that babies with detectable arsenic were 223.8 g heavier on average than those without ($P = 0.043$), although gestational ages were similar ($P = 0.813$).

During the time of pregnancy, inorganic arsenic exposure (<50 μg/L) affects the uterus and placental growth results in progeny birth weight. In an ecologic study conducted in Chile, a 70% increase in the risk for stillbirth (RR: 1.7) was reported for residents of Antofagasta where drinking water arsenic contamination was recorded for 50–860 μg/L compared to <5 μg/L of arsenic in Valparaiso (Hopenhayn-Rich et al., 2003). The average urine total inorganic arsenic concentration was approximately 10-fold higher among women residing in Antofagasta (54.3 μg/L) than in Valparaiso (5.3 μg/L). A nonsignificant −26 g (95% CI: −85, 31) decrease in newborn birth weight was reported per 1 μg/L increase in maternal inorganic arsenic in urine (Hopenhayn-Rich et al., 2003). Conducting a hospital-based case-control study following some 286 incident cases of spontaneous abortion matched by 1391 live births, Aschengrau et al. (1989) detected an association between arsenic exposure and occurrence of spontaneous abortions in eastern Massachusetts, USA. Women who used to drink arsenic-contaminated water with of 0.8–1.3 μg/L were found 1.2 times (OR: 1.2) or 1.4–1.9 μg/L were 1.7 times (OR: 1.7) at risk of high frequency of spontaneous abortion (Aschengrau et al., 1989). Examining maternal urinary arsenic levels at gestational weeks 24–28 and levels of inflammatory biomarkers in plasma from 563 pregnant women and 500 infants' cord blood in New Hampshire, USA, Farzan et al. (2017) found that arsenic exposure during pregnancy may affect markers of vascular health and endothelial function in both pregnant women and their children.

Altered cell proliferation, apoptosis, and abnormal DNA methylation patterns are the major mechanisms affecting embryogenesis due to arsenic exposure (Abdul et al., 2015; Bjørklund et al., 2017). In utero and early childhood, exposure to arsenic could hinder lung function in children (Recio-Vega et al., 2014; Steinmaus et al., 2014). However, Kwok et al. (2006) investigated statistically a slight association between arsenic exposure and birth defects (OR: 1.005; 95% CI: 1.001, 1.010), but they did not find any association between arsenic exposure and stillbirth (OR: 0.999), low birth weight, childhood stunting, and child underweight in a cross-sectional investigation in three arsenic-endemic areas of Bangladesh. Myers et al. (2010) detected no association for preterm delivery with arsenic ingestion of more than 50 µg/L compared to <50 µg/L in Inner Mongolia. In a prospective cohort study among 78 pregnant women receiving prenatal care in gestational weeks 9–40 and analysis of total arsenic in spot urine specimens (mean: 76.9 µg/g creatinine and range: 9.81–1603) in a nonarsenic-endemic area of Tokyo in Japan, Shirai et al. (2010) reported no associations with birth weight. Moreover, a very poor association was detected for maternal arsenic exposure and low birth weight (OR: 1.02, 95% CI: 0.78, 1.34) in a prospective cohort research of Chou et al. (2012) among 309 Taiwanese women.

4.3 CARCINOGENIC EFFECTS

Arsenic has been treated as a human carcinogen since the late 17th century (Kligerman and Tennant, 2007). Epidemiological studies indicate that the genotoxic effects of arsenic lead to carcinogenesis, which established arsenic as a carcinogen (Fernández et al., 2015; Gamboa-Loira et al., 2017; Li et al., 2017; Melak et al., 2014). Trivalent arsenic compounds can be the cause of human carcinogens in terms of tracheal and bronchogenic carcinomas and various skin cancers, such as intraepidermal carcinomas (Bowen's disease), BCC, and SCC (Marshall et al., 2007). Chronic arsenic exposure can induce tumorigenesis in humans in skin (Gentry et al., 2014; Surdu, 2014), lung (Celik et al., 2008), bladder (Radosavljević and Jakovljević, 2008), liver (Wang et al., 2014b), and prostate (Benbrahim-Tallaa and Waalkes, 2008) (Table 4.2). However, the mechanisms behind development of cancer in the human body still remain subtle and remain under investigation (Abdul et al., 2015). It is evident that high levels of chronic arsenic exposure (>300 µg/L) are linked with skin and internal cancers (Yoshida et al., 2004). The carcinogenic effects are related to the ability of arsenic to act as a cocarcinogen and its potential as a genotoxin (Hsu et al., 2015; Son et al., 2015).

4.3.1 Skin Cancer

Chronic arsenic exposure from drinking water leads to the development of skin cancer. Although chronic arsenic poisoning damages many organ systems, it usually first presents in the skin with manifestations including hyperpigmentation and hyperkeratosis. It is evident that skin cancer is more frequent in an area of chronic arsenicism. There are significant associations between these dermatological lesions and risk of skin cancer (Gentry et al., 2014; Hunt et al., 2014; Karagas et al., 2015; Zaldívar et al., 1981). The most common malignancies found in patients with long-term exposure to arsenic are Bowen's disease (intraepithelial carcinoma or carcinoma *in situ*), BCC, and SCC (Boonchai et al., 2000; Hunt et al., 2014). Merkel cell carcinoma (MCC), an uncommon but highly aggressive cutaneous neoplasm, has also been documented at a lower frequency (Martinez et al., 2011; Wong and Wang, 2010).

Arsenic-induced Bowen's disease (BD) is able to transform into invasive BCC and SCC. Skin SCC can develop either *de novo* or progress from Bowen's disease, but BCC develops on sun-protected areas of the body with multiple and recrudescent lesions (Karagas et al., 2015; Surdu, 2014). Individuals with this Bowen's disease are considered for more aggressive cancer screening in the lung and urinary bladder. Arsenic-induced Bowen's disease provides an excellent model for studying the early stages of chemical carcinogenesis in human beings (Yu et al., 2006). Arsenic-induced Bowen's disease can appear 10 years after arsenic exposure, while other types of skin cancer can have a latency period of 20 or 30 years (Yoshida et al., 2004). A case-control study from Hungary, Romania, and Slovakia showed positive associations between long-term exposure to inorganic arsenic <100 µg/L in drinking water and BCC of the skin (Leonardi et al., 2012).

High prevalence rates of chronic arsenic poisoning in the southwest coast of Taiwan for skin cancer, hyperpigmentation, and keratosis were calculated at 10.6/1000, 183.5/1000, and 71.0/1000, respectively (Tseng et al., 1968). Tseng (1977) found a positive dose-response relationship between inorganic arsenic and skin cancer in Taiwan with high concentrations of arsenic (10–1820 µg/L with a mean concentration of 50 µg/L) in well water and a total of 428 cases of skin cancer were recorded

TABLE 4.2
Epidemiological Analysis of Arsenic Concentrations in Drinking Water and Its Ingestion with Different Cancers

Reference	Location	Study Method	Time	Comparison (Exposed vs. Reference)	Cancer Site	Outcome	Risk Estimate (95% CI)
Skin Cancer: Arsenic Is a Documented Carcinogen, and Ingestion of Inorganic Arsenic from Drinking Water Increases the Risk of Developing Skin Cancer							
Baastrup et al. (2008)	Denmark	Cohort	1993–2003	Arsenic exposure of 0.05–25.3 μg/L and 56,378 persons (1010 NMSC cases)	Skin	IRR	0.99 (0.94, 1.06)
Chen et al. (1985)	Taiwan (84 villages in 4 townships in the BFD endemic area from drinking water, including Peimen, Hsuechia, Putai, and Ichui in Taiwan population)	Ecological	1968–1982	Drinking water up to 1140 μg/L with an average of 780 μg/L of arsenic vs. general population	Skin	SMR	M: 5.34 and F: 6.52
Chen et al. (2003a)	Southwest Taiwan	Case-control	1996–1999	UMTA (43.71 ± 29.34 μg/L); CMA (8.14 ± 15.48 mg/L/year). Some 76 NMSC (BD: 29.2%; BCC: 33.3%; SCC: 47.2%) vs. 224 control	Skin	OR	CMA (>15,000 μg/L/Y): 2.99 (1.30, 6.87) with P value: 0.007
Chen and Wang (1990)	China and Taiwan (314 townships in mainland)	Ecological	1972–1983	Average arsenic concentration	Skin	SMR	M: 0.9 and F: 1.0
Gilbert-Diamond et al. (2013)	United States (New Hampshire)	Case-control	2003–2009	Urine (inorganic arsenic, MMA, DMA) and Median arsenic (ΣAs = 4.76; 2.94–8.10 μg/L). Some 470 invasive SCC cases vs. 447 controls	Skin (SCC)	OR	Σas: 1.37 (1.04, 1.80) iAs: 1.20 (0.97, 1.49) MMA: 1.34 (1.04, 1.71) DMA: 1.34 (1.03, 1.74)
Guo et al. (2001)	243 townships in Taiwan	Ecological	1980–1989	Arsenic concentrations in wells more than 640 μg/L	Skin	RD	M:0.027 and F:0.128
Hsueh et al. (1995)	Southwest Taiwan (Homei, Fuhsin, and Hsinming villages in Putai township)	Case-Control	1988–1994	Average arsenic exposure (AAE) >50 μg/L and cumulative arsenic exposure (CAE) >5000 μg/L in many years; some 1081 cases (66 skin cancer cases including BD)	Skin (BCC)	OR	AAE: 5.04 (1.07, 23.8); CAE: 13.74 (1.69, 111.64); and LAAC: 1.18 (1.08, 1.28)

(Continued)

TABLE 4.2 (*Continued*)
Epidemiological Analysis of Arsenic Concentrations in Drinking Water and Its Ingestion with Different Cancers

Reference	Location	Study Method	Time	Comparison (Exposed vs. Reference)	Cancer Site	Outcome	Risk Estimate (95% CI)
Hsueh et al. (1997)	Southwest Taiwan (Homei, Fuhsin, and Hsinming villages in Putai township)	Cohort	1989–1992	Average arsenic exposure (AAE) >700 μg/L and cumulative arsenic exposure (CAE) >17,700 μg/L. Some 654 persons (33 skin cancer cases, including BD)	Skin	RR	AAE: 8.69 (1.08, 65.50); and CAE: 7.58 (0.95, 60.33)
Karagas et al. (2001)	United States (New Hampshire)	Case-control	1993–1995	Toenails Geometric mean (0.094: 0.01–0.81 μg/g). Some 587 BCC and 284 invasive SCC—BD excluded vs. 524 controls	Skin	OR	BCC: 1.44 (0.74, 2.81); and SCC: 2.07 (0.92, 4.66) (considered range: 0.345–0.81)
Leonardi et al. (2012)	Hungary, Romania, and Slovakia (ASHRAM study)	Case-control	2003–2004	Inorganic arsenic (1.2: 0.7–13.8 μg/L) and 529 BCC vs. 540 controls	Skin	OR	DAE: 1.13 (1.07, 1.20); and CAE: 1.10 (1.01, 1.19)
Ranft et al. (2003)	Slovakia (Prievidza district)	Case-control	1996–1999	Urine (iAs, MMA, DMA) and median Σas (6.04: 1–4 μg/L). Some 264 NMSC (BCC: 91%) vs. 286 controls	Skin	Ratio of means and *P* value	Σas: 1.12 (0.03) iAs: 1.13 (0.001) DMA: 1.21 (0.045)
Rosales-Castillo et al. (2004)	Mexico (Lagunera Region)			Urine mean arsenic (34.7: 3.2–124.5 μg/L); Cumulative arsenic with geometric mean (769: 256–11,698 ppm/years). Some 42 NMSC vs. 48 controls.	Skin	OR	uAs: 1.01 (*P* value: 0.18); and cAs: 4.53 (*P* value: 0.11)
Smith et al. (1998b)	Region II northern Chile compared to the rest of Chile (440,000 people with chronic arsenic exposure)	Ecological	1989–1993	Average 43–568 μg/L (1950–94) of arsenic vs. general population. Exposure decreased from 569 μg/L (1955–69) to 43 μg/L (1990–94)	Skin	SMR	M: 7.7 and F: 3.2
Wu et al. (1989)	Taiwan (Mortality and population data in 42 villages on the southwest coast)	Ecological	1973–1986	High arsenic exposure (>600 μg/L) vs. general population	Skin	SMR	M: 32.41 and F: 18.66
Zaldívar et al. (1981)	Argentina (Antofagasta)	Ecological	1968–1971	As concentration from 580 μg/L in 1968–69 to 8 μg/L in 1971	Skin	—	M:145.5/100000 and F:168.0/100000

(*Continued*)

TABLE 4.2 (Continued)
Epidemiological Analysis of Arsenic Concentrations in Drinking Water and Its Ingestion with Different Cancers

Reference	Location	Study Method	Time	Comparison (Exposed vs. Reference)	Cancer Site	Outcome	Risk Estimate (95% CI)
Internal Cancer: High Mortality Risks from Lung, Bladder, Liver, and Kidney Cancers Are Documented Among Populations Exposed to Arsenic Through Drinking Water							
Bates et al. (1995)	USA (Utah)	Case-control	1978	Arsenic exposure of 19–53 µg/L (cumulative lifetime exposure) and 71 cases vs. 160 controls	Bladder	OR	1.56
Buchet and Lison (1998)	Belgium (Four districts with chronic arsenic exposure from drinking water, including Charleroi, LieÁge, Maaseik, and Turnhout)	Ecological	1981–1991	Rural high exposure (20–50 µg/L) vs. low exposure	Lung	SMR	M: 1.05 (0.74–1.16); F: 1.24 (0.74–1.16)
Cebrián et al. (1983)	Mexico (Two rural populations in Lagunera)	Ecological	—	High exposure (410 µg/L) vs. low exposure (5 µg/L)	BCC	RR	1.4% high exposure and 0% low exposure
Chen and Wang (1990)	China and Taiwan (314 townships in mainland)	Ecological	1972–1983	Average arsenic concentration	Lung Liver	SMR SMR	M: 5.3 and F: 5.3 M: 6.8 and F: 2.0
Chen et al. (1985)	Taiwan (84 villages in 4 townships in the BFD endemic area from drinking water, including Peimen, Hsuechia, Putai, and Ichu in Taiwan population)	Ecological	1968–1982	Drinking water up to 1140 µg/L with an average of 780 µg/L of arsenic vs. general population	Lung Liver Bladder Kidney Colon	SMR SMR SMR SMR SMR	M: 3.2 and F: 4.13 M: 1.7 and F: 2.3 M: 11.0 and F: 20.1 M: 7.72 and F: 11.19 M: 1.60 and F: 1.68
Chen et al. (1986)	BFD area (69 bladder, 76 lung, 65 liver cancer decedents in 1980–82. 65 live controls matched by age and sex)	Case-control	1980–1982	40+ years of use of arsenic averaging of 780 µg/L vs. no lifetime use of artesian well water	Lung Liver Bladder	OR OR OR	3.01 ($p < 0.01$) 2.0 ($p < 0.1$) 4.1 ($p < 0.01$)
Chen et al. (1992)	42 villages in the BFD endemic area	Ecological	1973–1986	Drinking water arsenic of ≥600 µg/L vs. <100 µg/L	Lung	RR	M: 2.42 and F:3.90
Chen et al. (2004a)	Taiwan (Arseniasis-endemic area)	Cohort	—	Drinking water arsenic of ≥700 µg/L arsenic vs. <10 µg/L	Lung	RR	3.29

(Continued)

TABLE 4.2 (Continued)
Epidemiological Analysis of Arsenic Concentrations in Drinking Water and Its Ingestion with Different Cancers

Reference	Location	Study Method	Time	Comparison (Exposed vs. Reference)	Cancer Site	Outcome	Risk Estimate (95% CI)
Chen et al. (2007)	China (Jiao-le Township)	Ecological	1992–2004	1394 arseniasis patients with chronic arsenic exposure from drinking water	All site cancer Liver Lung Bladder	SMR SMR SMR SMR	M: 1.47 and F: 0.78 M: 1.25 and F: 1.44 M: 2.84 and F: 2.33 M: 4.54 and F: 16.67
Chen et al. (2010b)	Northeastern Taiwan	Cohort	1991–2006	Arsenic concentration of ≥ 10 μg/L vs. <100 μg/L and 36 cases vs. 8086 control population at risk	Bladder	RR	2.19 (0.43–11.1)
Cheng et al. (2017)	Taiwan	Case-control	2009–2010	Participants who lived in areas with arsenic levels ≥ 50 μg/L had a higher risk of rapid progression of chronic kidney disease	Kidney	OR	1.22 (1.05–1.42)
Chiu and Yang (2005)	Taiwan (four townships in BFD endemic area, Putai)	Ecological	1971–2000	Average of 780 μg/L of arsenic in 1960s vs. general population	Lung	SMR	M: 2.31 and F: 2.92
Chiou et al. (2001)	Taiwan (BFD-endemic area)	Cohort	–	>100 μg/L of arsenic ingestion in drinking water	Urinary Tract	RR	4.8
Chiou et al. (1995)	Taiwan (BFD-endemic area)	Cohort	–	≥ 20 μg/L × year vs. 0 μg/L × year cumulative arsenic exposure	Lung	RR	4.01
Chung et al. (2013b)	Taiwan (Putai Township)	Follow-up	1996–2010	1563 residents with chronic arsenic exposure from drinking water	All site cancer Liver Lung Bladder	SMR SMR SMR SMR	M: 3.40 and F: 3.39 M: 2.91 and F: 0.84 M: 6.05 and F: 7.18 M: 42.46 and F: 59.38
Fernández et al. (2015)	Antofagasta of northern Chile	Ecological	1958–1971	42 exposed to arsenic in Antofagasta and 137 in Santiago with no exposure to arsenic	Bladder	SMR	Higher Bladder cancer among arsenic-exposed population (14.3% vs. 0; $p < 0.001$)
Ferreccio et al. (2000)	Northern Chile	Case-control	1994–1996	Arsenic exposure of 200–400 μg/L vs. 0–10 μg/L	Lung	OR	Total: 8.9

(Continued)

TABLE 4.2 (Continued)
Epidemiological Analysis of Arsenic Concentrations in Drinking Water and Its Ingestion with Different Cancers

Reference	Location	Study Method	Time	Comparison (Exposed vs. Reference)	Cancer Site	Outcome	Risk Estimate (95% CI)
Guo (2004)	Taiwan (10 townships in the southwest coast)	Ecological	1971–1990	>640 µg/L vs. <50 µg/L	Lung	RD	M: 0.28 and F:0.18
Guo et al. (1997)	1972 cases in 243 Taiwan townships containing a total population of 11.4 million	Ecological	1980–1987	Arsenic concentrations in wells more than 600 µg/L and 11 cases vs. 8102 population at risk	Bladder	RR	M:14.21 and F: 19.25
Hinwood et al. (1999)	Victoria, Australia	Ecological	1982–1991	Median arsenic in drinking water vs. general population	Lung	SIR	1.0
Hopenhayn-Rich et al. (1998)	26 counties with arsenic exposure from drinking water in Córdoba Province, Argentina	Ecological	1986–1991	Average 178 µg/L of arsenic in Córdoba vs. general population in Argentina; highest measured concentration 533 µg/L	Lung Bladder Kidney Liver	SMR SMR SMR SMR	M: 1.77 and F: 2.16 M: 2.14 and F: 1.82 M: 1.57 and F: 1.81 M: 1.8 and F: 1.9
Huang et al. (2012)	Taiwan	Case control	—	Elevated level of arsenic exposure to drinking water and 132 cases vs. 245 population with cancer risk	Kidney	OR	0.92 (0.50–1.71)
Kurttio et al. (1999)	Arsenic contamination from Finland	Case-cohort	1981–1995	Arsenic exposure in drinking water >0.5 µg/L	Bladder	RR	RR: 2.44
López-Carrillo et al. (2014)	Mexico	Case control	—	Mean arsenic exposure to drinking water (15.76 ± 21.01 µg/L) and considered the inorganic arsenic exposure of >14.90 µg/L vs. ≤6.55 µg/L, %MMA of >13.30 vs. <7.01, and %DMA of >84.96 vs. <72.37	Breast	OR	Arsenic: 0.90 (0.65–1.24) %MMA: 2.63 (1.89–3.66) %DMA: 0.63 (0.45–0.87)
Melak et al. (2014)	Northern Chile	Case-control	2007–2010	Arsenic concentrations in drinking water with <200 µg/L vs. >200 µg/L; and 94 lung and 117 bladder cancer cases vs. 347 population-based controls.	Lung Bladder	OR OR	<200 µg/L: 2.48; and >200 µg/L: 6.81 <200 µg/L: 2.37; and >200 µg/L: 6.96
Nakadaira et al. (2002)	Niigata Prefecture in Japan: area with endemic arsenic poisoning	Cohort	1959–1992	Some 86 patients with diagnosed high exposure vs. general population	Liver Lung	SMR SMR	M: 0.80 and F: 0 M: 11.01 and F: 5.34

(Continued)

TABLE 4.2 (Continued)
Epidemiological Analysis of Arsenic Concentrations in Drinking Water and Its Ingestion with Different Cancers

Reference	Location	Study Method	Time	Comparison (Exposed vs. Reference)	Cancer Site	Outcome	Risk Estimate (95% CI)
Rivara et al. (1997)	Region II and VIII in Chile	Ecological	1976–1992	Exposed group (40–860 µg/L) vs. unexposed group	Skin Liver	SMR RR	All: 3.2 RR: 1.2
Smith et al. (1998b)	Region II northern Chile compared to the rest of Chile (440,000 people with chronic arsenic exposure)	Ecological	1989–1993	Average 43–568 µg/L (1950–94) of arsenic vs. general population; exposure decreased from 569 µg/L (1955–69) to 43 µg/L (1990–94)	Lung Bladder Liver	SMR SMR SMR	M: 3.8 and F: 3.1 M: 6.0 and F: 8.2 M: 1.1 and F: 1.1
Steinmaus et al. (2010)	Argentina	Case control	—	Arsenic exposure to drinking water between 4.8 and 112.3 µg/L and 109 cases vs. 141 population with cancer risk	Lung	OR	1.32 (0.49–3.54)
Tsai et al. (1999)	Four townships of arseniasis-endemic area from drinking water, including Peimen, Hsuechia, Putai and Ichu in Taiwan (BFD-endemic area)	Ecological	1971–1994	Averaging of 780 µg/L in well water vs. general population; drinking water up to 1140 µg/L	Lung Skin Liver	SMR SMR SMR	M: 2.64 and F: 3.50 M: 5.97 and F: 6.8 M: 1.8 and F: 1.9
Tsuda et al. (1995)	Residents of Niigata Prefecture in Japan: endemic arsenic poisoning	Cohort	1959–1992	High exposure (>1000 µg/L) vs. general population	Lung Bladder Liver	SMR SMR SMR	Total: 15.69 Total: 31.2 Total: 1.5
Tseng et al. (1968)	Taiwan (37 villages from southwest coast)	Ecological	—	Most of the wells contain 400–600 µg/L of arsenic	Skin	Mortality	Maximum 21.4 per 1000 population
Wu et al. (1989)	Taiwan (mortality and population data in 42 villages on the southwest coast)	Ecological	1973–1986	High exposure (>600 µg/L) vs. general population	Bladder Kidney Liver Lung	MR MR SMR MR	M: 92.7 and F: 111.3 M: 25.26 and F:57.98 M: 86.7 and F: 31.8 M:104.8 and F:122.16

ASHRAM: Arsenic Health Risk Assessment and Molecular Epidemiology; BCC: Basal-Cell Carcinoma; BFD: Blackfoot Disease; CAE: Cumulative Arsenic Exposure; cAs: Cumulative Arsenic; CI: Confidence Interval; CMA: Cumulative Mean Arsenic; DAE: Daily Arsenic Exposure; DMA: Dimethylarsinic Acid; iAs: Inorganic Arsenic; LAAC: Lifetime Average Arsenic Concentration; IRR: Incidence Rate Ratio; SCC: Squamous Cell Carcinoma; SMR: Standardized Mortality Ratio; SIR: Standardized Incidence Ratio; M: Male; F: Female; RR: Risk Ratio (Relative Risk); OR: Odds Ratio; MMA: Monomethylarsonic Acid; MR: Mortality Rate; NMSC: Nonmelanoma Skin Cancer; RD: Unit Rate Difference; ΣAs: Total arsenic summed from inorganic arsenic; uAs: Urinary Arsenic; UMTA: Urine Mean Total Arsenic; and NA: Not applicable.

in a population of 40,421, giving an overall prevalence of about 1%. Conducting a cross-sectional study, Hsueh et al. (1995) evaluated the prevalence of arsenic-induced skin cancer among residents in Taiwanese villages exposed to inorganic arsenic in drinking water (0–930 µg/L); a dose-response increase in skin cancer was found to be associated with arsenic, and lesions commonly observed are multiple BCC and multiple SCC. Similar trends were observed with incident skin cancer occurrences ($n = 33$) during the follow-up period (Hsueh et al., 1997).

A Taiwanese case-control study of 26 skin lesion patients with gender- and age-matched controls indicated that participants with a high MMA% (more than 15.5%) had an OR of skin lesions of 5.5, compared with those with low MMA%, and those with low DMA% (less than 72.2%) had an OR of 3.25, compared with those with high DMA% (Yu et al., 2000). In a hospital-based case-control study including 76 newly diagnosed Non-melanoma Skin Cancer (NMSC) cases with histologically confirmed Bowen's disease (29.2%), BCC (33.3%) or SCC (47.2%) and 224 hospital controls in southwest Taiwan, Chen et al. (2003a) found an association with NMSC and an increasing trend with a 30-year cumulative arsenic exposure ($p = 0.007$). In a case-control study from Taiwan conducted by Chen et al. (2009), a low DMA to MMA ratio was positively associated with skin cancer in participants with a high cumulative arsenic exposure.

In a case-control study from Bangladesh, Lindberg et al. (2008) found participants excreting higher proportions of MMA (>12%) but lower DMA proportions (<76%) to be at significantly higher risk of skin lesions. In another case-control study from Bangladesh, Kile et al. (2011) used path analysis to determine the odds of skin lesions that were significantly associated with log10 MMA%, but found no association with DMA. In a study from Bangladesh, some 2483 skin lesion cases and 2857 controls demonstrated independent association for both metabolism-related 10q24.32 SNPs (rs9527 and rs11191527) and skin lesion status, with the high-methylation capacity alleles being associated with decreased risk (Pierce et al., 2013). Evidence from India shows that malignant neoplasms like Bowen's disease with regard to skin cancers are the resultant effect of chronic exposure to arsenic (Saha et al., 1999). In contrast, data from 229 skin lesions cases and 199 controls sampled from an arsenic-exposed Indian population showed no association between AS^{3}MT SNP rs11191439 (Met287Thr) and skin lesion status (De Chaudhuri et al., 2008).

From a study in Thailand, Foy et al. (1992) confirmed the relationships between arsenic exposure and Bowen's carcinoma. Astolfi et al. (1981) conducted a study in Cordoba, northern Argentina, where they observed that a regular intake of drinking water with arsenic above 100 µg/L causes identifiable types of arsenic toxicity and eventually in some cases leads to skin cancer. In a study from the Huhhot Basin of China, Zhang et al. (2014b) found that the proportion of arsenic metabolites within urine samples to influence the risk of skin lesions, with lowered arsenic methylation capacity (i.e., lower DMA% and higher MMA%), suggested an increase in the risk of arsenic-induced skin lesions.

In a dermatological clinic-based case-control study from Lagunera, Mexico, involving 42 prevalent clinically diagnosed NMSC cases and 48 controls with individual historic arsenic exposure and urinary concentrations, Rosales-Castillo et al. (2004) hypothesized that human papillomavirus (HPV) infection contributed to NMSC risk among those chronically exposed to arsenic. Meanwhile, Farzan et al. (2013b) found no association with urinary arsenic or historic exposure among those seronegative for HPV 16 in Mexico. A higher odds ratio was found among those with high historic arsenic exposure who were HPV 16 positive (OR: 16.5; P: 0.001; 95% CI: 2.97, 91.75) when compared with those who were HPV 16 negative and had low arsenic exposure. In a study of arsenic-exposed Mexican individuals (some 71 skin lesions cases and 51 controls), Valenzuela et al. (2009) observed that metabolism-related 10q24.32 SNP (rs9527 and rs11191527) was associated with decreased DMA% and increased skin lesions risk.

A population-based case-control study (including 264 histologically confirmed first primary NMSC registered in Slovakia (of which 91% were BCC) and 286 randomly selected age and sex matched controls), Ranft et al. (2003) found that total urinary arsenic was higher in NMSC cases than controls ($p = 0.03$) as was inorganic urinary arsenic ($p < 0.001$) and DMA ($p = 0.045$). In a large case-control Arsenic Health Risk Assessment and Molecular Epidemiology (ASHRAM) study with 529 histologically confirmed skin BCC cases identified by hospital pathologists and 540 hospital controls in Hungary, Romania, and Slovakia, Leonardi et al. (2012) detected an OR of 1.18 (95% CI: 1.08, 1.28) that was associated with skin cancer and inorganic arsenic exposure. Conversely, analyzing 56,378 persons on the prospective diet, cancer, and health in Denmark, Baastrup et al. (2008) found no overall association with skin cancer with BCC and SCC (IRR: 0.99, 95% CI: 0.94, 1.06), and this was due to very low arsenic exposure (range: 0.05–25.3 µg/L with 95% of the population being ≤2.1 µg/L).

In a case-control study of incident basal cell ($n = 587$) and invasive squamous cell skin cancers ($n = 284$) and 524 age- and sex-matched controls from New Hampshire, USA, there was evidence of an increased risk of invasive SCC (OR: 2.07, 95% CI: 0.92, 4.66) and to a lesser extent BCC of the skin (OR: 1.44, 95% CI: 0.74, 2.81) among those with the highest levels of toenail arsenic concentration compared with those with the lowest levels of toenail arsenic concentration (Karagas et al., 2001). In a subsequent population-based case-control (470 cases and 447 controls) study from New Hampshire, USA, for the investigation of SSC of skin with urinary arsenic detection, MMA and DMA, and summed arsenic species (ΣAs), Gilbert-Diamond et al. (2013) observed a linear dose-related increase with total urinary arsenic (OR: 1.37, 95% CI: 1.04, 1.80) and each urinary fraction (e.g., MMA: 1.34, 95% CI: 1.04, 1.71; DMA: 1.34; 95% CI: 1.03, 1.74; and arsenic: 1.20; 95% CI: 0.97, 1.49). However, several epidemiological studies performed in the United States have not detected an increased frequency of skin cancer in small populations consuming water containing arsenic at levels of around 100–200 µg/L (Goldsmith et al., 1972; Harrington et al., 1978).

4.3.2 LUNG CANCER

Epidemiological studies show the susceptibility of human lungs to inorganic arsenic exposure and high mortality from lung cancers (Chung et al., 2013a; Gibb et al., 2011; Lamm et al., 2013; Smith et al., 2012). Human lungs are one of the major targets for tumorigenesis due to arsenic exposure, and arsenic-induced specific alterations of pathways can regulate tumor formation in lung cells (Hubaux et al., 2013). From an historical cohort study in the BFD-endemic area in Taiwan, Chen et al. (2004a) examined the dose-response relationship between arsenic exposure from drinking water and lung cancer with a RR of 3.29 for lung cancer mortality with high arsenic concentrations of ≥700 µg/L compared to <10 µg/L. Guo (2004) reported age-adjusted increases in lung cancer mortality rates of 0.28 and 0.18 cases per 100,000 person-years for males and females, respectively, comparing areas with arsenic in drinking water >640 µg/L to areas with <50 µg/L. In a case-control study of 152 lung cancer cases and 419 hospital-based controls in northern Chile from November 1994 to July 1996, Ferreccio et al. (2000) found a statistically significant increased lung cancer risk (OR: 8.9) comparing individuals consuming water containing 200–400 µg/L of arsenic to individuals consuming water containing <10 µg/L arsenic. Moreover, Steinmaus et al. (2013) examined lifetime exposure of inorganic arsenic in drinking water of >335 µg/L and identified a clear arsenic-induced lung cancer association with an OR of 4.35 (95% CI: 2.57, 7.36).

With an ecological study for lung cancer mortality statistics during the period 1973–1986 of residents and levels of arsenic concentrations in four townships in the BFD-endemic area in Taiwan, Wu et al. (1989) showed that age-adjusted SMR for lung cancer increased with increasing concentrations of arsenic in well water for residents aged 20 years or older. An ecological analysis during 1973–1986 in the BFD-endemic area by Chen et al. (1992) observed significantly elevated ratios for lung cancer mortality associated with high arsenic concentrations (≥600 µg/L compared to <100 µg/L) among both males (RR: 2.42) and females (RR: 3.90). With the information of cancer death occurring between 1986 and 1993 in a study population of 263 BFD-patients and 2293 healthy subjects in Taiwan, Chiou et al. (1995) found a statistically significant positive association between ingested inorganic arsenic (0–1140 µg/L) and cancer of the lung and bladder. Analyzing the cohort with urine samples and arsenic exposure in the BFD-endemic area of Taiwan during an average follow-up period of 17.8 years and 193 subjects with all-site cancer deaths, Chung et al. (2013b) observed 71 (37%) deaths from lung cancer.

In an historical cohort study of 454 individuals exposed to arsenic-contaminated well water for 33 years from 1959 to 1992 in Niiagata, Japan, Tsuda et al. (1995) found that individuals exposed to arsenic in well water >1000 µg/L had a significant number of excess deaths from lung cancer (SMR: 15.69) and urinary tract cancer (SMR: 31.18). Nakadaira et al. (2002) investigated an increased risk of dying from lung cancer among both men (SMR: 11.01) and women (SMR: 5.34) with arsenic-contaminated well water in Japan. Comparing the cancer mortality data for Region II in northern Chile (average 43–568 µg/L) during 1989–93 with age-adjusted mortality rates for the rest of Chile, Smith et al. (1998b) found elevated levels of mortality rates from lung cancer in Region II with SMR of 3.8 (95% CI: 3.50, 4.10) for men and 3.1 (95% CI: 2.70, 3.70) for women. Analyzing urinary arsenic metabolites in 94 lung cancer cases and 347 population-based controls from northern Chile with a wide range of arsenic concentrations in drinking water, Melak et al. (2014) showed the OR for lung cancer to be 2.48 (95% CI: 1.08, 5.68) where arsenic concentrations were <200 µg/L and 6.81 (95% CI: 3.24, 14.31) for >200 µg/L with a %MMA in the upper tertile (≥12.5%).

Lung cancer is one of the leading causes of mortality in North America and low to moderate levels of arsenic exposure (<100 µg/L) show an increased risk of lung cancer (small cell and SCC) in the US population (Heck et al., 2009). Analyzing age-adjusted SMR for lung cancer from 26 counties in the Córdoba province of Argentina during 1986–1991 and high arsenic levels in drinking water (mean 178 µg/L), Hopenhayn-Rich et al. (1998) found a significant incidence of lung cancer associated with increasing exposure to arsenic for both men (SMR: 1.77) and women (SMR: 2.16) per 100,000 person-years. However, a case-control study in the USA population revealed that drinking water arsenic concentrations near 100 µg/L showed no association with high risks of lung cancer (Dauphiné et al., 2013).

4.3.3 Liver Cancer

The liver is an important site of arsenic carcinogenesis (Liu and Waalkes, 2008), and a meta-analysis denotes that long-term arsenic exposure through drinking water increases the risk of liver cancer mortality (Wang et al., 2014b). In a case-control study with 65 cases of liver cancer mortality and 368 controls with an age-adjusted OR for those who consumed artesian well water for more than 40 years compared with those who never consumed artesian well water, Chen et al. (1986) examined an age-adjusted mortality rate for liver cancer (OR: 2.0 per 100,000 person-years). Conducting an ecological study of cancer mortality by using data on deaths from cancer during 1972–83 and on arsenic concentrations in well water of 1974–76 in 83,656 people in Taiwan, Chen and Wang (1990) found an increased mortality risk from liver cancer for both men (SMR: 6.8) and women (SMR: 2.0) per 100,000 person-years. Following an ecological analysis in the BFD-endemic area, Chen et al. (1992) found a significant dose-response relationship between arsenic level in drinking water and mortality with 202 people with liver cancer.

An ecological study by Wu et al. (1989) of cancer mortality statistics during 1973–1986 of residents and levels of arsenic concentrations from 42 villages in four townships in Taiwan's BFD-endemic area showed age-adjusted liver cancer mortality with a significant dose-response increase in relation to arsenic exposure in both male (SMR: 86.7) and female (SMR: 31.8). Analyzing liver cancer patients (802 male and 301 female) from 138 villages, Lin et al. (2013) showed significant cancer incidences in both genders that consumed a high level of arsenic in drinking water.

In another ecological study during 1976–1992, Rivara et al. (1997) investigated the liver cancer mortality with Relative Risk (RR) of 1.2 (95% CI: 0.99, 1.6) in arsenic exposed Region II compared with the control area of Region VIII in Antofagasta of Chile. Comparing the cancer mortality data for Region II in northern Chile (average 43–568 µg/L) during 1989–93 with age-adjusted mortality rates for the rest of Chile, Smith et al. (1998b) found elevated levels of mortality rates for liver cancer in Region II with SMR of 1.1 for both sexes using the national rate as the standard. An investigation of childhood liver cancer in Chile revealed that liver cancer mortality between ages 0 and 19 years was especially high due to exposure to arsenic in drinking water (Liaw et al., 2008). Analyzing the cohort with urine samples and arsenic exposure in the BFD-endemic area of Taiwan during an average follow-up period of 17.8 years and 193 subjects with all-site cancer deaths, Chung et al. (2013b) ascertained 29 (15%) deaths for liver cancer. Arsenic exposure accelerates a decrease in the mRNA levels and DNMT activity and reactivates the partially or fully silenced genes in liver cancer cells (Cui et al., 2006).

Analyzing mortality data for 1986–1991 and age-adjusted SMR for liver cancer in Córdoba Province, Argentina, with high arsenic concentrations (mean: 178 µg/L), Hopenhayn-Rich et al. (1998) found a significant incidence of liver cancer associated with increasing exposure to arsenic (SMR by 1.8 for men and 1.9 for women per 100,000 person-years). Lewis et al. (1999) examined a relationship between arsenic exposure from drinking water and cancer mortality in a cohort of residents in Millard County, Utah, with an observed SIR of 0.5 (95% CI: 0.3, 0.8) for liver cancer. In an historical cohort study of 454 individuals exposed to arsenic-contaminated well water for 33 years from 1959 to 1992 in Niiagata, Japan, Tsuda et al. (1995) found significantly elevated SMR for liver (7.17) and uterine cancers (13.47). However, Lin et al. (2013) showed cancer incidences are not prominent at exposure levels lower than 0.64 mg/L.

4.3.4 Bladder and Kidney Cancer

Arsenic is genotoxic to bladder cells and ingesting inorganic arsenic is an established cause of bladder malignancies (Melak et al., 2014; Radosavljević and Jakovljević, 2008; Steinmaus et al., 2013). Studies from Argentina, Taiwan, and Chile confirmed high bladder cancer SMR due to the presence of arsenic in drinking water. An increased risk of bladder cancer has been observed in Taiwan resulting from an association between secondary arsenic methylation index (SMI) and cumulative arsenic exposure (CAE, mg/L/year) (Chen et al., 2003b). In studies from Taiwan and Bangladesh with historic arsenic concentrations >200 µg/L, people with higher %MMA or MMA/DMA ratios have had relative risks of arsenic-induced bladder cancer that were 2–5 times higher than people with lower levels of these factors (Huang et al., 2008; Lindberg et al., 2008; Pu et al., 2007).

In a case-control study from the BFD-endemic area, Chen et al. (1986) examined an age-adjusted mortality rate for bladder cancer (OR: 4.1 per 100,000 person-years). Chen and Wang (1990) reported that the lifetime risk of kidney cancer of men exposed to inorganic arsenic at a dose of 1.0 µg/kg/day (body weight) was 0.042% and that of women was 0.048%. Following an ecological analysis in the BFD-endemic area, Chen et al. (1992) observed a significant dose-response relationship between arsenic level in drinking water and mortality in 202 bladder and 64 kidney cancers. Chiang et al. (1993) investigated a high annual incidence of bladder cancer (23.53 per 100,000 person-years) in the BFD-endemic area. Analyzing the cohort with urine samples and arsenic exposure from the BFD-endemic area and 193 subjects with all-site cancer deaths, Chung et al. (2013b) ascertained for 43 (22%) deaths from bladder cancer mortality hazard ratios (HR) of 3.53 (95% CI: 1.16, 10.77) and with %MMA for 1.77 (95% CI: 0.72, 4.36). Using 50 µg/L of arsenic as the cutoff in drinking water in Taiwan, Cheng et al. (2017) found a positive association between inorganic arsenic and the progress of chronic kidney disease, i.e., a higher level of arsenic in the drinking water was associated with an OR of 1.22 (95% CI: 1.05, 1.42, $p < 0.01$) for the rapid progression of chronic kidney disease (eGFR decline >5 ml/min/1.73 m^2/year), after adjusting for other independent risk factors including hypertension, DM, proteinuria, and anemia.

Clusters of bladder cancer in Chile were linked to arsenic contamination in the 1950s. People who had grown up in Antofagasta in the 1950s and 1960s were at high risk of cancer (Fraser, 2012). The study found a steady increase in bladder cancer mortality rates over the past 60 years, peaking for men at 26·7 per 100,000 in 1991 and for women at 18·7 per 100,000 in 2001. Between 1985 and 2000, the mortality rate for men in Antofagasta was 2·9 to 5·8 times the Chile national average. Since 2006, it has been between 2·8 and 3·3 times the national average. Cuzick et al. (1992) describe the highest risk for bladder cancer as the effect of chronic arsenic exposure measured from a cohort of 478 patients in England during 1945–1980. Analyzing the mortality data during 1986–1991 and age-adjusted SMR for bladder and kidney cancers and high arsenic in drinking water (mean 178 μg/L) in the Córdoba Province of Argentina, Hopenhayn-Rich et al. (1998) found significant incidence of bladder and kidney cancers associated with increasing exposure to arsenic (SMR by 1.8 for male and 1.9 for female per 100,000 person-years for bladder cancer and by 1.57 for male and 1.81 for female for kidney cancer). Moore et al. (1997) in their cross-sectional biomarker study in a Chilean male population chronically exposed to high and low arsenic levels in drinking water show a strong epidemiological association between arsenic ingestion and bladder cancers as well as inducing genetic damage to bladder cells at drinking water levels close to MCL of 50 μg/L for arsenic.

Exposure to inorganic arsenic provides evidence of an increased risk of bladder and kidney cancer in a register-based cohort of all Finns who had lived outside the municipal drinking water system during 1967–1980 (Kurttio et al., 1999). In an epidemiologic study in an endemic arsenic poisoning area in Cordoba, Argentina, Hopenhayn-Rich et al. (1996) found that the male SMR of renal carcinoma was 0.80, 1.42, and 2.14 in low, medium, and high arsenic exposure groups, respectively, and the female SMR of kidney cancer was 1.21, 1.58, and 1.82, respectively. Comparing the cancer mortality data for Region II in northern Chile (average 43–568 μg/L) during 1989–1993 with age-adjusted mortality rates for the rest of Chile, Smith et al. (1998b) found elevated levels of mortality rates for bladder cancer in Region II with SMR of 6.0 (95% CI: 4.8, 7.4) for men and 8.2 (95% CI: 6.3, 10.5) for women. A case-control study during 2007–2010 from northern Chile with patients of kidney cancer revealed that exposure to drinking water arsenic may cause renal pelvis and ureter cancer (Ferreccio et al., 2013). In utero and early life, arsenic exposure markedly increased the rate of lung and bladder cancer manifestation in the adult population of northern Chile (Steinmaus et al., 2014).

Analyzing urinary arsenic metabolites in 94 lung and 117 bladder cancer cases and 347 population-based controls from northern Chile with a wide range of arsenic concentrations in drinking water, Melak et al. (2014) showed the OR for bladder cancer to be 2.37 (95% CI: 1.01, 5.57) who had arsenic concentrations <200 μg/L and 6.96 (95% CI: 3.27, 14.79) for >200 μg/L of arsenic with a %MMA in the upper tertile (≥12.5%). Examining information from 179 bladder cancer patients from northern Chile having 42 patients exposed to arsenic during 1958–1971 in Antofagasta and 137 with no exposure to arsenic in Santiago, Fernández et al. (2015) investigated significantly higher bladder cancers among the arsenic-exposed population (35.7% vs. 20.4%; $p = 0.037$ and 14.3% vs. 0; $p < 0.001$, respectively). However, there is contradictory evidence in the association of low level arsenic exposure with the risk of bladder and kidney cancers. Lewis et al. (1999) indicated a slightly elevated, but not statistically significant, mortality from kidney cancer in both males (SMR: 1.75) and females (SMR: 1.60). Certain case studies in the U.S. population showed no association between arsenic related mortality for exposures below 100 μg/L and bladder cancer (Lamm et al., 2004; Meliker et al., 2010).

4.3.5 Prostate Cancer

Prostate cancer is the most common type of cancer that develops in men and is considered a major cause of increased mortality (Yang et al., 2008). A positive association was investigated between arsenic exposure from drinking well water and mortality from prostate cancer (Aballay et al., 2012; Chen and Wang, 1990; Chiou et al., 1995; Molina et al., 2014; Shearer et al., 2015). Epidemiological studies show that the prostate gland is considered to be a potential site for arsenic-induced carcinogenesis (Benbrahim-Tallaa and Waalkes, 2008). Moreover, chronic exposure to arsenic causes rectal cancer (Tsai et al., 1999) and colon cancer (Chen et al., 1985; Molina et al., 2014). Arsenic is also known as a clastogenic/aneugenic carcinogen and chronic exposure to arsenic causes cytogenic damage to humans (Gonsebatt et al., 1997). Besides, polyneuropathy (a peripheral neurological disturbance) also appeared in a study of Saha et al. (1999). Guo et al. (1997) identified associations for urinary cancers of various cell types and arsenic ingestion indicating that the "carcinogenicity of arsenic may be cell type specific."

4.4 SUSCEPTIBILITY AND LATENCY PERIOD

The biological systems affected by arsenic are controlled by varied individual genetically-controlled susceptibilities toward arsenic-induced skin lesions and skin cancer and differences in arsenic metabolism. Humans metabolize inorganic arsenic to MMA and DMA, which are excreted in urine (Ameer et al., 2017). A line of literature indicates that incomplete arsenic metabolism, with higher fractions of inorganic arsenic and MMA in the urine, is a marker for increased susceptibility to arsenic-related skin lesions and skin cancer (Antonelli et al., 2014; Engström et al., 2015), cardiovascular disease (Chen et al., 2013b), and genotoxicity (Ameer et al., 2016). Furthermore, nutritional factors can play important roles in arsenic methylation and elimination (Basu et al., 2011; Pierce et al., 2012). Low folate is associated with increased risk of skin lesions (Hall and Gamble, 2012; Naujokas et al., 2013; Pilsner et al., 2009). Low socioeconomic status along with low dietary intake of calorie, protein, and micronutrients may increase the risk of arsenic-induced malignant and nonmalignant diseases.

Arsenical skin lesions are an indicator of susceptibility to arsenic-related disease and a precursor to arsenic-induced skin cancers. Studies from Bangladesh show dose-response relationships with mortality and risk of arsenical skin lesion in populations with low-to-moderate arsenic exposure over many years. Once individuals are chronically exposed to arsenic, risk for arsenic-related diseases and mortality remains high for several decades even after cessation of exposure (Chang et al., 2004). These skin lesions gradually develop five to ten years after exposure commences, although shorter latencies are possible (Arain et al., 2009b). Table 4.3 shows the level of exposure to inorganic arsenic and exposure duration and its impacts on human health.

The literature shows that bladder cancer mortality peaks 25–36 years from the initiation of exposure (Marshall et al., 2007) and kidney cancer MRR peaks at 21–25 years from initiation of exposure and is highest for women (Yuan et al., 2010). Moreover, latency periods can be extended over 50 years including for skin cancer (Haque et al., 2003), urinary cancers (Chen et al., 2010a), and lung cancer (Su et al., 2011). Exposure to inorganic arsenic during pregnancy and childhood can be associated with an increased occurrence and severity of lung disease, cardiovascular disease, and cancer in childhood and later in life (Dauphiné et al., 2013). Childhood liver cancer MRR can be 9–14 times higher for those exposed as young children as compared with controls (Liaw et al., 2008). Research also shows that skin cancers can appear after a latency of about 10 years, while internal cancers, particularly bladder and lung, can appear after 30 years at a concentration of 50 µg/L of arsenic (Brown and Chen, 1995; Tsuda et al., 1995).

4.5 MECHANISMS OF ARSENIC TOXICITY

The mechanism of arsenic toxicity on human health has been under intense investigation in recent decades. There are a number of modes of action of arsenic genotoxicity and carcinogenesis, e.g., chromosomal aberrations, induction of oxidative stress and DNA damage, altered DNA repair, altered DNA methylation, altered growth factors, altered cell proliferation, altered cell signaling, the suppression of p53, and the induction of gene amplification (Faita et al., 2013; Flora, 2011; Jomova et al., 2011; Mandal, 2017; Minatel et al., 2018; Muenyi et al., 2015; Shi et al., 2004; Yan et al., 2011). Presently, three modes (chromosomal abnormality, oxidative stress, and altered growth factors) for arsenic carcinogenesis have a degree of positive evidence. There are indications that arsenic-induced epigenetic changes are associated with arsenic-related diseases, and epigenetics is described as the heritable changes in gene expression without any dependence on the DNA sequence (Abdul et al.,

TABLE 4.3
Levels of Significant Exposure to Inorganic Arsenic Ingestion and Exposure Duration and Its Impacts on Human Health

Impacts on Human Health	Systems	Exposure (Years)	NOEAL[a] (μg/kg/day)	LOEAL[a] (μg/kg/day)		References
				Less Serious (μg/kg/day)	Serious (μg/kg/day)	
Death		5	–	–	1000 (Increased death)	Tsuda et al. (1995);
		12	–	–	240 (Death in 4/208)	Zaldívar and Guillier (1977)
		1–39	–	–	130 (Death in 5/337)	
Systemic	Dermal	3–7	–	0.8 (Hyperkeratosis, hyperpigmentation)	–	ATSDR (2007); Borgoño et al. (1980)
	Dermal	11–15	–	10 (Hyperpigmentation, hyperkeratosis)	–	
	Dermal	Continuous	0.4	22 (Pigmentation changes, hyperkeratosis)	–	
	Cardio	3–7	–	–	110 (Blackfoot disease)	Foy et al. (1992); Chen et al. (1988a); Zaldívar and Guillier (1977)
	Cardio	1–39	–	–	60 (Arterial thickening, Raynaud's disease)	
	Cardio	Continuous	–	–	64 (Blackfoot disease)	
	Gastro	2–7	–	24 (Chronic diarrhea)	–	ATSDR (2007); Cebrián et al. (1983); Huang et al. (1985)
	Gastro	5–15	–	50 (Abdominal pain)	–	
	Gastro	Continuous	0.4	22 (Gastrointestinal irritation, diarrhea)	–	
	Hepatic	1–11	–	20 (Hepatomegaly)	–	ATSDR (2007); Chakraborty and Saha (1987)
	Hepatic	12	–	–	20 (Arterial thickening, cirrhotic changes)	
	Hemato	1–20	–	7 (Anemia)	–	
Immunological		2–7	–	24 (Splenomegaly)	–	Chakraborty and Saha (1987); Rosenberg (1974)
		1–11	–	20 (Spleen palpable in 3%)	20 (Arterial thickening in spleen)	
		12	–	–	110 (Weakness, anorexia)	Foy et al. (1992)
Neurological		3–7	0.8	0.007 (Tingling of hands and feet)	–	Mazumder et al. (1988); Hindmarsh et al. (1977)
		1–20	–	0.019 (Electromyographic abnormalities)	–	
		Continuous	0.7	–	40 (Functional denervation)	
Cancer		>10	–	–	1.0 (Liver, lung, bladder, and kidney cancer risk)	Chen et al. (1986, 1992)
		Continuous	–	–	64 (Bladder, lung, and liver cancers)	

[a] NOEAL—No observable adverse health effect; LOEAL—Lowest observable adverse health effect.

2015). Some of these major regulatory changes in human epigenetics include DNA methylation, microRNA expression, and post translational histone modifications (Argos, 2015; Bailey and Fry, 2014; Banerjee et al., 2017; Bjørklund et al., 2017; Heerboth et al., 2014; Marsit, 2015). Epigenetic changes have been linked with cardiovascular (Baccarelli and Ghosh, 2012), cancer (Mandal, 2017; Sarkar et al., 2013; Sandoval and Esteller, 2012), neurological complications (Fee, 2016; Kleefstra et al., 2014), and metabolic disorders (Mendez et al., 2016; Sterns et al., 2014).

Arsenic metabolism is significant for its toxicity. Both the arsenate (As^{5+}) and arsenite (As^{3+}) have different modes of action in the poisoning mechanism. Once arsenate enters the human body it is reduced to arsenite mainly in the blood or liver by utilizing glutathione (GSH) and thioredoxin (TRX) as reductants (Figure 4.4). The arsenic biotransformation process involves alternating oxidative methylation with monomethylarsonic acid (MMA^{5+}) and dimethylarsinic acid (DMA^{5+}) (Ameer et al., 2017; Bhattacharjee et al., 2013b; Bustaffa et al., 2014; Hubaux et al., 2013; Lesseur et al., 2012; van Breda et al., 2015). This transformation occurs mainly in the liver and these metabolites are ultimately excreted in the urine. These urinary arsenic metabolites are often used as a biomarker of arsenic methylation capacity (Huang et al., 2012; Reichard et al., 2007). Arsenic is metabolized through repeated reduction and oxidative methylation. There are two pathways of arsenic metabolism: (a) the classical pathway that proposes the DMA^{3+} and DMA^{5+} are the intermediate products of arsenic metabolism (Cullen and Reimer, 1989) and (b) an alternative pathway that suggests that both the DMA^{5+} and MMA^{5+} are the end products of arsenic metabolism (Bhattacharjee et al., 2013a; Dheeman et al., 2014). It is evident that an individual with a poor urinary arsenic profile having low ratios of DMA^{5+} and MMA^{5+} combined with high cumulative arsenic exposure (CAE) has a high risk of arsenic-related skin and bladder cancer (Medeiros and Gandolfi, 2016; Sarma, 2016) and peripheral vascular disease (Barchowsky and States, 2016; Sidhu et al., 2015).

Methylation is an important step in arsenic biotransformation, and carcinogenic effects induced by arsenic exposure are mostly generated due to its biotransformation process, having effects at genetic and epigenetic levels. Arsenic can cause hypo-methylation of DNA leading to the inhibition of DNA methyltransferases (Bailey and Fry, 2014). The methylated metabolites of arsenic are contributed to its disease outcome (Gomez-Rubio et al., 2011), and the alteration of DNA methylation is involved in arsenic-induced skin cancer (Seow et al., 2014; Ren et al., 2011). DNA methyltransferase plays an important role in regulating arsenic-mediated epigenetic disruption (Reichard and Puga, 2010; Rudenko and Tsai, 2014). Arsenic interferes with DNA methyltransferases, resulting in inactivation of tumor suppressor genes through DNA hyper-methylation. Arsenic-induced tumorigenesis (i.e., formation of a cancer, whereby normal cells are transformed into cancer cells) and malignant transformation are influenced by epigenetic modifications (Abdul et al., 2015) that are linked to DNA hypo-methylation subsequent to depletion of s-adenosyl-methionine (SAM), which results in aberrant gene activation including oncogenes (Roy et al., 2015; Suzuki et al., 2013). Epigenome studies show a positive association between arsenic exposure and gene specific differential white blood cell DNA methylation (Argos et al., 2015).

Cancer cells are characterized by an unstable genome, and genomes of certain cells are susceptible in accumulating DNA damage at an accelerated rate, developing a wide range of genetic abnormalities, and ultimately becoming cancerous. Genomic instability can be CIN (chromosome instability) and MIN (microsatellite instability) (Bhattacharjee et al., 2013c; Hudler, 2012). CIN is associated with mitotic errors such as chromosomal rearrangements or segregational anomalies and is most commonly observed in human cancer, while MIN is associated with DNA level instability seen in tumors (Bhattacharjee et al., 2013c) (Figure 4.5). As a nonmutagenic human carcinogen, arsenic has very strong clastogenic properties that can lead to its carcinogenic potential (Bustaffa et al., 2014; Collotta et al., 2013; Engström et al., 2011). The most prominent example of genetic variation influencing arsenic metabolism capacity is the 10q24.32 locus, which harbors the As^3MT (arsenite methyltransferase enzyme) gene. Genetic variants have shown consistent association with arsenic metabolism capacity. Studies have reported associations between 10q24.32 metabolism-related SNP and skin lesion risk (Agusa et al., 2011).

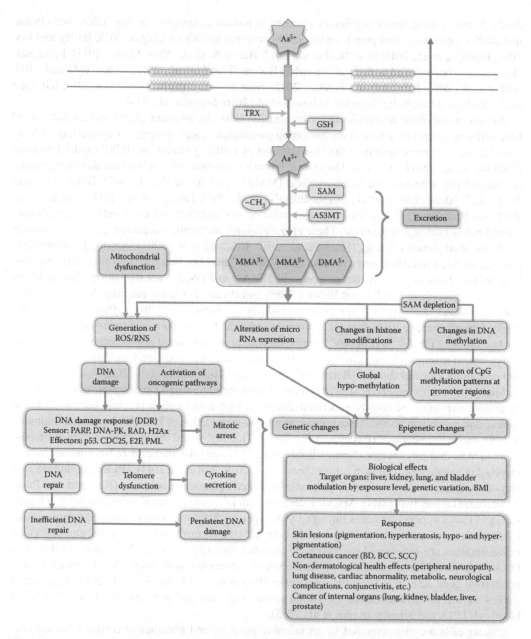

FIGURE 4.4 Schematic representation of arsenic-induced carcinogenic mechanisms and various disease outcomes that are mostly generated due to its biotransformation process, having effects at genetic and epigenetic levels. (Modified from Bailey, K. A. and R. C. Fry. 2014. *Current Environmental Health Reports* 1:22–34; Bhattacharjee, P. et al. 2013b. *International Journal of Hygiene and Environmental Health* 216(5):574–586; Brocato, J. and M. Costa. 2015. *Journal of Trace Elements in Medicine and Biology* 31:209–213; Hubaux, R. et al. 2013. *Molecular Cancer*, 12:20 (doi: 10.1186/1476–4598-12-20).)

Arsenic-induced (ROS) and Reactive Nitrogen Species (RNS) including nitric oxide (NO$^-$), hydrogen peroxide (H_2O_2), hydroxyl radical (OH), and superoxide anion (O_2^-) are known to be important in mutagenesis and carcinogenesis (Flora, 2011; Pachauri et al., 2013). Mutagenesis and DNA damage caused by ROS could contribute to the initiation of cancer. Exposure to arsenic can generate NO and O_2^- that is subsequently converted to more damaging reactive species such as OH

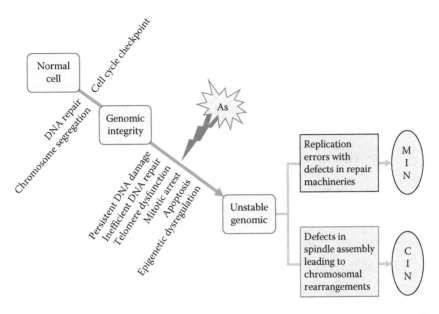

FIGURE 4.5 Genomic integrity in normal cell and its disruption are due to arsenic exposure, leading to MIN (microsatellite instability) and CIN (chromosomal instability). (After Bhattacharjee, P. et al. 2013c. *Environment International* 53:29–40.)

radical and ONOO⁻ (Shi et al., 2004). The reaction and interaction of different reactive species (e.g., O_2^- and H_2O_2) with target molecules lead to oxidative stress, DNA damage, and tumor promotion (Figure 4.6). Arsenic exposure results in the formation of ROS that could lead to DNA adduct formation like 8-OHdG (8-hydroxy-2′-deoxyguanosine), which is used as a common marker of oxidative DNA damage (Ding et al., 2005; Luna et al., 2010).

Oxidative stress is important in the molecular mechanism of arsenic-induced toxicity and carcinogenesis. Oxidative stress occurs if the equilibrium between the generation of ROS and the efficiency of detoxification is disrupted. The production of ROS and RNS by arsenic usually results in oxidative stress in cells. Oxidative stress has been involved in mediating many deleterious effects of arsenic. It is demonstrated that ROS and RNS are directly involved in oxidative damage to lipids, proteins, and DNA in cells exposed to arsenic, which can lead to cell death (Bhattacharjee et al, 2013c; De Vizcaya-Ruiz et al., 2009). It is also suggested that free radicals are generated during arsenic metabolism in cells, and enhanced oxidative stress might be associated with the development of arsenic-related diseases including cancers (Shi et al., 2004). Moreover, ROS increases the transcription of nuclear factor kappa B (NF-κB) (Barchowsky et al., 1996), tumor suppressor protein (p53) (Chanda et al., 2006), and activating protein-1 (AP-1) (Cavigelli et al., 1996; Simeonova et al., 2000), affecting cell signaling, transcription factor binding to DNA, cellular proliferation, apoptosis, and development (Bhattacharjee et al, 2013b).

Human bones have the capacity to accumulate pentavalent arsenic for a long period (Dani, 2013). The skeletal accumulation of pentavalent arsenic is due to the resemblance of arsenate to phosphate, so arsenate may substitute for phosphate in the hydroxyapatite crystal, thus forming arsenate apatite and other calcium-arsenic crystals (Bothe and Brown, 1999) that are relatively stable under the prevailing bone pH conditions. Under more acidic conditions during bone resorption, apatites are dissolved and high levels of calcium, phosphate, and arsenate are released into the extracellular fluid as osteoclasts tunnel into mineralized bone, breaking it down and releasing calcium, phosphorus, and arsenic from hydroxyapatite and arsenate apatite. This process results in a transfer of calcium, phosphorus, and arsenic from bone fluid to blood, from the blood to the cells of the body, and eventually to the urine (Dani, 2013).

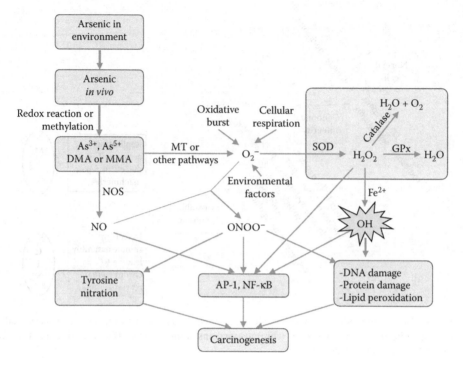

FIGURE 4.6 Mechanism of arsenic toxicity and carcinogenesis with ROS. (Modified from Hartwig, A. 2013. *Free Radical Biology and Medicine* 55:63–72; Shi, H. et al. 2004. *Molecular and Cellular Biochemistry* 255(1–2):67–78.)

Individuals with skin lesions show significant retention of arsenic in nail and hair and have a lower amount of urinary arsenic elimination capacity compared to the group without any skin lesions (Ghosh et al., 2007). The availability of biomarkers of exposure, metabolism, and biological effects can be helpful in analyzing arsenic-induced health consequences (Pimparkar and Bhave, 2010). Low arsenic metabolism capacity is thought to be a risk factor for arsenic toxicity because individuals with low-capacity genotypes will retain more arsenic in the body. More specifically, higher MMA% and lower DMA% in urine have been related to reduced methylation capacity and an increased risk of skin lesions (Agusa et al., 2014). It is notable that hypermethylation shows a dose-response relationship with arsenic measured in drinking water (Collotta et al., 2013). Gomez-Rubio et al. (2011) reveal a strong association between body mass index (BMI), genetic variation in arsenic (+3 oxidation state) methyltransferase (As^3MT), and arsenic methylation efficiency in adult women from the USA and Mexico.

4.6 CONCLUDING REMARKS

Drinking arsenic-contaminated water develops various pathological manifestations in the human body. This chapter has examined dose-related trends for arsenic-related skin lesions and a number of arsenic-induced nonmalignant and malignant diseases at levels below 100 μg/L in the populations of Argentina, Bangladesh, Chile, India, Taiwan, China, and some other arsenic-contaminated areas around the globe. The toxicity of arsenic has been known for centuries, and research during recent years has shown that arsenic is a potent human carcinogen. The evidence for arsenic exposure and different cancers in terms of skin, liver, lung, bladder, and kidney suggests a similar dose-related trend in South Asia, Mexico, Taiwan, Eastern Europe, and the USA. The relevant database indicates a role for genetic variation in arsenic metabolism.

5 Risk from Groundwater Arsenic Exposure

Epidemiological and Spatial Assessment

All substances are poisons: there is none which is not a poison. The right dose differentiates a poison and a remedy.

Theophrastus Philippus Aureolus Bombastus von Hohenheim
Known as Paracelsus, Swiss Physician and Father of Toxicology, 1493–1541

In recent years, there has been a remarkable interest in research on epidemiological and spatial risk for hazardous elements or natural events. The term "risk" here refers to the combination of likelihood or chance of an occurrence of a hazardous event or exposure(s) and the severity of injury or ill health that can be caused by the event or exposure(s). The elevated level of groundwater arsenic concentrations in drinking water can be considered a hazardous element. Rowe (1977: 24) defines risk as the "potential for realization of unwanted negative consequences of an event." USEPA (2001a) defines risk as the "probability of some adverse health effect that results from exposure to some contaminant in the environment." Is it possible to eliminate such adverse health effects within the framework of existing definitions of environmental health risk? Since risk itself cannot be eliminated, the only option is "risk aversion," which can be defined as an action taken to control risk or to manage the risk properly. Risk can remain after the implementation of risk treatment, when it is termed "residual risk." Risk management in this connection refers to the reduction of threats posed by known hazard(s).

Groundwater arsenic is a known carcinogen and elevated levels of arsenic concentrations have emerged in drinking water and food products in more than 100 countries in recent years (Amini et al., 2008; Singh, 2017). Long-term ingestion of arsenic-enriched water is associated with an increased risk of cancer (liver, lung, kidney, bladder, and skin) and nerve tissue damage (Bates et al., 1992; Cheng et al., 2017; Lin et al., 2013; Medeiros and Gandolfi, 2016; Sherwood and Lantz, 2016). It is evident that skin cancers can appear after a latency of about 10 years; internal cancers, particularly affecting the bladder and lung, can materialize after 30 years at a concentration of 50 µg/L of arsenic (Brown and Chen, 1995; Çöl et al., 1999). In addition, long-term exposure to inorganic arsenic is also associated with noncarcinogenic health effects in the form of melanosis, leuco-melanosis, and keratosis (Chakraborti et al., 2016b,c; Sarma, 2016; Singh, 2017; Tollestrup et al., 2005; Wei et al., 2017).

In risk assessment, an estimate is made of the severity or magnitude of risk to human health posed by exposure to an environmental hazard (Achour et al., 2005; Wen et al., 2006). The epidemiological evidence for risk assessment of inorganic arsenic comes from populations chronically exposed to elevated levels of arsenic in drinking water (>50 µg/L) in several countries, including Argentina (Hopenhayn-Rich et al., 1998), Bangladesh (Hassan and Atkins, 2007), Chile (Smith et al., 1998b), Taiwan (Chen et al., 2010b), and so on. Prolonged exposure to arsenic at 10 µg/L can still cause cancer (Smith et al., 2000a; WHO, 1981). Nevertheless, due to the socioeconomic and political conditions and lack of resources to combat arsenic poisoning, most developing countries still follow the previous World Health Organization (WHO) standard of 50 µg/L of arsenic (Singh and Stern, 2017; Singh and Vedwan, 2015). Risk assessments have traditionally focused on quantifying the probability of negative consequences from one or a number of identified or unknown sources (Allen et al., 2006a: 56).

Traditional nonspatial models for risk characterization are thought to be unreliable and potentially misleading (Liu et al., 2006). Therefore, the potential for environmental damage, and the resulting threats to human life from arsenic poisoning, demand an assessment of "spatial risk mapping." The use of spatial techniques in analyzing spatial risk mapping can overcome the error of nonspatial procedures. Risk mapping, in this chapter, can be defined as the process of estimating the spatial magnitude of risk to human health posed by exposure to hazardous arsenic. Spatial risk mapping for arsenic toxicity involves plotting the areas of affected people and those likely to be affected in the future as a result of ingesting different levels of arsenic concentrated in tubewells. The recent surge of research interest in "spatial risk mapping" for hazardous elements using Geographical Information Systems (GIS) is a new dimension in the field. GIS can be used for modelling in epidemiology and risk or exposure assessments. It can be applied in decision-making systems with mapping capabilities to geographically referenced information in mapping aspects of spatial risk (Achour et al., 2005; Berke, 2004; Hassan and Atkins, 2007). In view of increasing concerns about arsenic-related health risk issues, this chapter focuses on epidemiological risk assessment as well as spatial arsenic risk mapping with the interpolation of regional estimates of arsenic exposure.

5.1 CHRONOLOGICAL DEVELOPMENT OF RISK

The historical perspectives on risk and risk analysis applications in society date back more than 5200 years. The Chinese word for risk, *wēijī*, is composed of two characters, one representing "danger" and the other is "opportunity." Both terms show that risk is not a purely negative concept and uncertainty usually involves some balance between profit and loss (Smith and Petley, 2009: 50). In the past, it was seen that natural catastrophes were thought of as "Curses of Nature." This perspective viewed damaging events as a divine punishment for demoralized behavior, rather than a consequence of human use of the earth (Smith and Petley, 2009: 4). Around 3200 BC, in the valley of the Tigris-Euphrates in the Mesopotamia Civilization (modern-day Iraq), a group of scholars and practitioners of diagnosis and treatment called "Asipu" (Kuiper, 2010) served as risk analysis consultants for people making risky, uncertain, or difficult decisions.

The term "risk" has multiple conceptions and meanings, and it is a common concept that risk is closely related to "vulnerability" and "resilience." The ancients used to institutionalize prophylactic behaviors to prevent diseases. It is noted that procedures for regulating the consumption of potentially dangerous foods are mentioned in the book of Leviticus in the Old Testament. The dietary laws of the ancient Hebrew people, commonly known as Mosaic Law, were a form of risk management in response to food-borne hazards (Robson and Ellerbusch, 2007). The ancient Greeks were able to estimate risk, and Thucydides (431 BC) (1954) stated,

> We Athenians… take our decisions on policy and submit them to proper discussion. The worst thing is to rush into action before the consequences have been properly debated..... We are capable at the same time of taking risks and estimating them beforehand.

Greeks and Romans developed a system for environmental health risk where they detected causal relationships between exposure and disease, e.g., Hippocrates (4th century BC) correlated occurrence of diseases with environmental exposures; Vitruvius (1st century BC) noticed lead toxicity; and Agricola (16th century AD) noticed the correlation between occupational exposure to mining and health (Molak, 1997). Laws including risk management principles were present in legislation from the Roman Empire and the Middle Ages. English laws were issued in the 13th and 14th centuries for the prevention of health risks by unwholesome or damaged foods (Seip and Heiberg, 1989).

Historically, businesses have viewed risk as a necessary evil that should be minimized or mitigated whenever possible. In recent years, increased regulatory requirements have forced businesses to pay significant resources to address risk to scrutinize whether businesses had the right controls in place (Derr, 2016). John Evelyn, in 1661, discussed health problems caused by smog in London. In 1855, the question of how to evaluate a clean environment was expressed by Chief Seattle in his famous letter to the American president in which he states, "If we do not own the freshness of the air and the sparkle of the water, how can you buy them?" (Seip and Heiberg, 1989). Rachel Carson's *Silent Spring*, published in 1962, cited examples of indiscriminate use of pesticides, industrial pollution, water pollution, and so on and their impacts on environment and human health.

Despite the early attempts at risk assessment and risk management, it was only a couple of decades ago that such procedures became more generally recognized as useful tools in decision making at local, regional, and national levels. Recently, governmental agencies and scientific institutions in many countries are using well-established procedures to assess risk for decision optimization. This approach, for example, has led to the procedure for carcinogenic risk assessment that is presently used in many advanced countries. Conceptual development of risk analysis in recent times started with the establishment of the US Environmental Protection Agency (EPA), Occupational Safety and Health Administration, Centers for Disease Control and Preventions, Pan American Health Organization, National Institute for Occupational Safety and Health, Food and Drug Administration, and Agency for Toxic Substances and Disease Registry (ATSDR) in the United States and equivalent governmental agencies in some developed countries.

Modern risk analysis has its roots in probability theory and the scientific methods for causal links between different hazardous elements and adverse health effects. The probability theory of Blaise Pascal in 1657, life-expectancy tables of Edmond Halley in 1693, and a quantitative risk analysis of Pierre Simon de LaPlace in 1792 with the calculations of probability of death with and without smallpox vaccination are the background of modern probability theories for risk analysis (Molak, 1997). For noncarcinogenic chemicals, it is supposed that an adverse health effect occurs only if exposure to any chemical exceeds a threshold limit. It is also assumed that there is no probability of harm or injury if the exposure is below such a threshold limit. The probability of developing cancer as a result of exposure level to a concentration of a chemical is derived by modelling from animal data.

Despite public awareness and some attempts to reduce environmental pollution, the problems seem to increase on a global scale. The accident in the nuclear reactor in Chernobyl, the spread of deadly methyl isocyanate from the plant in Bhopal, and spills of chemicals in the Rhine River are recent catastrophes. In these cases, the damage is obvious and the effect is immediate (Seip and Heiberg, 1989). The adverse effects of chemicals may become apparent after a long period of time, for example, chronic arsenic poising from drinking water that could cause fatality with cancer. Under such circumstances, it may be very difficult to establish casual relationships with certainty.

Epidemiological risk analysis deals with establishing correlations or causal relationships between exposures to chemicals or physical agents and diseases. Because of the large uncertainty in estimating exposure, the results of the epidemiological studies are combined with studies in animals in order to confirm the causal relationship between exposures to an agent (carcinogen) and cancer (Molak, 1997). Over the last couple of decades, public health scientists have developed approaches to realize the extent of exposures to environmental agents, the nature of potential hazards to health, and the magnitude of such impacts on exposed populations. Recently, the epidemiological risk assessment from the exposure of inorganic arsenic from groundwater sources of drinking water (Bretzler et al., 2017; Liang et al., 2017; McBean, 2013), dietary staples (Ciminelli et al., 2017; Cubadda et al., 2017), and fish (Ahmed et al., 2015; Alamdar et al., 2017; Li et al., 2013b; Wei et al., 2014) is recognizable.

5.2 CONCEPTUAL FOCUS OF RISK

5.2.1 Defining Risk

The word "risk" is used with diversified meanings. It is a complex terminology that is best understood in context. Physical scientists prefer a quantitative view in defining risk, but social scientists favor inclusion of qualitative social elements in defining risk (Fjeld et al., 2007: 4). In the investment world, risk is typically equated with reward, while in the insurance industry risk is equated with loss. For public health perspective, risk is usually framed as a potential harm to human health or the environment. In the public health context, risk may be based on a science and policy construct. Science is used to estimate the likelihood of risk, while policy helps to define which level of risk is acceptable (Robson and Ellerbusch, 2007).

In its simplest form, risk is the probability of an adverse health effect from specified exposures (Omenn, 2007). It is a function of hazard and exposure. Without either hazard or exposure, risk is zero. Robson and Ellerbusch (2007) cited a good example with containers of cleaner and its use for risk concept. The cleaner substance in the container is hazardous and typically composed of caustic material that is corrosive to skin if contact is made. If the container of cleaner substance is left unopened and no contact is made with the contents, then the risk associated with the contained hazard would be zero. On the other hand, if the container is opened, the risk associated with using the cleaner can be greater than zero. How much greater than zero will depend on the exposure of the cleaner substance.

Generally, risk associated with technical equipment (e.g., a chemical plant) is often meant to include both the "probability" and the "consequences" of an accident. Usually, risk is considered the "likelihood" or "probability" of hazard occurrence of a certain magnitude. It is the "probability" that a substance or situation may produce harm under specified conditions. Risk is the quantitative probability that a health effect will occur after an individual has been exposed to a specified amount of a hazard (Yassi et al., 2001: 53). The most common definition of risk is that of the likelihood of an event occurring multiplied by the consequences of that event (Ansel and Wharton, 1992), where likelihood for quantitative measure is expressed either as a probability (e.g., 20%) or a frequency (e.g., 1 in 5 years) of occurrence of a disaster, whichever is appropriate for the analysis being considered (Coppola, 2011: 141). Lowrance (1976: 8) defines risk as "a measure of probability and severity of harm," and by severity, it means expected loss, vulnerability, or consequences of occurrence of an extreme event. Last (1995) defines risk as "the probability that an event will occur, e.g., that an individual will become ill or die within a stated period of time or before a given age; the probability of an unfavourable outcome." Figure 5.1 shows the basic concept of risk.

The definition of risk is different in physical science than in epidemiology or environmental health science. Van Dissen and McVerry (1994: 69) define risk as the likelihood of a hazard occurrence multiplied by consequence with vulnerability, and Twigg (1998) supports this definition.

FIGURE 5.1 Risk is possible when three basic elements (vulnerability, threat, and opportunity) exist at the same time.

Risk from Groundwater Arsenic Exposure

Vulnerability, here, implies high risk combined with an inability to cope. Human vulnerability was viewed by Timmerman (1981) as the degree of resistance offered by a social system to the impact of a hazardous event. Blaikie et al. (2004: 11) defined vulnerability as "the characteristics of a person or group in terms of their capacity to anticipate, cope with, resist and recover from the impact of a natural hazard." In turn, resistance depends on either resilience or reliability. Alexander (2000: 10) defines risk "as the likelihood, or more formally the probability, that a particular level of loss will be sustained by a given series of elements as a result of a given level of hazard impact." However, the difficulty in expressing risk of a hazard using probability and consequence dimensions lies in the fact that identical values may represent either high probability–low consequence or low probability–high consequence risks (Paul, 2011: 96). Kaplan and Garrick (1981) suggested a quantitative definition of risk in terms of the idea of a "set of triplets" and the definition is extended to include uncertainty and completeness. This definition is sometimes called the "Kaplan-Garrick risk triple" and it has been mentioned with an equation

$$R_i = \langle S_i, P_i, C_i \rangle, \tag{5.1}$$

where S_i is the scenario i, P_i the probability of scenario i, and C_i is the consequence of scenario i. In this construct, the scenario represents what can happen (or the set of conditions), the probability represents how likely it is, and the consequence represents the impacts.

Crichton (1999: 102) considers risk as the probability of a loss that depends on the hazard event itself, vulnerability, and exposure. If any of these three elements in risk increases or decreases, so does the risk probability. Thywissen (2006: 39) defines risk using four components "as a function of hazard, vulnerability, exposure, and resilience." Sometimes, risk probability is confused with hazard probability, and some think that they are analogous. Alwang et al. (2001), for example, use these two terms as synonyms. However, risk is usually considered a component of a hazard event; thus, these two terms are not synonymous. Sandman (1987) defined risk as a hazard with an outrage, and he stated that "the risk that kill you are not necessary the risk that angers and frighten you." In addition, Derr (2016) defines risk as the potential for an adverse outcome assessed as a function of threats, vulnerabilities, and consequences associated with an incident, event, or occurrence (Figure 5.2).

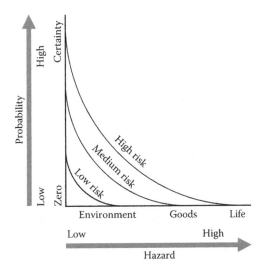

FIGURE 5.2 Theoretical relationships between the severity of environmental hazard, probability, and risk. (After Smith, K. and D. N. Petley. 2009. *Environmental Hazards: Assessing Risk and Reducing Disaster*. London: Routledge.)

5.2.2 Expression of Risk

Risk or likelihood of hazard can be expressed in both the objective or quantitative or statistical and the subjective or qualitative or perceived views (Coppola, 2011: 141). The objective perception occurs when the risks are scientifically assessed in a dispassionate way and are therefore not very susceptible to the effects of risk perception. In objective perception, all the risks and their consequences are assumed to be accurately assessed without bias and the objective risk assessment follows a highly specialized and formal procedure (Smith and Petley, 2009: 59). The objective perception is measured with numeric form of measurement that can be straightforward and incomparable. Likelihood, in objective perception, can be derived as either a frequency or a probability. Frequency, in risk assessment, refers to the number of times an event will occur within an established sample size over a specific period of time, i.e., how frequently an event occurs. For instance, in the United States, the frequency of auto accident deaths averages around 1 per 81 million miles driven (Dubner and Levitt, 2006). On the other hand, "probability" refers to a single-event scenario, and its value is expressed as a number between 0 and 1, with 0 signifying a zero chance of occurrence and 1 signifying certain occurrence. Using the auto accident example, in which the frequency of death is 1 per 81 million miles driven, it can be said that the probability of a random person in the United States dying in a car accident equals 0.000001 if he was to drive 81 miles (Coppola, 2011: 140).

Subjective or perceived risk assessment is not the result of a formalized numerical process and depends on a strong element of personal experience without any scientific validation of the results. Therefore, the resulting perception is not reproducible in a scientific sense, and this individual view may change greatly through time. The culture of a society or a community is critical to explaining differences in risk perception (Palm, 1998). Risk perceptions are socially constructed, and individual behaviors are driven by perceptions or beliefs about risks. Accordingly, some risks become "socially amplified," while others are "socially attenuated" (Paul, 2011: 101). The subjective risk or risk perception is vital to understanding people's decision making and adjustments before, during, and after a hazard event. With the growing importance of public involvement in hazard planning, risk perception can play a key role in shaping disaster policies at all government levels (Brody et al., 2003; Frewer, 1999).

The perceived risk can be measured qualitatively with three components: (a) risk ladder, (b) perceived exposure, and (c) relative severity (McClelland et al., 1990). In formulating a risk ladder, exposure levels and associated risk estimates are arranged with low levels at the bottom of a ladder and high ones at the top. Perceived exposure to risk involves individual views regarding the likelihood that their community will be affected by an extreme event of interest, and relative severity focuses on the individual's perceived severity of risk compared to their community's risk (Paul 2011: 102). Perceived risk involves neither estimating probabilities nor prediction of losses. This perceived risk contains defining various threats, determining the extent of vulnerabilities, and devising counter measures from an event that occurs. Coppola (2011: 141) suggests a qualitative representation of likelihood of a hazard event that could be expressed as

a. *Certain*: More than 99% chance of occurring in a given year, i.e., 1 or more occurrences per year.
b. *Likely*: 50%–99% chance of occurring in a given year, i.e., 1 occurrence every 1–2 years.
c. *Possible*: 5%–49% chance of occurring in a given year, i.e., 1 occurrence every 2–20 years.
d. *Unlikely*: 2%–5% chance of occurring in a given year, i.e., 1 occurrence every 20–50 years.
e. *Rare*: 1%–2% chance of occurring in a given year, i.e., 1 occurrence every 50–100 years.
f. *Extremely rare*: Less than 1% chance of occurring in a given year, i.e., 1 occurrence every 100 or more years.

In estimating the degree of risk, an important issue is related to determining an "acceptable risk." Acceptable risk is the degree of human and material loss that is perceived as bearable for actions taken to reduce disaster risk (Blanchard, 2005; Coppola, 2011: 165). In addition, "tolerable risk" is another issue, which represents temporarily acceptable risk. An individual may be willing to tolerate a risk because it is confined to a brief time period or associated with a short-term activity (Tobin and Montz, 1997). Acceptable risk is a policy-driven issue and many political, social, and economic factors influence the collective determination of what risks are acceptable and what risks are not. In a democratic society, it is a good sign to consider grass-roots concerns for any risk analysis; otherwise, the estimated risk characterization would be "irrational," the public may not accept it, and there would be problems in the risk management process. Two factors confounding the acceptability of risks are the benefits associated with certain risks and the creation of new risks by eliminating existing ones (Paul, 2011: 104). Acceptable risk is thought to be the best choice among alternatives. "How safe is safe enough?—the answer to this question depends on acceptable risk. "Acceptable risk is determined by what alternatives are available, what objectives must be achieved, the possible consequences of the alternatives, and the values to be used" (Derby and Keeney, 1981).

5.2.3 Risk Analysis, Risk Assessment, and Risk Management

Environmental risk analysis for human health is a systematic analytical process for assessing, managing, and communicating the risk to human health from contaminants released to or contained in the environment in which humans live (Fjeld et al., 2007: 1). In risk analysis, detailed examinations including risk assessment, risk evaluation, and risk management alternatives can be performed. Concepts of risk assessment and risk management have a long history, and they have been applied in past cultures. However, they were not formally defined or accompanied by detailed technical and scientific methodologies. Unfortunately, the nomenclature in the field of risk assessment and risk management is not universal. However, formal methods for assessing risk are helpful in making decisions on problems concerning health or environment, and over the last few decades, considerable progress has been made in the development of methods of risk assessment and risk management. Risk assessment can be carried out objectively, and risk management is a subjective activity, which involves preferences and attitudes. Risk assessment is the use of the factual base to define the health effects of exposure of individuals or populations to hazardous materials and situations. Risk management can be defined as the process of weighing policy alternatives and selecting the most appropriate regulatory action, integrating the results of risk assessment with engineering data and with social, economic, and political concerns to reach a decision (NRC, 1983). The term "decision analysis" may be used to cover both risk assessment and risk management.

Risk assessment is the beginning for risk management, the process of estimating the magnitude of risk to human health posed by exposure to an environmental hazard (Figure 5.3). Risk assessment examines the potential human health challenge from exposure to toxic contaminants in various environmental media. It is the procedure of obtaining the level of risk measured with quantitative or qualitative computation. The assessment of arsenic risk is based on a combination of information on the amount of arsenic that people are exposed to and its toxicity. Risk assessment is meaningful in that it allows us to understand the impact and probability of each risk (Allen, 2016a: 56). Omenn and Faustman (2002) defined risk assessment as the "systematic scientific characterization of potential adverse health effects resulting from human exposures to hazardous agents or situations." Risk assessment, in this chapter, refers to the determination of risk or a formal method for establishing the degree of risk that an individual or a community faces from groundwater arsenic poisoning.

Risk assessment is considered the process of estimating the nature, likelihood, and severity of adverse effects on human health or the environment. It is the use of the factual base to define health effects of exposure of individuals or populations to hazardous materials and situations (Fjeld et al., 2007: 3). Generally, in the environmental risk assessment process (ERA), a technical procedure can be implied through which quantitative estimates of risk are obtained. Risk assessment, in this

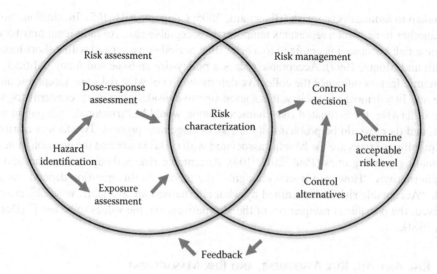

FIGURE 5.3 Risk assessment and risk management process. Dose-response assessment is the main process for risk characterization (https://toxtutor.nlm.nih.gov/06–001.html).

chapter, can be defined as the process of measuring a quantitative estimate of the human health risks resulting from ingesting arsenic-contaminated drinking water from groundwater sources.

Risk management is considered the process of identifying, evaluating, selecting, and implementing actions to reduce risk to human health and ecosystems. Risk management is thought to be a broader process of balancing risks, costs, and social values. Wu and Farland (2007) defined risk management as the "process of integrating the results of a risk assessment with social, economic, political, regulatory, and other information to make decisions about how to manage the risk." Risk management is best informed by the risk assessment process (Rodricks and Taylor, 1983). Risk management refers to both mitigation and preparedness (Figure 5.3). In mitigation, suitable actions should be taken to reduce the threats to life, property, and environment posed by extreme events; preparedness is ensuring the readiness of individuals and communities to forecast, take precautionary measures, and respond to an impending disaster (Christoplos et al., 2001). Risk management brings together the evaluation and perception of risk to control exposure to hazards. According to Plate (2002: 211),

> Risk management is a methodology for giving rational considerations to all factors affecting the safety or the operation of large structures or systems of structures. It identifies, evaluates, and executes, in conformity with other social sectors, all aspects of the management of a system, from identification of loads to the planning of emergency scenarios for the case of operational failure, and of relief and rehabilitation for the case of structural failure.

Since the 1970s, environmental hazards have continued to be noticeable as a social issue (Chaffe, 1985; Mintzberg, 1988), and risk management, in recent times, has become a foremost concern in public policy. Risk management is partly a scientific and quantitative exercise, in which the results of a risk assessment are compared to standards, guidelines, or comparable risks. In essence, risk management seeks to minimize, distribute, or share the potentially adverse consequences caused by hazardous events. Starting in the 1990s, perceived risks associated with genetically modified foods, Bovine Spongiform Encephalopathy, and variant Creutzfeldt Jakob Disease, and complex information about appropriate nutrition, have been the foci of public fear and skepticism about how food risks are managed (Frewer et al., 1997).

Disaster and health risk reduction measures include mitigation and preparedness aspects of the emergency cycle. These measures are undertaken to make individuals, households, and communities or society as a whole more resilient to disasters (Paul, 2011: 105). After the risk is evaluated and the exposure is controlled as appropriate, the risk must be monitored to ensure that it remains under control. The risk management process is an interactive one in which the risk must be reassessed and community perception re-evaluated on a continual basis. In reality, this interactive process means that the different steps in risk assessment and management may be carried out simultaneously (Yassi et al., 2001: 144). A number of actions can be implied for environmental risk management framework, and they are (Fjeld et al., 2007: 4) (a) define the problem; (b) analyze the risks associated with the problem in context; (c) examine options for addressing the risks; (d) make decisions about which options to implement; (e) take actions to implement the decisions; and (f) conduct an evaluation of the actions.

The quest for safety is a balancing act: how can people use risk to get more of the good and less of the bad (Wildavsky, 1988)? The controversy over risks and their management inevitably involves a confusing mixture of science and politics, including debates about which chemical substances and technologies present risks, which margins of safety are achievable, and how much money is needed for prevention (Graham et al., 1988). The known and potential risks can be considered with top priority if there is public fear about high risk and instability. Perhaps, the possibility of an increase in the risk of cancer, no matter how small or speculative, is sufficient to justify severe regulatory restrictions on the use of a suspect substance and technology (Sato, 2010: 3).

Risk communication is an important aspect of risk analysis. The public perception of risk is crucial for its communication (Rimal et al., 2005; Wong, 2009). Risk perception is subjective and based on community or individual culture. It refers to interactions among stakeholders, risk assessors, and risk managers. Generally, risk communication refers to an interactive process of exchange of information and opinion among individuals, groups, and institutions. The WHO defines risk communication as "an interactive process of exchange of information and opinion on risk among risk assessors, risk managers and other interested parties" (Gamhewage, 2016). It is a "two-way communication" between all the interested parties. It is "a social process by which people become informed and empowered about hazards and risks and subsequently are influenced towards behavioral changes, therefore, they can participate in the decision-making process about risk issues" (Malavé-Llamas and Cotto-Maldonado, 2010). It involves multiple messages about the nature of risk and other messages, but not strictly about risks (Covello, 2006).

Some authors think that risk communication is a community-based adaptation that addresses the societal, cultural, environmental, political, and economic characteristics of a population (Basu and Dutta, 2008; Ebi, 2009; Ford et al., 2009). In the risk communication process, people could accept involuntary exposure (e.g., exposure to pollution) and risk if they are involved in the process (Cai and Hung, 2005). Mass media play an important role in risk communication by sharing information between all parts because they have great influence in the perception of the risk (Bränström and Brandberg, 2010). This awareness develops from the individual's own values, beliefs, and experiences (Beacher et al., 2005). In risk communication, we should consider the "3Cs": context, complexity, and consequences (Glik, 2007; Stryker et al., 2008; Ford et al., 2009). This "3C" system covers all the issues in the risk assessment and management process.

Risk communication to the general populace can be influential in a better understanding of the decision-making process. Risk communication relates to sustainable development, and it plays major roles in achieving changes in attitudes that contribute to environmental awareness in society (Goosen et al., 2010: 3) as well as contributing to proper environmental management. Risk communication is the process by which persons or institutions with information of the risk at hand choose to communicate the risk to other people (Wu and Farland, 2007). Effective risk communication requires both effective transmission and reception of information; it is not merely a means for presenting the results of a risk analysis to stakeholders (Fjeld et al., 2007: 3). Risk communication focuses on finding communication methods that will enable others to make optimal decisions. Risk communication is likely to be more effective if it addresses the actual concerns of the public regarding groundwater arsenic hazard.

The most important issue during the risk communication process is the knowledge of cultural behavior (Mortenson et al., 2006). Taylor et al. (2009) investigated the effectiveness of risk communication for the massive death of wild birds in the coastal town of Esperance in Australia in 2006. When the native birds began to die, the community was found to be very concerned about the possibility of environmental pollution in the area as well as the health of the children of the community. The community received many reports regarding the causes of this disaster. This case demonstrates that quantity and level of data do not determine if the information is accepted and has a meaning for the community. Effective risk communication approaches require mutual understanding of the interested or affected parties regarding participant perceptions and the expected levels of concern, worry, fear, hostility, stress, and outrage (Malavé-Llamas and Cotto-Maldonado, 2010).

5.3 RELEVANT RISK TERMINOLOGIES

Is groundwater arsenic a hazardous element? Is there any risk to human health in ingesting toxic levels of arsenic from drinking water? If yes, how much arsenic causes what kind of harm? In an attempt to answer the questions, we need to deal with the terminological issues of risk, hazard, toxicity, threat, uncertainty, and probability since there is conceptual ambiguity. Without an understanding of these terminological issues, there would be a lack of clarity in assessing the real risk of chronic arsenic poisoning from drinking water.

5.3.1 Risk vs. Hazard and Disaster

Some see the term "risk" as analogous to "hazard." But the term hazard is not a synonym for either risk or toxicity, and there are conceptual uncertainties in "risk" and "hazard." Risk can be considered the possibility of suffering harm from a hazard; a hazard is a situation that poses a level of threat to humans or the environment (Cohrssen and Covello, 1989; Kates, 1985); and toxicity refers to the inherent potential of arsenic to cause systemic damage to humans. A highly toxic substance can damage an organism even if only very small amounts are present in the body. A substance of low toxicity cannot produce an effect unless the concentration in the target tissue is sufficiently high. Lee (1981) defines "hazard" as a situation or activity involving events where consequences are undesirable to some unknown degree and where future occurrence is uncertain. Last (1995) defines hazard as "a factor or exposure that may adversely affect health."

A hazard is a potential source of danger and hazards are a normal part of everyday experience (Fjeld et al., 2007: 5). Risk encompasses impacts on human health and on the environment that can arise from hazard. Mitchell (1990: 133) conceptualizes hazards as a multiplicative function of risk, exposure, vulnerability, and response. Risk does not exist if exposure to a hazardous substance does not occur. If hazard can be ranked, the probability of an event can be placed on a theoretical scale from zero to certainty (0 to 1). The relationship between a hazard and its probability can then be used to determine the overall degree of risk. A direct threat to life is a more serious risk faced by humans than damage to economic goods or the environment (Smith and Petley, 2009: 13).

The term hazard (or cause) is best viewed as a naturally occurring or human-induced process or event with the potential to create loss, that is, a general source of future danger, or it is the "potential threat to humans and their welfare," while risk (or consequence) is the

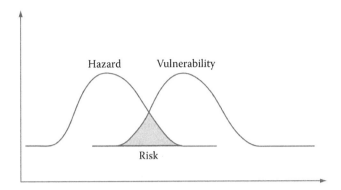

FIGURE 5.4 Graphical position of risk with relation to hazard and vulnerability. (After Menoni, A. 2004. *Natural Disasters and Sustainable Development*, ed. R. Casale and C. Margottini, 165–182. Berlin: Springer: 173.)

"probability of a hazard occurring, and creating loss" (Smith and Petley, 2009: 13). Menoni (2004: 173) very simply distinguished the differences between risk and hazard as well as risk and vulnerability (Figure 5.4). It is noted that there is a risk if there has been exposure to a hazard—not if a hazard is contained or if there is no opportunity for exposure. This distinction was illustrated by Okrent (1980) who considered two persons crossing an ocean, one in a passenger liner and the other in a rowing boat. The main hazard (deep water and large waves) is the same in both cases, but the risk (probability of capsize and drowning) is much greater for the person in the rowing boat. Likewise, an arsenic hazard can exist in any region, but the risk (probability of getting noncarcinogenic and carcinogenic symptoms for the human population) increases if people ingest arsenic-contaminated water from any arsenic-prone area for an extended period.

When a large number of people die with arsenic-induced cancers or are affected with dangerous levels of arsenic-related diseases, the event is termed a "disaster." Unlike hazard and risk, a disaster is a real or tangible or actual happening rather than a potential threat. Therefore, a disaster can simply be defined as the "realization of hazard" (Smith and Petley, 2009: 14). If a community suffers an exceptional, unusual, and non-routine level of stress and disruption from severe arsenic poisoning in terms of health effects and social problems, disasters can be social phenomena. Apart from health disasters, severely arsenic-affected patients in Bangladesh also suffer from a number of social problems that turn their lives into a social disaster.

5.3.2 Risk vs. Threat

There is a tendency to use the term "threat" similarly to "risk," but there is a sharp contrast between a risk and a threat. A threat is thought to be a "low probability" event with very large negative consequences. Derr (2016) mentioned that a threat is what we are trying to protect against. It is difficult to assess the probability of a threat. A threat is the specific use or attempted use of a risk, while a risk, on the other hand, is the probability that it may happen. Risk is a function of threats exploiting vulnerabilities to obtain, damage, or destroy assets. Thus, threats (actual, conceptual, or inherent) may exist, but if there are no vulnerabilities, then there is little or no risk. Similarly, if there is vulnerability, but no threat, then there is little or no risk (Allen, 2016b: 26). A risk can be defined as a "high probability" event, where there is enough information to make assessments of both the probability and the potential impact or consequences. Therefore, risks have two basic components: probability and potential impact. These two components in combination create the severity of risk, i.e., how much danger a risk poses with an event. Potential impact generally measures the damage occurring from the risk of a hazard.

5.3.3 Risk vs. Uncertainty

Risk and uncertainty are often treated as synonymous, but there is a distinction which hinges on the probabilities of occurrence. One dimension to the distinction is that the uncertainty does not imply risk if there are no direct consequences to the environment or human health. A risky situation, for example, is one where the set of possible outcomes, and a probability distribution for these outcomes, are well known. An uncertain situation occurs when either the set of outcomes is unknown or agreement as to a probability distribution cannot be reached. If there is no uncertainty, there is no risk. Holton (2004) notes that a person jumping out of an airplane without a parachute faces no risk since he is certain to die (i.e., there is no uncertainty).

The uncertainty, according to Derr (2016) surrounds "actual" events and outcomes for future events that may or may not take place. Wilson and Crouch (1987) mentioned risks of death and the estimated uncertainties to justify risk reliability. In the quantitative risk assessment process, the level of uncertainty associated with the estimate is usually high. Keeney (1995) stated that a sound approach to risk requires both good science and good judgment. Uncertainties in risk characterization are generally much larger in the estimates of exposure and in dose-response relationships. It is important that uncertainties originating from all steps in the analysis are included in the presentation of the final risk characterization (Suter et al., 1987). However, uncertainties often are large, but both exposure and related health effects can be determined objectively.

How should we think about risk and uncertainty? Frank Knight (1921) established the distinction between risk and uncertainty. The term "risk" is a state of uncertainty where some of the possibilities involve a loss, catastrophe, or other undesirable outcome, while "uncertainty" is the lack of complete assurance or certainty, i.e., the existence of more than one possibility and the "true" outcome is not known. Uncertainty without risk is possible but risk is meaningless without uncertainty. Knight (1921) also noticed that objective probability is the basis for risk, while subjective probability underlies uncertainty. Uncertainty is a necessary condition for risk. While some definitions of risk focus only on the probability of an event occurring, more comprehensive definitions incorporate both the probability of the event occurring and the consequences of the event. Thus, the probability of a severe earthquake may be very small, but the consequences are so catastrophic that it would be categorized as a high-risk event (Damodaran, 2008).

5.4 RISK ASSESSMENT THEORIES: ARSENIC THROUGH DRINKING WATER

How should we analyze the environmental health risk posed by groundwater arsenic poisoning? The assessment process of health risks associated with groundwater arsenic is a complex issue. Rich information is a prerequisite for understanding the mechanisms that lead to cancer and other long-term health effects. Moreover, low-potency carcinogens may easily escape detection in tests aimed at revealing carcinogenicity (Trønnes and Heiberg, 1989). There is a growing concern about levels of arsenic in the environment because of its toxicity to the human body. Hazard assessment and exposure assessment for environmental contaminants are two fundamental components of ERA (Calow, 1998), and ERA is the process by which hazards are identified, exposure quantified, and dose-response relationships determined for risk characterization (NRC, 1983). "Risk perception" (Durant, 1997) is another recent concern for ERA.

Risk assessment is a key part of the risk analysis process, involving the evaluation of the significance of a risk, either quantitatively or qualitatively. Risk identification is the foundation for risk assessment because it provides a mechanism for both risk identification and the opportunities for risk occurrence (Allen, 2016a: 56). When looking at "risk identification," what a risk is and how it can be identified are examined. The terminology and numerical outputs for ERA sometimes tend to be confused, and both experts and nonexperts are occasionally unable to understand the results of assessments. Risk analysis is a systematic process of determining the uncertainties and risks encountered in environmental health. In terms of risk reduction, the main practical process is risk management, which aims to lower the threats from known hazards whilst maximizing any related benefits. Risk management is the identification, assessment, and prioritization of risks (Derr, 2016).

The nature of risk depends on three elements: hazard, toxicity (vulnerability), and exposure. If any of these three risk elements increases or decreases, so does the pattern of risk. To answer questions relating to the safety of tubewell water requires performing a toxicological risk assessment, with an exposure assessment and a toxicity assessment. Is it safe to drink tubewell water? How much arsenic is an individual or population exposed to? Will anyone develop arsenicosis symptoms if s/he drinks that water? In answering these questions, a toxicological risk assessment is required, with an analysis of the exposure assessment and a toxicity or dose-response assessment. The results of three steps (hazard identification, exposure assessment, and toxicity assessment) are combined to produce an estimate of risk (Figure 5.5).

Since there are different susceptibilities and exposures, the risk may vary in both space and time. Risk assessment has four major elements, now commonly known as the risk assessment paradigm (NRC, 1983). The paradigm is central to an understanding of how human data are used in carcinogen risk assessment. The four elements for risk assessment are as follows (Gibb, 1997):

1. Hazard or agent identification (i.e., does the agent cause the adverse effect?)
2. Toxicity (dose-response) assessment (i.e., what is the relationship between dose and incidence in humans or how are quantity, intensity, or concentration of a hazard related to adverse effect?)
3. Exposure assessment (i.e., what exposures are currently experienced or anticipated under different conditions? who is exposed? to what and how much? how long? other exposures?)
4. Risk characterization (i.e., what is the estimated incidence of the adverse effect in a given population?)

Precision and accuracy are important issues for risk characterization, and some five factors determine the reliability of a risk characterization: (a) specification of the problem (scenario development), (b) formulation of the conceptual model (the influence diagram), (c) formulation of the computational model, (d) measurement or estimation of parameter values, and (e) calculation and documentation of results, including uncertainties (IAEA, 1989). In such a framework, there are many sources of uncertainty and variability—including lack of data, natural-process variation, incomplete or inaccurate data, model error, and ignorance of the relevant data or model structure (NRC, 1999a: 126). Table 5.1 shows different steps in arsenic-induced health risk characterization.

5.4.1 Hazard Identification

It is a significant aspect in risk analysis to determine whether exposure to an agent has the potential to cause an increase in the incidence of an undesirable effect such as ecological damage or human disease (Wu and Farland, 2007). Hazard identification is involved in describing how a chemical substance behaves in the human body, including its interactions at the organ, cellular, and molecular levels. The elevated level of arsenic concentrations in groundwater can be considered a hazardous element or "Contaminant of Concern" (CoC). Groundwater arsenic contamination in many countries around the world is already considered one of the worst environmental health disasters, but groundwater will continue to be widely used as drinking water until the availability of alternative

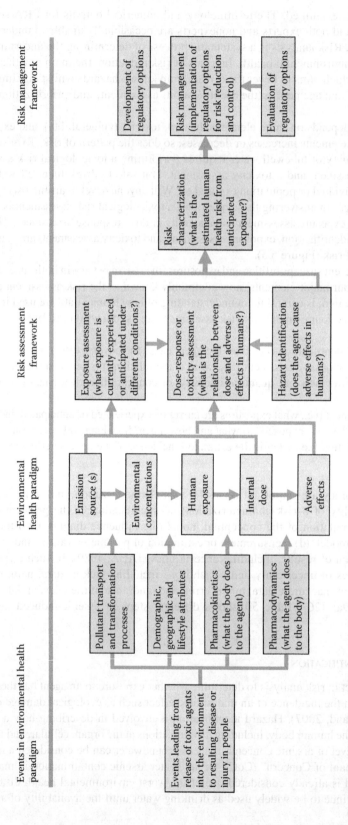

FIGURE 5.5 Sequence of exposure and risk as well as an environmental health paradigm with relationship to risk assessment and risk management framework. (Modified from Seip, H. M. and A. B. Heiberg. 1989. *Risk Management of Chemicals in the Environment*, ed. H. M. Seip and A. B. Heiberg, 1–10. New York: Plenum Press (Published in cooperation with NATO Committee on the Challenges of Modern Society): 6; Sexton, K., D. Kleffman, M. Callahan. 1995a. *Journal of Exposure Analysis and Environmental Epidemiology* 5:229–232.)

TABLE 5.1
Consecutive Steps in Arsenic-Induced Health Risk Characterization

Sl.	Steps	Description
(a)	Exposure	Pollutant concentration multiplied with exposure duration (or it is directly measured by integrated sampling).
(b)	Dose	Exposure multiplied with dosimetry factors (absorption rate, inhalation rate, etc.) divided by body weight or surface area.
(c)	Lifetime individual risk	Dose multiplied with risk characterization factor (carcinogenic potency, noncarcinogenic threshold, e.g., NOEAL), with uncertainty factors.
(d)	Risk to exposed population	Individual risk multiplied with number of exposed population (this should take into consideration age, other susceptibility factors, population activities, etc.).

Source: Yassi, A. et al., 2001. *Basic Environmental Health.* New York: Oxford University Press, 138.

options. In arsenic-contaminated countries, most people use the contaminated groundwater with different levels of arsenic for their drinking and cooking purposes. One line of literature shows that ingesting arsenic can be the cause of different arsenicosis diseases, with nonmalignant effects as well as skin and internal cancers (details in Chapter 4).

The identification of a hazard aims to determine the qualitative nature of the potential adverse health consequences of the contaminant. This can be done, for inorganic arsenic, by drawing from the results of the sciences of toxicology and epidemiology as well as from relevant scientific literature. Hazard identification is a complex process in which adverse effects are determined. This identification process should be based on well-designed toxicity studies. In this process, it needs to determine which adverse effects are toxic endpoints of concern, and once the endpoint is identified, the next step is to determine the highest "no-observable-adverse-effect level" (NOAEL).

There are a number of methods for determining a hazard, but a common method is to score both the "likelihood" of the hazard (turning into an incident) and the "seriousness" of the incident (if it occurs) on a numerical scale. This score can then be used to identify which hazards may need to be mitigated. A low score on the likelihood of occurrence may mean that the hazard is "dormant/inactive," whereas a high score would indicate that it may be an "active" hazard. The Toxicity-Concentration Screen (TCS) for selecting CoC is an important procedure for the identification of a hazardous element. The TCS is based on the scoring of contaminants according to their maximum observed concentration and toxicity properties (USEPA, 1989). The TCS does not consider the spatial variability of contaminant concentrations nor is it applicable for large sites; it is mainly used for small contaminated sites or portions of larger sites (Carlon et al., 2008). In order to apply for large sites, a modified TCS can be used and complemented with other criteria. The procedure implies the application of two criteria: (a) regulatory criteria and (b) ranking of contaminants.

In "regulatory criteria," the permissible limits in groundwater for drinking purposes can be applied. The "regulatory criterion" allows for excluding a certain level of arsenic concentrations that are not likely to pose any significant adverse effect to human health and the environment (Carlon et al., 2008). The "ranking of contaminants" by the TCS applies to the maximum and the mean observed concentrations. The objective is to identify the chemicals in a particular medium based on the concentration and toxicity that are most likely to contribute significantly to the risks for an exposure scenario involving that particular medium all over the site (USEPA, 1989). The procedure implies the calculation of a hazard score (HS), for each chemical in water (medium), following two equations: Equation 5.2 for noncarcinogenic substances and Equation 5.3 for carcinogenic substances (USEPA, 1989).

$$HS = \frac{CS}{RfD} \tag{5.2}$$

$$HS = CS \times SF \tag{5.3}$$

where CS is the contaminant concentrations at any site in the analyzed medium, in this case the groundwater arsenic concentration; RfD is the Reference Dose (mg/kg/day); and SF is the cancer slope factor (mg/kg/day). RfD is "an estimate of a daily exposure level for the human population, including sensitive subpopulations, that is likely to be without an appreciable risk of deleterious effects during a portion of a lifetime" (USEPA, 1989). The SF, on the other hand, is defined as "a plausible upper-bound estimate of the probability of a response per unit intake of a chemical over a lifetime" (Asante-Duah, 2002: 162). The slope factor is used to estimate an upper-bound probability of an individual developing cancer as a result of a lifetime exposure to a particular level of a potential carcinogen (USEPA, 1989). Chemical-specific HS are summed to obtain the total hazard score (THS) for all chemicals of potential concern in a medium. A separate THS is needed to calculate for carcinogenic and noncarcinogenic effects, and the ratio of HS to THS can be approximately regarded as the relative contribution of that chemical to the overall risk at the site. Chemicals are ranked according to their HS/THS ratio and excluded if the ratio is less than 1% (Carlon et al., 2008).

Apart from hazard identification methods, there is some literature regarding groundwater arsenic and its poisonous nature. The USEPA has classified inorganic arsenic as a Group "A" carcinogen (a known human carcinogen). The USEPA, in 2001, established a primary drinking water standard for a maximum contaminant level (MCL) of 10 μg/L for inorganic arsenic based on skin cancer incidence in humans. The American Conference of Governmental Industrial Hygienists (ACGIH) considers inorganic compounds of arsenic (except arsine) as confirmed human carcinogens (ACGIH, 1995). Exposure to inorganic arsenic in humans via drinking water (contaminated wells) can lead to skin cancers. There is substantial evidence that inorganic arsenic compounds can lead to skin and lung cancers in humans. Moreover, the International Agency for Research on Cancer (IARC) evaluated arsenic and classified "arsenic and arsenic compounds" in Group 1: "The agent is carcinogenic to humans" (IARC, 1990, 2004).

5.4.2 Toxicity (Dose-Response) Assessment

Toxicity assessment is vital for risk assessment. How much arsenic causes what kind of harm? Toxicity assessment can provide the answer to this question by investigating the potential for arsenic to cause harm. Toxicity to humans is not usually measured directly. This method is used to extrapolate data (e.g., from high to low exposure levels, from animal studies to humans, or from acute to chronic exposure) (Yassi et al., 2001: 105). It refers to the investigation of the potential for a contaminant to cause harm and how much that contaminant causes what kind of harm. Toxicity assessment is also known as dose-response assessment. Toxicity to humans is not usually measured directly. The dose-response concept is the basis of all toxicity assessments: as the dose (exposure) increases, the response (toxicity) increases (Kim et al., 2004) (Figure 5.6). Dose-response assessment is the process or analysis of the relationship between the dose of an agent administered or received and the incidence of an adverse health effect (responses) in the exposed population, and estimating the incidence of the effect as a function of human exposure (dose provided) to the agent (Holsapple and Wallace, 2008; IPCS, 2004). Typically, "dose" is used to indicate the amount of the agent, while "response" refers to the effect of the agent once administered. In toxicity assessment, it is necessary to determine exactly how high a dose causes what kind of a response or effect. The smaller the dose needed to cause an effect, the more potent (toxic) the substance is. Back in the 16th century, Paracelsus (Swiss Physician and Father of Toxicology) said, "the dose makes the poison."

Risk from Groundwater Arsenic Exposure

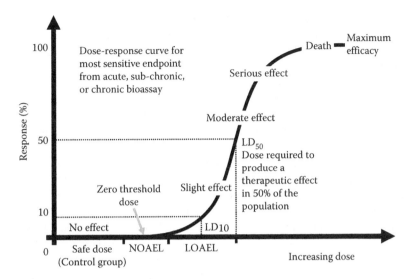

FIGURE 5.6 Dose-response relationship in epidemiology—the greater the toxic chemical exposure (dose) an individual has, the greater their response to the poison will be.

Hazard is not an intrinsic property of any chemical substance or agent but depends on the level of the exposure or the susceptibility of the receptor.

Yassi et al. (2001: 106) identified several factors that are considered when assessing the risk posed by a toxic substance, and they are the quantity of the substance actually absorbed (i.e., the dose), how the body metabolizes the substance, and nature and extent of the induced health effect at a given level of exposure (dose-response or dose-effect relationships). The term "dose-response" and "dose-effect" are occasionally used interchangeably. A "dose-response" relationship is one between the dose and the proportion of individuals in an exposed group that demonstrate a defined effect, and a "dose-effect" relationship describes that between the dose and the severity of a health effect in an individual (or a typical person in the population). The dose, in turn, depends on the route of exposure and the length, duration, and frequency of exposure. Generally, a dose-response relationship suggests causality between the degree of exposure and the adverse effect, and a dose-response curve is useful for developing basic dose-response relationships (Figure 5.6). A dose-response curve defines the relationship between dose and response based on two assumptions: (a) response increases as dose increases and (b) there is a threshold dose, i.e., a dose below which there is no effect. The dose-response curve has become a critical component of risk assessment and regulatory toxicology. In 1983, the National Research Council (NRC) detailed steps for hazard identification, dose-response assessment, exposure assessment, and the characterization of risks (Holsapple and Wallace, 2008). The "dose-response" step in risk assessment is frequently known as "hazard characterization" (Renwick et al., 2003).

The term "dose-response" can be used to describe either an "exposure-response," a "concentration-response," or other specific exposure conditions. Typically, as the dose increases, the response increases, and at low doses there may be no response. Dose-response assessment is a two-way calculation process: (a) nonlinear dose-response assessment and (b) linear dose response assessment. In nonlinear dose-response assessment, a range of exposures from zero to some finite value can be tolerated by the organism with essentially no chance of a toxic effect, and the "threshold of toxicity" is where the effects begin to occur. It is often prudent to focus on the most sensitive members of the population; therefore, regulatory efforts are generally made to keep exposures below the population threshold, which is defined as the lowest threshold of the individuals within a population (Keith, 1992: 22). If the toxicity has a threshold of low dose in which no harmful effect is expected to occur, then the type of assessment is referred to as a "nonlinear" dose-response assessment. In "nonlinear" assessment, the slope is zero and there is no response at a dose of zero.

The NOAEL is described as the highest dose (exposure level) that exerts no significant difference between the exposed population and unexposed control population. In cases in which a NOAEL has not been demonstrated experimentally, the term "lowest-observable-adverse-effect level" (LOAEL) is used. The LOAEL is the lowest dose that results in the appearance of some statistically significant mild biological effect. Regulatory toxicology has evolved around various mathematical derivations based on these basic values derived from a consideration of the dose-response (Faustman and Omenn, 2008; Holsapple and Wallace, 2008). There are alternatives to a NOEAL of effect levels for mathematical modelling for dose-response assessment known as a "Benchmark Dose" (BMD), or the "Benchmark Concentration" (BMC) or the "Benchmark Dose Lower-confidence Limit" (BMDL). A BMD or BMC is a dose or concentration that produces a predetermined change in the response rate of an adverse effect (called the benchmark response or BMR) compared to the background. The BMDL is a statistically lower confidence limit on the dose at the BMD. In developing the BMDL, a predetermined change in the response rate of an adverse effect (i.e., BMR) is selected. A BMR is an adverse effect, used to define a BMD from which an RfD or RfC can be developed. When the nonlinear approach is applied, the LOAEL, NOAEL, or BMDL is used as the point of departure for extrapolation to lower doses (USEPA, 1992).

Owing to existing uncertainties on dose-response relationships, the threshold mechanisms can be assumed for the adverse carcinogenic health effects for inorganic arsenic from drinking water (Cohen et al., 2013). Therefore, exposure levels for inorganic arsenic with no appreciable health risk, i.e., a tolerable daily intake, cannot be identified (Cubadda et al., 2017). As an alternative, reference points for health protection are currently based on BMR of a given percentage of extra risk from human data. A BMDL for 0.5% excess risk of lung cancer has been established by the FAO/WHO (BMDL $0.5 = 3$ µg/kg bw/day) (FAO/WHO, 2011), whereas the European Food Safety Authority (EFSA) identified a range of BMDL values for 1% excess risk of cancers of the lung, skin, and bladder, as well as skin lesions (BMDL $1.0 = 0.3–8.0$ µg/kg bw/day) (EFSA, 2014). Therefore, for risk characterization, an assessment of the margins of exposure (MOE) between the identified reference points and the estimated daily exposure to inorganic arsenic is required.

It is evident that being chronically exposed to inorganic arsenic via drinking water at moderate to low levels (e.g., <50 µg/L) continues to create adverse health effects for exposed populations (D'Ippoliti et al., 2015; García-Esquinas et al., 2013; Leonardi et al., 2012). Such evidence has yet to nourish into a new risk assessment. Furthermore, evidence also exists in negative impacts on fetal and infant development (Gilbert-Diamond et al., 2016; Raqib et al., 2009; Vahter, 2008) and impaired cognitive function in pre-school-aged children (Hamadani et al., 2011). Therefore, there is a need for further data supporting identification of dose-response relationships and critical exposure times (including in utero exposure) for these outcomes (Cubadda et al., 2017).

If the toxicity of a chemical does not have a threshold, then this type of estimation is referred to as a "linear dose-response assessment" (LDRA). In the case of carcinogens, linear extrapolation is typically used as the default approach for dose-response assessment. Theoretically there is no level of exposure for any chemical that does not pose a small and limited probability of generating a carcinogenic response. In LDRA, there is no use of an uncertainty factor (UF) for extrapolation; rather a straight line is drawn from the point of departure for the observed data (i.e., typically the BMDL) to the origin (where there is a zero dose and zero response). The slope of this straight line, called the "slope factor" (SF) or "cancer slope factor" (CSF), is used to estimate risk at exposure levels that fall along the line. The total cancer risk is calculated by adding the individual cancer risks for each pollutant in a pathway of concern (i.e., inhalation, ingestion, and dermal absorption), then summing the risk for all pathways (USEPA, 1992).

The acceptable daily intake (ADI) is a value used to define the daily intake of a chemical, which during an entire lifetime appears to be without appreciable risks on the basis of all known facts at the time; RfD or RfC is the estimate of a daily exposure to an agent that is assumed to be without adverse health impact in humans; tolerable daily intake (TDI) can be used to describe ingestions for chemicals that are not "acceptable" but are "tolerable" because they are below levels thought to cause adverse health effects (Faustman and Omenn, 2008). The TDI is an estimate of the amount of a substance in drinking water or in food staples, expressed on a body weight basis (mg/kg or, μg/kg) that can be ingested daily over a lifetime without appreciable health risk. The RfD, ADI, and related values are typically calculated by dividing NOAEL values by uncertainty factors (Barlow et al., 2006). In general, the RfD is defined as an estimate of a daily oral exposure of a chemical to the human population that is likely to be without an appreciable risk of deleterious effects during a lifetime. The RfD would be unlikely to cause adverse health effects even after a lifetime exposure (Barnes and Dourson, 1988). The RfD is generally expressed in units of milligrams per kilogram of bodyweight per day: mg/kg/day (USEPA, 1992). The RfD is derived from the NOAEL, LOAEL, or BMDL by application of a generally order-of-magnitude UF and MF. The RfD is determined by use of the following equation (Molak, 1997):

$$\text{RfD} = \frac{\text{NOAEL or LOAEL or BMDL}}{\text{UF} * \text{MF}} \quad (5.4)$$

where UF is the "uncertainty factor" to account for the type of study used to determine NOAEL or LOAEL and MF is the modification factor (1 to 10), which depends on the quality of the toxicological database for a chemical. The establishment of MF is often rather subjective.

The high-dose group in arsenic exposure shows a clear increase in skin lesions and is therefore designated a LOAEL. Tseng et al. (1968) investigated increased incidences of hyperpigmentation and keratosis with age and dose, and Tseng (1977) showed an increased incidence of blackfoot disease. There is some question of whether the low dose is a NOAEL or a LOAEL since there is no way of knowing what the incidence of skin lesions would be in a group where the exposure to arsenic is zero. Southwick et al. (1983) showed a marginally increased incidence of a variety of skin lesions in the individuals exposed to arsenic.

Arsenic is toxic in quantity, but the mere presence of arsenic does not automatically imply harm. Therefore, toxicity assessment is concerned with the type and degree of harm caused by differing amounts of arsenic. Chronic effects happen only after repeated long-term exposure. In determining how high a dose causes a response, the smaller the dose needed to cause an effect, the more potent (toxic) the substance is. We have examined the relationships between arsenicosis patients, their habits of consuming water, and the length of exposure to arsenic-contaminated drinking water.

5.4.3 Exposure Assessment

The term exposure refers to the "contact between an agent and a target," where an agent is defined as "a chemical, biological, or physical entity that contacts a target" (IPCS, 2004). Exposure assessment is "the process of estimating or measuring the magnitude, frequency, and duration of exposure to an agent, along with the number and characteristics of the population exposed" (Zartarian et al., 2006). The basic concepts used in exposure assessments were developed in the early 1980s by Duan (1982) and Ott (1982). Their introduction of the term "human exposure" (more simply exposure) emphasizes that the human being is the most important receptor of pollutants in the environment. Ott (1982) defines exposure as "an event that occurs when a person comes in contact with the pollutants." The exposure assessment includes the magnitude, frequency, duration, and routes of exposure, as well as evaluation of the nature of the exposed populations.

Exposure assessment is used in epidemiological studies to relate exposure concentrations to adverse health effects (Ott, 1990). Exposure assessment is an integral component of risk assessment, the process that provides scientific information for risk management. Exposure assessment is based on an "exposure scenario," which is defined as "a combination of facts, assumptions, and inferences that define a discrete situation where potential exposure may occur. These may include the source, exposed population, microenvironments, and exposure duration" (IPCS, 2004). When the duration of exposure is taken into consideration, the result is an "integrated exposure," calculated by integrating the concentration over time (Monn, 2001). Mathematically, the magnitude of exposure (E) can be defined as (Sexton et al., 1995b):

$$E = \int_{t_1}^{t_2} C(t)\, dt \qquad (5.5)$$

where $C(t)$ is the exposure concentration as a function of time that varies with time between the beginning and end of exposure and t_1 and t_2 are exposure durations. It has dimensions of mass time divided by volume. This quantity is related to the potential dose of contaminant by multiplying it by the relevant contact rate with different exposure routes. The contact rate itself may be a function of time (USEPA, 2009).

The definition of exposure refers to levels of pollutants in the ambient media, but it is not necessary that an individual inhales or ingests the pollutant. However, once the pollutant has crossed a physical boundary (e.g., skin, alveolar epithelial cells), the concept of "dose" is used (Ott, 1982). Dose is the amount of material absorbed or deposited in the body for an interval of time and is measured in units of mass (or mass per volume of body fluid in a biomarker measurement). The dose can be determined as an internal dose or as a biologically effective dose (NRC, 2001). Figure 5.7 shows the domain of exposure assessment in relation to an environmental health paradigm.

Exposure quantification aims to determine the amount of a dose (contaminant) that individuals and populations will receive. This is done by examining the results of exposure assessment. Exposure assessment mainly describes "the sources, pathways, routes, and the uncertainties in the assessment process" (IPCS, 2004). In exposure assessment, a description is given of how an individual or population comes in contact with a contaminant, including quantification of the amount of contact across space and time (Zartarian et al., 2005). Contact takes place at an exposure surface over an exposure period. Exposure assessment is the third step in the process of risk assessment.

Exposure assessment is an important analytical tool for evaluating the likelihood and extent of actual or potential exposure of receptors to the source of a chemical hazard. Apart from the mathematical models for estimating exposure factors, there are a number of methods in this connection, for example, simulation studies and dose-reconstruction studies for filling in data gaps regarding historical exposures. The probabilistic techniques (e.g., Monte Carlo analysis and Bayesian statistics) have been used recently for exposure assessment (Nieuwenhuijsen et al., 2006; Polya et al., 2009). Exposure assessment consists of quantifying the level of chemicals to which individuals and human populations are exposed, in terms of magnitude, duration, and frequency (RATSC, 1999). There are two main approaches for quantifying human exposures, and they are direct methods and indirect methods.

In the direct approach, exposures to pollutants are measured by scanning the pollutant concentrations reaching a person or population. The pollutant concentrations are directly monitored on or within the person through point of contact and biological monitoring. The point of contact approach indicates the total concentration reaching the host, while biological monitoring and the use of biomarkers infer the dosage of the pollutant through the determination of the body burden (Ott et al., 2007). In the point of contact approach, there is a continuous measure of the contaminant reaching the target through all the routes. In biological monitoring, the amount of a pollutant is measured within the body in various tissue media such as adipose tissue, bone, or urine. Biological monitoring measures the body burden of a pollutant but not its source (Lioy, 1990). An advantage of the direct approach is that exposures through multiple media are accounted for through one study technique. The disadvantages include the invasive nature of the data collection and associated costs (Ott et al., 2007). Direct methods generally tend to be more accurate but more expensive. Although arsenic can enter the human body through several pathways, all other intakes of arsenic (inhalation

Risk from Groundwater Arsenic Exposure

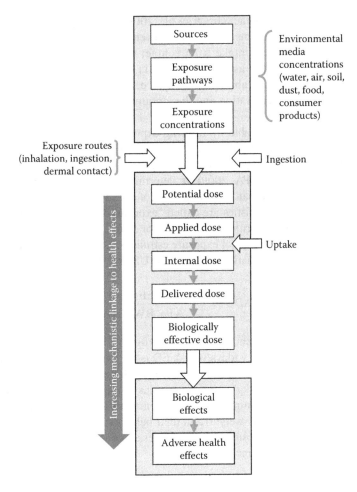

FIGURE 5.7 The domain of exposure assessment in relation to an environmental health paradigm. (Modified from Sexton, K. et al., 1995c. *Environmental Health Perspectives* 103 (Supplement 3):13–29.)

and dermal) are usually negligible in comparison to the oral route (ATSDR, 2000). Exposure pathways describe the means by which a receptor is exposed to a contaminant or chemical of concern (Mauro et al., 2000).

The indirect approach focuses on the pollutant concentrations within microenvironments to predict the exposure distributions within a population rather than the concentrations directly reaching the respondents (Ott et al., 2007). The indirect approach or exposure modelling determines the estimated exposure distributions within a population rather than the direct exposure that an individual has experienced. An exposure model is "a conceptual or mathematical representation of the exposure process" (IPCS, 2004). It has been developed in an effort to quantify human exposures to chemicals via contact with the surrounding natural environment. It is "a logical or empirical construct which allows estimation of individual or population exposure parameters from available input data" (WHO, 2000a). Exposure models represent important tools for indirect exposure assessments. They are typically used where direct measurements of exposure or biological monitoring data are not available or where these techniques are not appropriate for the exposure assessment situation.

There are two broad categories of environmental exposure models: (a) environmental concentration models and (b) human intake models. Environmental concentration models simulate environmental processes in order to generate chemical concentrations in particular media with which humans may come into contact (Fryer et al., 2006), and they are typically sophisticated environmental mechanistic assessment tools with a temporal or spatial aspect. Human intake models quantify human chemical intake from contact with the relevant environmental media, and contact is typically modelled as taking place via one of three possible exposure routes: ingestion, inhalation, and dermal absorption (Fryer et al., 2006). The output of an exposure model can be an exposure concentration and the amount of a chemical that is absorbed into the body.

The control of risk is only possible by limiting exposure. The regulatory and nonregulatory approaches are important to use in order to decrease the exposure to contaminants, but in many cases, reduction of exposure is suitable through the change of people's activities rather than regulating a source of contaminants (USEPA, 2009). In determining the exposure of a population rather than individuals, indirect methods can often make use of relevant statistics about the activities that can lead to an exposure. These statistics are called "exposure factors." They are generally drawn from the scientific literature or governmental statistics or different authentic sources. Exposure factor values can be used to obtain a range of exposure estimates. For example, to calculate the average daily dose (ADD), the following equation can be used (USEPA, 1998):

$$ADD = \frac{CS \times IR \times EF \times ED}{AT \times BW} \tag{5.6}$$

where CS is the contaminant concentration in water (µg/L), IR is the daily water ingestion rate (L/day), EF is the exposure frequency (days/years), ED is the exposure duration (years) taken as 30 years for noncarcinogens and 60 years for carcinogens in Bangladesh, AT is the average life time (days), and BW is the body weight (kg). All the variables in this equation are considered exposure factors, with the exception of contaminant concentration. Each of the exposure factors involves the characteristics of humans (e.g., body weight) and their behavior (e.g., the amount of time spent in a specific location, which affects exposure duration). These characteristics and behavior can encompass a great deal of variability and uncertainty.

5.4.4 RISK CHARACTERIZATION

Risk characterization is the integration of hazard identification, toxicity assessment, and exposure assessment in the risk assessment process. Risk characterization usually produces a quantitative estimate of the risk in the exposed population or estimates of the potential risk under different plausible exposure scenarios. Typically, in risk characterization, a range of estimates is developed, using different assumptions and statistical methods that determine how sensitive the estimates are to basic assumptions in the model (Yassi et al., 2001: 106). The purpose of risk characterization for this chapter is to figure out the nature and extent of the risk of chronic exposure to groundwater arsenic, including (USEPA, 1991)

1. The qualitative ("weight-of-evidence") conclusions as to the likelihood that a chemical may pose a hazard to human health;
2. A discussion of the dose-response information considered in deriving the RfD, including the UF and MF used;
3. Data on the shapes and slopes of the dose-response curves for the various toxic endpoints, toxicodynamic (absorption and metabolism), structure-activity correlations, and nature and severity of the observed effects;
4. Estimates of the nature and extent of the exposure and the number and types of people exposed; and
5. Discussion of the overall uncertainty in the analysis, including the major assumptions made, scientific judgment employed, and an estimate of the degree of conservatism involved.

Risk from Groundwater Arsenic Exposure

In the risk characterization process, a comparison is made between the RfD and the estimated exposure dose (EED). The EED should include all sources and routes of exposure involved. If the EED is more than the RfD, the need for regulatory concern is likely to be significant.

Risk characterization is performed by comparing the exposure level with the NOAEL to establish the margin of exposure (Heinemeyer, 2008). Generally, risk assessment concepts are mainly focused on the assessment of exposure and modelling of exposure pathways. Therefore, risk assessment is a process of estimating the potential harmful effects of chemical exposure to the environment where chemical-related risk is a function of both exposure and toxicity:

$$\text{Risk} = \int (\text{Exposure}, \text{Toxicity}) \tag{5.7}$$

Exposure identifies the existence of a chemical in the environment and the potential receptors that would be exposed to that chemical, while toxicity is a function of the response of the receptor to a chemical and the existence of a toxicity mechanism. Therefore, the process of risk assessment links the exposure pathways with the toxicity and dose-response of receptors (Figure 5.8).

Risk characterization takes place in both human health risk assessments and ecological risk assessments. Risk can be assessed in different ways for different purposes. For carcinogenic risk, the option is different from that of noncarcinogenic toxic risk. Toxic risk has been defined for noncarcinogenic exposure. It is calculated in terms of a Hazard Quotient (HQ) by the equation below (Equation 5.8). If the calculated HQ is <1, then no adverse health effects are expected as a result of exposure. If the HQ is >1, then adverse health effects are possible (Liang et al., 2017).

$$\text{HQ} = \frac{\text{ADD}}{\text{RfD}} \tag{5.8}$$

where RfD is commonly known as the "Reference Dose," the oral toxicity reference value for arsenic equals 3.04E-04 mg/kg/day (or, 3.04E-01 µg/kg/day), and ADD is the average daily dose from ingestion (µg/kg/day). The equation of ADD was depicted in the previous section (Equation 5.6).

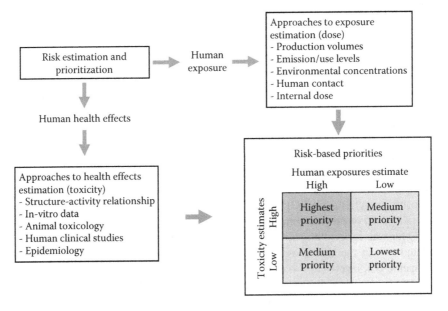

FIGURE 5.8 Basic elements in the estimation and prioritization of environmental health risks. (After Sexton, K. et al., 1993. *Toxicology and Industrial Health* 9:685–727.)

The oral RfD is an estimate of a daily exposure of the human population (including sensitive subgroups). The lifetime carcinogenic risk (LCR) is the probability of an incidence of cancer from chemical exposure and can be calculated by a different formula option:

$$LCR = 1 - \exp(-(SF \times ADD)) \tag{5.9}$$

where SF is the oral slope factor. SF is an upper bound (95% CL) on the increased cancer risk from a lifetime's exposure to an agent. This estimate, usually expressed in units of the proportion (of a population) affected per mg/kg/day, is generally reserved for use in the low-dose region of the dose-response relationship, that is, for exposures corresponding to risks of less than 1 in 100 (USEPA, 1992). The oral slope factor (SF) for arsenic is 1.5 mg/kg/day.

Risk assessments characterize carcinogenic and noncarcinogenic adverse effects by calculating the LCR and HQ, respectively, based on exposure concentration, duration, and pathways (Markley and Herbert, 2009). To undertake a risk assessment in a typical application, exposure scenarios are assessed to determine if the incremental risk of cancer/death exceeds one in a million (i.e., acceptable quality of water on the basis of increasing the probability of cancer by one in a million), or "de minimis risk" (USEPA, 2005). The concept of "de minimis risk" has its origin in the observation that some risks are so highly unlikely that they are ordinarily ignored, for example, the possibility of being struck by a meteorite (Mumpower, 1986). The term "de minimis" is a shortened version of the Latin phrase "de minimis non curat lex," which means "the law does not care about very small matters." This concept is widely used throughout Europe to set guidelines for acceptable levels of risk exposure to the general population (Coppola, 2011: 172; Fiksel, 1987: 4; Peterson, 2002). The concept of "de minimis risk" was derived from this legal principle by risk professionals in the early 1980s, and perhaps the first to use it was Comar (1979) who argued in favor of ignoring any hazard bearing a risk of less than 10^{-5} per person/year of death, unless it provided no benefit or could be easily reduced. De minimus risk is selected because it corresponds to circumstances which are considered part of normal living that represent a risk of one in a million.

By declaring a risk to be "de minimis," it is usually implied that the risk is so small that it is beyond concern or equivalent to no risk at all (Peterson, 2002). Fiksel (1987: 4) mentioned that a risk can be "de minimis" if the incremental risk produced by an activity is sufficiently small so that there is no incentive to modify the activity. Therefore, a "de minimis" risk level would represent a benchmark, below which any alleged problems or hazards could be ignored. De minimis risk is a "threshold concept" that postulates a threshold of concern below which it would be indifferent to change in the level of risk (Fiksel, 1987: 4). This "threshold of concern" is not easily identifiable, but is generally translated such that one new case of cancer is instigated above background per 100,000 people to one million people (i.e., a probability ranges from 10^{-5} or 1 in 100,000 to 10^{-6} or 1 in 1,000,000) as an incremental cancer death (McBean, 2013).

For chronic residential exposure, an acceptable carcinogenic risk level of 10^{-6}, which means that exposure would result in a 1 in 1,000,000 chance of developing cancer. For industrial purposes, the recommended carcinogenic risk level is 10^{-4} (i.e., 1 in 10,000). These risk levels are generally conservative when compared to the likelihood of developing cancer (USEPA, 2005). Four chronic exposure risk charts (CERC) were developed using the standard USEPA protocol to calculate arsenic risk based on arsenic concentration and speciation (Figure 5.9). These CERC visualize LCR and HQ for water ingestion pathways based on adult exposure (Markley and Herbert, 2009). The calculated risk in the CERC associated with chronic arsenic exposure includes the acceptable carcinogenic risk level $<10^{-6}$ (i.e., less than 1 in 1,000,000) and noncarcinogenic HQ is less than 1 (i.e., HQ is <1). In addition, related to "de minimis risk" is the concept of "de manifestis risk," or "obnoxious risk." With "de manifestis risk," there is a risk level above which mitigation is mandatory. In practice, this level is generally set at 1 in 10,000 per vulnerable individual (Coppola, 2011: 172).

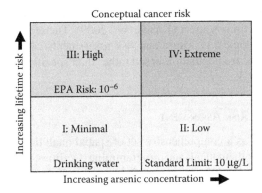

FIGURE 5.9 Theoretical categorization of chronic arsenic risk. (After Markley, C. T. and B. E. Herbert. 2009. *Water, Air, & Soil Pollution* 204:385–398.)

The LCR associated with chronic arsenic exposure in drinking water can be divided into four categories based on the calculated risk and arsenic concentrations (Figure 5.9). The first category for LCR represents the "minimal risk" and occurs where arsenic concentrations and associated risk are below the USEPA limits, while the fourth category is the "extreme risk" and occurs where both the arsenic concentration and the associated risk are above the USEPA limits. The third category is the "high risk" and occurs where arsenic concentrations are below drinking water standards, but associated risk is above the USEPA limit. The second category is considered to be the "low risk" and occurs where arsenic concentrations are above drinking water standards, but the associated risk is below the USEPA limit.

5.5 HUMAN EXPOSURE TO ARSENIC AND SPATIAL RISK

5.5.1 GIS AND HUMAN EXPOSURE

There has been an increased interest in GIS modelling for spatial risk assessment. Exposure of arsenic in the environment is inherently spatial, with real world coordinates. GIS can provide a methodological framework for exposure analysis and risk assessment since it has the ability to develop predictive models for the future consequences of a chemical contaminant in the environment. GIS can integrate different and diversified sources of relevant data for spatial risk assessment with the evaluation of human exposure in a quick and precise manner. The paradoxical nature of arsenic concentrations in groundwater requires an effective tool to compile information from different sources for risk characterization. ERA provides a qualitative and quantitative description of the exposure pathway by which a CoC travels from source to receptor, along with the interactions associated with the chemical and the transport media (Wilson, 1997).

Biomarkers have a great potential in epidemiological and toxicological research for assessing risk posed by exposure to chemicals (Caussy et al., 2003; Harrison and Holmes, 2006; Robson, 2003). There is a significant association between arsenic in drinking water and arsenic concentrations in urine, hair, and toenails. Adair et al. (2006) suggest that toenails are a better biomarker of arsenic exposure than urine because arsenic contents in toenails have been consistent among samples collected years apart. In other words, toenails provide a long-term integrator of arsenic exposure which cannot be said for urine. Exposure to arsenic in drinking water has produced dermatological disease in the arsenic-contaminated districts of Bangladesh, and therefore arsenic-induced skin lesions may also be considered a long-term biomarker of cumulative arsenic exposure (Khan et al., 2009). In GIS, the magnitudes of hazards are identifiable by looking at the spatial distribution of contaminant levels. The exposure or dose that an individual receives depends on the hazard present at the specific locations where most time is spent (Khan et al., 2009). Therefore, ERA has both spatial and time variability. Epidemiological exposure and risk assessments can allow for uncertainty since the exposure measurements are population-based rather than specific to individual people (Tristán et al., 2000).

5.5.2 GIS AND SPATIAL RISK ASSESSMENT

Gesler (1986) reviews GIS as a comprehensive set of spatial analytical tool used in analyzing risk patterns because of its mathematical and programming facilities. GIS is an automated spatial decision-making system used in the mapping of geographically referenced information. GIS has been widely used in risk assessment (Hiscock et al., 1995; Lovertt et al., 1997; Vine et al., 1997), and it has successfully been linked with risk assessment on a number of occasions. Wadge et al. (1993) used GIS for natural hazard assessment, and similar applications and approaches have been employed to identify indices for ecological risk and overlay analysis of groundwater vulnerability studies (Emmi and Horton, 1996; Engel et al., 1996). Risk mapping of potential pollution has also been accomplished by combining GIS, remote sensing (RS), and the DRASTIC model (Al-Adamat et al., 2003).

A GIS can be used as a powerful tool in assessing environmental risk (Kim et al., 2004; Morra et al., 2006) with its mathematical and programming facilities and spatial overlay capabilities that allow different map data to be combined in determining different "problem regions" of arsenic. A problem region represents a natural domain for the cause of a given disease and its spread over the area (Haining, 1996). The conventional ERA approaches for human health risk assessment have several spatial limitations that could be overcome with GIS (Tristán et al., 2000). GIS has been widely used to visualize, integrate, and analyze spatial data pertinent to evaluating changes in environmental ecological systems (Sydelko et al., 2000; Zandbergen, 1998).

GIS has the capacity to map arsenic concentration levels in groundwater, the location of tubewells based on safe and unsafe sources of drinking water, the distribution of arsenic-affected populations, and spatial relationships between arsenic exposure and the human health hazard. It is evident from recent studies that the present trend in arsenic risk assessments of human exposure is to integrate physical, socioeconomic, and biomarker samples into a GIS format database. Moreover, geostatistics can be used to evaluate and characterize risk, and the resultant risk is presented as a function of probability. A few attempts have been made to develop and incorporate human health risk assessment as a tool in GIS to assess and quantify risk (Bien et al., 2004; Chen et al., 2004b; Gay and Korre, 2006; Goovaerts et al., 2005; Hassan et al., 2003; Morra et al., 2006; Nguyen et al., 2009).

The most common static approach in GIS to characterize spatial risks is based on the overlay operation. The static risk modelling approach is used due to limited information being available on the dynamic and temporal variables of contaminant datasets (Khan et al., 2009). One application of GIS in the field of exposure and risk was made by Sabel et al. (2000) in which they modelled space time patterns of motor neuron risk using the kernel estimation calculation technique in the GRID component of the ESRI Arc/Info software. In analyzing GIS-based risk assessment of public water supply intake, Foster and McDonald (2000) looked at the applicability of various GIS techniques to identify and evaluate risk. In particular, they found that the GIS overlay technique is suitable for hazard identification, probabilistic techniques in GIS are capable of modelling risk, and GIS overall is suitable for the derivation of monitoring strategies.

Geostatistics have been used to investigate and map the pattern of spatial arsenic contamination. Based on a stochastic model, the geostatistical approach relies on both statistical and mathematical methods to create surfaces and to assess the uncertainty of predictions for regionalized variables (Bastante et al., 2008; Ghosh and Parial, 2014; Jalali et al., 2016; Liu et al., 2006; Uyan, 2016). Geostatistics represents one of the most powerful procedures for producing contour maps for regionalized variables and, thereby, indicates an appropriate method of prediction (Oliver and Khayrat, 2001). Gay and Korre (2006) presented a methodology to assess human health risk with the combination of quantitative probabilistic and spatial statistical methods from exposure to contaminated land.

Interpolation is the process of estimating the spatial arsenic concentrations at unsampled points from a surrounding set of measurements. When the local variance of sample values is controlled by the relative spatial distribution of these samples, geostatistics can be used for spatial interpolation and point interpolation is significant in GIS operation (Cinnirella et al., 2005). Yu et al. (2003) applied geostatistical methods to create arsenic concentration maps of Bangladesh by dividing the country into regions based on concentration and estimated the vertical arsenic concentration trends in these regions. Zhang et al. (2001b) used GIS methodology to identify the arsenic distribution pattern in Thailand. GIS interpolation was used to estimate that 72.3% of their study site had arsenic levels higher than 50 μg/L in shallow groundwater in Thailand.

Serre et al. (2003) suggested that a GIS-based holistochastic framework could be used to identify human health exposure due to arsenic contamination, based on Bayesian maximum entropy theory. Hassan et al. (2003) used GIS-based data processing to analyze and map groundwater arsenic concentrations and to identify risk zones in southwest Bangladesh. They used multiple methods combining spatial and attribute information. The pattern of arsenic concentration and its spatial variability was identified using the spatial kriging interpolation method. GIS has been identified by Foster and McDonald (2000) as a useful computing technique for pollution risk based on spatial and attribute data.

5.6 GROUNDWATER ARSENIC DOSE AND RISK RESPONSE

There is a clear dose-response relationship between the arsenic level in drinking water and the rate of skin lesions (Guo et al., 2001; Mazumder et al., 1998a; Sarma, 2016; Wei et al., 2017). But some researchers have reported contradictory results regarding the dose-response relationship between arsenic concentration in drinking water and cancer risks (Cheng et al., 2017; Lin et al., 2013; Karagas et al., 2015; Medeiros and Gandolfi, 2016; Sherwood and Lantz, 2016). It has not been identified unequivocally that the consumption of arsenic in drinking water at the current maximum contamination level (approximately 1.0 μg/kg/day) would cause cancer, although the WHO claims that a total daily intake of 2.0 μg/kg/BW of inorganic arsenic by humans may cause skin lesions within a few years (Khan et al., 2009). The maximum contaminant level in different countries for arsenic in drinking water (50 μg/L) is not safe for public health protection and therefore requires further analysis to assess the response to human health from the exposure of low levels of arsenic. It is reported that chronic exposure to inorganic arsenic in drinking water can be the cause of cancer and can increase risk even at very low exposures (WHO, 2001). Furthermore, death due to chronic arsenic exposure has been reported at lower concentrations (Zaldívar et al., 1981). Table 5.2 shows the level of arsenic exposure and its impact on human health.

5.6.1 Noncancerous Responses

Data from Taiwanese arsenic poisoning from drinking water and from studies of other populations reveal that there is a dose-response relationship for ingested drinking water arsenic and several noncancer toxic effects. Dose-response relationships between drinking water arsenic exposure and the dermatological manifestations in exposed populations are well-documented. Chronic exposure to low levels of arsenic causes different skin lesions in the form of melanosis, leuco-melanosis, and keratosis (Chakraborti et al., 2017a; Foster et al., 2002; Kile et al., 2011; Mazumder et al., 1998a, 2010; Nriagu et al., 2012; Saha, 2003; Sarma, 2016). The skin is quite sensitive to arsenic, and skin lesions (hyperkeratosis and pigmentation) have been observed even at exposure levels in the range of 5–10 μg/L arsenic in drinking water (Yoshida et al., 2004). Diffuse or spotted hyper-pigmentation

TABLE 5.2
Dose-Response Relationship of Arsenic Poisoning

References	Location	Dose Range (μg/L)	Frequency	Health Impact	Effects Prevalence	Remarks
Astolfi et al. (1981)	Argentina	>100	Daily	Skin cancer	–	Recognizable signs of skin cancer in some cases.
Baastrup et al. (2008)	Denmark	0.05–25.3	Cohort	Skin cancer	IRR: 0.99	Recognizable signs for skin cancer.
Chen et al. (1986)	BFD-endemic area in Taiwan	–	Ecological	Liver cancer	OR: 2.0	65 liver cancer patients were diagnosed.
Chen et al. (2004a)	Taiwan	≥700 vs. <10	Cohort	Lung cancer	RR: 3.29	Positive dose-response relationship.
Chen et al. (2010b)	Taiwan	≥10 vs. <100	Cohort	Bladder cancer	RR: 2.19	Significant association with bladder cancer risk.
Cheng et al. (2017)	Taiwan	≥50 μg/L	Case-control	Kidney cancer	OR: 1.22	Higher risk of chronic kidney disease.
Fernández et al. (2015)	Chile	–	Ecological	Bladder cancer	SMR	Bladder cancer among arsenic-exposed population.
Gilbert-Diamond et al. (2013)	USA (New Hampshire)	–	Case-control	Skin cancer (SCC)	OR: 1.37	Some 470 invasive SCC cases vs. 447 controls.
Guo and Tseng (2000)	Taiwan	>640	Daily	Bladder cancer	–	Significant association with bladder cancer risk.
Huang et al. (2012)	Taiwan	–	Case-control	Kidney cancer	OR: 0.92	Poor dose-response relationship was observed.
Lamm et al. (2004)	133 US Counties	3–60	Ecological	Bladder cancer mortality	SMR: 0.73	No dose-response relationship was observed.
Leonardi et al. (2012)	Hungary, Romania, and Slovakia	0.7–13.8	Case-control	Skin cancer	OR: 1.13	529 BCC vs. 540 controls.
Lewis et al. (1999)	Millard County, Utah	3.5–620	Ecological	Malignant cancer	SMR: 0.82	No dose-response relationship.
López-Carrillo et al. (2014)	Mexico	%MMA: <7.01 vs. >13.30	Case-control	Breast cancer	OR: 2.63	Significant association with breast cancer risk.
Melak et al. (2014)	Chile	–	Case-control	Lung and Bladder cancer	OR: 6.81 (Lung) & 6.96 (Bladder)	Significant association with lung and bladder cancer risk with arsenic exposure.
Moore et al. (2002)	Churchill County, Nevada	10–90	Ecological	Childhood cancer	SMR: 1.37	No evidence of childhood cancer.
Smith et al. (2000)	Chile	750–800	Daily	Skin lesion	–	Male and female are equally affected.
Steinmaus et al. (2003)	Nevada and California	0–1000	Daily	Bladder cancer mortality	SMR: 0.73	No association between bladder cancer.
Steinmaus et al. (2010)	Argentina	4.8–112.3	Case-control	Lung cancer	OR: 1.32	Significant association with lung cancer risk.
Watanabe (2001)	Bangladesh	1–535	Daily	Skin lesion	–	Males are more susceptible than females.

Note: IRR: Incidence Rate Ratio; OR: Odds Ratio; RR: Risk Ratio (Relative Risk); SMR: Standardized Mortality Rate.

may be seen after 6 months to 3 years by chronic ingestion of high doses of arsenic (40 µg/kg/day or higher) or 5–15 years of ingestion of low doses of the order of 10 µg/kg/day or higher (NRC, 2001).

By the dose-response relationship, as the arsenic exposure increases, both the frequency and the severity of toxic effects increase in the exposed population. The value of LOAEL in this aspect is the key determinant for the dose-response relationship. The LOAEL figures between 10 µg/kg/day and 18 µg/kg/day can lead to dermatological manifestations (Abernathy et al., 1999; Chakraborty and Saha, 1987). The levels of arsenic that most people ingest in food are not usually considered to be a health concern. The characteristic arsenical skin lesions may involve a latency period (the time from first exposure to manifestation of disease) of about 8 years (Brown et al., 1989), 10 years (Mazumder et al., 1998b), or 5–10 years (Tondel et al., 1999) depending on arsenic dose content and immunity level (Smith et al., 1992). There are some instances of patients with skin lesions in West Bengal (India), Taiwan, and Chile who were drinking water containing very low concentrations of arsenic (Chakraborty and Saha, 1987; Lu, 1990; Mazumder et al., 1998a; Smith et al., 1992).

Chronic arsenic exposure causes nonmalignant health effects, such as peripheral neuropathy, cardiovascular diseases, ischemic heart disease, hypertensive heart disease, lung function, and bronchitis (Abernathy et al., 1999; Ahmed et al., 2017; Chiou et al., 1997; Engel and Smith, 1994; Hei and Filipic, 2004; Lewis et al., 1999). It is not clear from the literature how much ingestion of arsenic causes which types of skin lesions. If arsenic builds up to higher toxic levels, organ cancers, neural disorders, and organ damage, often fatal, can result. A risk of mortality from hypertension and cardiovascular disease has also been associated with chronic exposure to arsenic (Yoshida et al., 2004). An association has been reported between chronic ingestion of inorganic arsenic and an increased risk of diabetes mellitus (Lai et al., 1994; Tondel et al., 1999). Some evidence suggests that the ingestion of arsenic can have effects on the immune and respiratory systems. Young children, the elderly, unborn babies, and people with long-term illness are at greatest risk of arsenic poisoning (Mondal et al., 2006).

5.6.2 Cancerous Responses

The cancer potency factor for ingesting arsenic is based on extensive studies from Taiwan, where the water supply naturally contains high levels of inorganic arsenic (Chen et al., 1992; Wu et al., 2001). Cancer risks from inorganic arsenicals in drinking water have been proven and reported. A few years of continued exposure to low levels of inorganic arsenicals cause different skin lesions, and after a latency period of 20–30 years, internal cancers, particularly of the bladder and lung, can appear (Byrd et al., 1996). The most common malignancies found in patients with long-term exposure to arsenic are Bowen's disease (BD) (intraepithelial carcinoma or carcinoma *in situ*), BCC, and SCC (Boonchai et al., 2000; Hunt et al., 2014).

Arsenic-induced BD can appear 10 years after arsenic exposure, while other types of skin cancer can have a latency period of 20 or 30 years (Yoshida et al., 2004). A case-control study from Hungary, Romania, and Slovakia showed positive associations between long-term exposure to inorganic arsenic <100 µg/L in drinking water and BCC of the skin (Leonardi et al., 2012).

In an analysis of the relative risks for lung and bladder cancer with the original data from southwestern Taiwan, Morales et al. (2000) showed that, although the shape of the exposure-response curve is uncertain at low levels of arsenic exposure over a lifetime, one out of every 100–300 people who consume drinking water containing 50 µg/L of arsenic may suffer an arsenic-related cancer (lung, bladder, or liver cancer) death. Foster et al. (2002) pointed out that the lifetime risk of death is 1 in 100 from consuming 50 µg/L and 1 in 50 from consuming 100 µg/L arsenic in drinking water. In a study of the risk of bladder and kidney cancer in Finland in a cohort of people who had been using arsenic-contaminated drinking water over a period of 13 years (1967–1980), Kurttio et al. (1999) found an increased risk of bladder cancer with increased arsenic intake during the third to ninth year prior to the cancer diagnosis, which reached statistical significance in the high-dose group.

Guo and Tseng (2000) detected a statistically significant association between high concentrations of arsenic in drinking water (>640 µg/L) and occurrence of increased bladder cancer risk, but they did not find any association of arsenic exposure with lower concentrations of arsenic in drinking water. Analyzing the cohort with urine samples and arsenic exposure from the BFD-endemic area and 193 subjects with all-site cancers deaths, Chung et al. (2013b) ascertained for 22% deaths from bladder cancer mortality hazard ratios (HR) of 3.53 (95% CI: 1.16, 10.77). Using 50 µg/L of arsenic as the cutoff in drinking water in Taiwan, Cheng et al. (2017) found a positive association between inorganic arsenic and the progress of chronic kidney disease. Examining information from 179 bladder cancer patients from Chile having 42 exposed to arsenic during 1958–1971 in Antofagasta and 137 in Santiago with no exposure to arsenic, Fernández et al. (2015) investigated significantly higher bladder cancers among the arsenic-exposed population.

Chronic arsenic exposure can induce tumorigenesis in humans in skin (Gentry et al., 2014; Surdu, 2014), lung (Celik et al., 2008), bladder (Radosavljević and Jakovljević, 2008), liver (Wang et al., 2014b), and prostate (Benbrahim-Tallaa and Waalkes, 2008). Brown et al. (1989) calculated the lifetime risk of skin cancer to be 1.3/1000 for males and 0.6/1000 for females per microgram of arsenic per day (µg/kg/day), while the lifetime risk of dying from cancer of the liver, lung, kidney, or bladder from drinking 1.0 Liter/day can be as high as 13 (Smith et al., 1992) or 100 (Smith et al., 2000b) per 1000 persons at the standard of 50 µg/L and 500 µg/L of arsenic, respectively. Thomas et al. (2001) pointed out the dose-response relationships for chronic exposure to arsenic as a toxin and a carcinogen, while Englyst et al. (2001) found a correlation between lung cancer risk and exposure to inorganic arsenic. Charlet et al. (2001) and Calderon (2000) also identified a potential health risk of chronic exposure to drinking well water arsenic. A significant dose-response relationship and age-adjusted mortality rates from cancers in southwestern Taiwan were reported (Wu et al., 2001). The maximum likelihood estimate (MLE) of skin cancer risk for a 70 kg/person drinking 2 L/day of water ranged from 1E-3 (0.001) to 2E-3 (0.002) for an arsenic intake of 1 µg/kg/day, and, expressed as a single value, the cancer unit risk for drinking water is 5E-5 (0.00005) per (µg/L) (USEPA, 2002).

Bates et al. (2004) examined the relationship between arsenic in drinking water and bladder cancer in a population in Argentina and analyzed multiple exposure scenarios and possible latency periods. They did not find an association between arsenic intake and cancer even at levels as high as 200 µg/L in drinking water. Astolfi et al. (1981) pointed out from their study from the Argentine case that the regular intake of drinking water containing >100 µg/L of arsenic leads to clearly recognizable signs of arsenic toxicity and ultimately in some cases to skin cancer. Tsuda et al. (1995) claim that exposure to 5 years of high dose of arsenic (>100 µg/L) can cause skin signs of chronic arsenicism for subsequent cancer development. A study from Finland found that people who regularly drank >5 µg/L of arsenic had more than a 140% increase in bladder cancer rates compared to those who consumed levels of less than 1 µg/L (Kurttio et al., 1999). The USEPA (2001a) has calculated that lifelong ingestion of 1 µg/kg/day (around 50 to 100 µg/day in an adult) is associated with a risk of skin cancer of about 0.1% (1/1000), and this dose level is comparable to drinking water containing <50 µg/L for a lifetime. On the contrary, Buchet and Lison (1998) concluded that a low to moderate level of environmental exposure to inorganic arsenic (20 to 50 µg/L) from drinking water does not have any dose-response relationship for arsenic and cancer.

Lewis et al. (1999) in their ecological study of an arsenic-exposed population in Utah, USA, looked for increased cancer and noncancer deaths with drinking water arsenic concentrations ranging between 3.5 and 620 µg/L, averaging 100 µg/L. Selecting the members of the Mormon Church who would be expected to have relatively low exposures to tobacco and alcohol, the study found no relationship between exposure to arsenic-contaminated drinking water and bladder and lung cancer. Analyzing some 75 million person-years of bladder cancer mortality data for the years 1950–1979 from 133 US counties with median groundwater arsenic concentrations ranging from 3 to 60 µg/L, Lamm et al. (2004) did not find any evidence of a dose-response relationship between arsenic intake and bladder cancer. In a population-based case-control study including 587 diagnosed BCC patients and 284 SCC cases in New Hampshire from 1993 to 1996, Karagas et al. (2001) suggested an association between toenail arsenic content and both types of skin cancer, but the relationship was not statistically significant, even for the highest exposure category (0.35–0.81 µg/g toenail arsenic content).

A number of researchers have shown that there is no dose-response relationship between chronic arsenic ingestion and cancer incidence. Bates et al. (1995) found no association between arsenic levels in water (range: 0.5–160 µg/L; and average: 5 µg/L) and bladder cancer, while a similar study conducted by Steinmaus et al. (2003) in Nevada and California, where arsenic drinking water concentrations ranged from 0 to 1000 µg/L, confirmed no dose-response relationship. Analyzing drinking water arsenic concentrations as high as 33 µg/L in Lane County, Oregon, Morton et al. (1976) found no evidence of increased skin cancer. Similarly, examining average arsenic levels in drinking water >5 µg/L (mean range: 5.4–91.5 µg/L) in 30 US counties, Engel and Smith (1994) found no association between arsenic levels and death due to malignant neoplasms.

Moore et al. (2002) investigated the relationship between childhood cancer incidence and arsenic exposure in drinking water in Churchill County, Nevada, from 1979 to 1989. Over 328,000 Nevada children were grouped into low, medium, and high exposure categories (i.e., >10, 10–25, and 35–90 µg/L, respectively). No statistically significant association between arsenic and any type of childhood cancer was found in any of the exposure groups, nor was any specific association with leukemia observed. It is noted that leukemia has not been associated with arsenic exposure, even in Taiwan, where exposure to arsenic was considerably higher. Moreover, it was reported in studies of the USA in the 1970s (Goldsmith et al., 1972; Harrington et al., 1978) that no clinical or hematological abnormalities were observed in the exposed population, despite the presence of higher arsenic concentrations in the groundwater (i.e., >50 µg/L).

5.7 GROUNDWATER ARSENIC EXPOSURE AND RISK: A BANGLADESH CASE STUDY

5.7.1 Epidemiological Risk

What are risk and uncertainty and how can we characterize them? The epidemiological modelling can be used to assess the risk characterization. It is a technique to utilize the entire range of input data to develop a probability distribution of risk rather than a single point value. Risk characterization combines information on exposure and toxicity to estimate the type and magnitude of arsenic risk faced by the exposed population. Combining the evaluation of arsenic toxicity with estimates of how much people are exposed to leads to an assessment of the risk pattern. In order to calculate both the carcinogenic and noncarcinogenic health risk, the assumption was made that residents in the study site are dependent on groundwater as their main source of water for drinking purposes. How much arsenic is an individual or population exposed to from groundwater source? The answer to this question denotes the exposure assessment, which depends upon (a) how much arsenic is present in the groundwater, (b) how long people have been exposed to arsenic, (c) whether arsenic exposure is continuous or intermittent, (d) whether there is an alternative source of safe drinking water, (e) how the people are exposed, and (f) how many people were exposed to arsenic from each tubewell.

What kind of health impacts are posed by arsenic? Arsenic poisoning occurs as a result of the consumption of drinking water that naturally contains a high amount of inorganic arsenic. Arsenic exposure for humans occurs via ingestion, inhalation, and dermal contact. The exposure pathways describe the means by which a receptor is exposed to a contaminant or chemical of concern (Mauro et al., 2000). The most deceptive and dangerous aspect of arsenic toxicity is its very slow and insidious development. There have been scientific discussions and debates about the evaluation of potential health risks associated with groundwater arsenic exposure at different exposure levels, particularly at low exposure levels (Schoen et al., 2004).

The main objective of health risk assessment is to estimate to what extent the health of people would be at risk through arsenic-contaminated drinking water. The health risk from both noncarcinogenic and carcinogenic exposure are documented from the intake of arsenic from drinking water. The USEPA has established a method to evaluate carcinogenic and noncarcinogenic effects related to regular consumption of arsenic (USEPA, 2001c). Four important parameters need to be determined for this health risk assessment: (a) the average daily dose (ADD), (b) the chronic daily intake (CDI), (c) the lifetime cancer risk (LCR), and (d) the hazard quotient (HQ). Health risk for carcinogenic exposure is evaluated based on the LCR index, which is expressed as the excess probability of contracting cancer over a lifetime of 70 years, and the health risk for noncarcinogenic exposure is evaluated based on the HQ index, which is defined as the ratio of the potential exposure to a level at which no adverse effects are expected. A value of LCR between 10^{-6} and 10^{-4} indicates that the carcinogenic risk is acceptable, and an HQ less than 1 indicates no significant risk of noncarcinogenic effects. According to the USEPA, for CDI calculations, the standard body weights for children (10 kg) and adults (70 kg) are fixed, but these standard weights may not apply to all populations (Singh and Ghosh, 2012). Researchers may need to derive a different body weight for their surveyed populations, for example, according to the biological factor of the Vietnamese population, an adult at the age of 26 years old has a weight of 55 kg (body weight) and an average lifetime of 50 years. The daily water consumption rate is 2.0 L/day (Nguyen et al., 2009; Pham et al., 2017).

Relevant information: The necessary information for health risk assessment was collected from Ghona *union* (the fourth order local government administrative unit in Bangladesh) of Satkhira district in southwest Bangladesh along the border with India. The area is 17.26 sq. km in area with a population density of 1026 per sq. km in 2011. A census survey for all the tubewells was conducted for this study. A minimum detection limit (MDL), the Bangladesh standard permissible limit of 50 µg/L, and the WHO permissible limit of 10 µg/L were considered for risk characterization. The collected samples were analyzed with the Flow Injection Hydride Generation Atomic Absorption Spectrometry (FI-HG-AAS) method by the School of Environmental Studies (SOES), Jadavpur University, Kolkata, India. In order to analyze the "risk factor" and "risk characterization," a dose-response analysis was conducted on the medical diagnoses of 11 patients with different confirmed levels of arsenicosis. Early symptoms of arsenic poisoning such as melanosis and keratosis were found in local residents, suggesting that the use of arsenic-contaminated groundwater results in potential health problems for local communities. No patients were identified as having cancer symptoms.

Arsenic concentrations in the untreated groundwater were found to be uneven over space, ranging between <3 and 600 µg/L, with a mean concentration of 238 µg/L and a standard deviation of 117 µg/L. The daily water ingestion rate (IR) was estimated at 3.0 L/day (2.0 L/day for drinking and 1.0 L/day for cooking), and it seems that the exposure duration (ED) of local people to this groundwater is a total of 30 years. Since there is no alternative drinking water option/source, they have been continuously using tubewell water. The study area is very badly affected with arsenic, with water from 99% (371 out of 375) of tubewells contaminated at the WHO standard (10 µg/L) and about 96.50% (358 out of 375) of tubewells contaminated at the Bangladesh standard daily permissible limit (50 µg/L). The mean arsenic concentration in the study area is five times higher than the Bangladesh standard limit and 25 times higher than the WHO permissible limit. The patterns of concentrations vary considerably and unpredictably and are therefore high over distances of just a few meters; here, about 46% of tubewells are located within 25 meters of another well within the settlement area of the study site (Hassan et al., 2003). According to the biology of the Bangladeshi people, an adult at the age of 25 years old has a body weight of about 60 kg and an average lifetime of 60 years (21,900 days). This information was collected directly from field visits (Table 5.3).

Assessing arsenic exposure: An exposure assessment is stated in terms of the likelihood that people are exposed to a given level of arsenic over a specified period. In the study, the average daily dose (ADD) of arsenic for different arsenic concentration categories and the calculated ADD values showing different exposure levels were calculated. At the WHO permissible limit, the ADD value was calculated at 0.099 µg/day, and for the DoE standard limit, the value was calculated at 0.498 µg/day. The average ADD for the whole study site is 5.95 µg/day (Table 5.4).

TABLE 5.3
Relevant Parameters Used in Groundwater Arsenic-Induced Health Risk Model

Parameter	Unit	Parameter Characteristics	Issues
ED (exposure duration)	(year)	30	ED is the exposure frequency for 365 days/year over 30 years (EF*ED = 10,950 days).
EF (exposure frequency)	(day/year)	350	EF can be considered how many days an individual is exposed to arsenic in a year.
IR (daily water intake rate)	(L/day)	3.0	2.0 L/day for drinking and 1.0 L/day for cooking.
BW (body weight)	(kg)	60	The standard BW for children (10 kg) and adults (70 kg) are fixed, but these standards may not apply to all populations.
AT (average time for carcinogenic exposure)	(days)	60 × 365 (21,900 days)	AT can be considered for 60 years for the exposure of carcinogenic risk, but this figure can be different in different countries.
RfD (reference dose)	(mg/kg/day)	3×10^{-4} (0.304)	USEPA (2002) developed the RfD value for noncarcinogenic health risk assessment.
CSF (cancer slope factor)	(mg/kg/day)	1.5	Markley and Herbert (2009) and IARC (2004) introduced the CSF value for carcinogenic health risk assessment.
CW (arsenic concentrations in the groundwater)	(μg/L)	<3.0–600 (Field data)	Different levels of arsenic concentrations from drinking water.

TABLE 5.4
Estimation of Lifetime Cancer Risk

Arsenic Concentration (μg/L)	Average Arsenic Concentration (μg/L)	Years of Arsenic Exposure	ADD (μg/day)	HQ	LCR (Probability)
<10	7	17	0.099	0.325	1.5E-4 (1 in 6734)
10–50	26	23	0.498	1.638	7.5E-4 (1 in 1339)
50–100	80	20	1.333	4.384	2E-3 (1 in 500)
100–300	198	18	2.97	9.769	3E-3 (1 in 339)
>300	366	25	7.625	25.082	115E-2 (1 in 87)
Total Average (<3–600)	238	30	5.95	19.572	8.9E-3 (1 in 112)

Source: RfD value (USEPA. 2002. *A Review of the Reference Dose and Reference Concentration Processes.* U.S. Environmental Protection Agency, Washington DC. [EPA/630/P-02/002F: December 2002]); and SF value (Markley, C. T. and B. E. Herbert. 2009. *Water, Air, & Soil Pollution* 204:385–398; USEPA. 2002. *A Review of the Reference Dose and Reference Concentration Processes.* U.S. Environmental Protection Agency, Washington DC. [EPA/630/P-02/002F: December 2002].)

Note: ADD: Average Daily Dose; HQ: Hazard Quotient; LCR: Lifetime Cancer Risk; RfD: Reference Dose; and SF: Cancer Slope Factor.

ADD (Total Average): $\frac{(238\,\mu g/L * 3L/day) \times (365\,days * 30\,years)}{(21900\,days * 60\,kg)} = 5.95\,\mu g/day$;

HQ (Total Average): $\frac{ADD}{RfD} = \frac{5.95}{0.304} = 19.572$;

LCR (Total Average): $1 - \exp(-(SF \times ADD)) = 1 - \exp(-(1.5 \times 5.95)) = 8.9E - 3 = 1\,in\,112$;

RfD Value: 0.304 μg/kg/day; and

SF Value: 1.5 mg/kg/day.

If the groundwater that is used for drinking and cooking purposes is found to have a mean arsenic content of 238 µg/L, a person having a 60 kg body weight who ingests 3.0 L/day of water for about 30 years in his 60 years of lifetime will have an exposure of 5.95 µg/day from this source. This figure is constantly changing due to changing exposure habits and amounts as well as changing concentrations of arsenic. Therefore, an exposure assessment is stated in terms of likelihood. In considering the toxic risk for noncarcinogenic exposure, it can be assessed with an HQ. Based on the HQ results, no adverse health effects are expected at the WHO permissible limit as a result of exposure since the HQ value is <1, while at the Bangladesh standard limit, the calculated HQ value is >1 and there is a probability of health effects. In the contamination categories (50–100 µg/L; 100–300 µg/L; and >300 µg/L), the calculated HQ values show the possibility of adverse health effects.

If arsenic builds up to higher toxic levels, organ cancers, neural disorders, and organ damage, often fatal, can result. In the most severe cases, cancer can occur in the skin and internal organs, and limbs can be affected by gangrene. A few years of continued exposure to low levels of inorganic arsenicals cause different skin lesions, and after about 10–15 years these turn into skin cancers (Byrd et al., 1996). Two patients were found with skin lesions, in particular, melanosis, which was the first visible symptom, who had been ingesting arsenic <50 µg/L for around 15–25 years. Two patients were also identified affected by skin lesions, with the symptoms of melanosis and keratosis, who had been ingesting arsenic between 50 µg/L and 100 µg/L for 20 years and two patients with keratosis and hyperkeratosis who had been ingesting arsenic between 100 µg/L and 300 µg/L for nearly 20 years. In addition, some five patients were found to be affected with hyperkeratosis, hyperpigmentation, and gangrene having ingested arsenic at more than 300 µg/L for 25 years (Table 5.5). One patient was also found with severe skin lesions, in particular hyperkeratosis, who had been consuming water containing arsenic at 446 µg/L for around 18 years and another with gangrene due to the impact of arsenic. This patient had been ingesting arsenic at 353 µg/L for about 26 years. During the field survey, advice was provided to the local people to use arsenic-safe water for their drinking and cooking purposes, which was available in only 17 tubewells as per the Bangladesh standard permissible limit.

Risk characterization: It is clear from this discussion that combining the uncertainties of toxicity assessment with the exposure assessment will lead to an overall risk assessment with greater uncertainty than that associated with either the toxicity or the exposure estimates. Thus, it is not possible to describe the pattern of exact risk, but it is possible to assess how high and how low it could possibly be. Despite the considerable uncertainties in the underlying data, the risks are "sobering" (Morales et al., 2000). The estimation of environmental health risk with uncertainties is described within a range of probabilities and should be seen as a "best guess" rather than an irrefutable statement of fact.

TABLE 5.5
Estimation of Risk Ratio for Arsenic Poisoning

Arsenic Concentration (µg/L)	Average Arsenic Concentration (µg/L)	Average Exposure Years	Identified Patients	Cumulative Frequency	Risk Ratio	Diagnosed Symptoms
<10	7	17	–	–	–	–
10–50	26	23	2	2	2/1 = 2	Melanosis
50–100	80	20	2	4	4/1 = 4	Melanosis, Keratosis
100–300	198	18	2	6	6/1 = 6	Keratosis, Hyperkeratosis
>300	366	25	5	11	11/1 = 11	Hyperkeratosis, Hyperpigmentation, Gangrene

1. The study shows that low exposures to inorganic arsenic in drinking water can be the cause of arsenicosis symptoms and can increase the health risk if the dose level contains <50 µg/L for a lifetime. In such cases, the risk of melanosis could be about 0.1% (1/1000). In considering the cancer risk, it is estimated that the LCR is 1.5E-4, i.e., 1 out of every 6734 people who consume drinking water containing <10 µg/L arsenic may suffer an arsenic-related cancer (Table 5.4).
2. It was calculated that the lifetime risk of death from cancer is 7.5E-4 (i.e., 1 in 1339) from consuming 10–50 µg/L of arsenic in drinking water. This figure increases to 2E-3 (1 in 500) if people consume arsenic at a rate between 50 and 100 µg/L. The research also shows that the lifetime risk for carcinogenic death would be about 3E-3 (i.e., 1 in 339) if people ingest arsenic-contaminated water at 100–300 µg/L. The figure will be about 115E-2 (i.e., 1 in 87) for ingesting water containing arsenic at >300 µg/L. It was estimated that the LCR of 8.9E-3 (1 in 112) is possible if the people of the study area drink water with an arsenic concentration of 238 µg/L; accordingly, there is a lifetime chance of about 119 people dying with cancer (Table 5.4).
3. The risk ratio, found by comparing the occurrence of arsenicosis symptoms with different toxic levels of arsenic, can be described as a process of estimating the environmental health risk from arsenic. A risk ratio close to 1 or <1 suggests that there is no health effect from arsenic, and a risk ratio of >1 suggests the increasing trend of risk of arsenicosis. It has been calculated that people who ingest arsenic between 10 and 50 µg/L daily are twice as likely to get arsenicosis symptoms as people who ingest at the safe level (<10 µg/L). Those who were ingesting arsenic at 50–100 µg/L daily were four times as likely to get arsenicosis symptoms, six times at between 100, and 300 µg/L and eleven times at >300 µg/L (Table 5.5). It was identified from fieldwork that a patient was affected with arsenicosis symptoms after an exposure level of 10 µg/L for 17 years, and this figure shows a chance of having health problems even at very low exposures.

There are a number of studies of arsenic concentrations in drinking water and arsenicosis patients that show the dose-response relationships between arsenic concentration in drinking water and the dermatological manifestations in exposed populations. The risk characterizations of these studies are supported by a number of studies that identify LOAEL values of 18 mg/kg/day (>200 µg/L) (Chakraborty and Saha, 1987), of 19 mg/kg/day (Hindmarsh et al., 1977), and of 14 mg/kg/day (Abernathy et al., 1997). The lifetime risk of skin cancer from arsenic has been calculated to be 1.3/1000 (1 in 769) for males and 0.6/1000 (1 in 1667) for females per microgram of arsenic per day (Brown et al., 1989). In analyzing arsenic data from a study in an arsenicosis-endemic area of Taiwan, Morales et al. (2000) concluded that, although the shape of the exposure-response curve is uncertain at low levels of arsenic exposure, over a lifetime, one out of every 100–300 people who consume drinking water containing 50 µg/L arsenic may suffer an arsenic-related cancer (lung, bladder, or liver cancer) death.

The NRC (2001) has reported that people who consume water daily containing 50 µg/L of arsenic have about a 1 in 1000 risk of developing bladder cancer. Following this result, it can be said that the Bangladeshis who are drinking water containing 50 to >1000 µg/L of arsenic are potentially exposed to a high risk of internal cancers in the long run. About 20% of the total population who are drinking arsenic-contaminated water above 200 µg/L of arsenic are potentially exposed to this health hazard. It is estimated that about 21 million people in Bangladesh are exposed to arsenic concentrations >50 µg/L (BGS/DPHE, 1999).

The visible carcinogenic impact of arsenic poisoning on human health will take decades to develop. The latency period (the time from first exposure to manifestation of disease) for arsenical skin lesions is about 8 years (Brown et al., 1989), 10 years (Smith et al., 2000b), or 5–10 years (Tondel et al., 1999) depending on the dose of arsenic (Mazumder et al., 1998a) and the immunity level of the exposed people. The impact of arsenic toxicity may be influenced by a number of issues: (a) pattern of arsenic concentration in drinking water and food, (b) amount of arsenic intake, (c) exposure duration, (d) nutritional status of exposed population, (e) genetic susceptibility to particular levels of arsenic exposure, (f) synergistic and antagonistic effect of other elements and dietary levels of interacting elements, (g) age and sex of exposed population, and (h) immunity level of the exposed population.

5.7.2 Spatial Risk and Mapping

Arsenic risk zones were mainly identified in a vector-base data analysis process by using GIS methods. A GIS was used as a platform enabling the management of the "criterion data" (Store and Kangas, 2001) for the spatial risk zoning. GIS technology has been applied to a wide range of environmental risks. A point-in-polygon operation can be performed in this regard. In a developed cartographic model, the data layer for arsenic risk zones was created by combining arsenic magnitudes with the threshold distances of tubewells. In addition, reclassification operations allow the transformation of attribute information, which represents the "recoloring" (Martin, 1991) of risk features in the map. A map of spatial arsenic concentrations within the buffer zones can be analyzed into different categories without reference to any other information.

Buffering and overlay: The spatial analytical capabilities of GIS with "buffer generation" and "overlay operation" can identify a spatial arsenic risk pattern (Figure 5.10a). "Buffer generation" was used in mapping the proximity area of arsenic, and overlay is the process of integrating different data layers (Martin, 1991). Reclassification allows the transformation of attribute information; it represents the "recoloring" (Martin, 1991) of features in the map. Thus, a map of spatial arsenic concentrations may be classified into categories such as "safe zones," "contaminated zones" or "severely contaminated zones" without reference to any other information.

A buffer is a zone of a specified distance around coverage features, and this analysis is used to identify areas surrounding geographic features. In this case study, a buffer zone or "buffer" is a polygon enclosing an area within a specified distance from a tubewell. Spatial risk zones were analyzed with tubewell buffer areas, clipping them from agricultural land. Here, buffer distances of tubewells were calculated based on the opinions of local people regarding the threshold distance in collecting water from the surveyed tubewells (Figure 5.10a). A GIS has strong spatial overlay capabilities that allow different map data to be combined in determining different "problem regions" of arsenic pollution. Spatial overlay is accomplished by joining separate data sets that share all or part of the same geographical area. GIS enables the combination and evaluation of different map overlays to provide new risk information. Therefore, overlaying the settlement within the buffer zone facilitates the generation of information on arsenic "problem regions" (Figure 5.10b).

Arsenic within threshold distances: The "threshold distance" for arsenic analysis refers to the areas from which people collect their drinking water. Travel time and travel distance are the two main factors in determining the buffer distance or proximity areas of tubewells. My field survey shows that most people are willing to collect their drinking water from a long distance if it comes from the safe hand-pump deep tubewells that the Government of Bangladesh has provided. But there are still many people who use contaminated tubewells within a very short distance. The buffer zones or proximity areas of tubewells were calculated from the threshold distances of tubewells having different degrees of arsenic magnitudes. The threshold distance of deep tubewells is estimated to be 500 meters, while tubewells with a high level of arsenic concentration have a much lower threshold distance. Accordingly, settlement areas within different buffer categories were calculated for risk zone quantification (Table 5.6).

Threshold distances of different tubewells were estimated on the basis of how far users are willing to travel to collect water from tubewells. Different people with different occupations and different levels of education had variable opinions about the threshold distance. The poor and the marginal farmers were not interested in collecting arsenic-safe water from a long distance since arsenic is not a priority issue to them. Five different threshold distances were calculated for different opinions of the local people and pattern of arsenic concentrations: (a) deep tubewells, 500 meter buffer distance; (b) tubewells with

Risk from Groundwater Arsenic Exposure

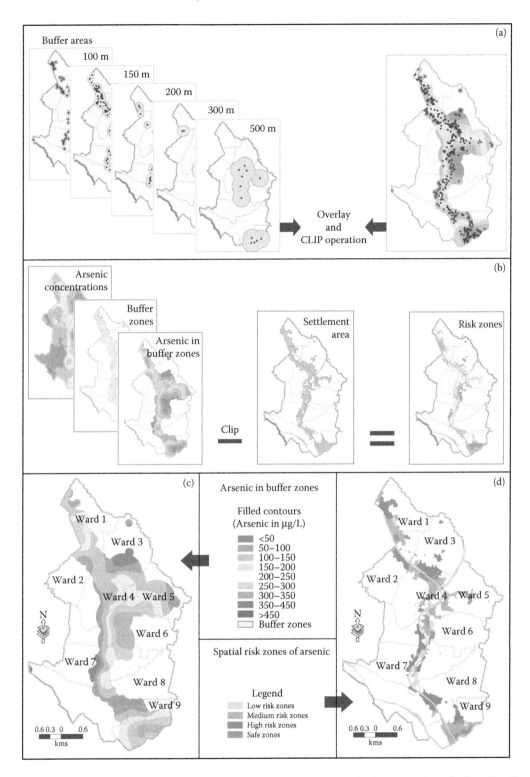

FIGURE 5.10 Spatial risk mapping with methodological concern: (a) buffer generation of tubewells; (b) extract operation for spatial risk zoning; (c) arsenic concentrations with the buffer areas; and (d) spatial risk zones of arsenic exposure. (Adapted from Hassan, M. M. and P. J. Atkins. 2007. *Journal of Environmental Science and Health, Part A* 42:1719–28.)

TABLE 5.6
Buffer Generation for Assessing Spatial Risk of Arsenic Poisoning

Buffer Distance (meters)	Average Arsenic (mg/L)	Tubewell Frequency	Buffer Area (hectares)[a]	Buffer Area (excluding overlapping)[a,b]	Settlement within Each Buffer (hectares)[a]
500	0.025	12	464.30	464.30 (66.78)	145.55 (60.06)
300	0.032	7	137.99	26.64 (3.83)	17.53 (7.23)
200	0.08	30	207.97	58.20 (8.37)	24.58 (10.14)
150	0.198	200	430.96	105.09 (15.11)	39.96 (16.49)
100	0.366	126	222.16	41.07 (5.91)	14.72 (6.07)
			Total:	695.30	242.34

[a] Figures in the parentheses indicate the percent of buffer area in hectares.
[b] People's preference was given priority. They want to collect their drinking water from any arsenic-free tubewell.

concentration of arsenic <50 µg/L, 300 meters; (c) tubewells with <100 µg/L of arsenic, 200 meters; (d) tubewells with <300 µg/L of arsenic, 150 meters; and (e) with <600 µg/L of arsenic, 100 meter buffer distance. In addition, primary, secondary or tertiary buffers were also estimated on the basis of different arsenic contents and threshold distances. It should be noted that a risk zone may exist outside such buffers. In identification of the pattern of arsenic concentrations within buffer zones or proximity areas, GIS was used as an integration of the layers of arsenic concentrations and buffer zones of different threshold distances (Figure 5.10c). An uneven concentration of arsenic was found within the buffer zones of tubewells in the study area. The kriging prediction map confirms that the safe zones are mainly concentrated in the north, central, and south part of the buffer areas in a scattered manner. The west and northeast part of the proximity areas are generally contaminated (Figure 5.10).

Spatial interpolation: The spatial pattern of arsenic concentrations was analyzed and interpolated in a GIS environment by using the kriging method. Interpolation methods can be used in the estimation of spatial continuity of variables to infer the values to nonsampling locations. This estimation method in geostatistics is known as kriging, an interpolation approach that provides optimal estimative of regionalized variables with minimum variance and without bias, using a theoretical variogram (Ahmadi and Sedghamiz, 2008; Bressan et al., 2009; Ghosh and Parial, 2014; Isaaks and Srivastana, 1989). Kriging is a means of local estimation in which each estimate is a weighted average of the observed values in the neighborhood (Patgiri and Baruah, 1995; Söderström and Magnusson, 1995). It is a distance weighting estimation method that takes into account the spatial characteristics of the local structure through a variogram function (Serón et al., 2001). The advantage of kriging is that the estimated values at observation sites are equal to the actual measurements (Rizzo and Dougherty, 1994; Uyan and Cay, 2013).

The arsenic interpolation map produced by kriging is constrained by the spherical semivariogram fits. The experimental variogram was computed from the raw data, and a "mathematical model" (Brooker et al., 1995; Mapa and Kumaragamage, 1996) was fitted to the arsenic concentration values by weighted least-squares approximation, using ArcGIS. The parameters of the variogram model (e.g., sill, nugget, and range) for arsenic concentrations were used with their values for estimating their concentrations over the area. We used Ordinary Kriging since the arsenic concentrations in groundwater are highly uneven. Ordinary Kriging is the most widely used type of kriging to estimate values when data point values vary or fluctuate around a constant mean value (Serón et al., 2001). It is applied for an unbiased estimate of the spatial variation of a component (Wang et al., 2001b).

Spatial risk zones: A point-in-polygon operation through kriged interpolation methods was performed to analyze the spatial arsenic concentrations of different magnitudes. The arsenic exposure data layer was created by combining the arsenic magnitudes map data, tubewell buffer area data, and the map data layer for tubewell installation years. The exposure data layer was then overlaid with the map data of the settlement area to yield a characterization of different risk zones (Figure 5.10d). The classified demarcated risk zones are (a) low risk zones, (b) medium risk zones, and (c) high risk zones (Table 5.7). The categories of risk zones were developed by poly-lines and these were converted to polygons using GIS in order to perform statistics. In addition, safe zones were also developed. Only the settlement areas were accounted for in the spatial risk zoning; agricultural land was not considered in this regard.

1. *Low risk zone:* Arsenic concentrations in this zone are mainly concentrated between 50 and 100 µg/L. A 200-meter buffer distance was considered for this category. The low risk zones cover about 27.88% (67.57 hectares) of the total settlement area and 3.92% of the total study area. They are found mainly in the northern, central, and lower central areas (Figure 5.10d). Some 9.68% (1286) of the population live with low risk of arsenic contamination. The first author identified two arsenic affected patients from this low risk zone.
2. *Medium risk zone:* Arsenic concentrations in this zone are mainly concentrated between 100 and 300 µg/L. A 150-meter buffer distance was selected to identify this risk zone. The medium risk zone covers slightly more than a quarter (63.38 hectares) of the total settlement area. About one-third (4216) of the total population live in this zone (Table 5.7). Two people here were found with symptoms suggesting the primary stage of arsenicosis. This zone is distributed in the northern and southern parts of the study area, covering about 3.67% of the total (Figure 5.10d). The first author found two arsenic-affected patients in this category. Both were in the primary stage of arsenicosis symptoms.
3. *High risk zone:* In high risk zones, arsenic is concentrated above 300 µg/L, covering about 5.44% (13.17 hectares) of the total settlement area and 0.76% of the total study area. The average buffer distance of each tubewell was calculated as 100 meters. They are found in the northern, central, and southern part of the study site (Figure 5.10d). About one-eighth (1588) of the total population use arsenic-contaminated water from this high-risk zone (Table 5.7). Some five arsenicosis sufferers were found to be living here.
4. *Safe zone:* Areas having concentrations of arsenic <50 µg/L are classed as safe zones. Buffer distances of 300 meters and 500 meters for arsenic-safe shallow tubewells and arsenic-safe deep tubewells were used in identifying safe zones. The isopleth map for this zone covers slightly more than two-fifths (98.22 hectares) of the total settlement area, in the northern, central, and southern part of the study area (Figure 5.10d). Arsenic-free tubewells are mainly concentrated in this zone. Almost half of the total population (46.46%) in the study site collects their drinking water from safe tubewells (Table 5.7).

TABLE 5.7
Spatial Risk Zones and Population Are at Risk

Safe and Risk Zones	Area (hectares)	% against Settlement	Net Area (%)	Population at Risk[a]
Safe	98.22	40.53	5.69	6197 (46.64)
Low	67.57	27.88	3.92	1286 (9.68)
Medium	63.38	26.15	3.67	4216 (31.73)
High	13.17	5.44	0.76	1588 (11.95)
Total:	242.34	100%	14.04%	13,287

[a] Figures in the parentheses indicate the percent of total population.

5.8 CONCLUDING REMARKS

Groundwater arsenic poisoning is considered to be a "natural calamity" in many countries because of its toxic nature. The assessed risk, spatial risk zoning, and risk factors show that, without any immediate mitigation action or awareness campaign, people of any arsenic-prone areas can be affected by mass poisoning and exposure to fatal disease with arsenicosis and in some cases cancer. Although the estimation of health risk in exposure to arsenic is uncertain, a low level of exposure to inorganic arsenic causes chronic toxicity in the body and will be related to health risks. High arsenic concentrations and high percentages of contaminated wells lead to a higher prevalence of arsenicosis patients. The validity of the risk calculation with arsenic concentrations in drinking water alone does not provide a good prediction of arsenicosis risk.

Which areas require further investigation with regard to groundwater arsenic poisoning? In this chapter, both the epidemiological and spatial risk for groundwater arsenic have been analyzed. Presently, GIS facilitates the overlying of several spatial variables in an attempt to identify the composite information in explaining spatial risk pattern. The current analytical system in GIS can perform spatial analysis using geostatistical methods and isoline mapping. The spatial risk zoning approach employed in this chapter could be used in planning and management for immediate arsenic mitigation, and, coupled with an awareness campaign, could be the basis of improved policy-making in arsenic-affected countries around the world.

6 Arsenic-Induced Health and Social Hazard and Survival Strategies
Experiences from Arsenicosis Patients

> Salcombe Hardy groaned: "How long, O Lord, how long shall we have to listen to all this tripe about commercial arsenic? Murderers learn it now at their mother's knee.
>
> **Dorothy L. Sayers**
> *British Author and Christian Humanist, 1893–1957*

Groundwater arsenic poisoning in Bangladesh is thought to be a major environmental health and social disaster. It is ironic that so many tubewells were installed in recent times for pathogen-free drinking water, but that water is now contaminated with toxic levels of arsenic (Hassan and Atkins, 2011). In a country with regular natural calamities (floods, cyclones, tidal surges, etc.), groundwater arsenic poisoning presents a new dimension of hazard. The extensive use of groundwater arsenic for drinking and cooking threatens the health of tens of millions of people in the world, of whom about 85 million are in Bangladesh alone.

The bodily marks of arsenic poisoning are found in the early stages on palms of the hands, soles of the feet, neck, and back. Arsenic poisoning also affects the central nervous system and the heart and blood vessels and causes a range of internal cancers, particularly affecting the bladder and lung (Barchowsky and States, 2016; Clewell et al., 2016; Kippler et al., 2016; Kumar et al., 2016a; Medeiros and Gandolfi, 2016; Sherwood and Lantz, 2016; Smith et al., 1998b). The stigmata on the hands are black patches known in Bengali as "zengoo" (Hassan et al., 2005). At first, blisters, sores, or swellings develop on palms and soles and there is itching. These swellings gradually turn into "zengoo" that develop slowly. Later the skin becomes dark and spotted due to the deposition of a black pigment. Eventually the spots become thickened and hard, the worst prognosis being a cancerous gangrene (Hassan et al., 2005).

Besides its toxicity, groundwater arsenic contamination creates widespread social problems for its victims and their families (Abedin and Shaw, 2013; Bhuiyan and Islam, 2002; Brinkel et al., 2009; Caldwell et al., 2003; Hanchett, 2004; Hassan et al., 2005; Sarker, 2010) as well as panicking the unaffected people in arsenic-prone areas in Bangladesh. There is, for instance, a tendency to ostracize arsenic-affected people, arsenicosis being thought of as a contagious disease. Within the community, arsenic-affected people are barred from social activities and often face rejection, even by their immediate family members. Women with visible arsenicosis symptoms are unable to get married, and some affected housewives are divorced by their husbands. Children with symptoms are not sent to school in an effort to hide the problem (Hassan et al., 2005). In addition, some patients are barred from access to the safe drinking water from deep tubewells because of a common misapprehension of infectivity. In Bangladesh, arsenic-affected women are more often victims of ostracism than arsenic-affected men. Moreover, domestic water collection and management in rural Bangladesh is predominantly undertaken by women and girls (Crow and Sultana, 2002; Sarker, 2010), and it is rare for men to participate in domestic water collection (Hanchett, 2004).

Social research is inevitably shaped by the research paradigm (Kuhn, 1962), and the choice of research paradigm dictates not only the research methods but also the interpretations of findings (Neuman and Kreuger, 2003). Therefore, it is important to reflect about social research conducted within different paradigms as a lens through which one can see and understand the reality of social difficulties (Shek and Wu, 2017) of arsenicosis patients. This chapter addresses a different departure—qualitative enquiry with verbatim data analysis. Qualitative methods are particularly suitable when describing phenomena from emic perspectives (Morse and Field, 1995). They are used to elicit in-depth material about culture, meanings, processes, and problems. Qualitative research involves an interpretive, naturalistic approach to its subject matter; it attempts to make sense of, or to interpret, phenomena in terms of the meaning people bring to them (Alasuutari, 2012; Denzin and Lincoln, 2005; Yang et al., 2017). Qualitative research is designed to help researchers to understand people and the social and cultural contexts within which they live (Myers, 2009).

There has been a plethora of studies on arsenic and its associated health and social problems, but the distress and survival strategies of arsenic-affected patients are an inspiration for this chapter. The chapter mainly addresses a number of issues: (a) exploring people's understandings about arsenic; (b) investigating the toxic impact of arsenic on health and social aspects; (c) outlining the survival strategies adopted by the arsenicosis sufferers, i.e., what arsenic-affected people think and do, classified into what we will call "coping strategies" and "adaptation strategies"; and (d) examining the pattern of health and social hazards developed by arsenic poisoning. The relevant data for this chapter were collected from different geophysical areas in Bangladesh in different times. The chapter dwells on the voices of arsenicosis patients, who are the best judges of their own experiences.

6.1 THEORIES OF QUALITATIVE RESEARCH

6.1.1 QUALITATIVE RESEARCH PHILOSOPHY

Man cannot discover new oceans unless he has the courage to lose sight of the shore (André Paul Guillaume Gide, 1869–1951, a French Nobel Laureate in Literature).

Qualitative research approaches help researchers to enter deeper into the waters to lose sight of the shore and diagram a new ocean (Varpio et al., 2017). The intention in this chapter is to consider a few qualitative modes of analyses that could uncover the inherent health and social difficulties of arsenicosis patients that they bear in their mind. Qualitative research philosophy is mainly based on the nature of research design and research method for nonquantifiable social and behavioral aspects. Research design assists to "plan, structure, and execute" research to maximize the "validity of the findings" (Burns and Grove, 2009; Polit and Beck, 2012), and research method is a strategy of enquiry that moves from underlying philosophical assumptions to research design, data collection, data analysis, and data interpretation (Fletcher et al., 2016; Myers, 2009). The research methodology chosen depends on the research questions and the philosophical perspectives from which the questions are to be investigated (Shepard et al., 1993). Qualitative research is especially useful for the exploration and discovery of inherent health and social problems. What is meant by qualitative research is somewhat a "contested terrain" (Johnson et al., 2007). Generally, qualitative research may be defined as an attempt to obtain an in-depth understanding of the meanings and "definitions of the situation" (Wainwright, 1997) presented by informants, rather than the "quantification" (Eisner, 2017: 5; Strauss and Corbin, 1998; Yang et al., 2017) of their characteristics. Qualitative analysis can be used to uncover and understand what lies behind arsenic poisoning in health and social concerns in which little is yet known, for instance, the intricate details of phenomena that are difficult to convey with quantitative methods. Qualitative research, according to Rich and Ginsburg, (1999) can be defined as

> Qualitative research is an ideal approach to elucidate how a multitude of factors such as individual experience, peer influence, culture, or belief interact to form people's perspectives and guide their behaviour.

Qualitative research examines the complex social world, especially meanings and behaviors in a social context (Powell and Single, 1996; Rich and Ginsburg, 1999). Qualitative inquiry is an umbrella term for various philosophical approaches to interpretive research (Eisner, 2017: 8; Glesne and Peshkin, 1992). Qualitative investigation offers insight into a vast array of human experiences and emotions in ways that are captured through qualitative enquiry rather than quantitative methods (Yang et al., 2017). It is argued by Alvesson and Deetz (2000: 1) about qualitative research that

> Qualitative research has become associated with many different theoretical perspectives, but it is typically oriented to the inductive study of socially constructed reality, focusing on meanings, ideas and practices, taking the native's point of view seriously.

A qualitative methodology is appropriate when the aim is to gain insight into the ways that people perceive, interpret, and explain their world (Stenner et al., 2016). Qualitative methods generate detailed and valid data with multiple forms of evidence (Eisner, 2017: 7) that permit formulation of new hypotheses and inform further study or practice (Bunne, 1999; Powell and Single, 1996; Strauss and Corbin, 1990). The qualitative approach is typically oriented to inductive logic (generation of theory), interpretivism, and constructionism (Berger and Luckmann, 1966; Boas, 1962). Qualitative enquiry is particularly suited to the combination of complex phenomena typically found in human behavior.

There is no universal regulation for qualitative research (Maxwell, 1998: 233), but there are several familiar approaches and practices for qualitative enquiry (Lofland et al., 2006) that can dictate proper qualitative methods to achieve quality research and rigor. Qualitative method is useful for broad and context sensitive fieldwork material as distinct from numerical data. Qualitative method offers a series of techniques for examining discursivity, dynamic processes, complexity, contextualization, relationality, and fine-grained detail (Fletcher et al., 2016). It is reported that qualitative methods are "powerful tools" for developing theory (Reay and Zhang, 2014: 5), and this tends to assume that there is "a fixed battery of methods" (Stewart, 2014: 77) that can be drawn upon to fill in gaps or explore unknown phenomena. In research practice, qualitative inquiry encapsulates "multiple practices and vocabularies which acquire different meanings in their use" (Patton, 2002: 76).

Qualitative research for arsenic poisoning and social implications includes a plethora of methodological issues, such as interviews (Rapley, 2007: 15), the oral history method (Bornat, 2007: 34), biographical research (Rosenthal, 2007: 48), the focus group method (Finch et al., 2013: 211; Macnaghten and Myers, 2007: 65), or the grounded theory (Dey, 2007: 80). In addition, there are various analytic frameworks for qualitative research, such as narrative research (Andrews et al., 2007: 97; Spencer et al., 2013: 295), feminist approaches (Kitzinger, 2007: 113), ethnomethodology (ten Have, 2007: 139), conversation analysis (Peräkylä, 2007: 153), discourse analytic practice (Hepburn and Potter, 2007: 168), critical discourse analysis (Wodak, 2007: 185), and so on. Qualitative methods for this chapter consist of field observations in the participants' natural environments, oral and written narrative, text, sounds, and visuals. Qualitative enquiry was mainly designed to understand the lives of arsenic-affected people and their social issues.

6.1.2 Interpretive Research Paradigm

The philosophy and strategies of the interpretive paradigm can be suitable for analysis of arsenic poisoning and social ostracism. The philosophical assumptions underlying this chapter come mainly from interpretivism. Interpretive approaches give the research greater scope to address issues of influence and impact of arsenic poisoning. The interpretive paradigm has the potential to generate new understandings of complex multidimensional human phenomena in groundwater arsenic poisoning, as well as the coping and adaptation strategies of the arsenic-affected people. Interpretive paradigm is based on the "epistemology of idealism" and encompasses a number of research approaches, which have a central goal of seeking to interpret the social world (Ajjawi and Higgs, 2007; Higgs, 2001; Putnam and Banghart, 2017). Epistemology is defined as "the philosophy and theory of knowledge, which seeks to define it, distinguish its principal varieties, identify its sources, and establish its limits" (Bullock and Trombley, 2000: 279), and in idealism, knowledge is viewed as a social construction (Ajjawi and Higgs, 2007).

The investigative approaches of Wilhelm Dilthey (1833–1911) and Max Weber (1864–1920) focused on interpretive understanding to access the meanings of participants' experiences as opposed to explaining or predicting their behavior, which is the goal of empiricoanalytical (or quantitative) research (Smith, 1983). In the interpretive paradigm, meanings are constructed by arsenicosis patients following their context and personal frames of reference as they engage with the world they are interpreting. The interpretive paradigm aims to explain the subjective reasons and experiences of individuals and the meanings that lie behind social action. It attempts to understand phenomena through the meanings that people assign to them (Deetz, 1996). It does not predefine independent and dependent variables, but focuses on the full complexity of human sense making as the situation emerges (Kaplan and Maxell, 2005). Within an interpretive hermeneutic research tradition, the "intent is not to develop a procedure for understanding, but to clarify the conditions that can lead to understanding" (Holroyd, 2007).

6.1.3 Hermeneutic Phenomenology: Philosophical Underpinnings

The philosophical base of interpretive research is hermeneutics and phenomenology (Boland, 1986; Koch, 1999; Maltby et al., 2016; Reiners, 2012; Van Manen, 2007). An interpretive hermeneutic phenomenological research approach is suitable in order to provide an understanding of arsenic affected people's beliefs and perceptions regarding arsenic poisoning and social situations as well as their survival strategies. Hermeneutic phenomenology is attentive to the philosophies underpinning both hermeneutics and phenomenology (Van Manen, 2003, 2007; Dowling, 2007). It aims at producing rich textual descriptions of experiencing of selected phenomena in the lifeworld of individuals that are able to connect with the experience of all of us collectively (Smith, 1997). The nub of phenomenology and hermeneutics was introduced into modern philosophy by Heidegger "1927" in opposition to Transcendental Phenomenology of Husserl (1962). Hermeneutic phenomenology is able to uncover the uniqueness of individuals' experiences with an emphasis on the individuals' historicity or backgrounds (Gadamer, 1960; Heidegger, "1927"). Hermeneutic phenomenology is a suitable qualitative method for investigating everyday social implications for the meanings of lived experiences (Van Manen, 2003) of a particular phenomenon through the interpretation of text and narratives and through actions by the informants (Mjørud et al., 2017).

Hermeneutics is a branch of the interpretive philosophy of Hans-Georg Gadamer (1976) and Paul Ricoeur (1981). It can be treated as both an underlying philosophy and a specific mode of analysis. Hermeneutics enhances the interpretive element to illuminate assumptions and meanings in the text that participants themselves may have trouble expressing (Crotty, 1998; Van Manen, 2003). Hermeneutics, in this chapter, has been used to interpret these meanings to reveal "hidden meaning" (Palmer, 1969: 147) beyond the "explicitness" of what participants themselves said in order to provide warranted, insightful interpretations (Nosek et al., 2016). This interpretation helped to deepen and strengthen understandings to promote new or different ways of viewing (Moules et al., 2015) the social pain of arsenic poisoning. Palmer (1969: 33) argued that hermeneutics has been interpreted in six distinct ways: (a) the theory of biblical exegesis; (b) general philological methodology; (c) the science of all linguistic understanding; (d) the methodological foundation of

Geisteswissenschaften (spiritual science); (e) phenomenology of existence and of existential understanding; and (f) the systems of interpretation, both recollective and iconoclastic, to reach the meaning behind myths and symbols.

The primary concern of hermeneutics is the "philosophy of understanding" (Geanellos, 1998). As an ontological inquiry and a method of conceptual-textual analysis, hermeneutics puts the text within its historical and sociocultural context. The term "hermeneutics," a Greek origin from "hermeneuo" translated to "interpret" or "understand" (Crotty, 1998), is philosophical in nature and can be used as a theory of interpretation and understanding (Mailloux and Mootz, 2017; Moules, 2002; Urmson and Jonathan, 2005). Hermeneutic or interpretive enquiry is a living tradition of interpretation with a rich legacy of theory, philosophy, and practice (Moules, 2002). The Latin "hermeneutica" was introduced in the 17th century by the theologian Johann Conrad Dannhauer (1603–1666) for biblical and theological textual interpretation, and it has grown into different schools incorporating the realms of the theological, juridical, and philosophical (Grondin, 1994). Hermeneutics has followed a changing course from pragmatism to philosophy.

Hermeneutics has been described as the practice and theory of interpretation and understanding in human contexts as well as a deductive process of reading and understanding the researched item (Bleicher, 1980; Chesla, 1995; Schleiermacher, 1985: 74). It is also treated as the science, art, and philosophy of interpretation as well as the "discipline of thought that aims at (the) unsaid life of our discourses" (Grondin, 1995). As a research methodology, hermeneutics focuses on understanding, interpreting, exploring, and clarifying the hidden meaning of texts or any other written or oral event. It is considered a reflective enquiry concerned with "entire understanding of the world and thus… all the various forms in which this understanding manifests itself" (Gadamer, 1976). "Hermeneutics deals only with the art of understanding, not with the presentation of what has been understood" (Schleiermacher, 1985: 74), and hermeneutics and rhetoric, in his opinion, are opposite sides of the same coin that is intimately related in that every act of understanding is the reverse side of an act of speaking, and there is a listener who receives it and understands it.

Hermeneutics attempts to understand human beings in a social context. Hermeneutics refers to the theory and methodology of interpretation, more specially the interpretation of biblical texts, wisdom literature, and philosophical texts (Audi, 1999: 377). As a mode of analysis, hermeneutics suggests a way of understanding the meaning or trying to make sense of textual data that may be unclear in one way or another. It started out as a theory of text interpretation of scripture or religious texts, but later it had been broadened to questions of general interpretation (Grondin, 1994: 2) including written, verbal, and nonverbal communication as well as semiotics, assumptions, and pre-understandings. Hans-Georg Gadamer (1960) in his "Truth and Method" deploys the concept of "philosophical hermeneutics" as it is worked out in the work of Martin Heidegger (1927) published as "Being and Time."

Hermeneutics is a means of interpretation, mainly of historical events, phenomena, or texts. It is the science of interpretation that has been developed with a history stretching back to the Ancient Greek and later established as a mechanism for Biblical exegesis in the Middle Ages. After that, it transformed into a philosophical and later sociological form of enquiry and, with the work of Martin Heidegger (1927), into an ontological principle. As a contemporary philosophy and research methodology, hermeneutics is concerned with the thick description that emanates from people's detailed stories of their experiences in their every-day understanding of "Being-in-the-world" (Stenner et al., 2016). Hermeneutics is described as a method that facilitates interpretation of texts within certain contexts, where "texts" refers to sources of information in addition to the written or spoken word (Honderich, 2005: 380).

Phenomenology refers to a philosophy with a set of principles that can be used to underpin a methodology that has descriptive and interpretive philosophical traditions (Finlay, 2012; Gullick et al., 2017; McWilliam, 2010; Stenner et al., 2016). "Descriptive phenomenology" seeks to describe rather than explain and starts from a researcher's perspective free from preconceptions (Giorgi,

1985; Husserl, 1962), while "interpretive phenomenology" aims to describe, consider, explain, and interpret participants' experiences (Dowling, 2007). Phenomenology sets out to gather "deep" information and perceptions through inductive qualitative research methods (Lester, 1999)—it describes the meaning of the lived experiences of several individuals about a concept or phenomenon (Creswell, 2009). Phenomenology is the science of ontology, which focuses on explicating the meanings of experiences to understand phenomena (Heidegger, 1993: 72). In this chapter, it is the intention to understand the meaning of participants' experiences regarding arsenic poisoning and social pain.

Phenomenology explores a systematic reflection on the lived experience of changing systems. It is guided by philosophical methodology and assisted by human and social science procedures and techniques (Kaivo-oja, 2016). Phenomenological strategies are particularly effective at bringing the forefront the experiences and perceptions of individuals from their own perspectives, and they therefore challenge structural and normative assumptions (Lester, 1999). In this work, qualitative methods bring forth in-depth realities about arsenic impact on social issues and people's survival strategies concerning arsenic poisoning. Phenomenology involves gathering context-specific, in-depth accounts of perceptions using inductive, qualitative methods such as interviews, participant observations, and other bespoke creative methods to present an interpretation of the meaning of the experiences of arsenic-affected patients.

Phenomenology strives to discover what it means to be human and to gain a deeper understanding of human experience within the context of their culture (Munhall, 2007). Phenomenology relies on the perceptiveness, creative insight, and interpretative sensitivity that are key elements of qualitative research practice (Van Manen and Adams, 2010). Phenomenology comprises its own "philosophical and theoretical approach premised on a phenomenological concept of experience as well as a research methodology consistent with this theoretical framework" (Cilesiz, 2011). Phenomenology is related to a number of different fields of philosophy: (a) ontology (study of being or what is), (b) epistemology (study of knowledge), (c) logic (study of valid reasoning), and (d) ethics (study of right and wrong action). All these philosophical pitches provide tools and methods for exploring arsenic-induced social ostracism.

6.1.4 META-SYNTHESIS: QUALITATIVE FOUNDATIONS

Meta-synthesis can be utilized to accommodate qualitative findings on arsenic poisoning and social issues with relevance and integrity. Meta-synthesis refers to a range of existing and emerging methods by which qualitative research in social and health issues are integrated into a larger concept that are relevant to practice (Paterson et al., 2009). Meta-synthesis is based on the interpretative paradigm, where the relevant data are research findings that have already been interpretive and integrated (Aguirre and Bolton, 2014; Jensen and Allen, 1996). Meta-synthesis is a systematic method, beyond the traditional literature review (Hong et al., 2017; Moeller et al., 2016), through which additional interpretive and inductive analysis are anticipated and the findings of a range of qualitative methods on a topic can be evaluated and presented (Edwards and Kaimal, 2016).

Usually qualitative research uses small sample sizes and is conducted with a view to depth rather than breadth. This is a caution when developing the meta-synthesis. The main target of meta-synthesis is to elicit novel understandings from comparison and synthesis of the findings of multiple studies. A meta-synthesis can provide a valuable overview of existing findings and new insights around social implications of arsenic poisoning. Barnett-Page and Thomas (2009) identified a number of methods for meta-synthesis, of which "narrative synthesis" and "critical interpretative synthesis" are recognizable for arsenic impacts on social aspects.

"Narrative Synthesis" or "Textual Narrative Synthesis" refers to a method that uses a text-based approach to systematic reviews and synthesis of findings (Edwards and Kaimal, 2016; Hong et al., 2017) from a qualitative thematic summary of findings (Barnett-Page and Thomas, 2009; Lucas et al., 2007). The findings in narrative synthesis are reported in a narrative format. Reviewing different studies in a narrative synthesis can be conducted with homogenous findings. Typically, study context, quality and findings are reported according to a standard format in narrative synthesis with comparisons and differences linked across arsenic and social studies. Narrative synthesis is likely to make transparent heterogeneity between studies and issues of quality appraisal (Barnett-Page and Thomas, 2009). This is because narrative synthesis makes the context and characteristics of each study clear.

"Critical Interpretative Synthesis" (CIS) was developed as a way to produce new findings from a large number of complex and diverse sources. Introduced by Dixon-Woods et al. (2006), Critical Interpretative Synthesis is an adaptation of meta-ethnography, which also borrows techniques from grounded theory (Edwards and Kaimal, 2016). CIS as an approach for synthesizing multi-disciplinary and multi-method evidence in the whole review process rather than just the synthesis component. It involves an iterative approach that is distinct in its "explicit orientation towards theory generation" (Dixon-Woods et al., 2006) in refining the research question from literature with relevance and quality (Gough, 2007).

CIS was developed to understand the complexity of issues in access to healthcare (Dixon-Woods et al., 2006). Later, CIS has explored the effectiveness of health interventions (Morrison et al., 2012), sexual and reproductive health (Walker et al., 2016), vision loss and risk in older people (Rudman et al., 2016), and difficulties in mental health problems (Markoulakis and Kirsh, 2013). Two analytical processes can be recognized in qualitative meta-synthesis: (a) content analysis with empirical foundation and (b) interpretive discussion (Aguirre and Bolton, 2014). The methods employed in this chapter are within the broad range of meta-synthesis approaches that "represent an inductive way of comparing, contrasting, and translating the original understanding of key metaphors, phrases, ideas, concepts, and findings across studies" (Kinn et al., 2013).

6.2 VERBATIM AND NARRATIVES

6.2.1 Qualitative Transcripts

The originality of this chapter lies in its exploration of evidence from detailed fieldwork organized to address the impact of arsenic exposure on social issues and the survival strategies of arsenicosis patients. The fieldwork employed qualitative methodologies, and surprisingly little effort has been made so far to use these valuable techniques in the social geography of Bangladesh. Observation and interviews are the two key data collection methods within phenomenology. The qualitative "tool kit" has many available techniques (Hay, 2000), and this study was designed using "multiple sources" of data derived from three methods: participatory rural appraisal (PRA); face-to-face in-depth interviews; focus-group discussions (FGD); and informal dialogues (Figure 6.1). Moreover, informal conversation and discussions with different people were carried out to "obtain a differentiated understanding of the community's attitudes, beliefs, and behaviours" (Mukherjee, 1995) toward the arsenic poisoning and social ostracism.

PRA is suitable to gain a quick understanding about the primary ideas of the overall physical and social conditions of different arsenic-contaminated areas. PRA approaches helped us to turn a theoretical awareness into reality (Chambers, 1997). Under a PRA conceptualization, triangulation, reconnaissance survey, and informal meetings with the local people were employed to get a quick picture of the arsenic situation in the study sites.

The participatory approaches were used for an epistemology of the social situation that relies on local understandings and perceptions. A central objective of the participatory approach is to ensure that the voices of local people or different groups figure prominently in the dialogue (Shaffer, 1996). The author of this book made a number of extended visits to many arsenic-contaminated sites all over the country, and he spent several years with the help of national and international NGOs as well as local NGOs. It is noted that the arsenic-affected people were not initially interested in providing information, but the presence of local NGO personnel made the situation possible to learn a great

FIGURE 6.1 FGD with people in different areas of Bangladesh: (a) Political leaders, Ghona, Satkhira; (b) Community people, Galachipa, Patuakhali; (c) Community people, Shivalaya, Manikganj; (d) Community people, Mongla, Bagerhat; and (e) NGO people, Kotalipara, Gopalganj.

deal about the background context of each arsenic-contaminated area and people easily provided us the relevant information. It should be mentioned that the perception of arsenicosis is also influenced by culture. There is a common belief in rural societies of Bangladesh that arsenicosis is the "curse of nature" because of its noncurative nature and confirmed fatality. Fieldwork by the present author indicates that culture imposes certain limitations on health and social issues of arsenicosis patients regarding their treatment and health care.

In-depth interviews and focus-group discussions were used to define people's own understandings about the impact of arsenic on social issues and subsequent survival strategies. These involved sufferers, nonsufferers, and groups containing different occupations. Interviewing is a highly personal process where meanings are created through personal interaction (Baxter and Eyles, 1999). The in-depth interviews were based upon open-structured questions so that a long discussion would be possible in each interview. Interviews were also digitally recorded and transcribed verbatim. Focus-group discussion employs an "interaction discussion" as a means of generating "rich details of complex experiences and the reasoning behind actions, beliefs, perceptions and attitudes" (Carey, 1995). This method was adopted in this study since the investigation of arsenic issues over health concerns was complex. A number of focus groups were selected from a spectrum running from very poor to rich people, illiterate to literate people, landless farmers to land holding farmers, local NGO officials to different local government officials, local social activists to local political leaders, and finally local elected administrators. Where possible, gender homogeneity and occupational segmentation were followed in this regard.

6.2.2 Rich Descriptive Narratives

Bogdan and Biklen (2007: 3) define qualitative data analysis as "working with the data, organising them, breaking them into manageable units, coding them, synthesising them, and searching for patterns." The aim of qualitative data analysis is to discover patterns, concepts, themes, and meanings. Qualitative modes have mainly been concerned with textual analysis, and the resulting data can be analyzed from multiple perspectives using different techniques (Merriam, 2009; Miles and Huberman, 1994; Wolcott, 1994). The data analysis procedure was conducted in conjunction with data collection. Data collection did not cease until no new information was obtained, which indicates that data saturation has been reached (Dong et al., 2016). The qualitative analytical modes seek to explore the rich descriptive narratives by developing valid and reliable concepts of groundwater arsenic poisoning and its impacts on social life. The thick description mode was used to consider the data without "interpretation and abstraction" (Geertz, 1973), and ethnographic representation was used to create a "rich descriptive narrative" (Strauss and Corbin, 1998) and a vivid presentation of new understandings.

Van Manen's circular process of hermeneutical writing, underpinned by Heidegger's hermeneutic circle of understanding, also influenced data analysis (Van Manen, 2007). The hermeneutical circle is a circle of interpretation that moves forward and backward between the parts and the whole (Chaplin et al., 2016). In keeping with the methodology adopted for this chapter, data analysis methods were developed from phenomenological and hermeneutic principles. Some six stages were employed (e.g., immersion, understanding, abstraction, synthesis and theme development, illumination and illustration of phenomena, and integration and critique of findings) for conducting hermeneutic phenomenological research as described by Allen and Jensen (1990) and Van Manen (2003). Throughout all stages of the data analysis, there was ongoing interpretation of the text and the phenomenon of learning to communicate social reasoning. By constantly crosschecking the interpretations with the original transcripts, we sought to maintain closeness (or faithfulness) to the participants' constructs, grounding interpretations in the data. This strategy to maintain authenticity was suggested by Lincoln and Guba (2000).

Theoretical memos (Glaser, 1978) are important during data coding to keep track of all categories, properties, and generative questions that evolved from the analytical process. Theoretical memos contain coding products, summary notes, and concepts that are potentially sensitive in possible story lines. The collected verbatim data from in-depth interviews, focus-group discussions, and informal dialogues were transcribed and analyzed using different techniques (Winters, 1997): (a) the transcribed interview data were divided into units designated by the subject matter being described; (b) individual units from each interview were coded using topical codes; (c) the codes were grouped into clusters of similar topics and recoded using interpretive codes; and (d) the interpretive codes were grouped to reflect the themes. The process of data analysis begins with the categorization and organization of data in search of patterns, critical themes, and meanings that emerge from the data. A process sometimes referred to as "open coding" (Strauss and Corbin, 1990) is commonly employed whereby the researcher identifies and tentatively names the conceptual categories into which the phenomena observed would be grouped.

In the hermeneutic context, the selected questions that emerge from studying arsenic phenomenon are open and allow respondents or participants to speak their experiences. This verbatim information was then transformed into text that would finally be used as themes and story. In hermeneutics, the act of interpretation itself represents a gradual convergence of insight on the part of the researcher and the text (Bontekoe, 1996). The phenomenological data analysis strategy is to "transform lived experience into a textual expression of its essence" (Van Manen, 2003). The application of phenomenology is to construct an animating, evocative description (text) of human actions, behaviors, intentions, and experiences as we meet them in the "lifeworld." Phenomenological descriptions are rich and evocative as well as simple and straightforward so that readers who have experienced the phenomenon may analyze their own reality with the identified themes (Swanson-Kauffman and Schonwald, 1988). Phenomenological themes may be understood as "structures of experience" (Van Manen, 2003) and offer a thick description of phenomena. A systematic method of thematic data analysis was adopted for this chapter, as informed by Titchen and colleagues' work (Edwards and Titchen, 2003; Titchen, 2000; Titchen and McIntyre, 1993). This method allowed for systematic identification of participants' interpretations and constructs (first order constructs), which were then layered with the researchers' own understandings, interpretations, and constructs (second order).

6.3 RIGOR AND TRUSTWORTHINESS

Generally, qualitative studies are criticized because of their lack of rigor and credibility (Baxter and Eyles, 1997; Decrop, 1999). There are strategies that can be used to enhance the quality and trustworthiness of interpretive research findings. Generally, smaller and nonrandom samples are utilized in qualitative research. Qualitative research is not based upon standardized instruments like the experimental studies. It is argued that the criteria used to ensure quality in interpretive research should be consistent with the philosophical and methodological assumptions on which the research is based (Koch, 1996; Koch and Harrington, 1998; Leininger, 1994; Pories et al., 2016).

Qualitative research is often criticized for lack of methodological rigor and research bias. Bias is inevitable in any research (Freshwater, 2005) and in the Gadamerian research approach, it should ensure research "trustworthy" in the sense of being able to demonstrate both rigor (process) and relevance (end product) (Stenner et al., 2016). Traditional criteria of rigor for interpretative phenomenology include trustworthiness, auditability and confirmability, credibility, and transferability (Koch, 1996). The term "trustworthiness" can be used as a measure of the quality of research in which relevant information and data analysis are believable and trustworthy. The concepts of validity, reliability, and generalizability provide the traditional framework for evaluation of quantitative research; however, these positivist concepts are inappropriate for qualitative research because they are concerned with measurement and representation, and in qualitative research, nothing of these evaluation principles is measured (Stenner et al., 2016).

Rigor refers to the trustworthiness of the research process, while trustworthiness is the believability or trueness of data and findings—equivalent to validity and reliability in quantitative research. Rigorous qualitative research involves using trustworthy data collection techniques and analytic procedures, and engaging in checks and balances (e.g., debriefing, triangulation) to ensure quality of the data and findings. Rigor is also established through clear documentation of the research focus, methods, data collection, analysis, and findings, which allows the reader to draw conclusions about the quality of a research process (Goodell et al., 2016). Scientific rigor is necessary for any research method to understand and accurately represent the phenomena it studies (Rich and Ginsburg, 1999). Ensuring quality of the study required the rigorous use of systematic methods of data collection and analysis, transparency in documenting these methods, and consistency in operating within the philosophical assumptions and traditions of the research paradigm and approach (Lincoln and Guba, 2000). Validity and reliability are important considerations in qualitative research. The conceptual meaning of validity may be applied by asking, Are the methods relevant for the aims and objectives and the research questions or to what contexts are the findings transferable? Such questions relate to the trustworthiness of the research findings.

A number of criteria are important to qualify trustworthiness of an interpretive hermeneutical study (Lincoln and Guba, 1985; Krefting, 1991; Creswell, 2013): (a) credibility (internal validity of the interpretation of the data)—how truthful the particular findings are; (b) transferability (external validity)—how applicable the research findings are to another setting or group; (c) reliability or dependability (the degree to which a research finding remains the same when data are collected and analyzed several times)—if the results are consistent and reproducible; and (d) confirmability (objectivity)—how neutral the findings are. Credibility in qualitative research refers to the faithfulness of the description to the phenomena (Koch and Harrington, 1998), or authenticity of the findings of the research (Denzin and Lincoln, 2005). Transferability is an important indicator of quality in qualitative research (Hammersley, 1992). It is analogous to external validity, and it can be achieved by detailing the research methods, contexts, and assumptions underlying the study so a reader can judge the applicability of the findings to other settings that they know. Dependability is analogous to reliability, that is, the consistency of observing the same finding under similar circumstances. It refers to the extent to which research findings can be "replicated" (Merriam, 2009) with similar subjects in a similar context. Two different techniques were employed to achieve reliability for this interpretive study: (a) explaining the assumptions and theory behind the study and (b) using multiple methods of data collection and analysis (triangulation). In addition, confirmability is the degree to which the research findings can be confirmed or corroborated by others.

To achieve trustworthiness, data from each key participant were collected several times. It has been mentioned earlier that field visits with local NGO personnel were helpful for establishing a rapport with the participants and gaining their trust, which gave them freedom to discuss their views and learning experiences, increasing the rigor and trustworthiness of the research findings. Interpretations about arsenic poisoning and experiences of arsenicosis patients are consistent with the philosophical underpinnings of the interpretive paradigm. The voices of arsenicosis participants are evident in the text, enhancing and ensuring authenticity about the meanings of their social issues (Lincoln and Guba, 2000).

Using the constant comparison process, the data analysis was continually adjusted to ensure (a) the degree of data fit, (b) its functionality, and (c) its relevance to the emerging theory (Glaser and Strauss, 1967). The degree of fit is the categories that are applicable to the research setting and directly derived from the collected data. Since the categories are generated directly from the collected data from field visits, the criteria of fit are automatically met (Sherman and Webb, 1988). Functionality refers to the ability of findings to explain the actions under study, i.e., for describing a theory that "works" (Kerlin, 1998). Functionality explained the variation in the data and interrelationships among the constructs in a way that produced a predictive element to the new concept. Relevance is the core categories that are meaningfully relevant to the research setting (Spaulding, 2000). Relevance evolves through the emergence of a core variable from the data in the way of the theoretical sensitivity of the arsenic milieu.

Triangulation is important to promote the credibility for qualitative enquiry of arsenic poisoning. Triangulation arose from an ethical need to confirm the validity of the research processes

(Yin, 2003). It is an important way to improve the trustworthiness of qualitative research findings. Triangulation reduces methodological biases and enhances the credibility of this study. Triangulation is the use of multiple sources of data, multiple settings, and multiple methods to increase confidence in research findings (Blaikie, 2000; Lincoln and Guba, 1985; Scandura and Williams, 2000) as well as to crosscheck (White and Taket, 1997). Multiple methods and sources of data collection provide multiple constructions of phenomena, thereby enhancing the depth and richness of the data and reducing systematic bias in the data (Denzin and Lincoln, 2005). Richardson and St. Pierre (2005) describe triangulation as a crystallization in which a story can be developed through data gathered from different data sources due to its multi-perspective sources and nature. This can be followed by a process that considers the data from various angles by highlighting different aspects, depending on different phases of the analysis. The triangulation for data collection entails the use of multiple-methods to collect relevant information for a single problem. Participant observation (with field notes), in-depth interviews, focus-group discussions, informal and dialectic interviews, repeat interviews, and field notes were employed as forms of triangulation for the field data.

6.4 TERMINOLOGICAL ISSUES AND PEOPLE'S UNDERSTANDINGS

What do the local people think about "arsenic," "health hazard" and "social hazard"? The respondents and the FGD participants identified different meanings of arsenic. The majority of the participants defined arsenic as a kind of "poison"; some considered it a "germ" contained in groundwater. Some participants knew that people could die if they take this poison continuously from their tubewell water. One common response was to see arsenic-related diseases as similar to eczema, skin lesions (sores), gangrene, and leprosy. A few participants assumed that cholera and diarrhea are the resultant effect of the chronic impact of arsenic. A majority of the arsenicosis patients thought that their skin lesions become worse during the winter because of the hard soil where they work. Some also thought that during the rainy season they got worse because of the mud in the fields. In addition, some assumed that arsenic means "iron," which is frequently found in local drinking water. When discussing the concentration of arsenic in tubewell water, they replied, "We know about arsenic, it is red in colour," which refers to the iron.

Perceptions of some unaffected people in rural Bangladesh are mainly confined to "poison" and "cancer." They consider arsenic to be a poisonous substance that is concentrated in tubewell water and is dangerous to human health. Some participants defined arsenic as a "fatal disease." Their understandings of the visible impact of arsenic are mostly confined to a diagnostic form: "Arsenic is a dangerous poison contained in tubewell water. If people drink this water, they could get many types of diseases. Arsenic is so dangerous that it can cause cancer. Once arsenic attacks a person, they will die."

People's perceptions about "social hazards" varied significantly. Some arsenic-affected respondents considered social hazard the "result of negligence." Some respondents recognized social hazard as "social degradation," and some considered it "social isolation." It is noted that many in-depth interviewees did not have any understanding about social hazard, but some respondents, in this regard said, "If people have difficulties living in society, then this could be called a social hazard." Some participants considered "social inequality" and "social injustice" social hazard, and they thought that loss of social norms and moral values due to "social injustice" are the resultant forms of a social hazard. Some respondents considered "social humiliation" due to arsenic poisoning social hazard. Such perception is mainly confined to the degradation and dishonor for arsenic-affected people. Some participants assumed that if people contract arsenic-induced diseases, they will be at risk of social isolation, social injustice, and social inequality.

6.5 ARSENIC EXPOSURE AND HEALTH EFFECTS

What do the arsenic-affected people think about arsenic and related diseases? In seeking to explore perceptions about health situations of the arsenic-affected people, their understandings of the problems and the difficulties they are experiencing from their diseases were examined. All arsenic-affected patients were interviewed about the chronological development of arsenicosis with symptoms. In an evaluation work with Caritas Bangladesh in Munshiganj district, the author found a family where two members had died with arsenicosis and another had been diagnosed with arsenicosis.

6.5.1 Symptom Recognition and Health Conditions

A sequential development of arsenicosis symptoms and the meanings that patients attached to them is the focus of this section. During the early stages of their illness, people ignore the symptoms, and they did not realize at first that their tubewell water could be contaminated with arsenic. Several years later, mainly at the beginning of this century, people have been getting information about arsenic in their tubewell water and its toxicity on human health. Since arsenic poisoning is new to them, participants denied the severity of the symptoms due to their unfamiliarity. A patient (48, male, marginal farmer, 2010) from Ghona union of Satkhira district, talking about this issue sounded complacent: "Almost everybody in this village has got black spots on their palms and soles; it is not a disease, if you take rest for a couple of days or if you do not toil in the paddy field, you will recover and no medical treatment will be needed." Generally, the poor people who work in agriculture get many types of skin diseases on their hands and feet. They consider this a normal aspect of their regular life, and they generally do not go to a doctor for skin diseases. One patient (40, male, casual laborer, 2001) in this connection said,

> We have heard something about arsenic, but it does not have any importance for us. If we get any sores on our palms, we always consider that it is due to ploughing the land or digging the soil with a shovel when working in marshy agricultural land. It is not a disease. We call it swelling spots on our hand and feet. We have never thought that arsenic could be the cause of it. Many people have got this type of swelling spots on their hands and feet.

There is a lack of awareness surrounding health issues, and some health problems caused by arsenic toxicity are of a low priority to many rural poor people. The health problems are mainly concentrated among poor people who are not health-conscious and are illiterate. They usually suffer from both malnutrition and undernutrition and most of them are unaware of the seriousness of their illness. Some patients try to ignore their health problems and are not so worried about them, since poverty has captivated them. One respondent (Ema, 47, housewife, 2014) from Magura union of Satkhira district told me that

> I am not worried about my rough skin. I have been experiencing this rough skin for about two years. I use medicine that was given to me by an NGO from Dhaka. They didn't take any money from me for the medicine. I hope this medicine will make me better soon.

Arsenicosis patients describe their disease as "black spots," and this is the most common symptom in some arsenic-prone areas of Bangladesh. Skin lesions caused by arsenic are considered mainly to be a skin disease in Bangladesh, and only a few people know about the relationship between drinking water arsenic and skin lesions. At the primary stage, some "swelling spots" develop on palms and soles and there is "itching." These "swelling spots" turn into "black spots," which develop slowly. Later the skin becomes dark in a spotted form due to the deposition of a black pigment. There is a lot of evidence of this in the interview texts, for instance,

> About six or seven years ago, there developed blisters on my whole body and there was a lot of itching. Few months later, these blisters turned into *zengoo* on my hands and legs. There was itching and some pain. A few years later, these black spots became hard and rough. Now they have turned into sores (40, male, casual laborer, 2001).

Some arsenic-affected people were found to be fearful of arsenicosis. Due to their declining health situation from arsenic related diseases, they also suffer from some social isolation. One patient (47, male, marginal farmer, 2011) said, "I got this disease few years ago. At the beginning, I went to a doctor and he gave me some medicines. I swallowed them, but I did not get any benefit. I am not taking any medicine now since it is expensive and does not work to cure me. My health condition is getting worse. Some people told me that I have arsenic related disease on my hands. They also told me that I will die and there is no curative medicine for arsenic."

6.5.2 Health within Illness

We are interested in the meanings that patients "attach to experience" (Bergum, 1989), i.e., how individuals experience their health problems caused by arsenic exposure. When people come to know that they are affected with arsenicosis and that no curative medicines have yet been invented, their attitude and behavior changes. However, micronutrients such as vitamins A, C, E, zinc, selenium, and folic acid have been shown to be effective in the treatment of arsenicosis (FAO/UNICEF/WHO/WSP, 2010). Patients with a long history of health problems are disheartened, particularly about their declining strength and "inability to do what they used to perform" (Winters, 1997)—in this case heavy physical labor—because of the worsening condition of their palms and soles. Some patients reported that their thinking has changed due to the current arsenic scare.

Some arsenicosis patients find it difficult to do any work with "black spots" on their palms and soles. Moreover, it is difficult to use their fingers if their palms are affected with sores. These black spots are painful, especially if they harden. A young woman patient (23, no occupation) affected with arsenicosis said, "I got blisters and swelling spots on my palms and soles. There is no itching, but a little bit of pain. When I work and write, I get more pain." Some patients thought that they would recover from these skin lesions, but their situation worsened and they are upset at their present health condition. A patient (19, female, no occupation) in this connection seemed depressed:

> My feet and hands are becoming harder and sometimes they are as hard as steel. It is bad looking and I hate to look at my own hands and feet. I'm continuously using ointment and swallowing medicines, but there is no improvement. The situation is getting worse.

Most of the patients know that arsenic is mainly concentrated in tubewell water and that people can die with gangrene or cancer if they are exposed to substantial amounts of arsenic for a long time. Some have heard about arsenic from the radio and television but took little notice. When they realized that they had been continuously ingesting arsenic-contaminated tubewell water and now have arsenic-related symptoms, only then did they become nervous. Some patients have come to know that arsenic poisoning can lead to a "terminal disease." Now even unaffected people are scared of arsenic and panic, thinking that, if arsenic attacks them, they will die with arsenic-related cancers.

Patients' perceptions regarding the symptoms of arsenic-related diseases are mainly confined to skin lesions, i.e., sores, blisters, boils, and swelling spots. In addition, black spots on palms and soles and skin roughness and skin hardness on palms and soles are the symptoms perceived by the indigenous patients. It is noted that interviews and talks were mainly about the kinds of health problems the arsenicosis patients had and the kinds of pain they experienced. Almost all the arsenic-affected patients are now leading constrained lives. In fear of possible social pressures, they are hesitant to talk of their illness and patients refused to talk about their health problems in front of others.

6.6 ARSENIC EXPOSURE AND SOCIAL IMPLICATIONS

Navon and Morag (2004) have referred to liminality as biographical disruption. They were researching adult males undergoing hormonal therapy for advanced prostate cancer and found several disruptive social side effects of the treatment. Many of their subjects were responding positively to the treatment and were therefore hoping for a return to something approaching normal social interaction with their friends and family. In contrast, arsenic patients in Bangladesh are faced with a total absence of effective medicines and, in some cases, very little choice in the water that they drink (Hassan et al., 2005). Arsenic-related disease creates immense social problems like social injustice, social isolation, social uncertainty, problematic family issues, and marriage-related problems (Abedin and Shaw, 2013; Brinkel et al., 2009; Bhuiyan and Islam, 2002; Hassan et al., 2005; Sarker, 2010).

6.6.1 OSTRACISM

If disease appears anywhere in rural Bangladesh, there is a tendency for people of that area to avoid and to isolate the affected people (Brinkel et al., 2009; Tsutsumi et al., 2004; Withington et al., 2003). There is a tendency to neglect arsenic-affected people in Bangladesh since it is thought that arsenicosis is like leprosy or some other contagious diseases (Hassan, 2000). Within the community, arsenic-affected people are barred from social activities and often face rejection, even by their immediate family members. Although a good number of villagers in different geophysical areas of Bangladesh, at different times of the fieldwork, had little or very little knowledge about arsenic, many feared it and assumed the disease to be contagious, even though they were unaware of its symptoms. As a result, some patients experienced social problems due to the visibility of "zengoo" (black spots) on their bodies. These "zengoo" are common among farmers and laborers in Bangladesh. The extreme stage of this "zengoo" makes some patients worry about terminal disease, and, when unaffected people come to know and see the extreme conditions of this "zengoo," they try to avoid these patients.

For instance, Kamal (22, male, farmer), an arsenic-affected patient, lives with his parents in Mollapara of Ghona Union at Satkhira district in Bangladesh adjacent to the Indian border. He discontinued his education due to financial constraints and is now working as a farmer and daily laborer like his father. He ploughs his own land and is a casual laborer in paddy fields of others near Dat-Bhanga Beel (marshy land). His whole family drinks water from highly arsenic-contaminated tubewells located there. He had heard some discussion of arsenic on the radio but did not attach any importance to it since he knew nothing about the subject. He is affected with arsenicosis himself, having had black spots on his feet and hands for six years. He has been to physicians several times for treatment and takes medicines and uses ointments as per their prescriptions, but there is no improvement. Some people no longer talk to him. One of his closest friends said, "Please don't come near me; if I touch you then the disease you have got will infect me." Also, he is marginalized in his own family: "my parents do not say anything directly, but I can understand their feelings and distance." Another seriously arsenic-affected patient (40, male, casual laborer) stated,

> Some people in the *Hatkhola* (periodic market) avoid me indirectly. When I go to any shop for my daily shopping and even to a tea-stall for a cup of tea, some people move away or try to leave. I don't know why they do this. They will not realise my problems until they get this disease themselves. I am very upset at this situation.

Some unaffected people are angry and aggressive. They think that patients should either stay in their homes or leave the village. One focus-group participant (45, male, political leader) summarized this combination of ignorance, prejudice, and fear: "If anybody is affected with gangrene, who will meet him? Who will go close to him? People will always make a safe distance from arsenic-affected patients because of arsenic panic. Everybody in this village is scared about arsenic." Patients complained that some tubewell holders misbehave toward them and do not give them access to their tubewells for collecting water. One was bitter:

> I used to go to a tubewell near to my home for collecting my drinking water. When the tubewell owner has come to know that I have got skin lesions on my body, he then told me not to collect water from his tubewell. He said that I could spread the disease to other people.

One tubewell holder is reported to have said, "Don't disrupt us, sink a new tubewell for yourself and tap your water from there." A respondent at a deep tubewell collecting water commented, "This is a government-owned tubewell. We have the right to access this deep tubewell, but the tubewell-holder and his family members always make problems for us to collect water. What can we do now"? It is noted that the tubewell is just outside the compound boundary wall of the holder, who is also the Ward Commissioner. This person contributed TK5000 (US$65.00) toward its sinking and feels a degree of ownership despite it being a public facility. When this issue was raised in a focus group of elected administrators, the tubewell-holder responded,

> No, I have never told them not to collect water from my (?) deep tubewell. They always quarrel during the collection of water. They collect water from early morning to midnight and we have to put up with noise from tubewell tapping and shrill unwanted sounds from them.

He commented that an additional deep tubewell was essential to reduce the pressure on water collection in this vicinity. It is worth noting that a keystone of government policy on public deep tubewells is to place them under the control of paid guardians. Because this gives social leverage to the holder, there is scope for inappropriate use. Anwar (2001a) pointed out an example of an 18-year-old girl who got arsenic lesions all over her body and could not get out of bed and her friends never visited her. Sarker (2010) investigated that nonparticipation by arsenicosis patients in social activities is because of not only self-restrained participation in the social activities but also there being dislike of other people to participate in their social activities. Due to the patriarchal system and lower sociocultural position of women in the society, unmarried women and women abandoned by husband and families live inhumanly (Chowdhury et al., 2006). Bearak (1998) unveils the life history of Pinjira Begum, 25, an arsenic-affected patient who was seriously ill and many indignities affected her life. It is interesting that most of the people do not know about arsenic and related diseases, but some of them considered the disease a contagious one, even though they do not know whether people are affected with arsenicosis or not.

6.6.2 In-Family Situation

There is an increasing tendency to avoid arsenicosis patients even within families—they are indirectly neglected and isolated. In most of the cases, arsenic victims are abandoned, not only by society but also by their family members. Some family members do not like to talk and hesitate to come close to their in-family arsenicosis patients. Studies found that social and economic loss for people in arsenic areas were acute and rapidly worsening (WHO, 2000b). As one respondent (26, male, farmer) said,

> My parents do not say anything to me directly, but I can understand their feelings and the distance they are making. One day, when I took rest on my bed, my mother asked me, why are you sleeping so much? Go to your work and earn money for the family.

Parents feel hesitant about being close to their children, and husbands keep a safe distance from their wives. A father (55, farmer) suffering from arsenicosis for four years said, "Two of my sons try to avoid me tactfully—they do not like to come close to me. I can understand their situation, but I never let them know about my health problem. It is an appalling situation in a family atmosphere." Korimon Bibi, an arsenicosis woman (60, housewife), expressed her situation with a big pain,

> When I visit someone's house, they don't take it easily now. Few years ago, my neighbours usually used to visit me, but after getting skin lesion on my hand, nobody visits me now. I can understand that why people like to avoid me. Also, my family members, their behaviour has also been changed and they don't talk with me as they did years ago. I don't know what can I do now.

Arsenicosis patients are facing in-family problems due to their financial constraints. It has been investigated that some patients are unable to contribute to their families due to their job loss and some are unable to work for their family. Moreover, some spend a large portion of their daily income for treatment. Abul Kashem (52, male, daily laborer) from Dumuria, Jessore district told his distress that

> Arsenic destroyed my dream. My dream was to educate my son, I thought he will do well in his life with higher education. My son was unable to continue his education due to my financial constraints. I spent a big amount of money for my arsenic treatment, but no improvement of my health. I lost my fingers with arsenicosis and I can't work now. There are so many problems now in my family and at present I can't afford my son's education. I don't know how can I overcome the situation.

Parvin, 17, a young woman who developed black spots on her palms and skin lesions on her whole body, is facing problems in her family. In desperation, she revealed that "My parents are rude to me. I have never seen this behavior before these sores appeared on my body. Probably, I am a burden to this family. I am really upset." Ahmad et al. (2007) and Zaman (2001) points out disastrous situations in the families of arsenic-affected patients.

6.6.3 Marriage and Conjugal Life

Apart from problems in social and daily life, arsenic poisoning also disturbs the marriage system. There is evidence of broken marriages and problems in getting married (Chowdhury et al., 2006; Keya, 2004; Sarker, 2010). Women afflicted with arsenic-induced skin lesion have been reported to be treated as contagious and often abandoned or denied marriage. For instance, wives were found to be divorced or separated from or sent back to their parents' house because of arsenic-related "zengoo." Sarker (2010) investigated that a happy marriage may end in divorce if the wife got visible arsenicosis symptoms on her body. Howard (2010) investigated from Bangladesh that about 8% of females with arsenicosis reported that they had been abandoned by their husbands. On the other hand, there is some evidence that wives left arsenic affected husbands because they were afraid of arsenicosis (Nasreen, 2003).

Problems before marriage are also notable. Generally, people are reluctant to establish marital relationships with those families suffering from arsenicosis. Women with visible signs of arsenicosis are facing difficulties in getting married and dowry is often demanded from women's families (Hanchett, 2004). Sarker (2010) explored that men having arsenicosis do not feel confident in getting married. Arsenic-affected women also experience socially undesirable events like dowry and physical torture (Chowdhury et al., 2006).

Women are socially the most vulnerable. Jarina, 31, who developed blisters and black spots on her body, is neglected by her husband. Her husband does not talk frequently to her now, and no longer asks her about her health situation. It is evident that some arsenic-affected housewives are divorced by their husbands and even forcibly sent to their parental home with their children. This latter extreme is a major problem in Bangladesh. Young women and their parents are certainly aware of the issues. A marginal farmer (46, male) in this connection stated,

> What can I do now? My daughter has got blisters on her whole body and it is gradually getting worse. If she does not recover quickly nobody will marry her. If she is in good health, she can help me in my housework. Now she is sick and she cannot do any work.

Korimon (32, housewife) from Srimongal, Moulvibazar district was facing severe problems in her conjugal life due to her recently developed blisters on her body. In addition, other family members ignore her and they do not like to talk. She stated,

> I have been suffering with my skin disease for the last 2–3 years. After getting my illness, my husband doesn't behave well with me—he gets angry if I want to talk with him. I was happy before getting this

disease. He got second remarried just four months ago. I don't know my future—where will I go. How can I manage my life with arsenic and family problems? Probably, I have to work for my own survival, I can only do housemaid, but I don't know whether I can get this opportunity or not.

Another patient, Lata Sheel (28, housewife) from Narayanganj, told of her problems within the family. She told of her existing situation with a lot of pain:

She has been ignored from her husband and parents in Law. I am now unable to visit my parent's home. They usually ask for bring the treatment costs from my parents. My mother-in-law tells me that your disease has contaminated from your parent's home, so you bring your treatment cost from there. How can I survive now?

During earlier fieldwork in Marua village, Jessore (June 1999), three wives (out of 37 affected women) were found to have been forced to return to their parents and two had been divorced as a direct or indirect result of their illness (Hassan, 2000). In an arsenic-prone area (e.g., Arihazar, Narayanganj district close to Dhaka City), we have investigated that four married women were forced to return to parents and one with visible signs of arsenicosis had been divorced. Young women and their parents are aware of social problems from having arsenicosis. Anwar (2001a) reveals the distrustfulness of parents about the health of their daughters. Families of arsenic-affected patients fear that a victim would be a burden to the family. Parents of a girl with arsenic-induced skin lesions told their problems:

People sometimes ask me, what has developed on your daughter's palms? Why don't you go to a doctor? You will face problems in the marriage of your daughter. We are upset at our daughter's present health condition.

Almost all of the arsenic-affected patients lead constrained lives. In fear of such social problems, they are hesitant to talk of their illness. Some refused to talk to us about their health problems in the presence of others. Others recalled their fear when they first realized that they had arsenicosis. One patient (41, male, mechanic) commented, "I don't show my hands to people, and I try not to tell my problems to anybody. If people come to know my health condition they will not be cordial with me."

6.6.4 Difficulties in Daily Activities

Some patients said that the difficulty of getting daily work or interruptions to daily labor are major consequences of arsenic poisoning. If an adult is affected with arsenicosis, there are subsequent problems in maintaining income stability, particularly if they are very poor, but sometimes they are the only earning members in their respective families. Most of the patients in arsenic-affected areas are engaged in work either in agriculture or as daily laborers. If they are absent due to sickness, they are not paid for days missed. Most patients thought that their skin lesions are the cause of getting work inconsistently. Some said that when people come to know that they are sick, then nobody is willing to provide them with any work. Some employers check the palms of arsenic-affected patients and refuse to provide "zengoo" sufferers with work. According to one informant (22, male, casual laborer),

My boss knows my health condition. One day, he told me that you are sick, you are not able to do any work. Go home and take a rest. When you recover then you can come for the work. I will give you the work then.

One arsenic-affected woman (40, housemaid) from Rajshahi district was found physically weak and she was not getting a job since she got visible skin lesions. She told her situation with distress:

I'm not literate. I manage my family with three children as a housemaid to someone's home. The tubewell water we use for our drinking and cooking purposes are arsenic contaminated. I know this water is not good, but there is no alternative—pond water is not tasty. My neighbour has got a safe tubewell, but they are barred us to collect water from that tubewell. I got *zengoo* on my body. At the early stage, I thought it was itchy and I used to visit a medical doctor and shallowed a lot of medicines, but there is no improvement. Day by day my health situation is getting worse and finally, I was given release from my maid job. They told me that when you come round, we will absorb you again. How can I survive with my children?

Public awareness is necessary to prevent arsenicosis. Unfortunately, poor people living in rural areas of Bangladesh are not adequately informed about arsenic contamination and arsenicosis (Hadi, 2003; Hassan et al., 2005; Khan et al., 2006; Paul, 2004). Parvez et al (2006) claimed that people with a higher socioeconomic status are more aware of the poisoning nature of arsenic than those with a lower socioeconomic status. As a result, some people believe in superstitions, prejudices, and fairy tales (Chowdhury et al., 2006). For instance, some people think that the disease is "an act of the devil" or "a curse of nature" or "the work of evil spirits" (Chowdhury et al., 2006; Hassan, 2000; Nasreen, 2003).

Arsenic-induced diseases not only are causing social difficulties for poor patients but also are creating serious concern among presently unaffected people. Unaffected people are generally scared of arsenicosis, and they tend to avoid and isolate arsenic victims (Chowdhury et al., 2006; Nasreen, 2003; Tsutsumi et al., 2004). Some unaffected people behave in a hostile manner and think that patients should either stay in their homes or leave the village. According to the participants of one focus group, "All the arsenic-affected patients are thinking about the recovery of their health, but we, the unaffected people, are not in a good situation either. We are worried about arsenic. If arsenic attacks us, we will face health and social problems like the arsenic-affected people." Moreover, arsenic-affected families are sometimes barred from taking baths in village ponds.

6.6.5 Schooling Children

In rural Bangladesh, if anybody gets any unknown disease, others consider the disease to be a contagious disease and they think that this disease could contaminate them if they are in physical proximity. Schoolchildren are also experiencing this situation. Talking to affected schoolchildren, it was clear that they also faced prejudice. In some cases, former friends keep their distance sitting close to them. They do not like to share books and pencils, and they do not play with affected children in school. In addition, in some cases, teachers restrict their access to school. One example is Taslima, aged 10, a girl who developed black spots on her palms and soles and is now having problems at school:

Nobody sits beside me in school. They do not like to talk with me, and do not share books. Nobody likes to play with me in school. When I play, some children shout—don't touch her, don't play with her, she's got arsenic. I will not go to school.

Some children hide their symptoms, as one girl (aged 13) confided, "I've got sores on my palms and if I show them or talk about it, my friends will not play with me in school." A line of literature shows evidence in support of this situation. Children with symptoms are not sent to school in an effort to hide the problem (World Bank, 1999), but their entrance to school is also restricted because of this illness. Children with arsenicosis symptoms are not allowed to attend social and religious functions as well as being denied taking water from a neighbor's tubewell and being barred from school (Keya, 2004). This situation is a serious impediment to the children getting an education.

6.7 ATTITUDES OF LOCAL LEADERS AND SERVICE PROVIDERS

Some patients focused their opinions on perceived social injustice and the negligence of their local village leaders. When patients go to them for help, some leaders play positive roles but others less so. Some leaders try to help patients by providing them with financial help, moral support, and advice, while others make commitments but then do nothing. For instance, one arsenic-affected woman had sought help from a local leader: "When I came to know that I am affected with arsenic, I went to our Ward Commissioner for help. I told him everything and he gave me some money for medicines and also told me that he will arrange a consultation with a doctor about my health. I am very pleased with him." Another remembered with distaste the attitude of a local leader who said,

> What can I do for you? I am not a medical doctor or general practitioner. When you have got a disease, go to a medical doctor for your treatment. Only a medical doctor can help you. If you are in political trouble or have other problems, then I can help you.

Patients also had mixed experiences from their own elected local administrators (UP Chairmen and Ward Commissioners). Generally, in rural Bangladesh, when people cannot get help from any other source, they go to their local representative. They are trusted more than any outside organization, but on the issue of arsenic, responses varied from insincere commitments to aggressive dismissal. One response is typical: "Why do you come to me? I cannot do anything for you. It is not my duty to deal with arsenic, I'm scared about it myself." Another member told an arsenicosis sufferer, "You did not cast your vote for me. Don't come to me for any help. I will help my men first." Another response shows a helpful attitude of some local administrators: "I requested my Ward Commissioner to tell the people in my vicinity not to make any problems for me. He replied, 'Oh, yes, I will do it for you, no problem, don't worry.'" He then asked me, "why don't you go to a doctor for treatment? It is not a good decision for not taking any medical treatment."

Some patients have sought credit from NGOs because they do not have any work to sustain their families and are at the stage of "distressed sales" of their assets in order to survive. Seeking financial help from a local NGO was a last resort, and patients generally thought that NGOs could help them because they are engaged in socioeconomic development as well as distributing relief to the flood-affected people. One patient (40, male, casual laborer) was optimistic but received a negative response: "Why do you need credit? How can we help you? You are a patient and you are so sick that you will not work hard. We don't know whether you will be able to repay the installments in time or not. When you recover, we will try to help you."

A very few of the village doctors in rural Bangladesh know a little bit about arsenicosis. This is not just a problem of diagnosis (Murshed et al., 2004) but a general shortage of information at this level of the medical profession. Therefore, arsenicosis patients have difficulty getting satisfactory treatment. One interviewee (40, male, casual laborer) described, "When I asked about my skin problems, three doctors explained the problems in three different ways and they prescribed different medicines for me. So, how can I rely on the prescriptions provided by the doctors?" The vast majority of local people cannot afford doctors' prescription fees and the cost of medicines over an extended period. They often present when their illness is already at a critical stage, when there is very little chance of recovery. Patients reported that some doctors have a tendency to prolong the treatment in order to boost their own incomes. We have no direct evidence that this allegation is true but it certainly seems that, in the absence of diagnostic certainty and therapeutic strategies, the village doctors have failed to reveal the full situation to their patients.

6.8 ARSENIC POISONING AND SURVIVAL STRATEGIES

The survival strategies adopted by arsenic-affected patients can be viewed as (a) coping strategies and (b) adaptation strategies. In a coping strategy, almost all of the patients take an immediate and temporary action for survival (WHO, 2000b). An adaptation strategy refers to the long-term and permanent attitudes of the arsenicosis patients in solving their health and social problems (Hassan, 2003). Some patients made decisions to solve their health and social problems quickly and others deployed a combination of adaptation strategies for the long term. Respondents from different geophysical areas of Bangladesh in different times talked of a combination of coping strategies that they employed during their critical health situations.

6.8.1 Coping Strategies

Frequent visits to physicians were thought to be necessary in order to get the health situation under control. Seriously affected patients usually go to a doctor, but most do not until their health condition is poor. Some patients with the worst symptoms went to several doctors but found no improvement in their health. One (55, male, farmer) told me that

> I went to several doctors several times. I've been taking medicines and ointments as per their prescriptions for the last five years, but there is no improvement of this disease, and the situation is getting worse slowly.

Some very poor patients take cheap treatments from quack doctors since they have lost their trust in mainstream doctors. Others rely on home remedies such as rubbing the skin with a small piece of garlic soaked with warm mustard oil. Some patients use traditional systems of treatment including amulets worn on their arms or waist or applying charmed oil or water to their wounds. Some rural poor people believe in these traditional treatment systems, but others do not. A patient who had been suffering from arsenicosis for about seven years was skeptical (40, male, casual laborer): "No, I don't believe it. It is meaningless to me to wear any amulet on my arm or on my waist. It is fake. How does an amulet work where the medicine does not work for this disease? What benefit will I get from an amulet?" Another female patient (23, no occupation) was more trusting: "When doctors fail, then amulets work well. I hope I will come round within a very short period of time. I've a trust in it."

In the context of social implications, the first strategy involves keeping a safe distance from the unaffected people in order to avoid social embarrassment. The most seriously affected patients do not feel able to go outside thinking that, if they leave their home, people will make hurtful comments to them. One patient (55, farmer) recounted his experience:

> One day I was at the local for my regular green vegetables. Somebody then started to talk about arsenic poisoning in my presence and at a certain point they made a criticism about my health. They even asked me why I was spreading out this disease in this society. I am very distressed about this situation. I have decided not go outside for any reason if I can avoid it.

Some patients decide not to attend social activities and functions and even not to continue with some personal relationships. One seriously affected patient (48, male, farmer) said, "… I went to a wedding and some people made problems there. I realised the situation and came back home. It was a really embarrassing situation for me and for the other guests as well." Moreover, very close friends may isolate arsenic-affected patients in different ways since they think that arsenic is a contagious disease. Keeping this in mind, some patients always avoid public situations. In addition, some arsenicosis patients thought that, since arsenic poisoning is new, participants denied the severity of the symptoms due to their unfamiliarity. Health issues and some health problems caused by arsenic toxicity are of a low priority to many rural poor people. A patient in this issue reveals his complacency:

> A good number of people in this village got black spots on their palms and soles, it is not a disease, if you take rest for a few days or if you do not toil in the paddy field, you will get recovery.

The second strategy covers coping with in-family problems. One affected person (26, male, farmer) experiencing such problems remarked, "after getting these sores on my palms, I am facing ignorance from my parents. I have decided not to talk with them and not to meet them. I think I am a burden to this family. Everybody in the family is rude to me." Some arsenic-affected children were found to keep a safer distance from their parents since they thought that arsenicosis is a contagious disease like AIDS. This is why they do not like to use and share the common objects of the family. One mother (49, housewife) stated, "My son seldom comes to me. He does not share the common plates and bowls—he uses his own." Some patients, especially young women, have problems since it is difficult to arrange a marriage for them. People are generally not interested in making new relationships with those from an affected family.

> I am about 19. My parents are always worried about my marriage. I have decided not to marry. I want to leave this village. I will work in a family as a maidservant in a different area. I hope that will make my parents happy (female, no occupation).

The third strategy covers children affected with arsenicosis, who have difficult access to school. They cannot play with their friends, and even some of their teachers neglect them. Some children now refuse to go to school, and they discontinue their education. They may have already missed a significant number of school days, or children's parents may decide to withdraw them from school. One parent of an 11-year-old child said,

> I have decided not to send my child to school. If there is not a tolerable environment and the teachers do not take care of them, why should I send my child to that school? If he stays at home, it is better for his mental health.

On the other hand, some children, especially girls, deliberately hide their arsenic symptoms. Such children want to continue with their education. The mother of a 10-year-old girl explained, "My daughter always avoids appearing in public. She goes to school covering herself (*borkha*) to make sure that no one sees the skin lesions that she has developed during the last two years." When asked about her situation, the girl added that "My mum strongly advised me not to show my skin lesions to anybody and not to say anything about my problems. My friends ask me why I wear a *borkha*. I cannot play with my friends if I am covered with this *borkha*."

6.8.2 Adaptation Strategies

Some seriously affected patients take medical treatment for their arsenic-related blisters and skin lesions. Although this is an expensive adaptation, they hope that if their health improves, they will be able to live as an accepted member of society. In a question concerning this measure and its effectiveness, one patient (41, male, mechanic) replied,

> What are the alternatives? I think this is the best possible way to save yourself from social injustice. If you continue the medication for a long time, you could get well and if you are well, why social isolation? People will do nothing if you are well.

Some patients thought that drinking filtered water or boiling water could remove their health problems. They came to know that arsenic-safe water was the only option to prevent arsenic poisoning. Boiling surface water and filtering were the obvious measures to take. One patient (32, male, farmer) stated, "If I can get arsenic-safe water by boiling pond water, I will do it. This arsenic-safe water could cure my skin lesions, and, if this happens, the social isolation that I am experiencing now will disappear." A few decades ago, boiling the surface water was a preventative measure against cholera and diarrhea, but, unfortunately, boiling the tubewell water is useless against arsenic toxicity. It is interesting that most respondents neither boil nor filter their water. Their main adapting strategy is to collect arsenic-safe water from the nearest deep tubewell. The following is a typical view: "I prefer a deep tubewell in place of using pond water, because people farm fish in their ponds, they bathe there and the water will not be clean afterwards."

We have seen that arsenicosis leads to changes in work responsibilities inside and outside the home. Patients are often physically unable to conduct laborious work in agriculture, and there is a reduction of income supporting the family. In such cases, degrees of reliance on other family members may increase in order to sustain the household economy of patients. As the wife (35) of one patient explained,

> My husband is unable to work in agriculture. His palms are full of *zengoo* and nobody wants him. So, I go to the fields and earn some money. My daughter [aged 11] also works and contributes to the family. Until he improves, we will continue to do that.

Some sufferers and their families think that if they can establish a relationship with well-known local people such as social activists, political leaders, and elected administrators, they can save themselves from social injustice. This policy can be seen as an adaptation mechanism at the community level. In addition, some organizations have planned awareness campaigns with the inclusion of arsenic messages in existing health and education programs. The impact so far, however, has been minimal in some arsenic-affected areas of the country. These are indirect survival measures. For example, one very poor man commented, "I do not have any access to the deep tubewell. I have told my wife and son to collect arsenic-safe water from a deep tubewell. I have come to know that the use of this water could cure my health problems." In addition, some patients have reduced the consumption by different family members of staple foods and other consumption items over the long term in order to reduce expenditure.

The final adaptation strategy might be called complacency. We found many people who, although they had heard of arsenic, claimed to be unconcerned. At different stages in public consciousness, it seems that arsenic patients are spatially marginalized, but most of the unaffected feel no need to mobilize for either prevention or mitigation. This attitude, sometimes fatalistic, is summed by one respondent (male, 30, farmer): "Why panic? Arsenic will not be a problem if God wants to keep me alive. Will God give you longer to live if you drink arsenic-free water?"

6.9 HEALTH AND SOCIAL HAZARDS

The pattern of arsenic exposure and its toxic effects can lead to an understanding of the pattern of environmental health hazards. A hazard is defined as the potential to cause harm or a general source of future danger (Gerrard, 2000). An environmental health hazard is therefore concerned with the nature and magnitude of harm to human health from a hazard event present in the environment. A hazard is not deemed to be synonymous with risk, although it can be a determinant of risk. Risk can be considered the possibility of suffering harm from a hazard, i.e., it is the likelihood of physical harm or adverse health effect due to any substance or technology or other processes (Beck, 1992). Health hazards may cause measurable changes in the body, and these changes are generally indicated by the occurrence of signs and symptoms in the exposed population. The chronic effect of arsenic is obviously a concern in the environment. It has been proved from epidemiological evidence that arsenic has measurable adverse health effects, especially the possible increased cancer risks as discussed in Chapters 4 and 5. Arsenic-affected people have already experienced health problems, and their health situation is getting worse. The worsening condition of patients' health year after year is noticeable. Some arsenicosis patients were found to be so ill that it was difficult for them to earn a living. Patients affected with arsenicosis for a long time are in pain and they are adopting various coping strategies for their health problems, but there is no improvement. They become anxious and depressed, and some panic about arsenicosis when they come to know that no curable medicines have yet been invented (Figure 6.2).

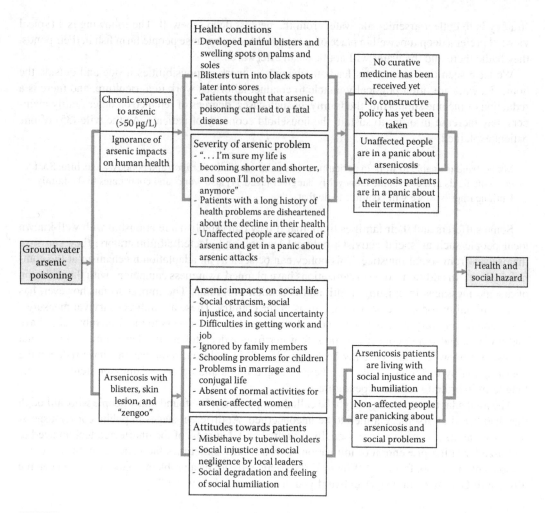

FIGURE 6.2 Health and social hazard posed by groundwater arsenic poisoning in arsenic-affected areas of Bangladesh.

Arsenicosis creates social implications in Bangladesh that are producing social stigmatization and discrimination. A social hazard is concerned with the characterization of the nature and magnitude of harm to people's social norms and social structure from a particular event (Hassan, 2003). Arsenic can be considered a social hazard if it represents a challenge to people's social status, their lifestyle, and sometimes their attitudes, whether measured in terms of "social degradation" or "social injustice." People in rural Bangladesh are concerned about arsenic poisoning, there are arsenicosis patients who are already experiencing many social problems, and a hazardous social situation is developing generally (Figure 6.2). Arsenic is not the only cause of toxicity to human health, but it results in major social dislocation for the affected people. Patients identified in Bangladesh are adopting various survival strategies for their social problems, but they face continuous hostility.

> What can I do now? I'm very upset about the social problems that I have been experiencing after getting this skin lesion. Everybody in this village treats me with disdain. They are rude and angry and I do not get any sympathy from anybody. I feel that it is unsafe to live here (40, male, casual laborer).

Most of the patients know that arsenic is mainly concentrated in tubewell water and that people can die with cancer if they ingest substantial amounts of arsenic for a long time. But some social problems caused by arsenic toxicity are a low priority to many rural poor people. Arsenicosis is found to be more prevalent among the poor (Argos et al., 2007; Chowdhury et al., 2006; Khan et al., 2006; Sarker, 2008) who suffer from dietary deficiency and who have no alternative sources of safe drinking water as well as who are unable to get proper care and treatment because of financial constraints. Most of the arsenicosis patients are unable to afford their treatment, which leads to social crisis. This lack of treatment further deteriorates the overall health and economic conditions of arsenicosis victims because poverty rises as the untreated poor victims are incapable of doing hard work and gradually lose strength to move. This disease is associated with social discriminations such as losing jobs, barriers to accessing new jobs, and social rejections (Brinkel et al., 2009).

There are some social risks that unaffected people are not aware of, but arsenicosis patients can weigh these risks on the basis of their practical experiences. Arsenic victims are often wrongly identified as leprosy patients and isolated from their close relations (Hassan, 2000). The medical risk of arsenic toxicity can be stated statistically, but the present concern of this chapter is how arsenic-affected patients are living within the context of their illness. The above discussion of the social problems reveals a picture of social hazard faced by arsenic-affected people, driven by loneliness, social injustice, and damage to social bonds. Some people affected by arsenicosis were found to be leading miserable lives. A national daily newspaper has reported that, frustrated by the treatment of the local doctors, a woman patient went to India for better treatment, but, failing to be cured, she attempted to commit suicide by taking poison (*The Daily Star*, 4 July 2001, Dhaka). In rural Bangladesh, a devastating situation arises when people die from arsenicosis—some local clerics are not burying them with Muslim rites (Hassan et al., 2005).

6.10 CONCLUDING REMARKS

Qualitative analytical procedures used here have provided insights into the lay understandings of the arsenicosis patients about arsenic and its toxic effects on health as well as the social implications of arsenic poisoning. Survival strategies of affected patients are the important aspects of this chapter. Participant observation, in-depth interviews, and focus-group discussions were used to explore people's perceptions concerning their health problems and their understandings about arsenic toxicity. Important grassroots opinions were expressed by the interviewees. The opinions of arsenicosis patients regarding health and social hazards and their survival strategies were collected through qualitative enquiry. In this chapter, the patients' own ideas about their health situation during their illness were also explored.

It was found that patients' experiences reveal severe negative social impacts. This involves living with social uncertainty, social injustice, social isolation, and problematic family issues. It was also found that a sharp difference of perceptions about arsenic and social issues between the arsenicosis patients and unaffected people exists. The unaffected group mainly focused on measures to prevent arsenic-induced diseases, such as the consumption of deep tubewell water, rather than on the existing social problems experienced by affected people.

Research on health aspects based on qualitative data remains important since it allows for a complementary understanding of the contextual aspects for people's narratives of their own lives (Williams et al., 1998b). This chapter has explored the patients' own ideas about their health and social problems and social management of them, i.e., what they think and do in terms of survival strategies and the solutions they envisage. The dissemination of arsenic information can change the behavioral pattern of people in arsenic-contaminated areas. If people get more information about the poisonous nature of arsenic and its preventive measures, there will be the possibility of minimizing arsenic risk to health and social issues. Madajewicz et al. (2007) argued that some 60% of people who learn that the well they used before the information campaign is unsafe change to another well within one year.

7 Policy Response and Arsenic Mitigation in Bangladesh

We have increased conservation spending, enacted legislation that enables us to clean up and redevelop abandoned brownfields sites across the country, and implemented new clean water standards that will protect us from arsenic.

Susan W. Kelly
Former Member of the US House of Representatives, 1936

In 2002, the United Nations Committee on Economic, Social and Cultural Rights (CESCR) recognized access to water as an independent human right:

> The right to water clearly falls within the category of guarantees essential for securing an adequate standard of living, particularly since it is one of the most fundamental conditions for survival

(WHO, 2003: 8)

The CESCR, in 2002 on the Right to Water, has noted that a core content of the right to water is that "water required for personal or domestic use must be safe, therefore free from micro-organisms, chemical substances, and radiological hazards that constitute a threat to a person's health" (CESCR, 2002: Article 12b). The CESCR has also mentioned that

> water should be treated as a social and cultural good, and not primarily as an economic good. The manner of the realization of the right to water must also be sustainable, ensuring that the right can be realized for present and future generations.

(CESCR, 2002: Article 11)

At the international level, Bangladesh has recognized the human right to safe drinking water and sanitation on several occasions. The country accepted the UN General Assembly resolution 64/292 (The human right to water and sanitation) of 28th July 2010, which "Recognizes the right to safe and clean drinking water and sanitation as a human right that is essential for the full enjoyment of life and all human rights" (UN, 2010). As a member of the Human Rights Council, Bangladesh adopted various resolutions affirming that the human right to safe drinking water and sanitation is derived from the right to an adequate standard of living (UNHRC, 2012). Among the Least Developed Countries, Bangladesh has taken this right most seriously. In 2013, it adopted a Water Act that declared water for drinking, sanitation, and hygiene as "the highest priority right" (GOB, 2013).

It was reported that about 68% of people in Bangladesh used to drink improved sources of drinking water in 1990, and the figure increased to 87% in 2015. This means that 41% of people have gained access since 1990 (JMP, 2015) as a result of the installation of about 12 million hand-pump tubewells and other improved sources of water. In addition, 61% of people are using improved sanitation facilities compared to 34% in 1990 due to access to improved sources of water (JMP, 2015). The availability of improved sources of water contributed to a remarkable decline in infant mortality, from 176 per thousand live births in 1960 to 100 per thousand in 1990, and the figure further reduced to 31 per thousand live births in 2015. Under-five mortality also fell from 264 per thousand live births in 1960 to 144 per thousand live births in 1990 and further to 38 per thousand in 2015 (IGME, 2015). Until the middle of the 1990s, the installation of hand-pump tubewells was the only element of national water policy that had widespread support and that did not present major technical difficulties (Black, 1990). The other strands, including the Flood Action Plan, sanitation, water pollution, irrigation, drainage, cyclone shelters, and fisheries, were all more or less problematic and their policies widely criticized (Wood, 1999).

Access to safe drinking water is considered a fundamental human right in protecting human dignity. The Sustainable Development Goals (SDG) of the United Nations are the framework for global development after the termination of the Millennium Development Goals (MDG). The SDG was adopted by the leaders of 193 countries of the world who unanimously adopted the post-2015 international development agenda for the period of 2016–2030. The SDG, also known as "global goals," build on the success of the MDG. With 17 goals and 169 targets, the SDG represent a new agenda to end poverty, fight inequality, tackle the adverse effects of climate change, and ensure a sustainable future for all. One of the SDG is directly linked with safe water—Goal 6 (clean water and sanitation), which mainly focuses on "achieving universal and equitable access to safe and affordable drinking water for all by 2030" (http://www.un.org/sustainabledevelopment/water-and-sanitation/).

With its vast, intersecting problems of poverty and environmental vulnerability, and a limited administrative capacity, the government of Bangladesh has struggled to formulate and implement a national water policy (Atkins et al., 2007b). This situation has been so comprehensive that Wood (1999) has commented acidly that "policy towards water is too important to be left only to those traditionally in charge of it." The legitimacy of the elected government itself is perhaps questionable when there is such neglect of both the national interest and the interests of water users and consumers, but some argue that the recent reshaping of environmental sovereignty in Bangladesh into a new form shared between local politicians, non-governmental organization (NGOs), and the aid industry means that we must reassess governance in terms of an ethics of distributed responsibility (Wood, 1997). Water is a well-known hazard in Bangladesh, with riverine floods causing annual disruption and death, but this chapter will investigate a different aspect of water policy and practice. Pollution by one of the trace elements in the groundwater has caused a major environmental health emergency. This chapter describes a number of issues about safe drinking water policy including improved sources, human health protection with water-related disease prevention, distribution or supply with proper management, and ensuring water rights.

7.1 ARSENIC TOXICITY: VICTIMS OF VENOM

In the 1980s, symptoms of what seemed at first to be a strange skin disease began to appear in the rural areas of Bangladesh and West Bengal of India (Saha, 1995). Cases of arsenic poisoning were noticed among people who had crossed into India from Bangladesh. Dhar et al. (1997) found a woman from Bangladesh adjacent to the Indian border with West Bengal, who had arsenical skin lesions that developed after her marriage. The woman also revealed that many of her relatives and neighbors living in two neighboring villages had similar skin lesions. In an International Conference in Kolkata in 1995 on arsenic poisoning, the School of Environmental Studies (SOES) from Jadavpur University, India, identified patients with arsenic-related skin lesions from both Bangladesh and West Bengal in India (Chakraborti et al., 2015). Later in 1998 and 1999, the British Geological Survey (BGS) conducted a rapid assessment survey of some 3500 tubewells used for domestic water supply across the country and reported that 25% of the surveyed tubewells were above 50 µg/L of arsenic (BGS/DPHE, 2001).

The numbers of people affected were small, but by the 1990s there was a flood of cases of hyperpigmentation (dark spots) and small hardened lumps (keratoses) on hands and feet (which were often disabling), as a result of proneness to fungal infections, and occasionally they became malignant (Atkins et al., 2007a). An epidemiological link was eventually made to water consumption because these symptoms were similar to those of arsenic poisoning in Taiwan, with its associated "blackfoot disease." The laboratory testing of Bengali patients' hair confirmed the diagnosis of arsenicosis, and there is no longer any doubt that a major problem exists. By analogy with Taiwan, consumption of arsenic-contaminated water in Bangladesh over periods of 5–20 years will lead to cancers of the skin (Fraser, 2012), lung (Sherwood and Lantz, 2016: 137), bladder (Medeiros and Gandolfi, 2016: 163; Melak et al., 2014), liver (Lin et al., 2013), and kidney (Hsu et al., 2013); hypertension (Stea et al., 2014); and cardiovascular disease and peripheral vascular disease (Barchowsky and States, 2016: 453), which is characterized by black skin discoloration, ulceration, and possibly dry gangrene (Clewell et al., 2016: 511; Kippler et al., 2016; Kumar et al., 2016a; Kuo et al., 2015; Sarma, 2016: 127; WHO, 2001).

Groundwater arsenic contamination in Bangladesh was first confirmed by the Department of Public Health Engineering (DPHE) in Chapai Nawabganj in late 1993. Since then a large number of local, regional, and national investigations have been conducted in the country. The Bangladesh Water Supply Arsenic Mitigation Project (BAMWSP) with the financial support from the World Bank completed screening of all the water supply wells in 190 arsenic-affected upazilas during 2000–2003. A survey was undertaken by DPHE and JICA, and it found that arsenic contamination is highest in the central part and lowest in the northern part of the country (DPHE/JICA, 2010). Arsenic is the single largest threat to safe water service provision, but others include low water table and bacteriological contamination of shallow tubewells.

The population exposed to drinking water with arsenic contamination above 50 µg/L was estimated to be 20.2 million by the BAMWSP (World Bank, 2007). It seems that about half of the exposed population is living in the severely affected areas, where more than 80% of the water sources are contaminated with arsenic. It has been estimated by the United Nations Children's Fund (UNICEF) (2011) that some 12.6% of the population is still consuming arsenic-contaminated water above 50 µg/L and some 3.1% of the population is drinking water with arsenic contamination higher than 200 µg/L, which is a huge threat to public health. If the World Health Organizaton (WHO) guideline value of 10 µg/L is considered, the population at risk increases to 23.1%. In a report from the DPHE/JICA (2010), it was suggested that some 3132 unions in 301 upazilas were found to be arsenic-contaminated with more than 82 million people exposed to arsenic in varying degrees. The report also revealed that there is a large gap between the number of installed safe water options and the areas urgently in need of safe water options.

Gradually, very gradually, in the 1990s it dawned on the Bangladeshi authorities that they had discovered perhaps "the largest mass poisoning of a population in history" (Smith et al., 2000a). About 28–35 million people regularly consume groundwater with levels of arsenic content that are considered unsafe (BGS/DPHE, 2001). About 37,000 arsenicosis patients were officially recognized by the Directorate General of Health Services (DGHS) survey in 2009. Yu et al. (2003) estimate a future likelihood of two million cases of arsenicosis, including 125,000 cancers, and there are already 9000 arsenic-related deaths a year (Lokuge et al., 2004). Ahsan (2000) estimates a probable 3 million cases of arsenicosis over a 30-year period. In the most heavily contaminated areas, Smith et al. (2000a) speculate that the long-term drinking of water containing 500 µg/L of arsenic may result in one in ten persons dying from arsenic-caused cancers, including those of the lung, bladder, and skin. This raises issues about the ethical responsibility of the experts who recommended the use of groundwater (Atkins et al., 2006).

7.2 ARSENIC POLICIES IN BANGLADESH

7.2.1 Relevant Policies and Strategies

"The global response to the arsenic crisis in the Bengal Delta has been marked by staggering inertia" (Meharg, 2004b: 19). Although the presence of arsenic in groundwater was known about in West Bengal in the 1980s, it was not until large-scale testing in the 1990s that it came to light in Bangladesh. The hazard itself had its origins decades ago with the tapping of aquifers that had arsenic naturally present. Millions of hand-pump tubewells have since been sunk into the soft deltaic sediments with the intention of providing an alternative to the microbe-abundant surface ponds and dug wells that were used traditionally in rural areas (Atkins et al., 2007b).

The water policies and strategies concerning groundwater arsenic issues include arsenic mitigation from drinking water and water supply in Bangladesh. A number of policies have been promulgated over the last two decades for groundwater arsenic mitigation (Table 7.1). The National Policy for Safe Water Supply and Sanitation (1998) represents the "parent policy," which stipulated the specific objective of facilitating access of all citizens to a basic level of services in water supply and sanitation. A few policies and strategies augment the national policy, and they focus on arsenic-safe water as well as water supply and sanitation promotion. They are (a) the National Water Policy 1999, (b) the National Water Management Plan 2001, (c) the National Policy for Arsenic Mitigation 2004 and the Implementation Plan for Arsenic Mitigation in Bangladesh, (d) the Sector Development Plan (2011–2025), and (e) the National Strategy for Water Supply and Sanitation 2014. In 2005, the Government developed the Sector Development Programme—Water and Sanitation Sector in Bangladesh (SDP-WSSB), which provides an avenue for the various policies to be incorporated on to a single platform. The Government targets were to provide access to pure drinking water to the entire population by 2011 and access to sanitation for all by 2013. Clearly these were aspirational, rather than realistic, achievable targets.

The National Water Policy 1999 in its water supply and sanitation section (Section 4.6), identifies surface water pollution, arsenic contamination in groundwater, lowering of the water table, seepage of agrochemicals into shallow aquifers, and salinity intrusion in coastal belts. The policy includes protection of water resources from pollution and facilitates the availability of safe and affordable drinking water supplies through various means, including rainwater harvesting and conservation as well as the necessity of increasing awareness on the issues to overcome water supply and sanitation problems (GOB, 1999).

The National Water Management Plan 2001 was formulated specifically to address the widespread groundwater contamination by inorganic arsenic toxicity. The policy focused on assessing the current and future extent of contamination, the implications for food safety of irrigating with arsenic-contaminated water, and the effectiveness of treatment methods for domestic water supplies. The policy includes measures to mitigate arsenic issues in the WSS (Water Supply and Sanitation), agriculture, and health sectors. The policy was given highest priority and urgent attention to achieve sustainable and affordable solutions in all areas where arsenic is a problem (GOB, 2001).

The National Policy for Arsenic Mitigation 2004 and the Implementation Plan for Arsenic Mitigation in Bangladesh stated goals of providing safe drinking water for domestic use in the arsenic affected areas. The government initiated several programs in such areas to reduce future adverse health impacts of groundwater arsenic poisoning and meet the policy goal. The Arsenic Mitigation policy attempted to incorporate new knowledge in arsenic issues and mitigation as well as an implementation plan.

The new National Water Management Plan was approved in March 2004 with a subsidiary National Policy for Arsenic Mitigation, which empowers and mandates the government to "facilitate availability of safe and affordable drinking water supplies through various means" and enunciates a policy that "access to safe water for drinking and cooking shall be ensured through implementation of alternative water supply options in all arsenic affected areas. All arsenicosis cases shall be diagnosed and brought under an effective management system." There are sections in this document on well-screening, identification and management of patients, mitigation, research, awareness raising, and alternative, safe supplies. Potentially the most significant move is "towards decentralized planning and delivery of safe water options and health services through the grass root level local government institutions," but whether this is a genuine move to empower the grassroots, or a fig leaf to cover the government's embarrassment at its own incapacity, remains to be seen (Atkins et al., 2007b).

The Sector Development Plan (2011–2025) for Water and Sanitation in Bangladesh strongly recognizes serious insufficient efforts in (a) arsenic investigation, assessment, monitoring, and management;

TABLE 7.1
A Timeline of Arsenic Issues in Bangladesh

Time	Events (Projects, Policy, Strategy, etc.)
1970s	UNICEF/DPHE begin installing TW to improve drinking water quality, but by 1990s private money is responsible for most new wells.
1976	First report of arsenicosis, in northwest India.
1978	Arsenic first found in West Bengal wells.
1982	First clinical case recognized in West Bengal.
1984	First patients from Bangladesh.
1993	BGS water quality report fails to mention arsenic.
1993	Arsenic found in groundwater at Chapai Nawabganj in northwest Bangladesh.
1995	Halcrow/DHV survey of baseline groundwater quality.
1995	International Conference at Kolkata hears about Bangladesh problem.
1996	BGS awarded DFID grant and confirms groundwater contamination.
1996	National Steering Committee, Scientific Research Committee, and Arsenic Technical Committee formed.
1997	World Bank fact-finding mission, followed by $44 million grant for the Bangladesh Arsenic Mitigation Water Supply Project (BAMWSP), implemented by DPHE: screening, community development, and mitigation program in 188 upazilas, 100 pourashavas; training of 2000 doctors and 11,000 health workers.
1997–99	DPHE/UNICEF test 51,000 wells using FTKs.
1998	Conference in Dhaka, followed by articles in *The Guardian* by Fred Pearce.
1998–99	Phase I survey of 3534 wells by Mott MacDonald Ltd, 6 volume Phase I BGS report.
1999	Phase I of BAMWSP National Emergency Screening Programme, testing of 49,000 wells in six upazilas.
1999	National Coordination Conference, WHO/GoB.
1999–06	Environmental Sanitation, Hygiene and Water Supply in Urban Slums and Fringes Project (abbreviated to the Urban Slums and Fringes Project) to help urban slum people in Bangladesh in water and sanitation sector, mainly in hygienic behavior with handwashing.
2000	Phase II of BGS project completed, 4 volume report published in 2001.
2000–01	Rapid Assessment of Household Level Arsenic Removal Technologies project, funded by WaterAid/DFID, implemented by WS Atkins International Ltd.
2000–03	Environmental Technology Verification—Arsenic Mitigation Project, CIDA funded.
2001	ICDDR-B research in Matlab, on the effects of exposure to arsenic, funded by WHO, SIDA, and AusAid.
2001–03	Phase II of BAMWSP in 15 upazilas, with BRAC, Grameen Bank, and other NGOs.
2001–04	Australian Arsenic Mitigation Programme for Bangladesh and West Bengal.
2001–04	DANIDA Arsenic Mitigation Pilot Project: 85,000 TW screened, 950 deep wells installed by DPHE.
2001–04	UNICEF capacity building of public health workers to identify and treat patients in 44 upazilas.
2001–05	USAID/World Vision screening, awareness raising, safe water provision.
2001–05	DFID funding for Arsenic Mitigation and Measurement Project, DPHE/UNICEF test 1.3 million wells in 45 "hot spot" upazilas, 15,000 health workers trained, awareness communication to 15 million people; Arsenic Policy Support Unit of the Local Government Division designing national strategy and action plan; National Arsenic Mitigation Information Centre opens. Emphasis 2002–05 on provision of safe water options.
2001–06	DFID supported Environmental Sanitation, Hygiene, and Water Supply in Rural Areas (ESHWSRA) Project to help rural poor people in Bangladesh in water and sanitation sector, mainly in hygienic behavior with handwashing.
2002	National protocol for the diagnosis of arsenicosis patients and recording system, Directorate General Health Services, funded by the WHO.
2002	International Workshop on Arsenic Mitigation in Dhaka, WHO/GoB.
2002	National Expert Committee on Arsenic.
2003	DGHS/DCH Trust develop patient treatment protocol and train doctors and health workers in 80 districts, UNICEF funding.

(Continued)

TABLE 7.1 (*Continued*)
A Timeline of Arsenic Issues in Bangladesh

Time	Events (Projects, Policy, Strategy, etc.)
2004	National Policy for Arsenic Mitigation.
2005–10	$40 million from World Bank for the Bangladesh Water Supply Programme Project (BWSPP) to help 21 villages worst affected with piped water system and install 7000 deep tubewells for arsenic safe water.
2007–13	Spent $100 million for the Sanitation, Hygiene Education and Water supply in Bangladesh (SHEWA-B) funded by Government of UK's Department for International Development (DFID), the Government of Bangladesh (GOB), and UNICEF to provide facilities for 21 million people from some areas in 19 districts and 18 municipalities.
2012–17	World Bank funded $75 million Bangladesh Rural Water Supply and Sanitation Project (BRWSSP) for building 28,000 new piped household water connections as well as constructing or rehabilitating 14,000 water points to provide safe water to villagers in 383 unions in 33 districts with acute arsenic contamination and low access to a safe water supply.

(b) protection of groundwater quality; (c) licensing of groundwater abstraction; (d) licensing for tubewell drilling; (e) assessing the sustainability of current and planned practices; and (f) creating the necessary legislative framework. The Sector Development Plan has given priority for action in arsenic mitigation to (a) removing exposure at contaminated water sources and (b) protecting presently safe and new wells in arsenic risk areas. The actions can be done in the medium-term (2016–20) and long-term (2021–25) planning periods of the Sector Development Plan (GOB, 2011).

The National Strategy for Water Supply and Sanitation 2014 addresses its target to give priority to arsenic mitigation. This National Strategy for WSS considers several strategic directions, and they are focused on (a) importance of arsenic screening and monitoring of all potential contaminated tubewells with the objective to identify arsenic-contaminated wells, arsenic patients, and populations at risk; (b) priority on arsenic-laden sludge disposal and research; (c) testing all new groundwater supply sources for arsenic before commissioning for installation; (d) preference given to arsenic mitigation technology; (e) carrying out public awareness campaign to warn about arsenic poisoning and remedial measures; (f) the process of lowering the standard for arsenic in drinking water to 10 µg/L; (g) promoting piped water supply in arsenic affected areas wherever feasible; and (h) coordinating arsenic-related activities between various ministries and divisions (e.g., Local Government Division (LGD), Ministry of Health and Family Welfare—MoH&FW, Ministry of Agriculture—MoA, and Ministry of Water Resources—MoWR) as well as between government agencies, NGOs, and the private sector at different levels, from national to union (GOB, 2014).

7.2.2 NATIONAL INSTITUTIONAL FRAMEWORK

The most striking feature of the institutional landscape in arsenic-related interventions is that of engaging a large number of agencies (governmental and non-governmental), but there is poor coordination in current institutional arrangements at the national level to face the challenges of the sector. In 2004, the Government of Bangladesh formulated a National Policy for Arsenic Mitigation, which was translated into the Implementation Plan for Arsenic Mitigation. An Arsenic Policy Support Unit (APSU) and a National Committee for the Implementation Plan for Arsenic Mitigation (IPAM) was established. However, APSU no longer exists. It is interesting that there is nothing about the provision for arsenic mitigation in the National Agriculture Policy, the National Water Policy, or the National Health Policy, although policies for these three different sectors are affected by groundwater arsenic contamination. In 2009, the Policy Support Unit (PSU) of the Ministry of Local Government, Rural Development, and Cooperatives (MoLGRD&C) called for a review of the IPAM. This represented an important step in arsenic mitigation efforts (UNICEF, 2010). Figure 7.1 shows the activities and roles of different stakeholders in groundwater arsenic mitigation in Bangladesh.

Policy Response and Arsenic Mitigation in Bangladesh

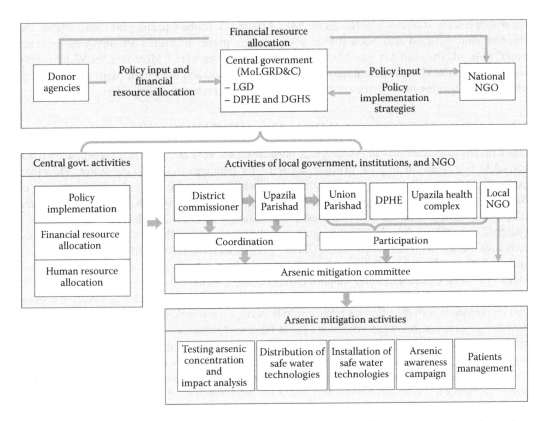

FIGURE 7.1 Activities of different stakeholders and their roles in groundwater arsenic mitigation in Bangladesh. (Modified from Khan, N. I., and H. Yang. 2014. *Science of the Total Environment* 488–489:493–504.)

At the national level, the LGD of the MoLGRD&C is responsible for the overall development of the Water Supply and Sanitation (WSS) sector. The DPHE, within the LGD under the MoLGRD&C, is the lead government agency and technical organization responsible for supporting the LGD and the local governments in planning and delivery of solutions and the management of the arsenic problem as well as sanitation services all over the country except for some urban areas, mainly outside the areas covered by the Water Supply and Sewerage Authority (WASA). In the urban areas, the DPHE was originally responsible for the WSS service, but gradually the Paurashavas (municipalities) and the city corporations are becoming more involved in planning, implementing, and managing the water systems. WASA is another government agency accountable for water supply in the metropolitan cities. In addition, the Local Government Engineering Department (LGED), also under the LGD, implements the water and drainage projects in the urban areas as part of the urban infrastructure development projects. Moreover, the National Forum for Water Supply and Sanitation (NFWSS) housed in the LGD is responsible for the national-level coordination among the government agencies, NGOs, development partners, and private sectors.

In the rural areas, the Local Government Institution (LGI) is accountable for promoting WSS facilities. The Union Parishad (UP-Council), the lowest level of local government in rural areas of Bangladesh, is responsible for management and maintenance of water points, preventing water point contaminations used for drinking water, and prohibiting the use of water points suspected to be dangerous to public health (Union Parishad Act, No. 61 of 2009, 2nd schedule, Articles 23, 24, and 25). In addition, the municipality (Pourashava) is responsible for providing or ensuring "sufficient wholesome water for public and private purposes" (Pourashava Act, No. 58 of 2009, Article 10) in urban areas governed by the Pourashava. The municipality controls, regulates, and inspects all private water supplies (Pourashava Act, No. 58 of 2009, Article 11). At the Upazila (sub-district) level, the Upazila administration also has a role in coordinating "proper measures for supplying drinking water" (Upazila Parishad Act, No. 24 of 1998, 2nd schedule, Article 6). Moreover, the Water Supply and Sanitation (WATSAN) committee of the Upazila Parishad and the Union Parishad are responsible for coordinating the activities of the DPHE, NGOs, and relevant stakeholders.

7.3 ARSENIC PROJECTS: SITUATION AND MITIGATION

Infant deaths from diarrheal and other water-borne diseases were at a peak in 1972 when UNICEF agreed to help with funding for drilling tubewells for pathogen-free drinking water. Therefore, expansion of tubewells came with the financial help of UNICEF after the independence of Bangladesh in 1971. At first sight, they seemed to be the perfect development tool—a cheap and effective technology that has been received enthusiastically by the users. Having a tubewell was a matter of convenience, but also a status symbol, and, as a result, people have been willing to invest their own money in private installations (Black, 1990). Later, these tubewells, installed in or very close to household compounds, were also very convenient, especially for women, who are the customary drawers of domestic water supplies (Atkins et al., 2007b).

The discovery of the vast scale of the arsenic pollution about three decades ago created a situation incompatible with the previous relationship between rural people and their groundwater. Since then the government has been very slow, but it tried to develop several nationwide programs to mitigate arsenic poisoning (Table 7.1). The most intensive water quality screening in Bangladesh was conducted by the BAMWSP funded by the World Bank and Swiss Agency for Development and Cooperation (SDC). A framework for this mitigation strategy had been provided in August 1998, when the World Bank invested $44.4 million in the establishment of the BAMWSP (Bangladesh Arsenic Mitigation Water Supply Project) principally to complete the emergency nationwide testing of all tubewells in Bangladesh. Second, there were to be rapid health surveys to identify arsenicosis sufferers and refer them to health services. Third, the foundation of a National Arsenic Mitigation Information Centre was to plug the yawning knowledge and information gaps that were hampering the implementation of policy (World Bank, 2007).

The BAMWSP was charged with a major emergency screening program. The preliminary target of the project was to cover 4000 of the worst arsenic-contaminated villages for new arsenic-safe drinking water sources, but the BAMWSP finally accomplished its target in 1808 villages (i.e., 122 villages in the first phase, 1026 villages in the second phase, and 660 villages in the third phase to provide 9,272 deep tubewells, 300 rainwater harvesting systems, 393 dug wells, and one piped water supply system (World Bank, 2007). A participatory process was used with villagers in deciding on the location and density of wells and contributing 10% of investment costs. By closing the BAMWSP on June 2006, about 3.04 million tubewells had been tested in 190 upazilas in place of the original target of 270 upazilas and 390 production wells in 100 Paurshavas. The screened well spouts were painted either red (unsafe) or green (safe) following the arsenic concentrations of Bangladesh standard permissible limit of 50 µg/L (Johnston and Sarker, 2007), and arsenic concentrations were detected at the field level using field testing kits (FTKs), which is questionable on grounds of validity and reliability.

The associated "Implementation Plan for Arsenic Mitigation in Bangladesh" initiates an ambitious "Emergency Water Supply Programme in Severely Arsenic Affected Areas," which aims to ensure at least one secure source of safe water "within a reasonable distance on an emergency basis" per 50 families. This project intervention criterion was set as an emergency response for villages where more than 80% of wells were contaminated with above 50 µg/L (GOB, 2004). Awareness raising was targeted at villages where more than 20% of tubewells are contaminated. Therefore, BAMWSP had developed mitigation measures under about 4331 Community Action Plans and 894 Ward Action Plans. In addition, the BAMWSP has trained 2300 physicians and 12,599 health workers in the diagnosis of arsenicosis symptoms as well as 54,000 field workers to use KTKs.

The BAMWSP also diagnosed about 29,500 confirmed arsenic-affected patients. As noted above, if more than 80% of wells in an area are contaminated with arsenic, an emergency mitigation procedure was triggered, in which alternative safe water sources were provided, but the commitment to awareness programs has been disappointing and by the end of the project, 190 upazilas were in place of the originally targeted 270 upazilas. In addition, following adjustment of the 4000 village target, project benefits accrued to between 2.0 million and 2.5 million people through BAMWSP with the installation of 9977 mitigation options in 1808 villages and wards and in one Pourshava (World Bank, 2007).

The design of the BAMWSP permitted flexibility to make changes in accordance with new knowledge emerging with regard to arsenic concentrations. Changes were made during implementation to refine the components, improve efficiency of implementation, and facilitate achievement of outcomes. Therefore, the fund closing date was extended four times, initially in 2002, and three more times, after management changes and improved implementation progress. Moreover, fund cancellation occurred twice: first, an amount of US$4.967 million was cancelled due to the slow progress of mitigation activities in 2002, and an amount of US$6.48 million of the unutilized credit balance was cancelled at project closure. Additionally, the SDC trust fund closed in 2005 due to extremely low disbursement (World Bank, 2007).

The Bangladesh Water Supply Programme Project (BWSPP) was launched on 17th June 2004 and ended on 31st December 2010. This was initially backed by $40 million of World Bank funding and was built on lessons learned by the BAMWSP. The development objective of the project was to contribute to achieving the Millennium Development Goals in water supply and sanitation by 2015 in selected rural villages and small towns affected by the 2007 floods and cyclones in Bangladesh. The project piloted some innovative service provision measures, including private sector participation in rural schemes. Due to unsatisfactory and poor performance in achievement of objective and implementation performance, the Mid-Term Review mission recommended that the project be revamped; accordingly, the project was restructured in 2008, and the grant was substantially reduced to US$18.6 million from the initially allocated budget of US$40 million (World Bank, 2011). Therefore, in terms of targets, the number of beneficiaries was cut by 54%. The goal became providing safe water to 570,000 people in villages and towns, compared to an original target of at least 1.25 million people. The rural piped water component in private-sponsor arrangements in a technically and financially sustainable manner was drastically cut from 300 villages to 21 villages, i.e., the rural piped water facilities reduced from 100,000 households to 25,000 households. The intermediate results indicators were greatly reduced, with the exception of those for the rural non-piped water component. However, the result indicator for rural deep tubewells was greatly expanded. The target of safe non-piped water points, that is the rural deep tubewells, went from 2000 to 7000 deep tubewells: 2000 as originally planned in three heavily arsenic-affected areas in three sub-districts (upazilas), plus 5000 additional deep tubewells in cyclone-affected areas. Around 70,000 households are getting access to these safe non-piped water points. The Pourashava (Municipality) water component was also cut. Instead, immediate priority civil works were undertaken in 5 Pourashavas and 19 other flood-affected areas. Some 20,000 households in 5 participating Pourashavas receive safe piped water supply. In addition, about 10,000 point sources of water were tested as part of water quality assessment (World Bank, 2011).

With the financial support from the World Bank, the Bangladesh Rural Water Supply and Sanitation Project (BRWSSP) was launched in March 2012 for increasing sustainable access to safe water supply and improved sanitation in the rural areas of Bangladesh, focused on supporting the government in mitigating against deteriorating water quality arising from arsenic, pathogens, salinity, and others (World Bank, 2016). Built on the experience of two earlier projects (i.e., BAMWSP and BWSPP), the BRWSSP follows a community participatory and strong monitoring process to provide safe water and hygienic sanitation to 1.2 million people in arsenic hot spots and salinity-prone coastal areas (World Bank, 2016). The target of the project was completion in 2017.

The BRWSSP is scaling up piped and non-piped water facilities to provide safe water to villagers in 383 unions in 33 districts with acute arsenic contamination and low access to a safe water supply with a total committed of $75 million. The BRWSSP is building 28,000 new piped household water connections as well as constructing or rehabilitating 14,000 water points, primarily deep tubewells along with pond sand filters, arsenic-iron removal plants (AIRP), and rainwater harvesting (RWH) units, where the piped water supply is not geographically or economically viable. With crosschecking by the DPHE, the BRWSSP follows a community participatory process, i.e., communities can decide on mitigation options, installation sites, and maintenance. The piped and non-piped water schemes, and the hygienic sanitation facilities, are implemented through local partnerships involving communities, local governments, non-governmental support organizations, local private operators, community-based organizations, and other entrepreneurs, following a Public-Private Partnership (PPP) model (World Bank, 2016).

The BRWSSP has strengthened water quality monitoring protocols where monitoring is carried out twice for installed water points. The monitoring activities include random quality checks (QC) at 10% of installed water points within three months of commissioning of the new wells under the project. These QC are being conducted in the DPHE laboratories. This monitoring ensures the water quality and reliability of any installed water points. The BRWSSP has already monitored almost 300,000 existing tubewells to ensure arsenic safety. If the water quality of a new well is found to be contaminated, an arsenic removal plant is added to ensure a safe water supply. The BRWSSP also works with DPHE to introduce a registration system for drillers to train and equip them to test for arsenic. Moreover, the World Bank has already introduced rapid evaluation of waterpoint location, water quality, and equity through Geographic Information Systems (GIS) mapping (World Bank, 2016).

Environmental Sanitation, Hygiene, and Water Supply in Rural Areas (ESHWSRA) was implemented over the period from 2001 to 2006 with the financial support from UNICEF. The project activities were conducted in 37 upazilas of 10 districts. The project trained 2,909 Community Hygiene Promoters (CHP) and installed 6,939 arsenic-safe new water points. About eight million people received benefits from this ESHWSRA project. At the same time, the Environmental Sanitation, Hygiene and Water Supply in Urban Slums and Fringes Project (commonly known as Urban Slums and Fringes Project, USFG) was implemented from 1999 to 2006 funded by UNICEF in collaboration with the DPHE. The USFG project contributed to a decline of diarrheal diseases by targeting one million people in 5 selected city corporations and 10 Pourashavas. The USFG project mainly focused on clean water supply and the improvement of personal hygiene.

Arsenic mitigation was also an integral component of a UNICEF-supported Sanitation, Hygiene Education and Water Supply in Bangladesh (SHEWA-B) project that was the largest and most integrated project of its kind in Bangladesh. The SHEWA-B project was a collaboration between the Government of Bangladesh (GoB), the UK Department for International Development (DfID), and UNICEF. The project was designed to reduce diarrheal disease and acute respiratory infection , the top two causes of post-natal under-five deaths in Bangladesh, by improving sanitation and hygiene behavior in a sustainable way, as well as ensuring the access to arsenic-safe water. Over its six years of implementation, running from January 2007 to December 2013, it operated on a budget of about US$100 million. SHEWA-B directly addressed arsenic along with other water problems and arsenic-safe technologies.

Built on the experience of two earlier DFID-funded projects (i.e., ESHWSRA and the Urban Slums and Fringes Project), the SHEWA-B directly targeted 21.4 million people for hygiene promotion in rural and urban communities in 19 districts across the country including 3 districts of Chittagong Hill Tracts (CHT) and an additional 10 million people outside the SHEWA-B intervention areas, bringing the total number of beneficiaries to 31.4 million. While, at its inception, SHEWA-B was originally designed to cover 30 million people, later, it targeted 20 million people in 19 districts and 19 municipalities.

In the SHEWA-B program, communication materials were used to raise awareness about sanitation, hygiene, and safe water, primarily through a network of 10,000 CHP with 70% females and 30% males who had at least 10 years of schooling. Arsenic mitigation activities in this project were fully integrated with hygiene and sanitation promotion. It provided arsenic-safe water supply facilities through the drilling of boreholes and the use of other technologies in both rural and urban settings to offer benefits of hygiene practice and prevent water-related diseases. It installed some 20597 tubewells in rural areas, 2351 water points and 403 mini piped water systems in urban areas in both the plain land and CHT districts, and 4551 water points in both primary and secondary schools for arsenic-safe drinking water (UNICEF, 2013).

Around 2.1 million people in both rural and urban areas are getting access to safe drinking water options as well as about 1.2 million school children have the access to safe water year-round. Moreover, some 137,690 beneficiaries were switched from an additional 1377 water points due to arsenic contamination to safe sources (UNICEF, 2013). Safe water sources included deep tubewells, dug wells, pond or river water filters, rainwater tanks, and arsenic removal systems. Following consultations with the government and key sectoral partners, a series of actions were considered in the SHEWA-B program. Accordingly, UNICEF and DPHE developed and implemented a joint arsenic rehabilitation project that included re-testing and re-painting (red/green) of all the SHEWA-B program wells and strengthening of laboratory procedures. It is noted that SHEWA-B installed 20,597 tubewells of which some 1,733 did not meet initially the Bangladesh water quality standard (50 µg/L) for arsenic, but UNICEF undertook the necessary actions to replace them and completed the project in September 2015 (HRW, 2016).

A Canadian International Development Agency (CIDA) funded project, Deployment of Arsenic Removal Technologies (DART), examined the effectiveness and social acceptability of some arsenic removal filters in some of the SHEWA-B area between 2005 and 2009. The DART project was implemented in 26 unions in 12 upazilas where arsenic was heavily contaminated, having lack of easy alternative water supplies. The Bangladesh Council of Scientific and Industrial Research has approved four arsenic removal filters (i.e., SIDKO, ALCAN, READ-F, and SONO) for Bangladesh. The DART project distributed these filters to communities through a participatory process and monitors their technical performance and social acceptability. Nearly 18,000 household filters and 50 community filters were installed and were regularly checked for arsenic removal effectiveness. These filters provided more than 100,000 people previously exposed to arsenic with safe water. The project, which ended in 2009, generated valuable information about the advantages and disadvantages of each of the filters for arsenic-safe drinking water (UNICEF, 2010).

The Hygiene, Sanitation and Water Supply (HYSAWA) project is under the supervision of the MoLGRD&C with the goal of reducing poverty through improved and sustainable public health and environment with improved water and sanitation. Established in 2007 under the Companies Act 1994, the HYSAWA Fund was responsible for mobilizing resources from the government and donor agencies and channeling those resources to capacity building support to local governments and communities, empowering them to manage decentralized Water, Sanitation and Hygiene (WASH) services in Bangladesh. Designed as a multi-donor support mechanism directed to local governments, the HYSAWA since its inception continues to contribute to the decentralization process and improve the quality of governance. HYSAWA is still operating in 16 districts of Khulna and Rajshahi Divisions. From its inception to 2016, the organization supported 1182 LGI out of a total of 4554 and provided them with greater financial and administrative power. The HYSAWA has facilitated the establishment of necessary institutional arrangements within LGI and engaged private sector management firms and NGOs to provide capacity support to LGI. With HYSAWA funding and technical assistance, LGI has successfully installed over 80,000 water points benefitting about 4 million people in hard-to-reach (H2R) areas, among them 46% of people are ultra-poor, and 237 rural piped water systems with over 15,000 connections benefitting over 230,000 people. In addition, more than 100,000 caretakers of tubewells received training on simple operation and maintenance of water points, of whom 50% were female (www.hysawa.org).

NGOs have also been involved with arsenic in one of two ways. Either they have been working independently with funding from foreign donors or they have been surrogates, locally delivering certain aspects of government policy. The largest contribution was made by the Bangladesh Rural Advancement Committee (BRAC) in water sector under the banner of the WASH program. Since its inception in 2006, BRAC completed their WASH program in 2015 for a total of 250 sub-districts (out of a total of 489) of Bangladesh in several phases. In May 2006, BRAC launched its WASH program in 150 sub-districts (152 after boundary changes) of Bangladesh with the financial support from the Embassy of the Kingdom of the Netherlands (EKN) having a population of about 38.8 million. The second phase (WASH-II: 2011–2015) addressed these issues by expanding into 20 new sub-districts funded by EKN and 5 more new sub-districts funded by the Bill & Melinda Gates Foundation. The third phase (WASH-III, starting in mid-2012) expanded to a further 73 sub-districts (mainly H2R) with the support of the Strategic Partnership Agreement supported by Britain's DFID and Australia's Department of Foreign Affairs and Trade. In 2014, the WASH in Schools component expanded to work in 250 secondary schools in rural areas of Chittagong and Khulna districts, and the international NGO Splash started work with BRAC WASH in urban schools to provide safe water, sanitation, and hygiene education among the poorest children in Dhaka and Chittagong cities. A total of US$155 million was spent in all the phases of the BRAC WASH program (BRAC, 2016). Between January 2007 and April 2015, in the 172 sub-districts supported by EKN, a total of 21.2 million people gained access to improved sources of water. Of these, 1.6 million received direct support from BRAC WASH. The water interventions were focused on high arsenic and salinity prone areas, and the technologies to reach households were (a) piped water systems where tubewells were not suitable to the hydrogeological context, (b) pond sand filters (PSF) or solar desalination in the coastal belt, and (c) two- or three-headed deep tubewells where boring was difficult for wells. Only the water points directly provided by BRAC WASH were tested for water quality at the point of installation and confirmed as a safe water source (BRAC, 2016).

Apart from BRAC, the NGO Forum for Public Health (it was known as NGO Forum for Drinking Water Supply and Sanitation) conducted a number of WASH projects where arsenic-safe water points were an important component. Since its inception in 1982, the NGO Forum installed many water points under different projects in unserved and underserved areas in association with different partner NGOs. The NGO Forum has installed so far 427,063 safe water options in 369 sub-districts all over the country that covered a total of 28,540 villages with the financial support from the government, donor agencies, and international NGOs from several projects. A total of about US$21 million (around BDT1.45 billion) was spent in this connection. The NGO Forum installed 280 safe water points in 114 villages in three sub-districts covering about 5360 households between April 2007 and September 2011 under the "Columbia University Arsenic Mitigation Project." In the HYSAWA Project, some 6584 water points were installed from May 2008 to December 2011. With the financial support from the Swedish International Development Cooperation Agency (SIDA), the Forum installed 300 arsenic-safe tubewells from April 2008 to May 2013 in Matlab upazila of Chandpur district as well as some 335 deep tubewells and 15 rainwater harvesting systems (RWHSs) on some off-shore islands of Galachipa Upazila of Patuakhali district from January 2009 to December 2011 funded by the Terre des Homes, the Netherlands. Moreover, in association with Bangladesh Centre for Advanced Studies and funded by the SDC, the NGO Forum installed 404 arsenic-safe water points and repaired 400 water points during December 2011 and December 2016 covering 47 H2R areas in 15 sub-districts of 13 districts, benefitting around 0.7 million vulnerable people.

BRAC employs local women as village health workers to operate FTKs, identify arsenicosis patients, and raise awareness, the hope being that communities will listen to their own members (Patel, 2001). Other NGOs have been involved in a smaller way, such as the Dhaka Community Hospital, the Dhaka Ahsania Mission and World Vision, and funding has come from Danish International Development Agency (DANIDA) and Britain's DfID. NGOs have been innovators in mitigation options, trying RWH, dug wells, PSF, and some of the chemical treatment methods, but none of these seem to be as popular with consumers as deep tubewells and rural piped water.

7.4 EXISTING MITIGATION OPTIONS: EFFECTIVENESS

Health problems caused by arsenic poisoning have a low priority to many rural poor people, and they see the issues surrounding health and illness as "non-threatening" (Gibbon, 2000). In some areas of Bangladesh, people are continuously using arsenic-contaminated water, and they are still ignoring arsenic poisoning although most of them know about the poisoning nature of arsenic in their drinking water. In recent years, attention has turned to mitigation to prevent or reduce exposure. This includes (a) chemical and nonchemical options and (b) household and community options (Brewster, 1992; Chen et al., 1999; Hering et al., 1996; Jekel, 1994; Jones, 2000; Kartinen and Martin, 1995). There are several methods that are expensive, and some are low-cost (Table 7.2). The DPHE has approved both the surface water and chemical options as well as four nonchemical based technological options for a short-term mitigation program: (a) deep tubewells, (b) rainwater harvesting, (c) pond sand filters, and (d) dug wells for mitigation purposes. The choice between these options should take into account their cost-effectiveness in providing arsenic-safe and microbiologically safe drinking water (Figure 7.2). Moreover, awareness of arsenic poisoning on human health can be important for people in arsenic-prone areas to accept suitable mitigation options. It is estimated that about 14% of the exposed population had access to several arsenic-safe technologies (GOB, 2011).

7.4.1 DEEP TUBEWELLS

Arsenic-safe drinking water can be available from deep aquifers. Deep tubewells are a proven suitable technology for arsenic-safe water as they are affordable and user-friendly. It is true that a deep aquifer is much less contaminated than a shallow one. A hydrogeological study conducted by the British Geological Survey (BGS) tested 280 tubewells deeper than 200 meters and found only 2 contaminated with arsenic (BGS/DPHE, 1999). The DPHE has also tested many deep tubewells and found only a very small number of tubewells with arsenic contamination (UNICEF, 2010). Therefore, use of deep tubewells is being prescribed as an arsenic-safe option. The DPHE along with a number of stakeholders have already made the recommendation to install deep tubewells in arsenic-contaminated areas with proper tests. The Government of Bangladesh is introducing deep tubewells in many arsenic-prone areas.

There are some hand-pump wells available for households and communities: (a) shallow hand-pump tubewells (depth: <70 meters in depth) and (b) deep tubewells (>100 m depth), but the required depth of wells depends on the depth of water table. To provide arsenic-safe water from the same tubewell to a bigger community or large population groups, some two-headed and three-headed tubewells have been introduced by BRAC WASH Programmes (BRAC, 2016). In rural Bangladesh, women mainly collect their drinking water by travelling for a long distance, sometimes 2–3 kilometers. These multi-headed tubewells make their water collection process easier.

People in Bangladesh abandoned their pond water practices about four decades ago, and they are now fully dependent on tubewell water rather than other sources of water. The installation cost of deep tubewells is high in the context of Bangladesh, but they are considered a cost-effective option for their sustainability and compared to other arsenic mitigation technologies (Ravenscroft et al., 2013). People in contaminated areas found that deep tubewells are the most preferred arsenic mitigation option (Inauen et al., 2013) because of the good water quality of the technology, especially, the taste and temperature of the water. Many are not motivated to take any preventive measures except the deep tubewells, and they want to confine themselves to deep tubewell water (Figure 7.2). Khan and Yang (2014) found that some 95% of the stakeholders prefer deep tubewell, and Hossain

TABLE 7.2
Suitability and Affordability Analysis of Existing Water Technology for Arsenic-Safe Portable Water

Technology	Description	Financial	Institutional	Environmental	Technical	Social	Suitability and Affordability
Shallow Tubewell (STW)	Water lifting option from a shallow aquifer. The depth of aquifer can range 6–70 meters. It is popular and widespread in Bangladesh.	Depending on depth and drilling method, a STW installation cost ranged US$120–225.	Due to simple installation and maintenance, facilities can be managed locally.	STW is very low ecological footprint technology. Water tapping with this technology is found to be arsenic-contaminated and different levels of saline concentration. STW cannot work properly during the dry season.	No need to train to operate. Maintenance includes replacement of washers, cup-seals, and bearings, is easy and can be managed without any external support.	People are aware of STW. Water from STW is well acceptable to the people, but the recent arsenic contamination causes a public health crisis.	Not suitable due to the concentration of arsenic, iron, saline, and other heavy metals in shallow aquifer.
Deep Tubewell (DTW)	Very popular option in Bangladesh. Water extracted from deep aquifer ranged between 100 and 400 meters or more. Believed to be arsenic safe.	Depending on depth and drilling method, a DTW installation cost ranged US$1000–1500.	Technology is not simple like STW and installation is inflexible. Maintenance can be managed locally, but is a little bit expensive.	The water tapping with this technology is found to be arsenic-safe. But, in coastal belts of Bangladesh, there are elevated level of saline concentration.	No need to instruct for operation. Maintenance cost is a little bit high and may require external support.	People are aware of using safe water with DTW. Water from DTW is well acceptable to the people, but the recent arsenic contamination causes a public health crisis.	Suitable for safe water, but check after installing the technology. Difficult to bore in some areas with subsurface geology. Moreover, are not suitable in some coastal areas.

(Continued)

TABLE 7.2 (Continued)
Suitability and Affordability Analysis of Existing Water Technology for Arsenic-Safe Portable Water

Technology	Description	Financial	Institutional	Environmental	Technical	Social	Suitability and Affordability
Pond Sand Filter (PSF)	Used for removal of algae, microorganisms, suspended solids, iron and manganese with a combination of physicochemical and biological processes.	Installation and commissioning costs are low. Installation cost ranged US$800–1200 (depends on size and capacity).	Applied at community level with limited labor. Can be used with other pre-treatment processes. Training required for operation and maintenance.	Contamination risk with open installations. Removal of contaminated top layer materials requires careful handling and disposal.	Technical expertise is required for construction. Operation includes flow control and cleaning for proper infiltration.	Viable but not guaranteed option for high quality drinking water delivery. Social acceptance level is still challenging.	Pathogen-free water technology, but very low acceptance at the community level. No water availability during the dry months.
Rainwater Harvesting (RWH)	Climate-dependent option. Water is usually collected from rooftops made of galvanized iron sheets during rainfall.	Investment for proper roofing costs around US$500. More costs are for treatment and storage facility.	Managed at household-level. Can be shared with families in neighboring households. Year-round consumption is limited.	Viable option for areas with rainfall abundance. High at risk of contamination with corroded roof, dust, leaves, insects, bird-dirt, etc.	Easy to maintain. Regular cleaning of roofs is needed.	Awareness is required to popularize the system. It is very traditional technology in coastal Bangladesh.	Depends on rainfall availability. Alternate water source is required in case of long dry-period.
Settling Water	Container filled with water is left undisturbed for more than 24 hours to settle solid particles down. Clean water is poured in clean container. Extra disinfecting steps may be required.	Various types of containers can be used for the settling. Containers can be purchased at minimum costs.	Suitable for household level. This method can easily be replicated and thus requires minimal training.	Used for both surface and groundwater. Proper care is needed in maintaining the containers, absence of which may contaminate water.	Can be applied with two or three pots. Water is kept for 24 hours in a pot and then poured into a new cleaned one.	Potential health risk for the users requires careful attention.	Affordable for household level, but not confirmed and established for safe water.

(Continued)

TABLE 7.2 (*Continued*)
Suitability and Affordability Analysis of Existing Water Technology for Arsenic-Safe Portable Water

Technology	Description	Financial	Institutional	Environmental	Technical	Social	Suitability and Affordability
Arsenic-Iron Removal Plants (AIRP)	A low-cost option based on the co-precipitation principle.	Installation and commissioning costs are low. Installation cost ranged US$800–1250 (depends on size and capacity).	Can be applied at community level with limited labor. Use with other pre-treatment processes. Training required for O/M.	Used for surface water. Proper care is needed in maintaining the containers.	Technical expertise is required for construction. Operation includes flow control and cleaning for proper infiltration.	Viable but not guaranteed option for high quality drinking water delivery. Problems in water availability during the dry season.	Suitable for pathogen-free water, but has not been accepted by the community people.

FIGURE 7.2 Arsenic-safe water points in different areas of Bangladesh and sharing the existing arsenic-safe tubewells as a preventive measure to reduce arsenic poisoning.

and Inauen (2014) investigated that more than 75% of the end-users prefer deep tubewell as one of the most suitable arsenic mitigation options in Bangladesh. One respondent (male, 52, retailer) from southwest Bangladesh said in this regard that

> I have learned that tubewells installed at a depth between 100–150 feet are concentrated with high levels of arsenic; while concentrations are very low in deep tubewells. I have been using water from a deep tubewell about one kilometer away from here from the time when I came to know that tubewells of this area are contaminated with arsenic. Some people of this area are collecting water from deep tubewells. I own a shallow tubewell, but I do not use this well water and do not allow others.

Bangladeshi people think that it is the responsibility of the government to help the poor. If the government installed deep tubewells for arsenic-safe water, there would be no arsenic problems as well as less water-borne diseases like cholera and diarrhea. Despite their popularity among the people as a long-term solution, it may be noted that deep tubewells are not as safe as sometimes assumed. Mandal et al. (1996) pointed out that, in 1990, the Indian Public Health Engineering Department installed deep tubewells to depths of 150 meters in Nadia, where the shallow aquifer was found to be arsenic-contaminated. At the outset, the water was arsenic-safe, but in the course of time, all of these deep tubewells have become contaminated with arsenic. A recent study mainly in coastal belts in Bangladesh is a source of concern. Arsenic concentrations in the deep aquifer were found to be higher than that of the shallow aquifer (Hassan, 2015). Installing so many deep tubewells is destroying the deep aquifer in Bangladesh; by contrast, in Colorado in the USA, one cannot even drill a water well on plots of less than 35 acres (14 hectares).

7.4.2 Rainwater Harvesting

The RWHS is an innovative alternative water supply approach for safe drinking water, and both the DPHE and UNICEF have recommended RWH to avoid arsenic poisoning. It is a recognized and successful water technology in use in many developing countries around the world including China, Pakistan, Sri Lanka, and Thailand. Properly stored rainwater is safe from bacteria and can be stored for many months (WHO, 2000b). Research by the International Centre for Diarrhoeal Disease Research in Bangladesh (ICDDR-B) confirms that rainwater can be a safe drinking water source (UNICEF, 2000a).

In China and Brazil, rooftop RWH is being used for drinking and domestic purposes. Gansu province in China and semi-arid zones in Brazil have the largest rooftop RWHS. In Beijing, some housing societies are now adding rainwater to their main water sources after proper treatment. In Senegal and Guinea-Bissau, houses are frequently equipped with RWHS. In the Irrawaddy Delta of Burma, communities rely on mud-lined rainwater ponds to meet their drinking water needs throughout the dry season. In New Zealand, RWH is the normal practice for rural housing. In Bermuda, the law requires all new construction to include RWH adequate for the residents (Low, 2016). The Virgin Islands of the USA have a similar law. Water rights laws almost completely restricted RWH in Colorado, USA, until 2009, but recently "rooftop precipitation collection system" has become permissible (Goodland, 2016). In New Mexico, rainwater catchment is mandatory for new dwellings in Santa Fe (Johnson, 2009).

Rainwater is collected using either a sheet material rooftop and guttering or a plastic sheet with the water being diverted to a storage container (WHO, 2000b). Users let the first few minutes of rainfall run off without collecting the water, to clean the roof and gutters. People can use rainwater for drinking and cooking through the dry season. This system has been used in coastal districts of Bangladesh for years and is being introduced in arsenic-affected areas. People of the southern districts in the coastal belt have been storing rainwater for their drinking purposes (Figure 7.2). RWHS has already become popular among the people of the coastal districts due to health-related concerns over tubewell water. The NGO Forum first started to install a RWHS in Patuakhali district in 1999 and now many stakeholders are installing more RWHS in areas where groundwater is contaminated with chemicals and heavy metals and surface water sources are contaminated with pathogens.

Since Bangladesh has a monsoon climate, people can preserve rainwater during the rainy season (June to September) for the dry months. This is a relatively low-cost mitigation option. Some of the respondents gave positive views on the use of rainwater, but mentioned that they need technological help, while other participants were not in favor of this measure because of its high installation costs. In addition, the rural poor have houses with straw rooftops. With rooftop harvesting, grass or palm leaf cannot be used to intercept the flow of rainwater and provide a household with high-quality drinking water and year-round storage. RWHS can range in size and complexity. All systems have basic components, which include a catchment surface or rooftop, conveyance system, storage, distribution, and treatment.

7.4.3 POND SAND FILTER

A PSF is a small-scale filtering device having manually operated treatment units used to treat the adjacent pond water based on the principle of slow sand filtration (SSF). PSF can remove bacteria from a nearby pond water by filtering it through a large tank filled with sand and gravel (Haq, 2001; UNICEF, 2000a). PSF is a community-based water treatment approach and it has been widely used in the coastal saline-prone areas of Bangladesh; now it is being installed in arsenic-contaminated areas of the country. The effect of SSF is to remove turbidity by straining and removing bacteria by biological action. The PSF was first introduced by the DPHE-UNICEF in Bangladesh in 1984 on a pilot basis.

Since the PSF operates on the principles of SSF, it has limitations in treating grossly polluted waters. Shakil and Martin (2000) revealed that the bacteriological quality of water from existing PSF does not satisfy the water quality standards for drinking. This is primarily because of the heavy pollution load of pond waters. Therefore, SSF in combination with pre-treatment by roughing filters may present a reliable and sustainable treatment process. Wegelin (1996) shows that roughing filters can achieve a particulate matter reduction of 90% or more. Roughing filters also reduced color and dissolved organic matter and other substances found in surface waters to some extent.

The initial design of PSF was developed under research and development activities of DPHE-UNICEF, and subsequently modifications were made by a number of international and national organizations. Subsequently, the construction of PSF was undertaken in development programs. PSF are being promoted by many organizations (DPHE-UNICEF, NGO Forum, Grameen Bank, etc.) in Bangladesh, as a method of surface water treatment in the coastal belt and arsenic prone areas. The NGO Forum, an apex NGO in Bangladesh, has been implementing PSF since 1997 in the coastal area of Bangladesh. In many arsenic-affected areas of the country, a significant number of PSF systems have been installed for safe water options. Microbial contamination is a major concern for PSF technology, and Alam and Rahman (2010) investigated that 77% of PSF suffered from microbial contamination. However, implementation of PSF technology alone may fail as hardware always has to be complemented by software and it requires adequate training of plant operators. However, rapid sand filters (RSF) can used for removal of arsenic from groundwater source. Gude et al. (2018) confirmed that arsenic-oxidising bacteria are able to grow and maintain their population on low trivalent arsenic (III) concentrations in RSF, either in presence, or absence, of other common groundwater bacteria and mineral precipitates, directly leading to an increased arsenic removal in the filter bed. It is noted that PSF is at times said to be a failed technology, and in most of the arsenic-prone areas the technology is socially unacceptable because of the poor water quality it produces and the availability of deep tubewells that have been installed recently.

7.4.4 DUG-WELLS

The hand-operated dug well is a nonchemical based short-term mitigation option. People in highly arsenic-contaminated areas can use water from dug wells. These are shallow hand-excavated wells and are a traditional source of water. Water from such wells is arsenic-safe and does not contain harmful chemicals and/or bacteria (UNICEF, 2000a); also, iron concentrations are quite low in dug wells (Chakraborti, 2001). Dug wells are safe with respect to arsenic contamination compared to hand-pump tubewells. It has been found that about 84% of dug wells are arsenic-free by the WHO permissible limit and 99% are within the Bangladesh standard permissible limit (Chakraborti, 2001). One hand-pump tubewell contained 1390 µg/L of arsenic, whereas a dug well located only 10 meters away from that hand-pump tubewell had only <3 µg/L of arsenic (Chakraborti, 2001). Before the hand-pump tubewell culture started decades ago, rural people were mostly dependent on dug wells. They drank dug well water, and this water was suitable at that time. Dug wells are now being re-introduced as a source of arsenic-safe water (UNICEF, 2000a). Dug wells can be used for drinking and cooking purposes after assessing the water quality (Alaerts et al., 2001). Problems arise with dug wells during the dry months because, since the green revolution in Bangladesh in the 1980s, extraction by machine-pumped deep tubewells for agricultural purposes has lowered the water table. People in arsenic-prone areas confirm that if water was available in dug wells, they would use it to reduce the risk of arsenic poisoning (Hassan et al., 2004).

7.4.5 Sharing Arsenic-Safe Tubewells

Well-switching or the sharing of any existing arsenic-safe tubewells is a community option for arsenic mitigation. The WHO regards this to be "the simplest and the most immediately achievable option" (WHO, 2000b), a view echoed by UNICEF (2000a), Van Geen et al. (2002, 2003b), and Caldwell et al. (2003). In any area where most of the tubewells are contaminated with arsenic, this option is suitable for drinking water. People are advised to collect water from arsenic-safe tubewells, and the owners of safe tubewells are encouraged to share their tubewells with their neighbors. Some people have decided to continue the collection of water from arsenic-safe tubewells until alternative options are available. Others are adamant that they want to collect water from arsenic-safe tubewells provided by the government rather than private tubewells. Many people are reluctant to use even these when they are in private hands because the sharing of such a scarce resource can cause tension and, in the words of one respondent, "It is embarrassing to collect water from a neighboring tubewell. I have had bitter experiences in collecting from different tubewells." Tubewell holders, on the other hand, claim that visitors may damage their tubewells, create a lot of noise, and make the tubewell platform dirty (Hassan et al., 2004).

7.4.6 Surface Water Usage: Digging Ponds and Boiling Water

In February 1999, the Minister for Local Government, Rural Development and Cooperatives (LGRD&C) promised in a National Conference on "Coordinated Action for Arsenic Mitigation Programme," which was co-organized by the Government of Bangladesh and UN agencies, that the government would resolve the arsenic problem within ten years by digging at least one pond as a reservoir in every union for arsenic-safe drinking water in arsenic affected areas. People were wary about such government policies, and they thought that the approach was untrustworthy and that it had a political "spin." They thought that the ponds would need to be renovated annually; otherwise, the banks would break during the rainy season and dirty water would enter the pond. Pond water is not pathogen-free—it is full of germs and dirt, is unhygienic, and its use for drinking can lead to different types of water-borne diseases. Pond water is used for washing cattle, bathing, and laundry. People are not at all interested in using pond water for drinking. They prefer deep tubewells to any alternative mitigation options. In their view, a deep tubewell is more economical to sink and maintain than to dig and manage a pond.

Boiled surface water is an important potential source for arsenic-safe drinking water, but people assume that boiling tubewell water can remove arsenic, whereas, in reality the risk is increased with boiling arsenic-contaminated water. Some respondents showed their willingness to drink boiled water until alternative options are available, but most people are reluctant since they cannot afford firewood for the regular boiling of water. They are mainly interested in using deep tubewells for their arsenic-safe water. A poor man (47, daily laborer) responded, "It is not a good decision to advise people to use pond water to boil; it is better to provide deep tubewells in each neighborhood."

7.4.7 Reflexive Sedimentation

A very simple, traditional technique for arsenic mitigation is to "pani basi kore khaoa," which means "to drink water after letting it settle overnight" (Alaerts et al., 2001). This "reflexive sedimentation" involves the storage of tubewell water for prolonged periods with no chemicals and the lower one-third of the water in a storage jar being discarded after settling for 12 hours (Jones, 2000).

Arsenic concentrations are reduced in the top layers. However, in many areas of southwest coastal Bangladesh (Satkhira, Khulna, and Bagerhat districts) almost all of the tubewells contain a remarkably high level of iron concentration and, if water is left overnight, it becomes viscous and yellowish and loses its original taste. So, unfortunately "reflexive sedimentation" in this region yields tasteless and smelly water.

7.4.8 Technology Options

There are several technological options for removing arsenic from the groundwater. Bold action has been discouraged by uncertainty about the most suitable approach (Hanchett, 2004). The key debate is between those who want high-tech solutions and those who want low-tech ones. Many water-treatment specialists are interested in resolving any environmental problem generated by technology with a further layer of technology. Their recommendation is to remove arsenic by chemical reaction including coagulation and filtration, ion exchange resins, activated alumina, and reverse osmosis; all are expensive and, in Bangladesh, likely to be restricted to urban situations. The use of aluminum-based technology, e.g., Alcan filter, is a grave health concern, and a positive statistical link has been found between high aluminum in drinking water and Alzheimer's disease (Martyn et al., 1989; Rondeau et al., 2009). The Steven technology for arsenic removal is based on coagulation and filtration, where they add iron salt (iron sulfate or iron chloride) as a coagulator with an oxidizing agent bleaching powder (Anwar, 2001a). Bucket treatment relies on flocculation after the addition of potassium permanganate and aluminum sulfate (alum). This technology results in high percentage reductions of arsenic from tubewell water. Arsenic removal using the bucket treatment method was tested by the DPHE and DANIDA as an alternative for the transitional period until a "permanent" solution is found (Jones, 2000). They may be "low-cost" in Western terms, but to rural people in Bangladesh, they remain beyond reach. Advocates of low-cost technologies are trialing, among other ideas, the three kolshi filter—three earthenware jars stacked on top of each other. Unfortunately, this and many of the other low-tech systems seem to have substantial disadvantages. Johnston et al. (2001) reviewed them and concluded that the delays caused by batch processing often outweigh the attractions of low cost. Several NGOs provided household-based arsenic removal filters to arsenicosis patients as well as to people living in arsenic-affected areas in Bangladesh to meet the immediate demand for safe water.

7.4.9 Piped-Water Systems

Many towns and cities in Bangladesh have arsenic-safe piped water systems. Treated piped water would be a suitable solution for inorganic arsenic poisoning, but the cost for this option in Bangladesh would be substantial. In 2004, the BAMWSP initiated pilot piped water supplies for villages, and the scheme was demand driven but the BAMWSP finally completed only one piped water system for a village. Later, the BWSPP installed piped water systems for 21 villages covering 25,000 households (World Bank, 2007). For a piped water supply system to reach the rural areas, it is necessary to consider the clustered form of rural settlement in Bangladesh. To minimize costs, treated water could be stored in reservoirs at some point of optimum distance from users and then supplied through a piped system to each settlement cluster or community for easy access from a standpipe. Piped water supply systems are suitable in areas where deep tubewells are difficult to install, such as hilly areas and some areas with a hard subsurface geology. In Bangladesh, water from deep aquifers is being delivered to households via an elevated water tank. The piped water is thought to be safe since the existing perception is that deep aquifer is arsenic-safe. There is no option for treating the water before supplying it to households. A number of stakeholders are enthusiastic about piped water supply systems, and they have already provided arsenic-safe water in many rural villages in Bangladesh. Experience shows that a majority of the people are interested in paying around US$1.25 (i.e., equivalent to BDT100.00) monthly for arsenic-safe piped water and more than two-thirds of women are prepared to walk for around five minutes to collect it from any standpipe. Piped water supplies have been implemented in Taiwan, Argentina, and Chile to mitigate arsenic contamination in drinking water.

7.5 NATURAL ARSENIC MITIGATION

7.5.1 Irrigation with Surface Water

Apart from a number of chemical and nonchemical low cost technologies for arsenic mitigation from groundwater, there are suitable preventive measures for arsenic poisoning. There is a possibility of adopting these natural mitigation options that could be environmentally supportive. Bangladesh is an agrarian economic country with a multiple cropping intensity. Despite rain in winter, irrigation is necessary to grow the "Rabi" (winter) crops. Irrigation by mechanized means (e.g., mainly the shallow tubewell and deep tubewell) began in the early 1960s and at that time only surface water sources (rivers, canals, marshy lands, and lakes such as beels) were drawn upon using Low Lift Pumps (Hossain, 1991). Since a sufficient quantity of surface water was not available during the dry months, mechanized tubewells of various bores were introduced to tap groundwater heavily for irrigation (Rashid, 1991). It has been observed that, during the dry months, most hand-pump tubewells all over the country do not have any water available for regular use due to the heavy withdrawal of groundwater by mechanized shallow tubewells and deep tubewells for irrigation.

Government aims and policies for agricultural development are to increase food production in order to reduce food imports. The government slogan was to build "Sonar Bangla" (Golden Bangladesh) during 1972–1975 and "Shobuz Biplab" (the green revolution) between 1975 and 1982 (Hassan, 1997). The government undertook a strategy of using technologies with chemical fertilizers and mechanized shallow tubewells and deep tubewells to increase food production. Arsenic from contaminated water of shallow aquifers through irrigation will penetrate through the roots to crops, vegetables, and fruits and finally come to humans through the food chain. If arsenic contaminates the food chain, people will ingest arsenic from both the contaminated drinking water and contaminated foods, and this will increase the health risk.

Since Bangladesh is a riverine country, there is an opportunity to utilize the surface water sources for irrigation. Moreover, there are "closed" and "open" water bodies, and most of them have water availability year-round. These water sources can be considered for irrigation rather than using shallow and deep aquifer sources. During 1977–81, the Government of Bangladesh undertook a principal strategy for agricultural development under the banner of the "green revolution." Apart from the use of mechanized means of irrigation, a "Canal Excavation Policy" is possible. This policy could be applied for irrigation to avert groundwater arsenic poisoning in soils and the food chain. River water is already being used in Ganga-Kabodak Irrigation Project and in Teesta River Barrage Project with a canal water-supply system. The proper excavation of canals could lead to the increased use of surface water for irrigation.

In considering the oxidation hypothesis, we may speculate whether it is possible to make a policy for using alternative water sources for irrigation rather than groundwater. The heavy withdrawal of groundwater is the cause of arsenic concentration. Therefore, use of water for the "Rabi" crops surface water sources (rivers, canals, and many closed water bodies) is desirable rather than groundwater. Thus, a policy for switching to surface water for irrigation can avert arsenic poisoning.

Arsenic can enter the food chain when groundwater is used for irrigation. Irrigation of dry season rice (boro) with arsenic-contaminated groundwater in Bangladesh is leading to increased arsenic levels in soils and rice (Dahal et al., 2008; Huq et al., 2006; Khan et al., 2010b; Roberts et al., 2011; Shrivastava et al., 2017). Arsenic concentrations and speciation in soil pore water are strongly influenced by redox conditions, and thus by water management during rice growth (Roberts et al., 2011). Moreover, inorganic arsenic in dietary staples (i.e., yams and rice) may have substantially contributed to exposure and adverse health effects observed in an endemic Taiwanese population historically exposed to arsenic in drinking water (Schoof et al., 1998). People in arsenic hot spots in Bangladesh eat an average of 300 g rice a day. Therefore, in addition to drinking water, dietary intake of arsenic from rice is another potential source of exposure and a new disaster for the human population (Rahman and Hasegawa, 2011). What will the government do if arsenic attacks humans through the food chain rather than drinking water? Is the existing government policy helpful in preventing human health risk from arsenic ingestion from food staples? The existing government policy regarding groundwater arsenic poisoning is not linked with arsenic from food staples. Therefore, the existing policy would not work to minimize the health risk from arsenic-rich food staples.

7.5.2 Switching Drinking Water

Generally, surface water in most places has no risk of arsenic contamination other than microbiological contamination. Before the advent of tubewells decades ago, people in rural areas of Bangladesh used to drink and cook with pond water and live with micro-organisms. Some used to purify pond water with camphor, some used to boil it before drinking, and others used to filter it. There was a tradition that a particular pond was used for drinking and cooking purposes, and there should be no bathing, washing of clothes, or other activities that might contaminate that pond water. People used that water only for their drinking and cooking purposes. For any emergency policy, where there are no alternatives, people could switch back from their present tubewell water culture to their previous pond water culture. Ahmad et al. (2002) and Van Geen et al. (2003b) are in favor of switching drinking water in rural areas of Bangladesh. Ahmad et al. (2002) found that 72% of their sample of 1331 arsenic-affected households preferred this solution. This traditional concept and practice could be transformed to a modern "water reservoir system" with sufficient safety measures. Rainwater can be a source of this reservoir water.

Recently, people have developed a tubewell culture in order to reduce incidence of water-borne diseases. They may find it difficult to shift back to past habits that have been altered over three decades, although they are still practiced in some parts of the country. People have to boil water before using it, which is still a traditional practice in many parts of the country, e.g., in Mongla area of coastal Bangladesh were groundwater is heavily contaminated with arsenic and sodium chloride. However, some researchers point out the importance of chemical methods, and their opinion is that, before using the surface water, it is necessary to remove certain toxic chemicals.

During the dry months, most of the tubewells remain dry, there is no water availability in most of the small water bodies, and the dug wells also remain dry. Villagers in different parts of the country say that before the Liberation War (1971) water was available in all ponds, dug wells, and hand-pump tubewells since there were no mechanized deep tubewells and shallow tubewells for irrigation. Tapping water from aquifers could make pressure to recharge these aquifers for making the water table balance.

7.6 POLICY ISSUES: VIGOR OR LETHARGY?

7.6.1 Government Vision

SDG Bangladesh is committed to fulfilling international commitments in achieving the targets of the SDG. SDG Goal 6 is specifically related to ensuring access to clean water and sanitation for all by 2030. "Sustainable water and sanitation services for all" remains a massive challenge in Bangladesh. Despite the significant increased water supply coverage from nearly 10% in the early days of independence to almost 97% in the early 1990s, it has now dropped back down to 74% due to the detection of toxic levels of arsenic in groundwater. The PSU of the LGD under the MoLGRD&C prepared a "Sector Development Plan (SDP) (FY 2011–2025) for the Water and Sanitation Sector in Bangladesh" in providing water for all by 2011 and sanitation for all by 2013 as well as basic services projected in Vision 2021 (GoB, 2011).

There are still limitations to fully reaching the hard-core poor in the H2R areas, settlement areas of CHT, coastal belts, offshore islands, haor (marshy lands), and char (point bar) for interventions to improve the water supply. In addition, there is a lack of comprehensive viable service opportunities for the people of urban slums and squatter settlements. Moreover, people in coastal Bangladesh do not have the access to safe drinking water due to elevated levels of salinity concentration in both the groundwater and surface water. Rural water supply also faces a host of problems including the lack of appropriate solutions with regard to hard-to-reach areas. In addition, groundwater levels are going down continuously in urban areas, thus restricting access to safe drinking water in a cost-effective manner.

The SDP (FY 2011–2025) for the Water and Sanitation Sector spells out the investment requirements divided into three major strategic timeframes (i.e., short-term, medium-term, and long-term), which is aligned with the government's planning systems. For reasons of conformity with broader national planning, the SDP was consulted with the Planning Commission to feed its views into the government's 15-year Perspective Plan. It has been reflective of the government's political commitment and international pledges, for example, MDG, Vision 2021, and SDG.

The DPHE is responsible for overseeing the installation of new water points in rural Bangladesh. Most that the DPHE installs are shallow tubewells and deep tubewells, while there are several water point technologies. Analyzing some 125,000 government water points installed between 2006 and 2012, it was found that DPHE water points have not targeted areas where the risk of arsenic contamination is high (DPHE/JICA, 2010; DPHE/UNICEF, 2013; HRW, 2016) and some 5% of these water points were found to be contaminated with the Bangladesh Standard Permissible Limit of 50 µg/L (HRW, 2016).

7.6.2 Safety Standards: 50 µg/L or 10 µg/L?

Demeritt (1998: 176) has discussed artefactual constructivism and argued that it "provides a way out of the dead-end debate about scientific truth." This approach sees social constructions of science and technological applications as joint achievements of society and nonhumans: equipment, geological strata, and pollutants. The results are interwoven, entangled, and only decipherable if the pretense of clean divisions between the social and the natural is dropped. To take the relationship one step further, one might also say that arsenic has changed Bangladesh because, in the words of Latour (2000), artefacts have the capacity to construct social order: "they are not 'reflecting' it, as if the 'reflected' society existed somewhere else and was made of some other stuff. They are in large part the stuff out of which socialness is made."

Arsenic pollution can be estimated with the means of testing and of pragmatic standards. Much of the literature accepts both as given, without discussing the crucial difference that they make to estimates of the pollution and of its health impact. As Schiappa (1996) has ably demonstrated, experts and politicians are often at odds about definitions and about threshold standards of value or quality. Standards for drinking water quality for Bangladesh were published in 1997 under the provision of the Environment Protection Act 1995, based on the WHO Guidelines (1993) for Drinking Water Quality (GOB, 2011). There is confusion in the international community, and in Bangladesh, about which water standards should be adopted in calculating the hazard. Should it be the 10 µg/L of the WHO or the much laxer 50 µg/L of the GoB? Cynics might say that the latter standard is particularly convenient for the GoB because it is five times more lenient and therefore reduces the scale of the problem in the eyes of both domestic consumers and international commentators (Atkins et al., 2007a).

Elevated levels of arsenic in groundwater are linked with cancer and other environmental illnesses, but concerns are shifting to the health effects of much lower doses of arsenic from drinking water in many countries. It is believed that arsenic risks from drinking water are more extensive than previously recognized, particularly during vulnerable periods such as pregnancy and childhood (Schmidt, 2014). Evidence for low-dose effects of inorganic arsenic ingestion is controversial. One view holds that arsenic has a dose threshold below which exposures are not harmful. Professor

Samuel Cohen explored this idea and that arsenic is carcinogenic only at doses high enough to induce cytotoxicity followed by regenerative cell proliferation, and if prolonged, mechanism can spawn tumors in the bladder, lungs, and skin (Cohen et al., 2013). However, a line of literature suggests that this threshold may not exist, so any exposure—no matter how small—could boost risks for diabetes, heart disease, immunological problems, and cancer (Gilbert-Diamond et al., 2013; Moon et al., 2013; Mostafa and Cherry, 2013; Navas-Acien et al., 2008).

It has been known since antiquity that arsenic is lethal, but lethal doses of arsenic are difficult to quantify (Schmidt, 2014). It was not noticed that arsenic was an environmental health threat until studies in Taiwan and Chile linked groundwater arsenic with skin cancers (which is rarely fatal) and a condition called black foot disease (Schmidt, 2014). It was suggested by the ATSDR (2010) that the minimal lethal exposure in humans ranges from 1000 µg/L to 3000 µg/L with death resulting from cardiovascular collapse and hypovolemic shock. The US Environmental Protection Agency (USEPA) developed a regulatory assumption that following exposure to a carcinogen, no matter how small, cancer risk increases to some degree. Therefore, National Research Council (NRC) designates arsenic levels below 50 µg/L as low, between 50 µg/L and 150 µg/L as moderate, and beyond 150 µg/L as high (NRC, 2014).

The levels believed to be "low" in early environmental health research on arsenic were much higher than what is considered "low" today. Studies from Taiwan up until the 1980s described groundwater levels of up to 300 µg/L as low, of up to 600 µg/L as moderate, and beyond that as high (Chen et al., 1988b; EPA, 2013). These delineations were based on a view that consuming arsenic in groundwater, while harmful, was not fatal in the long run. Chen et al. (1985, 1986) showed that arsenic could boost risks for fatal malignancies at groundwater concentrations far less than 600 µg/L. Chen et al. (1985) reported statistically significant associations between chronic exposure to artesian well water in southwestern Taiwan and elevated mortality from cancers of the lung, bladder, and other internal organs. In their follow-up study, Chen et al. (1986) reported that this relationship was dose-dependent, i.e., that cancer rates grew with higher arsenic exposure.

But linear assumptions drive considerable risk even at low exposures. The US Public Health Service aimed to protect against the arsenic-related skin problems seen in Taiwan when it set a 50 µg/L standard for arsenic in drinking water in 1942, which was then adopted by the USEPA in 1975 (EPA, 2013). Extrapolating from high-dose human data, the NRC predicted that the 50 µg/L water standard could induce cancer in as many as 1 in 100 people (NRC, 1999b). Evaluating more than 300 studies, the USEPA (2012) dropped the standard from 50 µg/L to 10 µg/L in 2001—a level that the NRC estimated might lead to a cancer risk of approximately 1 in 300 for people exposed over a lifetime (NRC, 2001).

One of the most interesting aspects is how regulators have interpreted the science in terms of safety standards for drinking water. Smith et al. (2002) and Smith and Smith (2004) show that there is a long and complex story about this internationally. In the USA, a 50 µg/L limit was established in 1942 lasting until 2001, but as early as 1962 there was advice from the US Public Health Service that this was too high and that 10 µg/L was a safer limit. In 1986, Congress instructed the USEPA to revise standards, but there were further delays. Similarly, the WHO safety threshold was 50 µg/L for 30 years until 1993, but this was apparently dictated more by the pragmatic consideration of what was possible in the field in terms of testing than by any clinical factors. The first WHO recommended limit was 200 µg/L set in 1958. In 1993, WHO reduced the guideline value for arsenic in drinking water from 50 µg/L to 10 µg/L, but the limit was reaffirmed in 2004 (WHO, 1993, 2004). The WHO considered the guideline provisional until reaffirmed in 2004 because of measurement difficulties and practical difficulties in removing arsenic from drinking water. Several countries have since lowered their permissible limits to 10 µg/L, and some have established lower standards. However, the limit in Bangladesh (since 1989) and many other countries has remained at the previous standard of 50 µg/L (BGS/DPHE, 2001). Still, the evidence for considerable death and illness from exposure to arsenic in drinking water between 10 µg/L and 50 µg/L in Bangladesh is increasing (Ahsan et al., 2006).

The University of Chicago-based Health Effects of Arsenic Longitudinal Study (HEALS) team has assembled a cohort of tens of thousands of individuals living in Araihazar, Bangladesh, where arsenic levels measured in well water have ranged from undetectable to more than 900 µg/L. The HEALS team first reported an association between arsenic and high blood pressure in 2007 at tubewell water concentrations of 10–40 µg/L. Since then, HEALS has produced papers associating arsenic at levels below 50 µg/L with health conditions including heart disease, hematuria (blood in the urine), and impaired lung function (McClintock et al., 2014; Parvez et al., 2013). Studies also showed that increased total urinary arsenic was associated with skin lesions, which are known precursors to skin cancer (Sober and Burstein, 1995; Woolgar and Triantafyllou, 2011). In their estimate on the excess mortality from arsenic exposure, Flanagan et al. (2012) noted that "the excess deaths among people exposed to arsenic concentrations of 10–50 µg/L represent from 45 percent to 62 percent of all arsenic-related deaths."

Allan H. Smith, one of the scientists most responsible for bringing the arsenic crisis in Bangladesh to the attention of the international community, espouses a solution of classic utilitarian logic when he argues that lowering the threshold from 50 µg/L to 10 µg/L may well in theory reduce the long-term cancer risk from one in 100 to one in 500, but in practice such a policy would be self-defeating because action is likely to be postponed until such a time that the technical and administrative capacity is in place to achieve that limit (Smith and Smith, 2004). Meanwhile, many people would be exposed in the absence of a credible standard. The implication of this style of argument is that policy should be incremental and derived from the real world of practical implementability (Atkins et al., 2007a). Such a conclusion seems likely to appeal to many local people, some of whom are already ignoring the red paint warnings on their tubewells and continue to consume contaminated water (Caldwell et al., 2003; Quamruzzaman et al., 2003).

7.6.3 Opinion of Things: Arsenic Data Accuracy

At the stroke of a pen, it is possible to manipulate the figures of the number of people identified as being at risk of arsenic poisoning. Leave aside the number of authors who uncritically cite the estimates of others, and there are many of these in the literature on Bangladesh. We are dependent, first, upon the screening technology used; second, the sampling methodology; and third, astonishing microspatial variability of contamination that means a strong component of indeterminacy and this can be analyzed with GIS-based smoothed cartographic models. These models therefore do considerable violence to the data and are of questionable predictive value on the ground (Atkins et al., 2007a).

When in the 1990s it became apparent that there was a problem with arsenic poisoning, the wheels of scientific realism began to turn. A method of testing was needed that could be used under field conditions in all parts of Bangladesh. It had to be chemically sensitive to a range of possible levels of arsenic concentration, physically robust, simple and safe to use, and affordable (Deshpande and Pande, 2005). This was an epistemological dilemma, a trade-off between precision and practicability: in other words, what version of the truth was knowable within certain constraints? The alternative of laboratory analysis simply was not available at that time, but later the capital investment was made to enhance indigenous laboratory capacity. The FTKs that were introduced in the 1990s have been heavily criticized. First, technically, they have been shown to give unreliable and varying results. This was due to a design problem of miniaturizing the chemistry to a portable size, combined with the difficulty of developing a test outside the laboratory that is accurate to below 100 µg/L (Atkins et al., 2007a). By comparison, atomic absorption spectrophotometry, the main laboratory method, yields accuracy to 3 µg/L. Second, there has been, apparently, significant operator variability in the field, probably due to inadequacies of training (Pearce and Hecht, 2002) and to the perverse incentive of completing a maximum number of tests in a day, which leads to skimping on demanding aspects of the process, such as leaving the test paper for a standard period to develop its full color. Most of the FTKs currently in use are based on the Gutzeit method, which involves the reduction of arsenite and arsenate by zinc to give arsine gas. This gas then produces a colored stain on mercuric bromide paper (Pande et al., 2001). The color (yellow, brown, or black) is interpreted using a standard reference chart in order to gauge the amount of arsenic present, but it seems that this is not a straightforward procedure.

Rahman et al. (2002) have written a particularly damning assessment of the four most commonly used FTKs. When calibrated against laboratory-based flow injection hydride generation atomic absorption spectrometry (FIHGAAS), they were reliable only at high concentrations (>100 μg/L), and correctly identified the binary status (acceptable or contaminated) of the water in only 49.3% of the 2866 tubewells tested. If true, this totally unacceptable result throws the whole status of the nationwide testing program into disarray and deracinates the government's policy. It is important to note here that Van Geen et al. (2005) published a more reassuring account of the accuracy of the Hach kit, one of the FTKs most widely used at present. They found a 12% discrepancy between field and laboratory tests. Unless new FTKs can be found immediately that are more accurate by an order of magnitude, the only credible solution seems to be to replace FTKs altogether with a large-scale laboratory testing program. Certainly, a short-term goal must be to establish sustainable testing regimes to retest every tubewell in Bangladesh on a regular basis.

In the present campaign of painting tubewell spouts red or green in accordance with the results from arsenic analysis, the success of such an information campaign depends upon people believing that the screening method is valid and accurate (Ahmed et al., 2006). In truth, several FTKs presently used to determine arsenic content are under a cloud because of the high potential for operator error in field conditions (Pande et al., 2001). The alternative of using laboratories is now realistic. If the laboratory method is used, for instance, the accuracy premium is potentially tenfold over most FTKs. But most alarming of all is the operational variability of results that has been found between laboratories (Foster and Tuinhof, 2004; Kinniburgh and Kosmus, 2002).

7.6.4 Arsenic Awareness: Shortcomings

Awareness campaigns are potentially an important aspect of arsenic mitigation. Before implementing any relevant policy, public awareness is needed about arsenic poisoning that could be helpful in reducing suffering. The approaches to an arsenic awareness campaign should emphasize the "communication process" (Hanchett et al., 2000) to the people. The need for information concerning arsenic poisoning and its mitigation options is an important aspect of any such campaign. People continue to need information and support when there are changes in their health situation due to arsenic poisoning. Arsenicosis neither develops in a day, nor do arsenic-affected patients seek help in a day, nor does an arsenic-affected person die in a day. Thus, the best possible way is to prevent arsenic poisoning in the first place, and the best measure is to make people aware of arsenic poisoning and related diseases.

Awareness with posters and stickers is important to alert people to the poisoning nature of arsenic. Generally, the advice is not to use red-labelled tubewell water but rather to rely on green-labelled tubewell water for drinking and cooking purposes. Moreover, the posters and stickers also referred to the extreme level of arsenic-related diseases, such as gangrene and lost fingers. Most people in rural areas have never seen that type of patient in their vicinity, and they therefore ignore the advice on drinking and cooking with green-labelled tubewell water. Some arsenic awareness posters and stickers focused on the advice that arsenicosis is not a contagious disease, but some rural people consider arsenicosis to be a curse of nature (Hassan, 2000).

Awareness campaigns through the media are not strong enough to make people aware; even the government awareness campaigns through the media do not make sense to most of the public since a small number of patients have been diagnosed and in most areas no tubewells have been painted with red or green colors. What do the local people think about this arsenic-safe water? Most of them still confuse arsenic with iron. When they find iron concentrations in their tubewell water, they think it is arsenic. Many stakeholders recommend drinking surface water after boiling and cooling it, since surface water is arsenic-free and it will be pathogen-free if boiled properly. But some rural people boil their arsenic-contaminated tubewell water before using it for drinking purposes. Ironically, boiling arsenic-contaminated water increases the level of arsenic concentration.

Messages concerning arsenic issues can be communicated to people to make them aware of arsenic poisoning and preventive measures. It is important to make people aware first about the toxic nature of arsenic because they can then adopt preventive measures until the appearance of a long-term sustainable mitigation option. Arsenic poisoning might result from continued use of arsenic-contaminated tubewell water for drinking and cooking purposes. A few years of continued exposure to low levels of arsenic causes different skin lesions, and after a latency period of 15–20 years, internal cancers, particularly in the bladder and lung, can appear. The messages let them know about the source of arsenic and its associated health risks. Moreover, photographs of severely arsenic-affected patients help people to understand the toxic nature of arsenic.

What do rural people think about arsenic poisoning? Many had the habit of drinking and cooking with surface water until a few decades ago. Recently, they have changed their practice to tubewell water in order to save themselves from water-borne diseases. But this tubewell water is now contaminated with arsenic toxicity. Which water will they drink now? People can understand about the extreme level of danger of arsenic impact on human health from media-based communication, but they do not understand about the early stage of arsenicosis symptoms. According to one respondent (farmer, 56) from Araihazar upazila in Narayanganj district near Dhaka City, "Nobody knows the initial symptoms of arsenic-related diseases. It is important to emphasize the initial health symptoms so that people can understand the problems from the very beginning."

Painting tubewell spouts with green or red based on arsenic-safe or arsenic-contamination, respectively, has been fundamental in arsenic awareness campaigns in Bangladesh. People are advised to use green-labelled tubewells and to avoid the red-labelled ones for drinking and cooking purposes. But different conditions can hamper the overall situation; for example, one national NGO installed arsenic-safe deep tubewells, but they are red in color and people are skeptical in using this arsenic-safe tubewell water. If people identified the green-labelled tubewells, they collected water from those tubewells rather than the red-labelled tubewells. Another issue is how to let people know about arsenic concentrations in their tubewells where no analysis has yet been done, as in almost 50% of the total tubewells all over the country. People assume that painting arsenic-safe tubewells green would help children not to drink water from red-colored ones. Children drink water anytime from any tubewell close to them, but they are more aware of arsenic than their elders and have already learned some ideas about it. Moreover, schoolchildren can easily come to know about arsenic and related issues if it is included in the academic curriculum, just as population problems, floods, and cyclones have already been included in different academic curricula for permanent proliferation.

Training for different groups of people, for example, health workers, NGO workers, and village doctors, would be productive. After receiving training on arsenic issues, they could work in rural areas to make people aware as part of their job. People prefer health workers because they are well known to rural poor people, and in recent years, they have played contributory roles in reducing diarrhea and have organized family-planning activities, and national schemes for immunization against polio. Moreover, teachers from different schools should be trained regarding arsenic issue because children listen to their teachers—whatever a teacher teaches pupils concerning the arsenic issue, they will try to follow it, and they may share their knowledge with their parents. In addition, *Imams* (religious leaders) of mosques could contribute to inform people about arsenic poisoning and its preventive measures. Local leaders and social activists could also advise their people to use green-labelled tubewells in place of red-labelled tubewells.

Dissemination concerning arsenic poisoning through electronic media is important. Proper means of communication are vital to preventing arsenic problems. It is necessary to propagate and circulate widely information about the toxic effects of arsenic. Most of the rural poor people cannot afford radios or televisions, but in recent times they have gotten mobile phones. They can use them in getting information from the government. A movie clip can make people aware quickly. Theater staging is another interesting and potentially important method for arsenic awareness.

7.7 MITIGATION ACTIONS: JUSTIFICATION AND VALIDATION

There is a disappointing lack of coordination among the agencies working to address the groundwater arsenic poisoning in Bangladesh. Very few organizations have shared data or information with others or with agencies working on the same problem. Despite repeated calls for coordinated efforts to address the issue, many are not cooperating. There have been many projects on arsenic in recent years, but few have been effective at pointing out the real problems and solutions of arsenic poisoning. Moreover, outcomes of any arsenic mitigation project of the World Bank, UNICEF, the WHO, and the DANIDA funded programs have reached the poor people only a little. The government implementing agencies, NGOs, and many stakeholders are working on arsenic mitigations (Figure 7.1), but their roles are insufficient in combating arsenic poisoning.

7.7.1 GOVERNMENT ACTIONS

The government has taken various measures to provide arsenic-free drinking water to the people through alternative sources, but these are not enough as either an urgent or a permanent solution. Since arsenic is a national problem, the government is ultimately responsible. However, after the first detection of arsenic concentrations in groundwater in 1993, the government has initiated several projects for detecting arsenic contamination and providing adequate treatment for affected people, but they were not sufficient to tackle the problem completely. Mitigation measures undertaken so far by successive governments are inadequate. Only in some severe cases has the government undertaken a comprehensive program for sinking tubewells in the rural affected areas.

The government could take quick action through different international donor agencies and deal with arsenic problems within years. People thought that, where people are affected with waterborne diseases like cholera, diarrhea, and abdominal problems as well as recent arsenic poisoning and cancer, the government should take quick action in this regard. One respondent (54, political leader) from Satkhira stated, "Many organisations or many professionals can make people aware of arsenic, but the mitigation policy should come from the government."

7.7.2 DPHE ACTIVITIES

The DPHE has a contributing role in mitigating arsenic poisoning, and it has been instrumental in a number of large- and small-scale arsenic initiatives. These include arsenic analysis in tubewell water, assessing the extent of arsenic concentrations, testing arsenic removal technologies and alternative options, and implementing mitigation measures. The DPHE was the first to uncover groundwater arsenic contamination in Bangladesh, and this government institution is now playing a contributory role in providing deep tubewells for arsenic-safe water. In association with UNICEF, it has taken the mitigation initiative of installing a significant number of water points all over the country.

There are incompatibilities in the DPHE plan and activities in terms of tubewell distribution, installation, repair and maintenance. A significant number of deep tubewells are provided with political drive (Van Geen et al., 2016). In severe arsenic-contaminated areas, DPHE provided a negligible number of safe water points, while in areas with low arsenic concentrations more tubewells were installed. In a question concerning repairing of a government tubewell, one respondent (49, social activist) from Satkhira district replied that "It is a troublesome matter to make any contact to the DPHE engineers for repairing damaged tubewells. They do not repair it until the tubewell owner provides them with money. The best way is to repair any damaged tubewell is on one's own." Some people have alleged that the DPHE collects money from poor people for arsenic-safe deep tubewells, but then fails to provide them. One elected administrator from Ghona union in Satkhira district in this regard said without any hesitation that "DPHE engineers take more against their officially assigned contribution money for each deep tubewell without having any proper money receipt, and there is no date fixed in providing the promised deep tubewell. People were worried if this engineer transfers elsewhere, they will get neither the deep tubewell nor their money back."

The DPHE analyzes water samples at a nominal cost, but people reported that they are not cordial and sincere in this connection. Some people pointed out their negligence in analyzing water samples. One respondent (45, retailer) from Satkhira district shared his experience of a testing of his tubewell water samples by the DPHE: "A few months ago, some of us went to Satkhira DPHE office with iron-rich red colored water samples from our own tubewells for testing. One officer of the DPHE took the samples, and a few minutes later told us that all the water samples are safe and that there is no problem." Some respondents claim that the DPHE misinforms people about the real situation of the water quality of tubewells. Anwar (2001b) pointed out that two new tubewells were drilled under the BAMWSP arsenic mitigation program in Faridpur, and both of them were found to be highly contaminated with arsenic and one tubewell contained 1760 µg/L of arsenic and another tubewell was measured at 400 µg/L of arsenic. Unfortunately, people were found to be drinking water from these tubewells as they were certified to be safe by the BAMWSP.

7.7.3 HEALTH AND FAMILY PLANNING DEPARTMENT

The Health and Family Planning department under the Ministry of Health and Family Planning Welfare is involved in the mitigation process of arsenic poisoning. Through their satellite camp program, they campaign on arsenic poisoning and safe drinking water. When people first came to know years ago that arsenic is concentrated in tubewell water, the family planning department arranged a course on arsenic to train some local health workers, local village doctors, schoolteachers, and so on for an awareness campaign. The concept was that the trained people would pass the arsenic message to others and those to others again. Health workers are working on arsenic issues apart from rural sanitation and family planning. They have grassroots contacts with the people, and they can contribute more in arsenic awareness campaign.

Most of the people do appreciate health workers for their role in many aspects of health. Health workers do go to the doorstep of poor people to provide them with health services, and they could therefore contribute to combating arsenic problems. They could visit every housewife in every household in rural areas and discuss the impact of arsenic and how to combat it. On the positive role of health workers, one respondent (62, political leader) from Bagerhat district cited an example of diarrhea: "A few years ago when there was a serious diarrhea problem in this area, the health workers went to every household and they taught them how to make a saline solution with salt and sugar to treat diarrhea. This saline saved many lives. The poor people who could not afford to buy saline, made it by mixing salt and sugar."

The Union Health Complex (the primary or lowest level health care establishment in Bangladesh) is contributing to providing health services to local poor people. If the government adopts a policy to give special medication through this institution, arsenic-affected people could benefit. Doctors have contributory roles in treating arsenicosis patients, but in rural areas village doctors are not properly trained and they sometimes provide wrong prescription in confusing arsenicosis with sores on palms and soles. If they get proper training on arsenic issues, they could contribute efficiently in reducing arsenic-related health risk.

7.7.4 LOCAL ELECTED ADMINISTRATORS

Rural people always appreciate their own local elected representatives since they have direct contact with them. People thought that the elected administrators could raise the arsenic problems in their monthly meeting with the Upazila Nirbahi Officer at the Upazila Headquarters. However, in a question concerning the role of Union Parishad Administration in the arsenic mitigation issue, one elected member from Ghona union in Satkhira district added here, "We do not have any direct contribution to arsenic mitigation. If the government takes any decision in this regard, we can send it to the doorsteps to our people." Some Union Council members criticized the role of the government for not having done enough to combat arsenic poisoning. They pointed out that the government and donor agencies have the tendency to implement their policies through NGOs rather than involving the people's representatives. Some people think that Union Parishad administrators can contribute to arsenic awareness campaigns. They could advise people about arsenic poisoning and its mitigation options in their own vicinity.

7.7.5 NGO Activities

NGOs can play positive roles in arsenic mitigation action in providing community-based arsenic removal technologies to their association members through their micro-credit program. NGOs such as the BRAC, Grameen Bank, and the NGO Forum are working on arsenic mitigation with safe water options. Many NGOs in Bangladesh are continuously providing their efforts to strengthening arsenic preparedness and mitigation activities. The activities are mainly focused on the use of arsenic-safe drinking water, arsenicosis patient management, and awareness-raising campaigns about arsenic toxicity. Caritas Bangladesh, for instance, has advised people about RWH and dug wells in their program areas, and they have provided a good number of alternative options, such as deep tubewells, ring wells, and RWH systems. The NGO Forum for Public Health prefers other options, e.g., AIRPs, deep tubewells, PSFs in their intervention areas, mainly in the coastal belts of Bangladesh. As a result, there has been some noticeable change in the health condition of some patients with early arsenicosis symptoms. Unfortunately, less improvement was found in seriously affected patients. In addition, a remarkable number of local community and political leaders, local government administrators, religious leaders (*Imams*), and teachers have been trained in planning and decision making in arsenic issues. This awareness of arsenic issues has discouraged the use of red-labelled tubewells for cooking and drinking purposes, and the campaign has made people more aware of arsenic poisoning.

Some NGOs analyze arsenic concentrations in tubewell water, and they paint tubewells red or green to let people know about arsenic concentrations. NGOs have grassroots links to poor people—their workers work at the field level and they are well known to the local people. People were found to be optimistic but reluctant with NGOs, and a respondent (54, male, political leader) from Satkhira district said against NGO activities that "The main target of an NGO is to earn more profits from poor women. They provide micro-credit to very poor and illiterate women. After providing them with credit, NGO workers are reported to go to those women's doorsteps once a week or several times in a week to collect instalments of the credit provided by NGOs at high annual interest rate."

Many NGOs are allegedly using arsenicosis patients for their own interests. They collect biomarker samples from arsenicosis patients without mentioning the purpose. They took photographs of the victims. Some NGOs, while taking samples, promise to the victims that they will notify them of the test results, but it has never been done (Haq, 2000). Locally, people thought that NGOs were doing business in the name of "socioeconomic development." One respondent (57, businessman) from Satkhira district told me stridently that "Actually, the main objective of NGO activities is the development of the socio-economic conditions of rural poor people, but the reality is different. They will not provide us with anything except high-interest credit. NGOs now mean credit programs and this is their prime business. They are continuing all of their money-making activities under the banner of socio-economic development."

7.7.6 Donors and International Organizations

UNICEF has become an important contributor to reducing high mortality rates from cholera, dysentery, diarrhea, and other water-borne diseases by providing pathogen-free tubewell water decades ago. The assistance of UNICEF to the DPHE in promoting groundwater is an important aspect for the drinking water. These tubewells are now contaminated with toxic levels of arsenic. UNICEF did not monitor the quality of drinking water regularly in terms of its toxic chemical content. Amidst their enthusiasm to drill tubewells in Bangladesh in the 1970s, UNICEF forgot about the Taiwan experience, but the enquiry will focus on how it was possible for the deadly water to remain untested for four decades.

The British Geological Survey (BGS) carried out studies on behalf of the Bangladesh Government in the mid-1980s and early-1990s. However, the British scientists "failed to detect dangerous levels of arsenic in the supply of drinking water implicated in the biggest mass poisoning in history" (Connor and Pearce, 2001). Two studies of groundwater quality in Bangladesh carried out by British hydrologists failed to monitor natural arsenic levels. The BGS in its report in 1992 did not mention arsenic in groundwater in Bangladesh although the WHO recommends testing for arsenic for drinking water quality tests. During 1991–92, a BGS team surveyed the quality of groundwater in thousands of tubewells in central and northeastern regions of Bangladesh and in its report said that the water was "safe for drinking." A few years later, it was found that water in many parts of the regions studied had high levels of arsenic. More dangerous is their recommendation for deep tubewells, as they reported that "available data shows that aquifers deeper than 150–200 m are essentially arsenic-free over much of Bangladesh. Systematic sampling showed only 2 out of 280 wells deeper than 200 meters to be contaminated" (BGS/DPHE, 1999). On the basis of this report, donor-aided arsenic mitigation projects installed thousands of deep tubewells, but a significant number of them in different areas are contaminated with arsenic.

7.8 POTENTIAL POLICY STRATEGY IN BANGLADESH

The policy strategy in Bangladesh should aim at ensuring arsenic-safe water, awareness programs, and finally, arsenic patient management. Sarker (2010) identified a lack of resources and medical treatment facilities as major obstacles in overcoming the arsenic health risk. Hassan (2015) shows that a significant number of arsenicosis patients in rural areas have access to alternative health care including homeopaths, village doctors, ayurvedic doctors, and local pharmacists. Provision of safe water options, periodic screening of water sources for arsenic, availability of trained doctors, regular availability of medicine, doorstep treatment, follow-up on severe patients are components for an arsenic mitigation strategy. Under the banner of "water governance" (Bakker and Morinville, 2013), multi-level actions are needed to be taken by the government, donors, NGOs, and other stakeholders at local, regional, and national levels in mitigating arsenic poisoning. In mitigating arsenic poisoning, many aspects need to be considered in terms of socioeconomic and sociocultural aspects for proper policy development. Both short-term and long-term policies are needed on an urgent and sustainable basis.

7.8.1 SHORT-TERM STRATEGY

a. A comprehensive and separate Arsenic Mitigation Project is needed to minimize drinking water arsenic exposure. As a government-implementing agency, the DPHE should implement the mitigation plan in areas with high arsenic contamination. Minimization of arsenic exposure is possible immediately by installing suitable alternative water sources for different geophysical areas. The DPHE can conduct this assignment efficiently with its experienced and skilled labor as well as extensive knowledge and practice in this sector. In doing so, it is obligatory to end any malpractice in the distribution of water technology and its installation—from local political pressure to bribery. The distribution will be based on the actual need of people and communities, more specifically, for the hard-core poor in H2R areas. There must be a strategy of "zero tolerance" of any attempts to influence the process of allocating new water points, and these must be maintained properly.

b. Formulating and implementing improved surveillance and treatment facilities is needed for chronic arsenicosis diseases, including skin lesions, hypertension, respiratory disease, diabetes, cardiovascular disease, and cancers. The DGHS should be responsible for these treatment facilities. Likewise, basic symptomatic treatment for skin lesions is required at village-level community healthcare centers, so poor people in rural areas can get healthcare facilities.
c. The drinking water standard for arsenic ingestion should be lowered from 50 μg/L to 10 μg/L to minimize high health risk to exposed people. Otherwise, all the planning and mitigation activities will be ineffective. The WHO and USEPA as well as many countries are currently maintaining 10 μg/L of drinking water arsenic ingestion for minimizing arsenic-related health risk.
d. Awareness campaigns regarding arsenic poisoning and mitigation are important since most of the people are not conscious of the impact of chronic arsenic exposure. The health sector neglects arsenic poisoning. Our experience shows that existing arsenic awareness campaigns are not appropriate since most of the people in rural Bangladesh cannot read newspapers and cannot afford TV or radio. Therefore, it is important to launch a national awareness campaign on the health problems caused by arsenic. Dissemination of relevant information through media, mobile phones, and more innovative forms of communication such as village theater and music is a prerequisite to minimizing health risk of people from arsenic poisoning. As a government-implementing agency, the DGHS in association with the DPHE can implement this health-related national awareness campaign. Information on water quality testing, improved medical testing for arsenicosis, and healthcare facilities for arsenicosis patients can be integrated with the national awareness campaign. In addition, village-level health camps can improve awareness levels of the people in arsenic-contaminated areas as well as sufferers from arsenic-related diseases.
e. Switching back from present tubewell dependence to the previous pond water culture is an important option to people in areas where there are no safe water points. This option can be maintained until new technology can be installed in the area. In Bangladesh, decades ago before the advent of tubewells, people used surface water that was contaminated with microorganisms, but they are now habituated to the tubewell culture. Many find it difficult to shift back to the past habits that have been altered over three decades. It is essential in this regard to make a policy to increase awareness and influence people to use pond water by either boiling it or otherwise purify it before drinking. The government should take the necessary action for the maintenance of safe pond water, which is a prerequisite for public health.
f. Short training courses for different working groups (e.g., schoolteachers, religious leaders, health workers, village doctors, village development committee, etc.) in rural areas of Bangladesh would be productive. After receiving training on different aspects of arsenic issues, these people could subsequently contribute to the arsenic awareness campaign as well as to arsenic preventive and mitigation measures.

7.8.2 Long-Term Strategy

a. It is proven that heavy withdrawal of groundwater for irrigation leads arsenic into the groundwater. Bangladesh is a riverine country, and her huge surface water sources in terms of "closed" and "open" water bodies could be used for irrigation rather than the use of groundwater. Apart from this, it is possible to use river water flowing through connecting canals with nearby rivers. The "canal excavation policy" for irrigation could avert groundwater arsenic poisoning in soils that would prevent arsenic poisoning in the food chain. The proper excavation of canals could lead to the appropriate use of surface water for irrigation. If the government formulates a policy to eliminate arsenic from groundwater but not ending its use for irrigation, in the long run this will be environmentally disastrous.

Therefore, the DPHE in association with the MoA and the MoWR could develop an implementation policy regarding this issue.

b. Deep tubewells connected to piped water supply systems are thought to be a safe and viable mitigation option, especially in high-density villages. This system would meet arsenic-safe drinking water demands for the people living beyond the municipalities. Considering the clustered form of rural settlement in Bangladesh, treated water can be stored in reservoirs at some point of optimum distance from users and then supplied through a piped system to each settlement cluster or community with easy access to a standpipe.

c. Surface water treatment technology is one of the most suitable in arsenic mitigation options. Since surface water is pathogenic, it is possible with a simple and sustainable technology to treat that water and supply every household with a piped water system or any settlement cluster or community with easy access to a standpipe. Borgoño et al. (1977) observed a reduction in prevalence (from 23% to 7%) of common respiratory illnesses after providing arsenic-safe water, reinforcing the effect of arsenic on chronic respiratory illness in Chile. Moreover, Zaldívar and Ghai (1980) found a reduction of cough from 38% to 7% after an arsenic removal plant was installed in an area of Chile.

d. Deep tubewells account for more than 80% of arsenic mitigation interventions in Bangladesh. These deep aquifers are not properly assessed for long-term risk of pollution by salinity and arsenic. The deep tubewell option has been the most popular and cost-effective, but very little has been done to monitor its status and provide warning of deleterious trends (GoB, 2011). A line of literature shows that deep tubewells are not as safe as assumed. Mandal et al. (1996) pointed out that a once safe deep tubewell can be contaminated with arsenic in the course of time. Moreover, subsurface geology in some parts of Bangladesh is unsuitable for installing deep tubewells. Therefore, care is needed in providing deep tubewells.

e. Dug wells and pond water are generally not contaminated with arsenic, but they carry microbiological risks. Therefore, these technologies need to be properly maintained and protected from pollution. Moreover, all new groundwater supply sources need to be tested for arsenic before commissioning by the individual or the agency installing the water source.

f. Water quality of all the water points should be monitored regularly with laboratory methods since FTKs are of unreliable accuracy. The existing DPHE laboratories as well as the laboratories from different universities can be utilized for regularly monitoring the water quality.

g. The DPHE should exchange contaminated water points information with DGHS so that improved surveillance and treatment facilities for chronic arsenicosis diseases can be targeted to these areas. Moreover, all community healthcare centers at different administrative levels need access to the DPHE zonal laboratories for testing the levels of groundwater arsenic contamination so that quick decisions can be possible for any treatment for arsenicosis patients. The current lack of coordination between inter-ministerial bodies regarding water quality and arsenic issues hampers the government's vision.

h. The deposition of arsenic sludge is a serious threat to the environment. The unplanned disposal of highly toxic arsenic sludge can contaminate nearby waterways and can infiltrate into the groundwater. Therefore, it is urgent to arrange a sludge management program before treating arsenic-contaminated water from aquifers.

i. Water quality and arsenic-related issues can be included in the school level academic curriculum so that pupils can easily get information. A number of natural disasters and population problems have already been included in different academic curricula for permanent proliferation.

j. For the transparency of any project implementation, "bulletin boards" are needed in public places to inform local communities about water quality with arsenic concentration as well as installation positions and distribution of arsenic-safe water points and technologies. The DPHE can conduct this in association with the DGHS and the Union Parishad. Likewise, the Nationwide Public Water Point Mapping database including GPS coordinates of all water points, owners, installation year, funding sources, depth, and water quality should be publicly available. This system can be further used for research purposes, and better recommendations would then be possible with the provided information.

k. An inter-ministerial committee should be developed to oversee national arsenic mitigation activities through an independent third party regulatory body other than the DPHE and the DGHS to monitor and evaluate the relevant data on mitigation progress and arsenic patient management. This regulatory body could independently test all public and private water points, check the health situation of arsenicosis patients, and provide reports to parliament including about progress on the national plan.

l. Arsenic mitigation-related activities between various ministries and divisions (e.g., LGD, MoH&FW, MoA, and MoWR) as well as between government agencies, NGOs, and the private sector at different levels, from national to union should be coordinated.

m. Without research, it is not possible to mitigate arsenic poisoning. A continuous research program on arsenic poisoning would be effective at improving surveillance and better treatment for arsenicosis patients.

7.9 CONCLUDING REMARKS

There are several challenges for arsenic mitigation in Bangladesh, which include (a) the lack of adequate coordination among stakeholders; (b) the lack of research regarding arsenic poisoning, mitigation options, and safe water technologies suitable for Bangladesh; (c) the lack of knowledge and experience about arsenic mitigation of stakeholders; and (d) incompetent arsenic mitigation policy. Many organizations are campaigning repeatedly and frequently about arsenic awareness and about drinking safe water through media, but the public do not give any credence to this awareness program since most of the people are more worried about their poor economic status. On the other hand, a group of people is aware of arsenic and its toxicity. Moreover, some people are changing their water-use habits and are drinking safe water from deep tubewells.

Among the low-cost technologies, most are not affordable by poor people, but another factor is the reluctance among the population generally to adopt unfamiliar innovations when they are so fully adapted to the tubewell culture (Hassan et al., 2004). The roles of stakeholders, mainly the DPHE activities, should be reorganized based on transparency in any implementation activities for arsenic mitigation—there will be no hidden issues in mitigating arsenic problems. It is essential in this regard to increase awareness and to influence people to try safe options, for instance, to use pond water by either boiling it or purifying it with a filter until a sustainable mitigation option is accessible. In our opinion, the best basis for such outreach is, first of all, to listen to the people's voices about their present constraints, understandings, and prejudices. A top-down information campaign might fail without such prior knowledge.

8 Arsenic Poisoning in Bangladesh and Legal Issues of Responsibility

As long as the world shall last there will be wrongs, and if no man objected and if no man rebelled, those wrongs would last forever.

Clarence Darrow
Prominent American Lawyer, 1857–1938

In 2002, the United Nations Committee on Economic, Social and Cultural Rights (CESCR) recognized access to water as an independent human right: "the right to water clearly falls within the category of guarantees essential for securing an adequate standard of living, particularly since it is one of the most fundamental conditions for survival" (WHO, 2003: 8). Among the Least Developed Countries (LDCs), Bangladesh has taken this right most seriously. Approximately 97% of its rural people have access to bacteriologically clean water (GOB, 2003), as a result of the installation of about 12 million tubewells. However, as we have seen in previous chapters, the irony is that millions of people have been exposed to the poisonous contamination of arsenic in this tubewell water. This raises the issue of who is responsible when well-meaning attempts at development go wrong. Are there legal consequences for the ill-health caused by exposure to toxic substances that people would not otherwise have encountered in their environment? This chapter draws mainly upon Atkins et al. (2006) that was published in the *Transactions of the Institute of British Geographers*.

8.1 SUTRADHAR v. THE NATURAL ENVIRONMENT RESEARCH COUNCIL

One exemplary case is that of Binod Sutradhar, a carpenter from Ramrail in Brahmanbaria district, 80 kms (50 miles) east of Dhaka, Bangladesh. He suffered from painful keratoses, hard lumps or papules, on his hands and feet, which he claimed were the result of consuming water contaminated with minute traces of arsenic from a shallow well installed in his village in 1983. In 2001, along with a large group of his fellow sufferers, he decided to take on the might of the Western science-based aid complex, in the surrogate form of the Natural Environment Research Council (NERC). They are the parent body of the British Geological Survey (BGS), who in the early 1990s undertook a survey of groundwater quality in his region (Davies and Exley, 1992), on behalf of the Overseas Development Administration (ODA). Because this did not check for arsenic, Mr. Sutradhar accused the NERC of negligence.

Mr. Sutradhar alleged a "tort," legally defined as a damage caused by someone's action or inaction. The point of law at stake was the controversial notion of "proximity": the nature of the relationship between the plaintiff and defendant in terms of geography and "duty of care." The NERC/BGS was alleged to have been liable because of its responsibility, through its water monitoring activities, to the water consumers who would have benefited from a fuller analysis of the samples they collected. This is a developing area of environmental law that has major implications for scientific consultancy and the application of expert knowledge in the aid industry, and it raises the issue of whether Western academics and researchers have a legal responsibility to their "clients" in LDCs.

8.2 WHAT SPACE FOR ENVIRONMENTAL JUSTICE?

Various lines of redress for damage are available in the international law protecting public environmental health. The first is application of the criminal law, where a state body may prosecute polluters for infringements of official regulations (Koenig and Rustad, 2004). This is most appropriate for companies producing hazardous waste and is not relevant for our discussion. Second, there is the arena of public international law, in which obligations are adduced for both state and nonstate actors. The International Law Commission of the United Nations is seeking to clarify and codify these, for instance, with respect to transboundary environmental damage and impacts upon the global commons.

Third, private international environmental law is a field that has grown in parallel with the global green agenda in the last twenty years, but cross-boundary toxic tort cases between private parties have so far had limited exposure. The International Court of Justice has taken on environmental cases since 1993 but only at the state-to-state level (Alkoby, 2003; Fitzmaurice, 2004). The potentially promising idea of an International Environmental Court for nonstate disputes has not progressed beyond the stage of speculation (Rest, 1998), and the bilateral and multilateral agreements within which cross-boundary environmental torts can be litigated remain inadequate for private cases and are institutionally thin (Rao, 2004). Plaintiffs thus have to fall back on testing their entitlement to environmental justice by suing for compensation in their own courts under domestic common law or in foreign jurisdictions, as with *Sutradhar v. NERC*.

In tort lawsuits, injured parties may sue to recover their position before the tort, through various forms of corrective justice such as monetary compensation or remediation. In the USA, the father of tort theory in the late 19th century was Oliver Wendell Holmes, and he distinguished three types of legal redress (Rosenberg, 1995). First was the application of the principle of "strict liability," where the perpetrator of unintentional damage may be held to account, for instance, for the use of an inherently dangerous or unpredictable technology, even without the need to prove fault. Second, negligence may be judged tortious, as may the third, and obvious category, of intentional wrongs. In Britain, the law of torts, a branch of the common law, has been gradually accumulating over the centuries, with scope for further development. For instance, there has been a rapid acceleration of compensation claims in recent years, following the example of personal injury litigation in the USA. As the name suggests, a "toxic tort" is caused by a noxious substance that damages health, a recent example, for instance, being mesothelioma, which is caused by exposure to asbestos (White, 2004).

The social movement literature on LDCs indicates that, although civil society remains weak in many of these countries, there are increasing numbers of NGOs pursuing legal channels for the resolution of environmental issues, thus producing democratic spaces that in the global North would more often be associated with formal state institutions (Stiles, 2002). Morgan (2005), for instance, has found widespread evidence of social protest on water issues. In Bangladesh, there is a rich variety of such movements with a legal or environmental emphasis, and their reach is gradually taking on an international dimension through links with overseas partners. Important in the present case study were Brotee, an indigenous NGO, and the Bangladesh Environmental Lawyers Association (BELA). Both were active in a range of environmental issues, and for the arsenic case they partnered with the London-based Bangladesh International Action Network and Leigh Day & Co, one of a new breed of environmental law companies that are actively involved in LDCs. There was also in Dhaka the NGOs' Arsenic and Information Support Unit, a joint venture between WaterAid Bangladesh and the NGO Forum for Drinking Water Supply and Sanitation, and many other organizations working on practical help for the victims of arsenic poisoning.

8.3 THE BODY-ENVIRONMENT NEXUS

According to Allan Smith, arsenic in Bangladesh has been responsible for "the largest mass poisoning of a population in history" (Smith et al., 2000a). But in court it is often uncertain to what extent such expert opinions are admissible as evidence (Browne et al., 1998). The notorious US Supreme Court judgment of *Daubert v. Merrell Dow Pharmaceuticals Inc.* (1993) insisted on a filter that has significantly influenced the law of torts in that country, with implications further afield (Edmond and Mercer, 2004). The justices required that judges in the future must screen the quality of experts called by both sides, with a view to admitting only reliable testimony. This sounds reasonable, especially when one considers that many of the so-called "experts" called to the witness stand have in the past often been poorly qualified to give an opinion (Huber, 1991), but the outcome in practice has been to disadvantage plaintiffs. Where the science is indeterminate, controversial, or unsettled or causal links are difficult to establish, it is not easy to persuade judges to take a case seriously, for instance, in complex diseases that have a long period of latency. Arsenicosis is one such disease, and we might add that the problems of plaintiffs are multiplied if they are poor, nonliterate, and perhaps from a country lacking a depth of experience in environmental law (Kanner, 2004). All of these disadvantages apply to the Bangladeshi arsenic sufferers, who are nearly all poor rural people. Due to inadequate nutrition and a high consumption of water in tropical heat, their bodies are more open than most to the insidious poisoning of arsenic (Smith et al., 2000a), yet their access to environmental justice is limited, both in Bangladesh and in foreign courts where Daubert-like attitudes prevail.

8.4 TOXIC TORTS: *SUTRADHAR v. NERC*

Denunciation of a guilty party has not been straightforward in the case of arsenic in Bangladesh. There were many stakeholders involved in the provision of clean water, from technical experts and funders to policy-makers and well engineers. Who among them is to blame? When it became clear in the 1990s that there was a major environmental health crisis emerging, it was UNICEF that the Ministry of Environment and the Department of Public Health Engineering (DPHE) first thought of suing (Anon, 1999). This was because UNICEF has provided funds for the sinking of up to one million tubewells since 1972 (Mahmud and Capella, 1999; Smith et al., 2000a). D'Monte (2004) has reported that a senior health official of the Government of Bangladesh (GoB) claimed at a meeting of Asian environmental journalists in Comilla that Bangladesh was "a victim of UNICEF aggression" and that the arsenic poisoning was a case of "criminal negligence." There is no hint, however, in UNICEF publications devoted to their contribution to water development in the region of any sense of guilt or culpability (UNICEF, 2000b). According to Pearce (2001), UNICEF's usual defense is that "at the time, standard procedures for testing the safety of groundwater did not include tests for arsenic [which] had never before been found in the kind of geological formations that exist in Bangladesh." In any case, 90% of tubewells are privately owned, so UNICEF and other funders only started and encouraged what became largely a popular movement of self-provisioning (BAMWSP, 2004), and, ultimately, UNICEF has global legal immunity as part of the United Nations family of institutions.

A possible model for litigation is the Bhopal disaster. This was a chemical leak in 1984 from a Union Carbide factory in India that was responsible for the deaths of thousands, immediately and in the twenty-year aftermath. The Indian government assumed responsibility for suing the transnational corporation, and this was done initially in the American courts (Xue, 2003). However, the GoB has not shown any resolve for mounting a similar action, and all of the legal proceedings have so far been privately initiated (Murshid, 2004). First, in 1999 the prominent Advocate, Rabia Bhuiyan, applied to the Bangladesh High Court for a writ to force the GoB to show cause why they should not halt the installation of further tubewells when they knew about arsenic-contaminated groundwater. Soon after, Brotee, a campaigning NGO in Dhaka, made a similar writ application, and in July 2001 the government was instructed to respond; then in August 2005 the Supreme Court in Dhaka directed the GoB to implement its own National Arsenic Mitigation Policy and Plan and to honor its legal duty to provide safe water (Anon, 2005).

These positive developments in Dhaka are part of a dual strategy adopted by Bangladesh's environmental justice organizations. In addition to legal action in Bangladesh, they spotted a potential opening in the international dimension of aid and consultancy that is intimately entwined with every aspect of development in that country. In May and July 2001, a team of solicitors from the British firm of Leigh Day & Co met fifteen arsenic victims in two villages of Chandpur district (Mortoza, 2003). Its partners in Bangladesh were Brotee, BELA, the Dhaka Community Hospital, and the Bangladesh Legal Aid and Services Trust (Anon, 2001, 2002). Brotee aimed for compensatory payments of at least £5000 per head for the victims they had identified (Anon, 2003). Legal action was started in London in August 2002 on behalf of 512 clients alleging negligence by the BGS in the execution of their 1992 survey (Lawson, 2003). The BGS report purported to comment on the quality of groundwater but, despite testing for 31 trace elements, did not look for a contaminant, arsenic, that had been found in other parts of the country and was listed by the WHO "Guidelines for Drinking Water Quality" as a hazard in drinking water. A sample case was issued in the name of Binod Sutradhar, asserting that he had suffered personal injury as a result of drinking the water, and in February 2003 a further case was lodged for Mrs. Lucky Begum.

In May 2003, the High Court in London gave permission for this case to go to trial (Bachtold, 2003). The Bangladeshis were hoping that eventually it would be possible to bring a class action involving hundreds or thousands of those who have the symptoms of arsenicosis. Much depended on whether the BGS (through the NERC) owed these water consumers "a duty of care." The British courts have hitherto dealt with such cases of environmental justice mainly in terms of nuisance and negligence (Pugh and Criddle, 2004). Counsel on behalf of the claimant here argued a failure of a duty of care. Davies and Exley were accused of not mentioning arsenic, not because that particular element was itemized in the brief, it was not, but because the study aimed to understand "the modes of occurrence of trace elements that may be toxic to biological systems" (Case No: A2/2003/1148: *Sutradhar v. NERC* [2004] EWCA Civ 175). According to Mr. Sutradhar, "in 1992 the possibility of arsenic being present in the groundwater should have been known to reasonably competent hydrogeologists." If the report was not intended as an analysis of the potability of water, then that should have been made clear. Murshid (2004) goes further:

> The arsenic crisis in Bangladesh is a classical example of negligence and distorted development policies. Both governments and international agencies must share the blame and must be made accountable for [their] actions.

In response, the NERC defended the BGS by stating that, as far as they were concerned, they owed no duty of care to Mr. Sutradhar and Mrs. Begum, nor, by implication, to any of the other consumers of water in Bangladesh (BGS, 2001; NERC, 2005). The 1992 report was prepared for the ODA, and there were no contractual arrangements with any organization in Bangladesh. The work was part of an agricultural irrigation project and, as such, had nothing to do with drinking water. The BGS had no relationship with the claimants, who were anyway unaware that the report existed and therefore could not have made any decisions about their water supply based upon it. The NERC also reminded the court that they were "not responsible for the presence of arsenic in the water and had no responsibility for removing the arsenic" and, anyway, that they had at no time certified the water as fit for human consumption. In their opinion "no reasonably experienced researcher would have tested for the presence of arsenic in this location without some special reason to do so" (NERC, 2005).

During the High Court hearing the claimant argued that the NERC had withheld material documentation. Mr. Justice Simon decided that this was true and that a fuller analysis of the case at trial was therefore justified.

In February 2004, the Court of Appeal disagreed with the lower court and struck out the claim, chiefly on the grounds of a lack of proximity between the parties (*Sutradhar v. NERC* [2004] EWCA Civ 175). Legally, proximity may involve closeness in space and time but mainly addresses other aspects of the relationship between parties, for instance, the close association between a parent and a child, a solicitor and a client, or a food manufacturer and a consumer (NERC, 2005). Because proximity, or "neighborhood" as it is sometimes called, is contingent upon the facts of each case, the courts have tended to rely incrementally upon the precedent of case law rather than any precise definition or foundational principle. Having said that, "foreseeability" of the damage is generally thought of as a necessary condition, in this case referring to the reasonable likelihood that the 1992 BGS report would be shown to the Bangladeshi authorities responsible for ensuring a safe water supply in the study region as a basis for action.

In the Court of Appeal, the three judges delivered technical judgments that had little factual or abstract moral content. Lord Justice Kennedy's commentary was divided into what he called "the construction issue" and "the proximity issue." On the former, he remarked that the report was clearly not intended as a comprehensive and definitive statement of water standards, and, on the latter, he concluded that the BGS had "no duty to provide the claimant or his fellow citizens with potable water. They had no power to do so, and they could not even warn him of any dangers." Lord Justice Wall agreed and said that these points were sufficient to strike out the appeal, but Lord Justice Clarke demurred in his minority judgment. He attached particular weight to the statement of Dr. Sara Bennett, a Canadian environmental specialist consulting on Bangladesh's Northeast Regional Water Management Plan, who pointed out that the GoB relies heavily upon foreign organizations for data gathering and analysis, with the implication that the 1992 report was therefore in close associative proximity to their decision-making processes. Lord Justice Clarke did not give any indication that he favored Sutradhar's case, but he perceived in its element of proximity sufficient novelty to recommend a trial in this developing area of jurisprudence. For him, the case of the NERC was weak in as much as

> The citizens of Bangladesh like the claimants were (at least potentially) so closely and directly affected by the negligent act or omission of the defendant in failing to test for arsenic and/or, having done so, in failing to make it clear to the reader of the 1992 report that there might be trace elements (including arsenic) not tested for which might pose a hazard to human health, that it ought reasonably to have had them in contemplation when deciding what to test for and how to report the results.

Lord Justice Wall said that at first he had agreed with Lord Justice Clarke but then had changed his mind on hearing the arguments of counsel. The basis for this was that the precedent cases were in his opinion not sufficiently favorable for a definition of proximity broad enough for Mr. Sutradhar's case.

8.5 HOUSE OF LORDS

Sutradhar v. NERC did not finish in the Court of Appeal. In January 2005, the Appellate Committee of the House of Lords granted permission for a further hearing, which followed in May 2006. Five Law Lords (Lord Nicholls of Birkenhead, Lord Hoffmann, Lord Walker of Gestingthorpe, Lord Brown of Eaton-under-Heywood, and Lord Mance) sat, usually taken as a measure of the significance of a case. They are not bound by precedent in quite the same way as lower courts and are therefore more open to new and innovatory ideas. In the event, it was Lord Hoffmann who gave the decisive judgment on behalf of the others. He was scathing about the case, which he described as "hopeless." It was not necessary to consider the facts because in principle the case did not meet the basic tests for judging whether a statement is negligent: foreseeability and proximity. Lord Hoffmann stated that "It seems to me that the alleged implied statement about arsenic in the BGS report is no different from a statement in an authoritative textbook on geology to the effect that the aquifers of Bangladesh are very unlikely to contain arsenic." He also observed that "the fact that one has expert knowledge does not in itself create a duty to the whole world to apply that knowledge in solving its problems."

8.6 CONCLUDING REMARKS

As a result of this judgment in the House of Lords, the legal position for the time being seems to be that consultants are not legally liable in development aid situations "because of the geography, chain of dealings, and supposedly the need to protect the future provision of development aid funds which may become restricted if legal claims arise in respect of services provided" (Michalowska, 2004). Beck (1999) predicted exactly this kind of outcome due to the "risk society's" "organized irresponsibility," where attribution of responsibility has become increasingly difficult in complex technological systems.

The case was important though, not only in the context of arsenic contamination of drinking water in Bangladesh but also as a commentary on the inter-relationships between expertise and funding from the global North and poor recipients in the global South. With the advance of globalization, such relationships are becoming ever more intense and important. The outcome of *Sutradhar v. NERC* should not be taken as an indication that international linkages are without legal consequences. On the contrary, the fact that the costs of £380K for the case were born by the UK's Special Cases Unit in the Legal Services Commission is proof that it is in the public interest to support such multi-party actions and to scrutinize closely all such alleged environmental torts.

9 Epilogue and Way Forward

> Arsenic sticks around and today it's easily found after death if somebody thinks of looking for it, because the problem with arsenic, it isn't looked for in the common tests for drugs.
>
> **Michael M. Baden**
> *Physician and Forensic Pathologist, 1934*

This book has explained one of the world's largest environmental health and social hazards, one that is hardly known by the public in Europe and North America. It therefore opens up an important environmental agenda and demonstrates another dimension of risk faced by the people living in Bangladesh. The book is therefore fascinating and appalling, while at the same being rigorous in its scientific approach. The social science approach to contamination has been responsible for the public understanding of food safety, the classic examples being mad cow disease and genetically modified foods. Arsenic toxicity is thought to have as high a profile as this, and the scale of the problem is huge and deserves to be discussed in detail. This book reviews sources of arsenic and its worldwide geographical distribution, clinical manifestations of arsenic exposure to human health, and risk assessment concepts.

As the "King of Poisons" or the "Poison of Kings" over the years, arsenic has been used since ancient times both as a poison and a curative. Arsenic has been used not only to commit homicide but also as a medicinal agent for a number of fatal diseases and an effective cancer chemotherapeutic agent. Given the known health consequences of arsenic exposure, there is an imminent need to develop arsenic mitigation programs. At present, people of more than 100 countries worldwide rely heavily on groundwater for drinking purposes. Groundwater development has been actively encouraged over the last few decades as a means of providing a pathogen-free alternative to polluted surface water in reducing the incidence of water-borne diseases. The United Nations Children's Fund, about a couple of decades ago, and later the World Bank suggested the tapping of groundwater for an immediate solution to the problem of untreated surface water. Accordingly, many tubewells have been installed in recent decades for pathogen-free drinking water, but the water pumped is often contaminated with toxic levels of arsenic. Arsenic contamination has been increasing at an alarming rate, and the risk is spreading all over the country. The discovery of groundwater arsenic poisoning in Bangladesh has been characterized as the "worst mass poisoning in human history."

9.1 PRÉCIS OF VENOMOUS ARSENIC: POLICY MAPPING AND LEGAL ISSUES

9.1.1 Arsenic Scenario and Mechanism

The occurrence of high concentrations of arsenic in drinking water has been recognized as a major public health concern in several parts of the world over the last couple of decades. In the last few decades a few major incidents of groundwater arsenic contamination have been reported in Argentina (Díaz et al., 2016), Bangladesh (Radloff et al., 2017), Chile (Díaz et al., 2015), China (Wei et al., 2016), India (Biswas et al., 2014), Mexico (Mendez et al., 2016), Taiwan (Hsu et al., 2015; Tseng et al., 1968), Vietnam (Wilbers et al., 2014), the Duero River Basin of Spain (Pardo-Igúzquiza et al., 2015), Nova Scotia in Canada (Dummer et al., 2015), Wisconsin in the USA (Luczaj and Masarik, 2015), the Águeda watershed area in the Portuguese district of Guarda, the Spanish provinces of Salamanca and Caceres (Antunes et al., 2014), and so on. Arsenic exposure has been well known for many years in the area of endemic "black-foot disease" on the southwest coast of Taiwan (Blackwell et al., 1961). In Cordoba Province of Argentina, "the illness of Bell Ville" was reported as endemic arsenical skin disease, and cancer was first recognized as long ago as 1917 (Goyenechea, 1913). In addition, about half of the countries around the world have arsenic contamination in their groundwater, and where remediation systems are not in place there is a hazard of poisoning.

The spatial pattern of arsenic concentrations does not show any uniformity corresponding to surface geology. Is geological variability the main cause of the differences? The variation of arsenic concentrations over time also raises the issue of the mechanism of arsenic in groundwater. What are the reasons for the variation of arsenic concentration with tubewell age: heavy withdrawal of groundwater or according to geological origin and lithology? Therefore, more research is needed on spatiotemporal analysis, more specifically depth-specific distribution in different geological settings to investigate the nature of arsenic mobilization. Moreover, there is a need to investigate the role of organic matter in the mechanism of arsenic release in groundwater. The natural biogeochemical processes and anthropogenic activities lead to the contamination of groundwater with elevated levels of arsenic. The primary source of arsenic in groundwater is predominantly natural (geogenic) and mobilized through complex biogeochemical interactions within various aquifer sediments (e.g., pyrite, arsenopyrite, and other sulfide minerals) and water (Herath et al., 2016). The geochemistry of arsenic is a function of multiple oxidation states, speciation, and redox transformation. Mobilization of arsenic in groundwater is controlled by adsorption onto metal oxyhydroxides and clay minerals. Geochemical processes including reductive dissolution, alkali desorption, sulfide oxidation, and geothermal processes can initiate the natural release of arsenic from aquifer materials into groundwater (Bundschuh et al., 2013). The mechanism of releasing arsenic into groundwater may occur in a variety of geochemical environments; however, an important occurrence of arsenic is in the young alluvial basins adjacent to active mountain belts (Herath et al., 2016).

9.1.2 Arsenic-Induced Health

Drinking water arsenic contamination is a global public health issue. The chronic deleterious effects of arsenic on human health have been well known for the last couple of centuries. Repeated low levels of exposure to inorganic arsenic from drinking water over an extend period of time can produce health effects. Inorganic arsenic has been recognized as a "human poison," and this inorganic arsenic was first classified as a carcinogen in 1980 by the IARC. Only a small quantity of arsenic at ppb to a few ppm can constitute a serious health hazard. Arsenic toxicity in acute form can appear in exposed populations ingesting arsenic at oral doses of around 0.02 mg/kg/day (about 1.0 to 1.5 mg/day for an adult). The symptoms of acute toxicity of arsenic include muscular cramps, facial edema, and cardiac abnormalities (Kamijo et al., 1998). Stomach pain and nausea may also lead to shock, coma, and even death (Saha et al., 1999). It is evident that oral exposure to arsenic in water at 60 mg/L can kill promptly (ATSDR, 2007).

Chronic exposure to elevated levels of arsenic is associated with substantial increased risk of a wide array of diseases including skin manifestations (Wei et al., 2017), neurological disorders (Fee, 2016), respiratory effects (Parvez et al., 2013), diabetes (Kuo et al., 2015), cardiovascular diseases (Barchowsky and States, 2016), and so on. The skin is quite sensitive to arsenic, and chronic exposure to low levels of arsenic can cause different skin lesions in the form of melanosis, leucomelanosis, and keratosis (Chakraborti et al., 2017a; Mazumder et al., 1998a; Sarma 2016). Skin lesions (hyperkeratosis and pigmentation) have been observed even at exposure levels of 5–10 µg/L arsenic in drinking water (Yoshida et al., 2004). Moreover, epidemiological investigations with chronic exposure to elevated levels of arsenic (i.e., 100–1000 µg/L) have found significant cancerous health risks (Clewell et al., 2016), including tumorigenesis in skin (Karagas et al., 2015), lung (Sherwood and Lantz, 2016), bladder (Medeiros and Gandolfi, 2016), liver (Wang et al., 2014b), kidney (Cheng et al., 2017), and prostate (Benbrahim-Tallaa and Waalkes, 2008).

The mechanism of arsenic toxicity on human health is still under intense investigation. There are indications that chromosomal abnormality, oxidative stress, and altered growth factors for arsenic carcinogenesis have a degree of positive evidence. Arsenic-induced epigenetic changes are associated with arsenic-related diseases, and epigenetics is described as the heritable changes in gene expression without any dependence on the DNA sequence. Epigenetic changes have been linked with cardiovascular (Baccarelli and Ghosh, 2012), cancer (Mandal, 2017), neurological complications (Fee, 2016), and metabolic disorders (Mendez et al., 2016).

9.1.3 Assessing Risk of Arsenic Toxicity

Risk assessment has become an important priority issue in recent years. The term "risk" has multiple conceptions and meanings, and it is a common concept that risk is closely related to "vulnerability" and "resilience." Risk is itself a complex concept incorporating at least three elements: probability, magnitude, and reliability (Alder and Wilkinson, 1999: 168). In its simplest form, risk is the probability of an adverse health effect from specified exposures—it is a function of hazard and exposure. Risk is the potential for realization of unwanted negative consequences of groundwater arsenic exposure. Without either hazard or exposure, there is no risk or, there is zero risk. Risk can be expressed in both the objective and subjective views (Coppola, 2011: 141).

Risk assessment is the "systematic scientific characterization of potential adverse health effects resulting from human exposures to hazardous agents or situations" (Omenn and Faustman, 2002). The risk assessment procedure can be conducted with four elements (i.e., hazard identification, toxicity assessment, exposure assessment, and risk characterization) that are known as the risk assessment paradigm. In assessing risk, the "de minimis risk" and "de manifestis risk" concepts should be considered. The "de minimis risk" represents a benchmark, below which any alleged problems or hazards could be ignored. The health risk associated with chronic arsenic exposure includes the acceptable carcinogenic risk level $<10^{-6}$ (i.e., less than 1 in 1,000,000) and noncarcinogenic hazard quotient (HQ) is less than 1 (i.e., HQ is <1). With "de manifestis risk," there is a risk level above which mitigation is mandatory. In practice, this level is generally set at 1 in 10,000 per vulnerable individual (Coppola, 2011: 172). The assessed LCR in my work is about 1.5E-4 for consuming <10 μg/L arsenic from drinking water, and the figure could be about 115E-2 for ingesting water containing arsenic at >300 μg/L.

9.1.4 Social Implications of Arsenic Toxicity

The inherent health situation and social problems as well as social stigma that arsenicosis patients experience during their illness are important foci for this book. The qualitative research philosophy was utilized to uncover the health issues of arsenicosis patients as well as their social implications in which little is yet known. Qualitative research philosophy is mainly based on the nature of research design and research method for nonquantifiable social and behavioral aspects. In qualitative research, an attempt can be made to obtain an in-depth understanding of the meanings and "definitions of the situation" (Wainwright, 1997) presented by informants, rather than the "quantification" (Eisner, 2017: 5) of their characteristics.

There has been a lot of work on arsenic poisoning on human health, but the social issues have so far been underplayed. Arsenic-affected people have already experienced health problems, and their health situation is getting worse. The worsening condition of patients' health year after year is noticeable. Some arsenicosis patients are so ill that it is difficult for them to earn a living. Patients affected with arsenicosis for a long time are adopting various coping strategies (e.g., visiting local doctors for medicines, using amulets on arm or waist, rubbing skin with mustard oil and charmed oil) for their health problems, but there is no improvement. They become anxious and depressed, and some panic about arsenicosis when they come to know that no curable medicines have yet been invented.

Arsenicosis creates social stigmatization and discrimination. Arsenic poisoning affects people's social status, their lifestyle, and sometimes their attitudes, whether measured in terms of "social degradation" or "social injustice." Arsenic toxicity results in major social dislocation for the affected people. As a social consequence of arsenic poisoning, there is, for instance, a tendency to ostracize arsenic-affected people since arsenicosis is thought of as a contagious disease. There are mixed attitudes from service providers and local leaders to the arsenicosis patients. Some are helping the patients, but some are not. Patients identified in Bangladesh are adopting various survival strategies for their social problems, but they face continuous hostility.

9.1.5 SPATIAL ARSENIC MAPPING AND PLANNING

Analyzing of arsenic concentrations in groundwater with spatio-temporal dimension is a very recent phenomena. Some literature relevant to this issue shows the variation of arsenic concentrations. There is a complex pattern of spatial discontinuity of arsenic concentrations in groundwater with differences between neighboring wells at different scales and changes with aquifer depth (Hassan and Atkins, 2011). Analysis of spatial discontinuity is very useful in spatial decision-support for water quality monitoring and identification of safe drinking water zones in groundwater sources. The widely applied traditional statistical techniques ignore spatial structures of information for geochemical patterns (Wang et al., 2015). Therefore, geostatistics (Luz et al., 2014), fractal and multifractal models (Zuo et al., 2015), and spatial autocorrelation (Delbari et al., 2016) have recently been introduced in geochemical analysis.

Apart from GIS and geostatistics, PPGIS, in recent times, is an important issue in spatial planning for arsenic mitigation. PPGIS has been developed in combination with a Participatory Rural Appraisal (PRA) and GIS modelling. Conventional GIS focuses on digital representations of social and environmental phenomena that best reflect their "expert viewpoint" (Weiner et al., 1995) rather than on "lay perceptions" or a "bottom-up approach." PPGIS is a "bottom-up" approach to view the GIS practices in social settings within the domain of "information-democracy" and "policy-right for the community." GIS methodologies with PPGIS can be very effective for spatial policy formulation for arsenic mitigation in any arsenic-prone areas on the globe. Participatory mapping can mark a distinction between grassroots perspectives and official arrangements of functions of land (Panek, 2015), and PPGIS provides a critical complement to grassroots efforts that are undertaken to empower communities (Bauer, 2009; Panek, 2015).

Spatial risk mapping can be possible with the application of GIS and geostatistics. The spatial risk assessment process involves plotting the areas of affected people and those likely to be affected in the future as a result of ingesting different levels of arsenic concentrated in tubewells. Spatial risk mapping can demarcate the areas: which areas are safe and which areas are at risk and to what extent? The answer to these questions can only be investigated with spatial risk mapping. Spatial risk mapping can easily monitor the groundwater arsenic concentrations and possible arsenic mitigation. Spatial risk assessment and risk management procedures have become useful tools in decision making at local, regional, and national levels. The governmental agencies and scientific institutions, in recent times, are using well-established procedures to assess both the probabilistic risk and spatial risk to decision optimization.

9.1.6 ARSENIC AND LEGAL RESPONSIBILITY

The environmental justice and legal issues of drinking water and its quality are also motivating issues of this book. Are there legal consequences for the ill-health caused by exposure to toxic substances that people would not otherwise have encountered in their environment? The identification and consequences of arsenic concentrations in groundwater have been a matter of international legal dispute in recent years. A case study from Bangladesh (*Binod Sutradhar v. Natural Environment Research Council-NERC*) is an important issue regarding environmental justice. Mr. Binod Sutradhar, a carpenter from Ramrail in Brahmanbaria district of Bangladesh, suffered from painful arsenicosis on his hands and feet, which he claimed was the result of consuming water contaminated with minute traces of arsenic. Mr. Sutradhar alleged a "tort," legally defined as a damage caused by someone's action or inaction. Beck (1999:6) recognized the "global market risk"

Epilogue and Way Forward

based on Asian economic crisis which is a new form of the "risk society's" "organized irresponsibility," where attribution of responsibility has become increasingly difficult in complex technological systems. The point of law at stake was the controversial notion of "proximity": the nature of the relationship between the plaintiff and defendant in terms of geography and "duty of care."

9.2 WAY FORWARD

The CESCR in 2002 recognized access to water as an independent human right, noting that "water required for personal or domestic use must be safe, therefore free from micro-organisms, chemical substances, and radiological hazards that constitute a threat to a person's health" (CESCR, 2002: Article 12b). As a member of the Human Rights Council, Bangladesh adopted various resolutions affirming that the human right to safe drinking water and sanitation is derived from the right to an adequate standard of living (UNHRC, 2012). In 2013, Bangladesh adopted a Water Act that declared water for drinking, sanitation, and hygiene as "the highest priority right" (GOB, 2013). This priority will ensure one of the 17 SDG that is directly linked with safe water—Goal 6 (clean water and sanitation), which mainly focuses on "achieving universal and equitable access to safe and affordable drinking water for all by 2030" (UN, 2015). Ensuring the clean water and sanitation (Goal 6) will ensure the healthy life and promote well-being for all at all ages (Goal 3). In order to ensure Goals 6 and 3, to achieve universal and equitable access to safe and affordable drinking water must be ensured for all.

There are several challenges for arsenic mitigation in Bangladesh, which include (a) the lack of adequate coordination among stakeholders; (b) lack of research regarding arsenic poisoning, mitigation options, and safe water technologies suitable for Bangladesh; (c) lack of knowledge and experience about arsenic mitigation of stakeholders; and (d) incompetent arsenic mitigation policy. Many organizations are campaigning repeatedly and frequently about arsenic awareness and about drinking safe water through media, but the public do not give any credence to this awareness program since most of the people are more worried about their poor economic status. Therefore, to determine the magnitude of groundwater arsenic poisoning in Bangladesh, a thorough review of existing literature and further research are essential. In mitigating arsenic poisoning, focus should be given on a number of areas: (a) extent of arsenic poisoning in terms of health and social consequences, (b) mitigation options with technologies, and (c) sustainability and future directions.

In the case of groundwater arsenic mitigation, a number of policies have been promulgated over the last two decades for groundwater arsenic mitigation in Bangladesh. The water policies and strategies concerning groundwater arsenic issues can focus on arsenic mitigation from drinking water. The prominent feature of the institutional landscape in arsenic-related interventions is that of engaging quite a large number of governmental and non-governmental agencies, but there is poor coordination in current institutional arrangements at the national level to face the challenges of the sector. Moreover, numerous arsenic projects were very unsuccessful for implying suitable mitigation options, and their success stories are very unsatisfactory. The existing arsenic mitigation options have a number of limitations that are not suitable for preventing arsenic poisoning. Therefore, the author, from his extensive experience in arsenic mitigation research, proposed several mitigation strategies:

- A comprehensive and separate Arsenic Mitigation Project is needed to minimize drinking water arsenic exposure. Among the existing arsenic mitigation technologies, some are not suitable in all the geophysical areas, some are not affordable in the context of underprivileged socioeconomic areas, and some are cost-effective, but not sustainable. Therefore, a suitable and appropriate technology for arsenic-safe water is a prerequisite.
- A surface water treatment technology is one of the most suitable of arsenic mitigation options. Since surface water is pathogenic, it is possible with a simple and sustainable technology to treat that water and supply every household with a piped water system or any settlement cluster or community with easy access to a standpipe.

- The drinking water standard for arsenic ingestion should be lowered from 50 µg/L to 10 µg/L to minimize high health risk to exposed people.
- Awareness campaigns regarding arsenic poisoning and mitigation are important since most of the people are not conscious of the impact of chronic arsenic exposure.
- Water quality of all the water points should be monitored regularly with laboratory methods.
- Water quality and arsenic related issues can be included in the school level academic curriculum so that pupils can easily get water quality related information.
- An inter-ministerial committee should be developed to oversee national arsenic mitigation activities through an independent third party regulatory body with monitoring and evaluating the relevant data on mitigation progress and arsenic patient management.
- A continuous research program on arsenic poisoning is essential, and this would be effective at improving surveillance and better treatment for arsenicosis patients.

9.3 CONCLUDING REMARKS

Arsenic toxicity affects millions of people in more than 100 countries across the globe. Unfortunately, arsenic still remains a neglected public health concern in many developing countries. Although the government of Bangladesh has made arsenic toxicity a priority agenda for access to safe drinking water, arsenic still remains in many tubewells at a dangerous level and most of the people are frequently using this contaminated water for their drinking and cooking purposes. Therefore, under the banner of "water governance," multi-level actions should be taken by the government, donors, NGOs, and other stakeholders at local, regional, and national levels in mitigating arsenic poisoning. The strategies would be applicable to other arsenic-affected countries.

Bibliography

Aballay, L. R., P. Díaz Mdel, F. M. Francisca et al. 2012. Cancer incidence and pattern of arsenic concentration in drinking water wells in Cordoba, Argentina. *International Journal of Environmental Health Research* 22(3):220–231.

Abbot, J., R. Chambers, C. Dunn et al. 1998. Participatory GIS: Opportunity or oxymoron? *PLA Notes* 33:27–34.

Abdul, K. S. M., S. S. Jayasinghe, E. P. S. Chandana et al. 2015. Arsenic and human health effects: A review. *Environmental Toxicology and Pharmacology* 40:828–846.

Abedin, M. A. and R. Shaw. 2013. Arsenic contamination in Bangladesh: Contemporary alarm and future strategy. In *Disaster Risk Reduction Approaches in Bangladesh*, eds. R. Shaw, F. Mallick, A. Islam, Tokyo: Springer.

Abernathy C. O., R. L. Calderon, W. R. Chappell. 1997. Arsenic exposure and health effects. *Proceedings of the Second International Conference on Arsenic Exposure and Health Effects, San Diego, California, (12–14 June)*, London: Chapman and Hall.

Abernathy, C. O., Y. P. Liu, D. Longfellow et al. 1999. Arsenic: Health effects, mechanisms of actions, and research issues. *Environmental Health Perspectives* 107:593–97.

Abhyankar, L. N., M. R. Jones, E. Guallar et al. 2012. Arsenic exposure and hypertension: A systematic review. *Environmental Health Perspectives* 120(4):494–500.

ACGIH. 1995. 1995–1996 Threshold Limit Values (TLVs™) for Chemical Substances and Physical Agents and Biological Exposure Indices (BEIs™). *American Conference of Governmental Industrial Hygienists*, Cincinnati, Ohio, USA.

Acharyya, S. K. 1997. *Arsenic in Groundwater-Geological Overview*. New Delhi: World Health Organization.

Acharyya, S. K., P. Chakraborty, S. Lahiri et al. 1999. Arsenic poisoning in the Ganges Delta. *Nature* 401: 545.

Achour, M. H., A. E. Haroun, C. J. Schult et al. 2005. A new method to assess the environmental risk of a chemical process. *Chemical Engineering & Processing* 44(8):901–909.

Adair, B. M., E. E. Hudgens, M. T. Schmitt et al. 2006. Total arsenic concentrations in toenails quantified by two techniques provide a useful biomarker of chronic arsenic exposure in drinking water. *Environmental Research* 101:213–220.

Adhikary, P. P., H. Chandrasekharan, D. Chakraborty et al. 2009. Statistical approaches for hydrogeochemical characterization of groundwater in West Delhi, India. *Environmental Monitoring and Assessment* 154:41–52.

Adhikary, P. P., H. Chandrasekharan, D. Chakraborty et al. 2010. Assessment of groundwater pollution in West Delhi, India using geostatistical approach. *Environmental Monitoring and Assessment* 167:599–615.

Adhikary, P. P. and C. J. Dash. 2017. Comparison of deterministic and stochastic methods to predict spatial variation of groundwater depth. *Applied Water Science* 7(1):339–348.

Adriano, D. C. 1986. *Trace Elements in the Terrestrial Environment*. New York: Springer-Verlag.

Affum, A. B., S. D. Osae, B. J. B. Nyarko et al. 2015. Total coliforms, arsenic and cadmium exposure through drinking water in the Western Region of Ghana: Application of multivariate statistical technique to groundwater quality. *Environmental Monitoring and Assessment* 187:1–23.

Afridi, H. I., T. G. Kazi, N. Kazi et al. 2011. Association of environmental toxic elements in biological samples of myocardial infarction patients at different stages. *Biological Trace Element Research* 141(1–3):26–40.

Ağca, N., S. Karanlık, B. Ödemiş. 2014. Assessment of ammonium, nitrate, phosphate, and heavy metal pollution in groundwater from Amik Plain, Southern Turkey. *Environmental Monitoring and Assessment* 186 (9):5921–5934.

AGMI. 2004. *Quarterly Report on Environmental Monitoring at the Puriton Closed Landfill, near Bridgwater, Somerset*. Cambridge (UK): Arcadis Geraghty and Miller International Ltd.

Aguirre, R. T. and K. W. Bolton. 2014. Qualitative interpretive meta-synthesis in social work research: Uncharted territory. *Journal of Social Work* 14(3):279–294.

Agusa, T., J. Fujihara, H. Takeshita et al. 2011. Individual variations in inorganic arsenic metabolism associated with AS3MT genetic polymorphisms. *International Journal of Molecular Sciences* 12(4):2351–2382.

Agusa, T., R. Kubota, T. Kunito et al. 2007. Arsenic pollution in groundwater of Vietnam and Cambodia: A review. *Biomedical Research on Trace Elements* 18:35–47.

Agusa, T., T. Kunito, J. Fujihara et al. 2006. Contamination by arsenic and other trace elements in tube-well water and its risk assessment to humans in Hanoi, Vietnam. *Environmental Pollution* 139:95–106.

Agusa, T., P. T. K. Trang, V. M. Lan et al. 2014. Human exposure to arsenic from drinking water in Vietnam. *Science of The Total Environment* 488–489:562–569.

Ahamed, S., M. K. Sengupta, S. C. Mukherjee et al. 2006. An eight-year study report on arsenic contamination in ground water and health effects in Eruani village, Bangladesh and an approach for its mitigation. *Journal of Health, Population, and Nutrition* 24(2):129–141.

Ahmad, S. A., D. Bandaranayake, A. W. Khan et al. 1997. Arsenic contamination in ground water and arsenicosis in Bangladesh. *International Journal of Environmental Health Research* 7:271–276.

Ahmad, J. K., B. N. Goldar, M. Jakariya et al. 2002. *Fighting Arsenic, Listening to Rural Communities: Findings from a Study on Willingness to Pay for Arsenic-Free, Safe Drinking Water in Rural Bangladesh.* South Asia, Dhaka: Water and Sanitation Programme, The World Bank.

Ahmad, S. A., M. Maharjan, C. Watanabe et al. 2004. Arsenicosis in two villages in Terai lowland Nepal. *Environmental Science* 11(3):179–188.

Ahmad, S. A., M. H. Sayed, S. Barua et al. 2001. Arsenic in drinking water and pregnancy outcomes. *Environmental Health Perspectives* 109(6):629–631.

Ahmad, S. A., M. H. Sayed, M. H. Khan et al. 2007. Sociocultural aspects of arsenicosis in Bangladesh: Community perspective. *Journal of Environmental Science and Health, Part A* 42(12):1945–58.

Ahmadi, S. H. and A. Sedghamiz. 2008. Application and evaluation of kriging and cokriging methods on groundwater depth mapping. *Environmental Monitoring and Assessment* 138:357–368.

Ahmadian, S. 2013. Geostatistical based modelling of variations of groundwater quality during 2006 to 2009 (in Tehran-Karaj Plain). *Journal of Basic Applied Science Research* 3:264–272.

Ahmed, M. F. 2003. Drinking water standard and guideline value for arsenic: A critical analysis. In *Arsenic Contamination: Bangladesh Perspective*, ed. F. M. Ahmed. Dhaka: ITN, BUET, Bangladesh.

Ahmed, M. F., S. Ahuja, M. Alauddin et al. 2006. Ensuring safe drinking water in Bangladesh. *Science* 314(5806):1687–1688.

Ahmed, S., E. Akhtar, A. Roy et al. 2017. Arsenic exposure alters lung function and airway inflammation in children: A cohort study in rural Bangladesh. *Environment International* 101:108–116.

Ahmed, K. M., P. Bhattacharya, M. A. Hasan et al. 2004. Arsenic contamination in groundwater of alluvial aquifers in Bangladesh: An overview. *Applied Geochemistry* 19:181–200.

Ahmed, M. K., N. Shaheen, M. S. Islam et al. 2015. Dietary intake of trace elements from highly consumed cultured fish (*Labeo rohita, Pangasius pangasius* and *Oreochromis mossambicus*) and human health risk implications in Bangladesh. *Chemosphere* 128:284–292.

Ahmed, M. K., N. Shaheen, M. S. Islam et al. 2016. A comprehensive assessment of arsenic in commonly consumed foodstuffs to evaluate the potential health risk in Bangladesh. *Science of the Total Environment* 544:125–133.

Ahoulé, D., F. Lalanne, J. Mendret et al. 2015. Arsenic in African waters: A review. *Water Air and Soil Pollution* 226(9):1–13.

Ahsan, S. M. 2000. Governance and Corruption in an Incomplete Democracy: The Bangladesh Case. Paper presented at *the Sixth Workshop of the European Network of Bangladesh Studies, University of Oslo*, 14–16 May, Oslo.

Ahsan, H., Y. Chen, F. Parvez et al. 2006. Arsenic exposure from drinking water and risk of premalignant skin lesions in Bangladesh: Baseline results from the Health Effects of Arsenic Longitudinal Study. *American Journal of Epidemiology* 163(12):1138–48.

Aitchison, J. 1982. The statistical analysis of compositional data. *Journal of the Royal Statistical Society. Series B (Methodological)* 44(2):139–177.

Ajjawi, R. and J. Higgs. 2007. Using hermeneutic phenomenology to investigate how experienced practitioners learn to communicate clinical reasoning. *The Qualitative Report* 12(4):612–638.

Akay, C., C. Thomas III., Y. Gazitt. 2004. Arsenic trioxide and paclitaxel induce apoptosis by different mechanisms. *Cell Cycle* 3(3):324–334.

Aksever, F. 2011. *Hydrogeological investigations of the Sandkl (Afyonkarahisar) Basin*. Doctoral Thesis (unpublished). Demirel University, Turkey.

Al-Abadi, A. M., H. R. Pourghasemi, S. Shahid et al. 2017. Spatial mapping of groundwater potential using entropy weighted linear aggregate novel approach and GIS. *Arabian Journal for Science and Engineering* 42:1185–1199.

Al-Adamat, R. A. N., I. D. L. Foster, S. M. J. Baban. 2003. Groundwater vulnerability and risk mapping for the Basaltic aquifer of the Azraq basin of Jordan using GIS, remote sensing and DRASTIC. *Applied Geography* 23:303–324.

Bibliography

Alaerts, G. J., N. Khouri, B. Kabir. 2001. *Strategies to Mitigate Arsenic Contamination of Water Supply*. Washington, DC: United Nations Synthesis Report on Arsenic in Drinking Water.

Alam, M. A. and M. M. Rahman. 2010. Comparative assessment of four alternative water supply options in arsenic affected areas of Bangladesh. *Journal of Civil Engineering* 38(2):191–201.

Alam, M. G. M., E. T. Snow, A. Tanaka. 2003. Arsenic and heavy metal contamination of vegetables grown in Samta village, Bangladesh. *Science of the Total Environment* 308:83–96.

Alamdar, A., S. A. M. A. S. Eqani, N. Hanif et al. 2017. Human exposure to trace metals and arsenic via consumption of fish from river Chenab, Pakistan and associated health risks. *Chemosphere* 168:1004–1012.

Alarcón-Herrera, M. T., I. Flores-Montenegro, P. Romero-Navar et al. 2001. Contenido de arsénico en el agua potable del valle del Guadiana, México. *Ingeniería Hidráulica en México* 16:63–70.

Alarifi, S., D. Ali, S. Alkahtani et al. 2013. Arsenic trioxide-mediated oxidative stress and genotoxicity in human hepatocellular carcinoma cells. *Onco Targets and Therapy* 6:75–84.

Alasuutari, P. 2012. The rise and relevance of qualitative research. *International Journal of Social Research Methodology* 13(2):139–155.

Alcaine, A. A., A. Sandhi, P. Bhattacharya et al. 2012. Distribution and mobility of geogenic arsenic in the shallow aquifers of the northeast of La Pampa, Argentina. In *Understanding the Geological and Medical Interface of Arsenic*, eds. J. C. Ng, B. N. Noller, R. Naidu, J. Bundschuh, P. Bhattacharya, 132–134. Boca Raton: CRC Press.

Alden, J. C. 1983. The continuing need for inorganic arsenical pesticides. In *Arsenic: Industrial, Biomedical, Environmental Perspectives*, eds. W. H. Lederer and R. J. Fensterheim, 63–71. New York: Van Nostrand Reinhold Company.

Alder, J. and D. Wilkinson. 1999. *Environmental Law and Ethics*. London: Macmillan.

Alexander, D. 2000. *Confronting Catastrophe*. Oxford: Oxford University Press.

Alkoby, A. 2003. Non-state actors and the legitimacy of international environmental law. *Non-State Actors and International Law* 3(1):23–98.

Al Lawati, W. M., J. S. Jean, T. R. Kulp et al. 2013. Characterisation of organic matter associated with groundwater arsenic in reducing aquifers of southwestern Taiwan. *Journal of Hazardous Materials* 262:970–979.

Allen, G. 2016a. Risk identification. In *Threat Assessment and Risk Analysis: An Applied Approach*, eds. G. Allen and R. Derr. Amsterdam: Elsevier and BH.

Allen, G. 2016b. Risk analysis. In *Threat Assessment and Risk Analysis: An Applied Approach*, eds. G. Allen and R. Derr. Amsterdam: Elsevier and BH.

Allen, M. and L. Jensen. 1990. Hermeneutic inquiry: Meaning and scope. *Western Journal of Nursing Research* 12(2):241–53.

Aller, L. T., T. Bennett, J. H. Lehr, G. Hackett. 1987. DRASTIC: A standardized system for evaluating ground water pollution potential using hydrogeologic settings. *Journal of the Geological Society of India* 29(1):38–57.

Al-Omran, A. M., A. A. Aly, M. I. Al-Wabel et al. 2017. Geostatistical methods in evaluating spatial variability of groundwater quality in Al-Kharj Region, Saudi Arabia. *Applied Water Science* [doi:10.1007/s13201-017-0552-2].

Altaş, L., M. Işik, M. Kavurmaci. 2011. Determination of arsenic levels in the water resources of Aksaray Province, Turkey. *Journal of Environmental Management* 92(9):2182–92.

Altunay, N. and R. Gürkan. 2017. Determination of sub-ng g-1 levels of total inorganic arsenic and selenium in foods by hydride-generation atomic absorption spectrometry after pre-concentration. *Food Additives and Contaminants: Part A* 34(3):390–403.

Alvesson, M. and S. Deetz. 2000. *Doing Critical Management Research*. London: Sage.

Alwang, J., P. B. Siegel, S. L. Jorgensen. 2001. *Vulnerability: A View from Different Disciplines*. Social Protection Discussion Paper Series, No. 0115. World Bank: Washington, D.C.

Amaro, A. S., B. C. V. Herrera, E. Lictevout. 2014. Spatial distribution of arsenic in the region of Tarapaca, northern Chile. In *Once Century of the Discovery of Arsenicosis in Latin America (1914–2014)*, eds. M. I. Litter, H. B. Nicolli, M. Meichtry, N. Quici, J. Bundschuh, P. Bhattacharya, R. Naidu, 54–55. Boca Raton: CRC Press.

Ameer, S. S., K. Engström, M. B. Hossain et al. 2017. Arsenic exposure from drinking water is associated with decreased gene expression and increased DNA methylation in peripheral blood. *Toxicology and Applied Pharmacology* 321:57–66.

Ameer, S. S., Y. Xu, K. Engström et al. 2016. Exposure to inorganic arsenic is associated with increased mitochondrial DNA copy number and longer telomere length in peripheral blood. *Frontiers in Cell and Developmental Biology* 4:87 (doi:10.3389/fcell.2016.00087).

Amini, M., K. C. Abbaspour, M. Berg et al. 2008. Statistical modeling of global geogenic arsenic contamination in groundwater. *Environmental Science and Technology* 42(10):3669–3675.

Amiri, V., M. Rezaei, N. Sohrabi. 2014. Groundwater quality assessment using entropy weighted Water Quality Index (EWQI) in Lenjanat, Iran. *Environmental Earth Sciences* 72:3479–3490.

Amundson, S. A., T. G. Myers, A. J. Fornace Jr. 1998. Roles for p53 in growth arrest and apoptosis: Putting on the brakes after genotoxic stress. *Oncogene* 17(25):3287–3299.

Andra, S. S., K. C. Makris, C. A. Christophi et al. 2013. Delineating the degree of association between biomarkers of arsenic exposure and type-2 diabetes mellitus. *International Journal of Hygiene and Environmental Health* 216:35–49.

Andrade, A. I. A. S. S. and T. Y. Stigter. 2013. The distribution of arsenic in shallow alluvial groundwater under agricultural land in central Portugal: Insights from multivariate geostatistical modelling. *Science of the Total Environment* 449:37–51.

Andrew, A. S., M. R. Karagas, J. W. Hamilton. 2003. Decreased DNA repair gene expression among individuals exposed to arsenic in United States drinking water. *International Journal of Cancer* 104(3):263–268.

Andrews, M., S. D. Sclater, C. Squire et al. 2007. Narrative research. In *Qualitative Research Practice*, eds. C. Seale, G. Gobo, J. F. Gubrium, and D. Silverman, 97–112. London: Sage.

Anon. 1999. *Bangladesh: Conflict over blame for arsenic poisoning.* SUNS, North-South Development Mirror. (http://www.sunsonline.org/trade/process/followup/1999/09020699.htm). Accessed 2 February 2006.

Anon. 2001. *Arsenic patients to take BGS to court.* The Daily Star (Dhaka), 27 July 2001.

Anon. 2002. *BIAN launches campaign for arsenic victims.* The Daily Star (Dhaka), 10 January 2001.

Anon. 2003. *British HC verdict: Each arsenic victim likely to get £5,000.* The New Nation (Dhaka), 11 May 2003.

Anon. 2005. *SC asks govt to implement arsenic mitigation policy.* New Age National (Dhaka), 28 August 2005.

Ansel, J. and F. Wharton. 1992. *Risk Analysis, Assessment, and Management.* Chichester: John Wiley & Sons Ltd.

Antman, K. H. 2001. Introduction: The history of arsenic trioxide in cancer therapy. *The Oncologist* 6 (Suppl. 2):1–2.

Antonelli, R., K. Shao, D. J. Thomas et al. 2014. AS3MT, GSTO, and PNP polymorphisms: Impact on arsenic methylation and implications for disease susceptibility. *Environmental Research* 132:156–167.

Antunes, I. M. H. R. and M. T. D. Albuquerque. 2013. Using indicator kriging for the evaluation of arsenic potential contamination in an abandoned mining area (Portugal). *Science of the Total Environment* 442:545–552.

Antunes, I. M. H. R., M. T. D. Albuquerque, M. F. M. Seco et al. 2014. Uranium and arsenic spatial distribution in the Águeda watershed groundwater. *Procedia Earth and Planetary Science* 8:13–17.

Anwar, J. 2001a. All quiet on arsenic front. *The Daily Star*, September 07, Dhaka, Bangladesh: The Daily Star Centre. [http://www.eng-consult.com/arsenic/as217.txt].

Anwar, J. 2001b. Arsenic mitigation: A costly delay. *The Daily Star*, January 08, Dhaka, Bangladesh: The Daily Star Centre. [http://www.eng-consult.com/arsenic/as217.txt].

ANZECC. 1992. *Australian water quality guidelines for fresh and marine waters.* ANZECC (Australian and New Zealand Environmental and Conservation Council), Canberra, Australia.

Appelo, C. A. J. and D. Postma. 1996. *Geochemistry, Groundwater and Pollution.* Rotterdam: Balkema.

Appleyard, S. J., J. Angeloni, R. Watkins. 2006. Arsenic-rich groundwater in an urban area experiencing drought and increasing population density, Perth, Australia. *Applied Geochemistry* 21:83–97.

Arain, M. B., T. G. Kazi, J. A. Baig et al. 2009a. Determination of arsenic levels in lake water, sediment, and foodstuff from selected area of Sindh, Pakistan: Estimation of daily dietary intake. *Food and Chemical Toxicology* 47(1):242–248.

Arain, M. B., T. G. Kazi, J. A. Baig et al. 2009b. Respiratory effects in people exposed to arsenic via the drinking water and tobacco smoking in southern part of Pakistan. *Science of the Total Environment* 407:5524–5530.

Argos, M. 2015. Arsenic exposure and epigenetic alterations: Recent findings based on the Illumina 450K DNA Methylation Array. *Current Environmental Health Reports* 2(2):137–144.

Argos, M., L. Chen, F. Jasmine et al. 2015. Gene-specific differential DNA methylation and chronic arsenic exposure in an epigenome-wide association study of adults in Bangladesh. *Environmental Health Perspectives* 123:64–71.

Argos, M., T. Kalra, B.L. Pierce et al. 2011. A prospective study of arsenic exposure from drinking water and incidence of skin lesions in Bangladesh. *American Journal of Epidemiology* 174:185–194.

Argos, M., T. Kalra, P. J. Rathouz et al. 2010. Arsenic exposure from drinking water, and all-cause and chronic-disease mortalities in Bangladesh (HEALS): A prospective cohort study. *Lancet* 376(9737): 252–258.

Argos, M., F. Parvez, Y. Chen et al. 2007. Socioeconomic status and risk for arsenic-related skin lesions in Bangladesh. *American Journal of Public Health* 97:825–831.

Argos, M., F. Parvez, M. Rahman et al. 2014. Arsenic and lung disease mortality in Bangladeshi adults. *Epidemiology* 25(4):536–543.

Armienta, M. A., R. Rodríguez, N. Aguayo et al. 1997a. Arsenic contamination of groundwater at Zimapán, Mexiko. *Hydrogeology Journal* 5(2):39–46.

Armienta, M. A., R. Rodríguez, O. Cruz. 1997b. Arsenic content in hair of people exposed to natural arsenic polluted groundwater in Zimapán, Mexico. *Bulleton of Environmental Contamination and Toxicology* 59:583–589.

Armienta, M. A. and N. Segovia. 2008. Arsenic and fluoride in the groundwater of Mexico. *Environmental Geochemistry and Health* 30:345–353.

Armstrong, C. W., R. B. Stroube, T. Rubio et al. 1984. Outbreak of fatal arsenic poisoning caused by contaminated drinking water. *Archives of Environmental Health* 39:276–279.

Aronoff, S. 1989. *Geographic Information Systems: A Management Perspective*. Ottawa, WDL.

Aronson, S. M. 1994. Arsenic and old myths. *Rhode Island Medicine* 77(7):233–234.

Arriaza, B., D. Amarasiriwardena, L. Cornejo et al. 2010. Exploring chronic arsenic poisoning in pre-Columbian Chilean mummies. *Journal of Archaeological Science* 37:1274–1278.

Arshad, N. and S. Imran. 2017. Assessment of arsenic, fluoride, bacteria, and other contaminants in drinking water sources for rural communities of Kasur and other districts in Punjab, Pakistan. *Environmental Science and Pollution Research International* 24(3):2449–2463.

Asante, K. A., T. Agusa, A. Subramanian et al. 2008. Contamination status of arsenic and other trace elements in drinking water and residents from Tarkwa, a historic mining township in Ghana. *Chemosphere* 66:1513–1522.

Asante-Duah, K. 2002. *Public Health Risk Assessment for Human Exposure to Chemicals*. Dordrecht: Kluwer Academic Publishers.

Aschengrau, A., S. Zierler, A. Cohen. 1989. Quality of community drinking water and the occurrence of spontaneous abortion. *Archives of Environmental Health* 44(5):283–290.

Ashraf, M., J. C. Loftis, K. G. Hubbard. 1997. Application of geostatistics to evaluate partial weather station networks. *Agricultural and Forest Meteorology* 84(3–4):255–271.

Astolfi, E., A. Maccagno, J. C. G. Fernández et al. 1981. Relation between arsenic in drinking water and skin cancer. *Biological Trace Element Research* 3(2):133–143.

Atkins, P., M. M. Hassan, C. E. Dunn. 2006. Toxic torts: Arsenic poisoning in Bangladesh and the legal geographies of responsibility. *Transactions of the Institute of British Geographers* 31:272–285.

Atkins, P., M. M. Hassan, C. E. Dunn. 2007a. Environmental irony: Summoning death in Bangladesh. *Environment and Planning A* 39(7):2699–2714.

Atkins, P., M. M. Hassan, C. E. Dunn. 2007b. Poisons, pragmatic governance and deliberative democracy: The arsenic crisis in Bangladesh. *Geoforum* 38(1):155–170.

ATSDR. 2000. *Toxicological Profile for Arsenic*. Atlanta: Agency for Toxic Substances and Disease Registry, U.S. Department of Health and Human Services.

ATSDR. 2007. *Toxicological Profile for Arsenic*. Atlanta: Agency for Toxic Substances and Disease Registry, U.S. Department of Health and Human Services.

ATSDR. 2010. *Arsenic Toxicity Clinical Assessment*. Atlanta: Agency for Toxic Substances Disease Registry, U.S. Department of Health and Human Services.

Audi, R. 1999. *The Cambridge Dictionary of Philosophy*. Cambridge: Cambridge University Press.

Auge, A. 2014. Arsenic in the groundwater of the Buenos Aires province, Argentina. In *One Century of the Discovery of Arsenicosis in Latin America (1914–2014)*, eds. M. I. Litter, H. B. Nicolli, M. Meichtry, N. Quici, J. Bundschuh, P. Bhattacharya, R. Naidu, 125–128. Boca Raton: CRC Press.

Axtmann, R. C. 1975. Environmental impact of a geothermal power plant. *Science* 187:795–797.

Ayerza, A. 1918. Arsenicismo regional endémico (keratodermia y melanodermia combinadas). *Boletín de la Academia Nacional de Medicina de Buenos Aires* 1:11–24.

Ayotte, J. D., B. T. Nolan, J. R. Nuckols et al. 2006. Modeling the probability of arsenic in groundwater in New England as a tool for exposure assessment. *Environmental Science & Technology* 40(11):3578–3585.

Azcue, J. and J. Nriagu. 1994. Arsenic: Historical perspectives. In *Arsenic in the Environment, Part 1: Cycling and Characterization*, ed. J. O. Nriagu, 1–9. New York: John Wiley & Sons.

Baastrup, R., M. Sørensen, T. Balstrøm et al. 2008. Arsenic in drinking-water and risk for cancer in Denmark. *Environmental Health Perspectives* 116(2):231–237.

Baccarelli, A. and S. Ghosh. 2012. Environmental exposures, epigenetics and cardiovascular disease. *Current Opinion in Clinical Nutrition and Metabolic Care* 15(4):323–329.

Bachtold, D. 2003. Arsenic victims to take British science body to court. *Science* 300:1066.

Backman, B., S. Luoma, T. Ruskeeniemi et al. 2006. *Natural occurrence of arsenic in the Pirkanmaa region in Finland*. Espoo: Geological Survey of Finland.

Bacquart, T., S. Frisbie, E. Mitchell et al. 2015. Multiple inorganic toxic substances contaminating the groundwater of Myingyan Township, Myanmar: Arsenic, manganese, fluoride, iron, and uranium. *Science of the Total Environment* 517:232–245.

Bae, M., C. Watanabe, T. Inaoka et al. 2002. Arsenic in cooked rice in Bangladesh. *Lancet* 360:1839–1840.

Bagshaw, N. E. 1995. Lead alloys: Past, present and future. *Journal of Power Sources* 53:25–30.

Baguet, J. P., J. M. Mallion, A. Moreau-Gaudry et al. 2000. Relationships between cardiovascular remodelling and the pulse pressure in never treated hypertension. *Journal of Human Hypertension* 14:23–30.

Baig, J. A., T. G. Kazi, M. B. Arain et al. 2009. Evaluation of arsenic and other physico-chemical parameters of surface and ground water of Jamshoro, Pakistan. *Journal of Hazardous Materials* 166:662–669.

Baig, J. A., T. G. Kazi, A.Q. Shah et al. 2010. Speciation and evaluation of arsenic in surface and ground water: A multivariate case study. *Ecotoxicology and Environmental Safety* 73:914–923.

Bailey, K. A. and R. C. Fry. 2014. Arsenic-associated changes to the epigenome: What are the functional consequences? *Current Environmental Health Reports* 1:22–34.

Baj, G., A. Arnulfo, S. Deaglio et al. 2002. Arsenic trioxide and breast cancer: Analysis of the apoptotic, differentiative and immunomodulatory effects. *Breast Cancer Research and Treatment* 73(1):61–73.

Bakker, K. and C. Morinville. 2013. The governance dimensions of water security: A review. *Philosophical Transactions of the Royal Society A* 371:20130116. [http://dx.doi.org/10.1098/rsta.2013.0116].

Baldwin, K. and H. Oxenford. 2014. A participatory approach to marine habitat mapping in the Grenadine Islands. *Costal Management* 42:36–58.

Balın, D. 2015. *Hydrogeological investigations of the Çöl (Haydarlı/Afyon) basin*. Master's Thesis (unpublished). Almaty: Suleyman Demirel University, Kazakhstan.

BAMWSP. 2004. *Implementation plan for arsenic mitigation in Bangladesh*. Dhaka: Bangladesh Arsenic Mitigation Water Supply Project. (www.bamwsp.org/Arsenic%20Policy%2004/Implementation.pdf). Accessed 1 February 2005.

Banerjee, N., A. K. Bandyopadhyay, S. Dutta et al. 2017. Increased microRNA 21 expression contributes to arsenic induced skin lesions, skin cancers and respiratory distress in chronically exposed individuals. *Toxicology* 378:10–16.

Banerjee, M., N. Banerjee, P. Bhattacharjee et al. 2013. High arsenic in rice is associated with elevated genotoxic effects in humans. *Scientific Reports* 3:2195.

Banerjee, M., P. Bhattacharjee, A. K. Giri. 2011. Arsenic-induced cancers: A review with special reference to gene, environment and their interaction. *Genes and Environment* 33:128–140.

Banerjee, S., A. E. Gelfand, A. O. Finley et al. 2008. Gaussian predictive process models for large spatial data sets. *Journal of the Royal Statistical Society: Series B (Statistical Methodology)* 70(4):825–848.

Barats, A., G. Féraud, C. Potot et al. 2014. Naturally dissolved arsenic concentrations in the Alpine/Mediterranean Var River watershed (France). *Science of the Total Environment*, 473–474:422–436.

Barchowsky, A., E. J. Dudek, M. D. Treadwell et al. 1996. Arsenic induces oxidant stress and NFxB activation in cultured aortic endothelial cells. *Free Radical Biology and Medicine* 21:783–790.

Barchowsky, A. and J. C. States. 2016. Arsenic-induced cardiovascular disease. In *Arsenic: Exposure Sources, Health Risks, and Mechanisms of Toxicity*, ed. J. C. States, 453–468. New Jersey: John Wiley & Sons.

Barlow, S., A. G. Renwick, J. Kleiner et al. 2006. Risk assessment of substances that are both genotoxic and carcinogenic report of an International Conference organized by EFSA and WHO with support of ILSI Europe. *Food and Chemical Toxicology* 44:1636–1650.

Barnes, D. G. and M. Dourson. 1988. Reference dose (RfD): Description and use in health risk assessments. *Regulatory Toxicology and Pharmacology* 8:471–486.

Barnett-Page, E. and J. Thomas. 2009. Methods for the synthesis of qualitative research: A critical review. *BMC Medical Research Methodology* 9:59 [doi:10.1186/1471-2288-9-59].

Bashir, S., Y. Sharma, M. Irshad et al. 2006. Arsenic induced apoptosis in rat liver following repeated 60 days exposure. *Toxicology* 217(1):63–70.

Bastante, F. G., C. Ordóñez, J. Taboada et al. 2008. Comparison of indicator kriging, conditional indicator simulation and multiple-point statistics used to model slate deposits. *Engineering Geology* 81:50–59.

Basu, A. and M. J. Dutta. 2008. The relationship between health information seeking and community participation: The roles of health information orientation and efficacy. *Health Communication* 23(1):70–79.

Basu, A., J. Mahata, S. Gupta et al. 2001. Genetic toxicology of a paradoxical human carcinogen, arsenic: a review. *Mutation Research* 488(2):171–194.

Basu, A., S. Mitra, J. Chung et al. 2011. Creatinine, diet, micronutrients, and arsenic methylation in West Bengal, India. *Environmental Health Perspectives* 119(9):1308–1313.

Bates, M. N., O. A. Rey, M. L. Biggs et al. 2004. Case-control study of bladder cancer and exposure to arsenic in Argentina. *American Journal of Epidemiology* 159(4):381–389.

Bates, M. N., A. H. Smith, K. P. Cantor. 1995. Case-control study of bladder cancer and arsenic in drinking water. *American Journal of Epidemiology*, 141(6):523–530.

Bates, M. N., A. H. Smith, C. Hopenhayan-Rich. 1992. Arsenic ingestion and internal cancers. *American Journal of Epidemiology* 135(5):462–476.

Bauer, K. 2009. On the politics and possibilities of participatory mapping and GIS: Using spatial technologies to study common property and land use change among pastoralists in central Tibet. *Cultural Geographies* 16:229–252.

Bauer, M. and C. Blodau. 2006. Mobilization of arsenic by dissolved organic matter from iron oxides, soils and sediments. *Science of the Total Environment* 354:179–190.

Baxter, J. and J. Eyles. 1997. Evaluating qualitative research in social geography: Establishing 'rigour' in interview analysis. *Transactions of the Institute of British Geographers* 22(4):505–525.

Baxter, J. and J. Eyles. 1999. The utility of in-depth interviews for studying the meaning of environmental risk. *Professional Geographer* 51(2):307–20.

BBS. 2011. *Bangladesh Population census (2011) Thana Statistics*. GOB: Bangladesh Bureau of Statistics, Dhaka.

Bearak, B. 1998. *Death by Arsenic: A special report - New Bangladesh Disaster: Wells that pump poison.* New York: The New York Times.

Beck, U. 1992. *Risk Society: Towards a New Modernity.* London: Sage.

Beck, U. 1999. *World Risk Society.* Cambridge: Polity Press.

Beck, R., M. Styblo, P. Sethupathy. 2017. Arsenic exposure and type 2 diabetes: MicroRNAs as mechanistic links? *Current Diabetes Reports* 17:18 (doi:10.1007/s11892-017-0845-8).

Bednar, A. J., J. R. Garbarino, M. R. Burkhardt et al. 2004. Field and laboratory arsenic speciation methods and their application to natural-water analysis. *Water Research* 38(2):355–364.

Beecher, N., E. Harrison, N. Goldstein et al. 2005. Risk perception, risk communication, and stakeholder involvement for biosolids management and research. *Journal of Environmental Quality* 34:122–128.

Beg, M. K., S. K. Srivastav, E. J. M. Carranza et al. 2011. High fluoride incidence in groundwater and its potential health effects in parts of Raigarh district, Chhattisgarh, India. *Current Science* 100(5):750–754.

Belkhiri, L. and L. Mouni. 2014. Geochemical characterization of surface water and groundwater in Soummam Basin, Algeria. *Natural Resources Research* 23(4):393–407.

Belkhiri, L. and T. S. Narany. 2015. Using multivariate statistical analysis, geostatistical techniques and structural equation modeling to identify spatial variability of groundwater quality. *Water Resources Management* 29:2073–2089.

Belkin, H. E., B. Zheng, R. B. Finkelman. 2000. Human health effects of domestic combustion of coal in rural China: A casual factor foe arsenic and fluorine poisoning. *Second World Chinese Conference on Geological Sciences, extended abstracts.* Stanford, CA: Stanford University; pp522–524.

Benbrahim-Tallaa, L. and M. P. Waalkes. 2008. Inorganic arsenic and human prostate cancer. *Environmental Health Perspectives* 116(2):158–164.

Benbrahim-Tallaa, L., R. A. Waterland, M. Styblo et al. 2005. Molecular events associated with arsenic-induced malignant transformation of human prostatic epithelial cells: Aberrant genomic DNA methylation and K-ras oncogene activation. *Toxicology and Applied Pharmacology* 206(3):288–298.

Benetos, A., M. Safar, A. Rudnichi et al. 1997. Pulse pressure: A predictor of long-term cardiovascular mortality in a French male population. *Hypertension* 30:1410–1415.

Benner, S. G., M. L. Polizzotto, B. D. Kocar et al. 2008. Groundwater flow in an arsenic-contaminated aquifer, Mekong delta, Cambodia. *Applied Geochemistry* 23:3072–3087.

Benowitz, N. L. 1992. Cardiotoxicity in the workplace. *Occupational Medicine* 7(3):465–478.

Benramdane, L., M. Accominotti, L. Fanton et al. 1999. Arsenic speciation in human organs following fatal arsenic trioxide poisoning: A case report. *Clinical Chemistry* 45(2):301–306.

Bentley, R. and T. G. Chasteen. 2002. Arsenic curiosa and humanity. *The Chemical Educator* 7(2):51–60.

Berg, M., C. Stengel, P. T. K. Trang et al. 2007. Magnitude of arsenic pollution in the Mekong and Red River Deltas - Cambodia and Vietnam. *Science of the Total Environment* 372(2–3):413–425.

Berg, M., H. C. Tran, T.C. Nguyen et al. 2001. Arsenic contamination of groundwater and drinking water in Vietnam: A human health threat. *Environmental Science and Technology* 35(13):2621–2626.

Berg, M., P. T. K. Trang, C. Stengel et al. 2008. Hydrological and sedimentary controlls leading to arsenic contamination of ground water in the Hanoi area, Vietnam: The impact of iron-arsenic ratios, peat, river bank deposits, and excessive groundwater abstraction. *Chemical Geology* 249(1–2):91–112.

Berger, P. L. and T. Luckmann. 1966. *The Social Construction of Reality: A Treatise in the Sociology of Knowledge*. New York: Anchor Books.

Bergum, V. 1989. *Woman to Mother. A Transformation*. Granby, Massachusetts: Bergin and Garvey Publishers.

Berke, O. 2004. Exploratory disease mapping: Kriging the spatial risk function from regional count data. *International Journal of Health Geographics* 3(1):18.

BGS. 2001. *Bangladesh claims against the British Geological Survey*. (www.bgs.ac.uk/scripts/showitem/view.cfm?item=279). Accessed 2 February 2006.

BGS/DPHE. 1999. Groundwater Studies for Arsenic Contamination in Bangladesh. *Phase I: Rapid Investigation Phase, Final Report*, Dhaka: British Geological Survey.

BGS/DPHE. 2001. Arsenic contamination of groundwater in Bangladesh. *Final Report. BGS Technical Report WC/00/19*. British Geological Survey and Department of Public Health Engineering, Ministry of Local Government, Rural Development & Co-operatives, Government of Bangladesh, British Geological Survey: Keyworth, UK.

Bhattacharjee, P., M. Banerjee, A. K. Giri. 2013c. Role of genomic instability in arsenic-induced carcinogenicity: A review. *Environment International* 53:29–40.

Bhattacharjee, S., S. Chakravarty, S. Maity et al. 2005. Metal contents in the groundwater of Sahebgunj district, Jharkhand, India, with special reference to arsenic. *Chemosphere* 58(9):1203–1217.

Bhattacharjee, P., D. Chatterjeea, K.K. Singhb et al. 2013b. Systems biology approaches to evaluate arsenic toxicity and carcinogenicity: An overview. *International Journal of Hygiene and Environmental Health* 216(5):574–586.

Bhattacharjee, P., N. Das, D. Chatterjee et al. 2013a. Association of NALP2 polymorphism with arsenic induced skin lesions and other health effects. *Mutation Research* 755:1–5.

Bhattacharya, S. 2017. Medicinal plants and natural products in amelioration of arsenic toxicity: A short review. *Pharmaceutical Biology* 55(1):349–354.

Bhattacharya, P., D. Chattargee, G. Jacks. 1997. Occurrence of arsenic contaminated groundwater in alluvial aquifers from Delta Plains, Eastern India: Options for safe drinking water supply. *International Journal of Water Resources Management* 13(1):79–92.

Bhattacharya, P., M. Claesson, J. Bundschuh et al. 2006. Distribution and mobility of arsenic in the Rio Dulce alluvial aquifers in Santiago del Estero Province, Argentina. *Science of the Total Environment* 358:97–120.

Bhattacharya, P., M. A. Hasan, O. Sracek et al. 2009. Groundwater chemistry and arsenic mobilization in the Holocene flood plains in south-central Bangladesh. *Environmental Geochemistry and Health*, 31:23–43.

Bhattacharya, P., M. Hossain, S. N. Rahman et al. 2011. Temporal and seasonal variability of arsenic in drinking water wells in Matlab, southeastern Bangladesh: A preliminary evaluation on the basis of a 4-year study. *Journal of Environmental Science and Health: Part A* 46(11):1177–84.

Bhattacharya, P., A. C. Samal, J. Majumdar et al. 2010. Arsenic contamination in rice, wheat, pulses, and vegetables: A study in an arsenic affected area of West Bengal, India. *Water Air and Soil Pollution* 214(1–4):3–13.

Bhattacharya, P., N. Tandulkar, A. Neku et al. 2003. Geogenic arsenic in groundwaters from Terai alluvial plain of Nepal. *Journal of Physics IV France* 107:173–176.

Bhowmik, A. K., A. Alamdar, I. Katsoyiannis et al. 2015. Mapping human health risks from exposure to trace metal contamination of drinking water sources in Pakistan. *Science of The Total Environment* 538:306–316.

Bhuiyan, M. A. H., M. Bodrud-Doza, A. R. M. T. Islam et al. 2016. Assessment of groundwater quality of Lakshimpur district of Bangladesh using water quality indices, geostatistical methods, and multivariate analysis. *Environmental Earth Sciences* 75(12):1020. [doi:10.1007/s12665-016-5823-y].

Bhuiyan, R. H. and N. Islam. 2002. Coping strategy and health seeking behavior of arsenicosis patients of rural Bangladesh: A case study of Ramganj Upzilla, Lakshmipur. Paper presented at *The International Workshop on Arsenic Mitigation*, Dhaka, 14–16 January.

Biagini, R. E. 1972. Hidroarsenicismo crónico y muerte por cánceres malignos. *La Semana Médica* 25:812–816.

Bibi, M., M. Z. Hashmi, R. N. Malik. 2015. Human exposure to arsenic in groundwater from Lahore district, Pakistan. *Environmental Toxicology and Pharmacology* 39(1):42–52.

Bidone, E., Z. Castilhos, R. Cesar et al. 2016. Hydrogeochemistry of arsenic pollution in watersheds influenced by gold mining activities in Paracatu (Minas Gerais State, Brazil). *Environmental Science and Pollution Research* 23(9):8546–8555.

Bien, J. D., J. ter Meer, W. H. Rulkens et al. 2004. A GIS based approach for the long-term prediction of human health risks at contaminated sites. *Environmental Modelling and Assessment* 9:221–226.

Billib, M., P. W. Boochs, J. Aparicio et al. 2012. *Hydro-geochemical Investigation of the Arsenic in the Coma rca Lagunera*. Report MEX 08-009, Hannover, Germany.

Bird, B. 1995. The EAGLE project: re-mapping Canada from an indigenous perspective. *Cultural Survival Quarterly* 18 (4): 23–24.

Biswas, B. K. 2000. *Groundwater Arsenic Poisoning in Bangladesh*. PhD Thesis (Unpublished), School of Environmental Studies (SOES), Kolkata: Jadavpur University.

Biswas, B. K., R. K. Dhar, G. Samanta et al. 1998. Detailed study report of Samta, one of the arsenic-affected villages of Jessore District, Bangladesh. *Current Science* 74(2):134–145.

Biswas, A., M. J. Hendry, J. Essilfie-Dughan. 2017. Geochemistry of arsenic in low sulfide-high carbonate coal waste rock, Elk Valley, British Columbia, Canada. *Science of The Total Environment* 579:396–408.

Biswas, B. K., J. Inoue, K. Inoue et al. 2008. Adsorptive removal of As(V) and As(III) from water by a Zr(IV)-loaded orange waste gel. *Journal of Hazardous Materials* 154:1066–1074.

Biswas, A., H. Neidhardt, A. K. Kundu et al. 2014. Spatial, vertical and temporal variation of arsenic in shallow aquifers of the Bengal Basin: Controlling geochemical processes. *Chemical Geology* 387:157–169.

Bjørklund, G., J. Aaseth, S. Chirumbolo et al. 2017. Effects of arsenic toxicity beyond epigenetic modifications. *Environmental Geochemistry and Health*, 2017:1–11. [https://doi.org/10.1007/s10653-017-9967-9].

Black, M. 1990. *From Handpumps to Health: The Evolution of Water and Sanitation Programmes in Bangladesh, India and Nigeria*. New York: UNICEF.

Blackwell, R. Q., T. H. Yang, I. Ai. 1961. Preliminary report on arsenic level in water and food from the endemic Blackfoot area. *Journal of the Formosan Medical Association* 60:1139–1140.

Blaikie, N. W. H. 2000. *Designing Social Research: The Logic of Anticipation*. Cambridge (UK): Polity Press.

Blaikie, P., T. Cannon, I. Davis et al. 2004. *At Risk: Natural Hazards, People's Vulnerability and Disasters*. New York: Routledge.

Blanchard, W. 2005. *Vulnerability Assessment Techniques and Applications (VATA)*. [http://www.csc.noaa.gov/vata/glossary.html].

Blanke, U., G. Troster, T. Franke et al. 2014. Capturing crowd dynamics at large scale events using participatory GPS-localization. In *Intelligent Sensors, Sensor Networks and Information Processing (ISSNIP). 1-7. IEEE Symposium on Participatory Sensing and Crowd Sourcing*. Singapore: IEEE. (DOI:10.1109/ISSNIP.2014.6827652).

Bleicher, J. 1980. *Contemporary Hermeneutics: Hermeneutics as Method, Philosophy, and Critique*. London: Routledge.

Bloom, M. S., I. A. Neamtiu, S. Surdu et al. 2016. Low level arsenic contaminated water consumption and birth outcomes in Romania-An exploratory study. *Reproductive Toxicology* 59(1):8–16.

Boas, F. 1962. *Anthropology and Modern Life*. New York: Courier Dover Publications.

Bode, A. M. and Z. Dong. 2002. The paradox of arsenic: Molecular mechanisms of cell transformation and chemotherapeutic effects. *Critical Reviews in Oncology/Hematology* 24: 5–24.

Bogdan, R. C. and S. K. Biklen. 2007. *Qualitative Research for Education: An Introduction to Theories and Methods*. California: Pearson (5th edition).

Bohling, G. 2005. *Introduction to Geostatistics and Variogram Analysis*. Kansas: Kansas Geological Survey. (http://people.ku.edu/~gbohling/cpe940/Variograms.pdf). Accessed on 22 February 2017.

Boland Jr., R. J. 1986. Phenomenology: A preferred approach to research on information systems. In *Trends in Information Systems*, eds. B. Langefors, A. A. Verrijn-Stuart, G. Bracchi, 341–349. Amsterdam: North-Holland Publishing.

Bolstad, P. 2008. *GIS Fundamentals*. Atlas Books: Minnesota (3rd edition).

Bondu, R., V. Cloutier, E. Rosa et al. 2017. Mobility and speciation of geogenic arsenic in bedrock groundwater from the Canadian Shield in western Quebec, Canada. *Science of the Total Environment* 574:509–519.

Bonnemaison, M. 2005. L'eau, facteur de liberation de l'arsenic naturel. *Geosciences* 2:54–59.

Bonsor, H. C., A. M. MacDonald, K. M. Ahmed et al. 2017. Hydrogeological typologies of the Indo-Gangetic basin alluvial aquifer, South Asia. *Hydrogeology Journal* 25(5):1377–1406.

Bontekoe, R. 1996. *Dimensions of the Hermeneutic Circle*. Atlantic Highlands, NJ: Humanities Press International.

Boochs, P. W., M. Billib, C. Gutiérrez et al. 2014. Groundwater contamination with arsenic, Región Lagunera, México. In *One Century of the Discovery of Arsenicosis in Latin America (1914–2014)*, eds. M. I. Litter, H. B. Nicolli, M. Meichtry, N. Quici, J. Bundschuh, P. Bhattacharya, R. Naidu, 132–134. London: CRC Press.

Boonchai, W., M. Walsh, M. Cummings et al. 2000. Expression of p53 in arsenic-related and sporadic basal cell carcinoma. *Archives of Dermatology* 136:195–198.

Borba, R. P. 2002. *Arsênio em ambiente superficial: Processos geoquímicos naturais e antropogênicos em uma área de mineração aurífera*. PhD Thesis, Instituto de Geociências, Universidade Estadual de Campinas, Campinas, São Paulo.

Borba, R. P., B. R. Figueiredo, J. Matschullat. 2003. Geochemical distribution of arsenic in waters, sediments and weathered gold mineralizes rocks from Iron Quadrangle, Brazil. *Environment Geology* 44(1):39–52.

Borgoño, J. M. and R. Greiber. 1971. Epidemiological study of arsenicism in the city of Antofagasta. *Revista Médica de Chile* 99:702–707.

Borgoño, J. M., H. Venturino, P. Vicent. 1980. Clinical and epidemiologic study of arsenicism in northern Chile. *Revista Médica de Chile* 108(11):1039–1048.

Borgoño, J. M., P. Vicent, H. Venturino et al. 1977. Arsenic in the drinking water of the city of Antofagasta: Epidemiological and clinical study before and after the installation of a treatment plant. *Environmental Health Perspectives* 9:103–105.

Bornat, J. 2007. Oral history. In *Qualitative Research Practice*, eds. C. Seale, G. Gobo, J. F. Gubrium, and D. Silverman, 34–47. London: Sage.

Börzsönyi, M., A. Bereczky, P. Rudnai et al. 1992. Epidemiological studies on human subjects exposed to arsenic in drinking water in Southeast Hungray. *Archives of Toxicology* 66(1):77–78.

Bothe, J. V. and P. W. Brown. 1999. The stabilities of calcium arsenates at 23+/−1 degrees C. *Journal of Hazardous Materials* 69(2):197–207.

Bowell, R. J. 1994. Arsenic speciation in soil porewaters from the Ashanti Mine, Ghana. *Applied Geochemistry* 9(1):15–22.

Boyer, J. L. 1999. Arsenic: An endothelial cell toxin in the liver? *Indian Journal of Gastroenterology* 18:141–142.

Boyle, D. R., R. J. W. Turner, G. E. M. Hall. 1998. Anomalous arsenic concentrations in groundwaters of an island community, Bowen Island, British Columbia. *Environmental Geochemistry and Health* 20:199–212.

BRAC. 1998. *Village Health Workers Can Test Tubewell Water for Arsenic*. Dhaka: Bangladesh Rural Advancement Committee.

BRAC. 2016. *Water, Sanitation and Hygiene: Nine Years of Scale and Innovation in Bangladesh*. (Programme Report 2006-2015). Dhaka: Bangladesh Rural Advancement Committee.

Brahman, K. D., T. G. Kazi, H. I. Afridi et al. 2013. Evaluation of high levels of fluoride, arsenic species and other physicochemical parameters in underground water of two sub districts of Tharparkar, Pakistan: A multivariate study. *Water Research* 47:1005–1020.

Brahman, K. D., T. G. Kazi, H. I. Afridi et al. 2016. Exposure of children to arsenic in drinking water in the Tharparkar region of Sindh, Pakistan. *Science of the Total Environment* 544:653–660.

Brahman, K. D., T. G. Kazi, J. A. Baig et al. 2014. Fluoride and arsenic exposure through water and grain crops in Nagarparkar, Pakistan. *Chemosphere* 100:182–189.

Braimoh, A. K. and T. Onishi. 2007. Geostatistical techniques for incorporating spatial correlation into land use change models. *International Journal of Applied Earth Observation and Geoinformation* 9:438–446.

Bränström, R. and Y. Brandberg. 2010. Health risk perception, optimistic bias, and personal satisfaction. *American Journal of Health Behavior* 34(2):197–205.

Bräuner, E. V., R. B. Nordsborg, Z. J. Andersen et al. 2014. Long-term exposure to low-level arsenic in drinking water and diabetes incidence: A prospective study of the diet, cancer and health cohort. *Environmental Health Perspectives* 122(10):1059–1065.

Bressan, G. M., V. A. Oliveira, E. R. Hruschka Jr et al. 2009. Using Bayesian networks with rule extraction to infer the risk of weed infestation in a corn-crop. *Engineering Applications of Artificial Intelligence* 22:579–592.

Breton, C. V., W. Zhou, M. L. Kile et al. 2007. Susceptibility to arsenic-induced skin lesions from polymorphisms in base excision repair genes. *Carcinogenesis* 28(7):1520–1525.

Bretzler, A., F. Lalanne, J. Nikiema et al. 2017. Groundwater arsenic contamination in Burkina Faso, West Africa: Predicting and verifying regions at risk. *Science of the Total Environment* 584–585:958–970.

Brewster, M. D. 1992. Removing arsenic from contaminated wastewater. *Water Environment and Technology* 4(11):54–57.

Brikowski, T. H., A. Neku, S. D. Shrestha et al. 2014. Hydrologic control of temporal variability in groundwater arsenic on the Ganges floodplain of Nepal. *Journal of Hydrology* 518(Part C):342–353.

Brinkel, J., M. M. H. Khan, A. Kraemer. 2009. A systematic review of arsenic exposure and its social and mental health effects with special reference to Bangladesh. *International Journal of Environmental Research and Public Health*, 6(5):1609–1619.

Brocato, J. and M. Costa. 2015. 10th NTES conference: Nickel and arsenic compounds alter the epigenome of peripheral blood mononuclear cells. *Journal of Trace Elements in Medicine and Biology* 31:209–213.

Brody, S. D., D. R. Godschalk, R. J. Burby. 2003. Mandating citizen participation in plan making: Six strategic planning choices. *Journal of the American Planning Association* 69(3):343–351.

Brooker, P. I., J. P. Winchester, A. C. Adams. 1995. A geostatistical study of soil data from an irrigated vineyard near waikerie, South Australia. *Environment International* 21:699–704.

Brown, K. G., K. E. Boyle, C. W. Chen et al. 1989. A dose-response analysis of skin cancer from inorganic arsenic in drinking water. *Risk Analysis* 9:519–28.

Brown, K. G. and C. J. Chen. 1995. Significance of exposure assessment to analysis of cancer risk from inorganic arsenic in drinking water in Taiwan. *Risk Analysis* 15(4):475–484.

Browne, M. N., T. J. Keeley, W. J. Hiers. 1998. The epistemological role of expert witnesses and toxic torts. *American Business Law Journal* 36:1–72.

Bryan, J. 2011. Walking the line: Participatory mapping, indigenous rights, andneoliberalism. *Geoforum* 42:40–50.

Buamah, R., B. Petrusevski, J. C. Schippers. 2008. Presence of arsenic, iron and manganese in groundwater within the gold-belt zone of Ghana. *Journal of Water Supply: Research and Technology-AQUA* 57(7):519–529.

Buchet, J. P., J. F. Heilier, A. Bernard et al. 2003. Urinary protein excretion in humans exposed to arsenic and cadmium. *International Archives of Occupational and Environmental Health* 76(2):111–120.

Buchet, J. P. and D. Lison. 1998. Mortality by cancer in groups of the Belgian population with a moderately increased intake of arsenic. *International Archives of Occupational and Environmental Health* 71(2):125–130.

BUET/NEMIP. 1997. *A Study on Arsenic Concentration in the Groundwater of the North-eastern Zone of Bangladesh*. Dhaka: Bangladesh University of Engineering and Technology (BUET) and North East Minor Irrigation Project (NEMIP).

Bullock, A. and S. Trombley. 2000. *The New Fontana Dictionary of Modern Thought*. London: Harper Collins.

Bundschuh, J., B. Farias, R. Martin et al. 2004. Groundwater arsenic in the Chaco-Pampean Plain, Argentina: Case study from Robles county, Santiago del Estero Province. *Applied Geochemistry* 19(2):231–243.

Bundschuh, J., M. I. Litter, P. Bhattacharya. 2010. Targeting arsenic-safe aquifers for drinking water supplies. *Environmental Geochemistry and Health* 32(4):307–315.

Bundschuh, J., M. I. Litter, F. Parvez et al. 2012. One century of arsenic exposure in Latin America: A review of history and occurrence from 14 countries. *Science of the Total Environment* 429:2–35.

Bundschuh, J. and J. P. Maity. 2015. Geothermal arsenic: Occurrence, mobility and environmental implications. *Renewable and Sustainable Energy Reviews* 42:1214–1222.

Bundschuh, J., J. P. Maity, B. Nath et al. 2013. Naturally occurring arsenic in terrestrial geothermal systems of western Anatolia, Turkey: Potential role in contamination of freshwater resources. *Journal of Hazardous Materials* 262:951–959.

Bunne, M. 1999. Qualitative research methods in otorhinolaryngology. *International Journal of Pediatric Otorhinolaryngology* 51:1–10.

Burges, W. and K. M. Ahmed. 2006. Arsenic in aquifers of the Bengal Basin: From sediment source to tube-wells used for domestic water supply and irrigation. In *Managing Arsenic in the Environment: From Soil to Human Health*, eds. R. Naidu, E. Smith, G. Owens, P. Bhattacharya, P. Nadebaum, 31–56. Victoria (Australia): CSIRO.

Burgess, W. G. and L. Pinto. 2005. Preliminary observations on the release of arsenic to groundwater in the presence of hydrocarbon contaminants in UK aquifers. *Mineralogical Magazine* 69(5):887–896.

Burguera, J. L., M. Burguera, C. Rivas et al. 1998. On-line cryogenic trapping with microwave heating for the determination and speciation of arsenic by flow injection/hydride generation/atomic absorption spectrometry. *Talanta* 45(3):531–542.

Burns, N. and S. K. Grove. 2009. *The Practice of Nursing Research: Appraisal, Synthesis, and Generation of Evidence.* St. Louis, MO: Saunders Elsevier.

Burrough, P. A. and R. A. McDonnell. 2000. *Principles of Geographical Information Systems for Land Resource Assessment.* Oxford: Oxford University Press.

Buschmann, J., M. Berg, C. Stengel et al. 2007. Arsenic and manganese contamination of drinking water resources in Cambodia: Coincidence of risk areas with low relief topography. *Environmetal Science and Technology* 41:2146–2152.

Buschmann, J., M. Berg, C. Stengel et al. 2008. Contamination of drinking water resources in the Mekong delta floodplains: Arsenic and other trace metals pose serious health risks to population. *Environment International* 34:756–764.

Bustaffa, E., A. Stoccoro, F. Bianchi et al. 2014. Genotoxic and epigenetic mechanisms in arsenic carcinogenicity. *Archieves of Toxicology* 88(5):1043–1067.

Byrd, D. M., M. L. Roegner, J. C. Griffiths et al. 1996. Carcinogenic risks of inorganic arsenic in perspective. *International Archives of Occupational and Environmental Health* 68:484–94.

Byrne, S., D. Amarasiriwardena, B. Bandak et al. 2010. Were Chinchorros exposed to arsenic? Arsenic determination in Chinchorro Mummies' hair by laser ablation inductively coupled plasma-mass spectrometry (LA-ICP-MS). *Microchemical Journal* 94:28–35.

Cadag, J. and J. Gaillard. 2012. Integrating knowledge and actions in disaster risk reduction: The contribution of participatory mapping. *Area* 44(1):100–109.

Cai, D. A. and C. J. F. Hung. 2005. How relevant is trust anyway? A cross-cultural comparison of trust in organizational and peer relationships. In *International and Multicultural Organizational Communication*, eds. G. Cheney and G. A. Barnett. Cresskill (NJ): Hampton Press.

Cai, L., Z. Xu, P. Baoe et al. 2015. Multivariate and geostatistical analyses of the spatial distribution and source of arsenic and heavy metals in the agricultural soils in Shunde, Southeast China. *Journal of Geochemical Exploration* 148:189–195.

Calderon, R. L. 2000. The epidemiology of chemical contaminants of drinking water. *Food and Chemical Toxicology* 38(1):S13–S20.

Calderon, J., M. E. Navarro, M. E. Jimenez-Capdeville et al. 2001. Exposure to arsenic and lead and neuropsychological development in Mexican children. *Environmental Research* 85(2):69–76.

Caldwell, B. K., J. C. Caldwell, S. N. Mitra et al. 2003. Searching for an optimum solution to the Bangladesh arsenic crisis. *Social Science and Medicine* 56(10):2089–2096.

Calow, P. 1998. Environmental risk assessment and management: The whats, whys and hows? In *Handbook of Environmental Risk Assessment and Management*, ed. P. Calow, 1–6, UK: Blackwell Science.

Campaña, D. H., A. Airasca, G. Angeles. 2014. Arsenic in groundwater of the southwestern Buenos Aires province, Argentina. In: *One Century of the Discovery of Arsenicosis in Latin America (1914–2014)*, eds. M. I. Litter, H. B. Nicolli, M. Meichtry, N. Quici, J. Bundschuh, P. Bhattacharya, R. Naidu, 148–150. Boca Raton: CRC Press.

Campbell, R. K. 2009. Type 2 diabetes: Where we are today: An overview of disease burden, current treatments, and treatment strategies. *Journal of American Pharmacists Association* 49(Suppl.1):S3–S9.

Campos, V. 2002. Arsenic in groundwater affected by phosphate fertilizers at São Paulo, Brazil. *Environmental Geology* 42(1):83–87.

Cano-Lamadrid, M., S. Munera-Picazo, A. Burgos-Hernández et al. 2016. Inorganic and total arsenic contents in rice and rice-based foods consumed by a potential risk subpopulation: Sportspeople. *Journal of Food Science* 81(4):T1031–T1037.

Carey, M. A. 1995. Concerns in the analysis of focus group data: Comment. *Qualitative Health Research* 5:487–495.

Carlon, C., L. Pizzol, A. Critto et al. 2008. A spatial risk assessment methodology to support the remediation of contaminated land. *Environment International* 34:397–411.

Carrera, P., P. J. Espinoza-Montero, L. Fernández et al. 2017. Electrochemical determination of arsenic in natural waters using carbon fiber ultra-microelectrodes modified with gold nanoparticles. *Talanta* 166:198–206.

Carson, R. 1962. *Silent Spring* (1st edition). Boston: Houghton Mifflin.

Carver, S. 2003. The future of participatory approaches using geographic information: Developing a research agenda for the 21st century. *Urban and Regional Information Systems Association (URISA) Journal* 15(APA I):61–71.

Caussy, D., M. Gochfeld, E. Gurzau et al. 2003. Lessons from case studies of metals: Investigating exposure, bioavailability, and risk. *Ecotoxicology and Environmental Safety* 56:45–51.

Ćavar, S., T. Klapec, R. J. Grubešić et al. 2005. High exposure to arsenic from drinking water at several localities in eastern Croatia. *Science of the Total Environment* 339(1–3):277–282.
Cavigelli, M., W. W. Li, A. Lin et al. 1996. The tumor promoter arsenite stimulates AP-1 activity by inhibiting a JNK phosphatase. *The EMBO Journal* 15(22):6269–6279.
Cebrián, M. E., A. Albores, M. Aguilar et al. 1983. Chronic arsenic poisoning in the north of Mexico. *Human Toxicology* 2:121–33.
Cebrián, M. E., A. Albores, G. Garcia-Vergas et al. 1994. Chronic arsenic poisoning in humans: The case of Mexico. In *Arsenic in the Environment. Part II: Human Health and Ecosystem Effects*, ed. J. O. Niragu, 93–107. New York: John Wiley & Sons.
Celik, I., L. Gallicchio, K. Boyd et al. 2008. Arsenic in drinking water and lung cancer: A systematic review. *Environmental Research* 108:48–55.
CESCR. 2002. *General Comment No. 15: The Right to Water*. (Articles 11 and 12 of the Covenant). UN Committee on Economic, Social and Cultural Rights (CESCR), (UN document: E/C.12/2002/11).
Cha, Y., Y. M. Kim, J. W. Choi et al. 2016. Bayesian modeling approach for characterizing groundwater arsenic contamination in the Mekong River basin. *Chemosphere* 143:50–56.
Chaffe, E. E. 1985. Three models of strategy. *Academy of Management Review* 10:89–98.
Chakraborti, D. 2001. *Dugwell Survey Report During February 2001*. Unpublished Report, SOES, Kolkata: Jadavpur University.
Chakraborti, D., B. K. Biswas, T. R. Chowdhury et al. 1999. Arsenic groundwater contamination and sufferings of people in Rajnandgaon, Madhya Pradesh, India. *Current Science* 77(4):502–504.
Chakraborti, D., B. Das, M. M. Rahman et al. 2009. Status of groundwater arsenic contamination in the state of West Bengal, India: A 20-year study report. *Molecular Nutrition & Food Research* 53(5):542–551.
Chakraborti, D., B. Das, M.M. Rahman et al. 2017b. Arsenic in groundwater of the Kolkata Municipal Corporation (KMC), India: Critical review and modes of mitigation. *Chemosphere* 180:437–447.
Chakraborti, D., S. C. Mukherjee, S. Pati et al. 2003. Arsenic groundwater contamination in Middle Ganga Plain, Bihar, India: A future danger? *Environmental Health Perspectives* 111:1194–1201.
Chakraborti, D., M. M. Rahman, S. Ahamed et al. 2016b. Arsenic groundwater contamination and its health effects in Patna district (capital of Bihar) in the middle Ganga plain, India. *Chemosphere* 152:520–529.
Chakraborti, D., M. M. Rahman, S. Ahamed et al. 2016c. Arsenic contamination of groundwater and its induced health effects in Shahpur block, Bhojpur district, Bihar state, India: Risk evaluation. *Environmental Science and Pollution Research* 23(10):9492–9504.
Chakraborti, D., M. M. Rahman, M. Alauddin et al. 2015. Groundwater arsenic contamination in Bangladesh - 21 years of research. *Journal of Trace Elements in Medicine and Biology* 31:237–248.
Chakraborti, D., M. M. Rahman, A. Chatterjee et al. 2016a. Fate of over 480 million inhabitants living in arsenic and fluoride endemic Indian districts: Magnitude, health, socio-economic effects and mitigation approaches. *Journal of Trace Elements in Medicine and Biology* 38:33–45.
Chakraborti, D., M. M. Rahman, B. Das et al. 2017a. Groundwater arsenic contamination and its health effects in India. *Hydrogeology Journal* 25(4):1165–1181.
Chakraborti, D., M. M. Rahman, S. Mitra et al. 2013a. Groundwater arsenic contamination in India: A review of its magnitude, health, social, socio-economic effects and approaches for arsenic mitigation. *Journal of the Indian Society of Agricultural Statistics* 67(2):235–266.
Chakraborti, D., M. M. Rahman, M. Murrill et al. 2013b. Environmental arsenic contamination and its health effects in a historic goldmining area of the Mangalur greenstone belt of Northeastern Karnataka, India. *Journal of Hazardous Materials* 262:1048–1055.
Chakraborti, D., M. M. Rahman, K. Paul et al. 2002. Arsenic calamity in the Indian sub-continent: What lessons have been learned? *Talanta* 58:3–22.
Chakraborti, D., M. K. Sengupta, M. M. Rahman et al. 2004. Groundwater arsenic contamination and its health effects in the Ganga-Meghna-Brahmaputra plain. *Journal of Environmental Monitoring* 6(6):74N–83N.
Chakraborti, D., E. J. Singh, B. Das et al. 2008. Groundwater arsenic contamination in Manipur, one of the seven North-Eastern Hill states of India: A future danger. *Environmental Geology* 56(2):381–390.
Chakraborty, M., A. Mukherjee, K. M. Ahmed. 2015. A review of groundwater arsenic in the Bengal Basin, Bangladesh and India: From source to sink. *Current Pollution Reports* 1(4):220–247.
Chakraborty, A. K. and K. C. Saha. 1987. Arsenical dermatosis from tubewell water in West Bengal. *The Indian Journal of Medical Research* 85:326–334.
Challis, J. R., C. J. Lockwood, L. Myatt et al. 2009. Inflammation and pregnancy. *Reproductive Sciences*, 16(2):206–215.

Chambers, R. 1997. *Whose Reality Counts? Putting the First Last.* London: Intermediate Technology.

Chan, K. 2002. Understanding the toxicity of Chinese herbal medicinal products. In *The Way Forward for Chinese Medicine*, eds. K. Chan, and H. Lee, 71–90. London: Taylor and Francis.

Chanda, S., U. B. Dasgupta, D. Guhamazumder et al. 2006. DNA hypermethylation of promoter of gene p53 and p16 in arsenic-exposed people with and without malignancy. *Toxicological Sciences* 89(2):431–437.

Chang, F. J., P. A. Chen, C. W. Liu et al. 2013. Regional estimation of groundwater arsenic concentrations through systematical dynamic-neural modeling. *Journal of Hydrology* 499:265–274.

Chang, C. C., S. C. Ho, S. S. Tsai et al. 2004. Ischemic heart disease mortality reduction in an arseniasis-endemic area in southwestern Taiwan after a switch in the tap-water supply system. *Journal of Toxicology and Environmental Health: Part A* 67(17):1353–1361.

Chanpiwat, P., S. Sthiannopkao, K. H. Cho et al. 2011. Contamination by arsenic and other trace elements of tube-well water along the Mekong River in Lao PDR. *Environmental Pollution* 159(2):567–76.

Chapagain, S. K., S. Shrestha, T. Nakamura et al. 2009. Arsenic occurrence in groundwater of Kathmandu Valley, Nepal. *Desalination and Water Treatment* 4:248–254.

Chaplin, J., J. Kelly, S. Kildea. 2016. Maternal perceptions of breastfeeding difficulty after caesarean section with regional anaesthesia: A qualitative study. *Women and Birth* 29:144–152.

Charlet, L., A. A. Ansari, G. Lespagnol et al. 2001. Risk of arsenic transfer to a semi-confined aquifer and the effect of water level fluctuation in North Mortagne, France at a former industrial site. *The Science of the Total Environment* 277(1–3):133–147.

Chatterjee, D., A. Kundu, D. Saha et al. 2017. Groundwater arsenic in the Bengal Delta Plain: Geochemical and geomorphological perspectives. *Procedia Earth and Planetary Science* 17:622–625.

Chen, Y., H. Ahsan, V. Slavkovich et al. 2010c. No association between arsenic exposure from drinking water and diabetes mellitus: A cross-sectional study in Bangladesh. *Environmental Health Perspectives* 118:1299–1305.

Chen, J. W., H. Y. Chen, W. F. Li et al. 2011b. The association between total urinary arsenic concentration and renal dysfunction in a community-based population from central Taiwan. *Chemosphere* 84(1):17–24.

Chen, C. J., C. W. Chen, M. M. Wu et al. 1992. Cancer potential in liver, lung, bladder and kidney due to ingested inorganic arsenic in drinking water. *British Journal of Cancer* 66(5):888–892.

Chen, J. G., Y. G. Chen, Y. S. Zhou et al. 2007. A follow-up study of mortality among the arseniasis patients exposed to indoor combustion of high arsenic coal in Southwest Guizhou Autonomous Prefecture, China. *International Archives of Occupational and Environmental Health* 81(1):9–17.

Chen, C. L., H. Y. Chiou, L. I. Hsu et al. 2010a. Arsenic in drinking water and risk of urinary tract cancer: A follow-up study from Northeastern Taiwan. *Cancer Epidemiology, Biomarkers and Prevention* 19(1):101–110.

Chen, C. L., H. Y. Chiou, L. I. Hsu et al. 2010b. Ingested arsenic, characteristics of well water consumption and risk of different histological types of lung cancer in northeastern Taiwan. *Environmental Research* 110:455–462.

Chen, C. J., Y. C. Chuang, T. M. Lin et al. 1985. Malignant neoplasms among residents of a blackfoot disease-endemic area in Taiwan: High arsenic artesian well water and cancers. *Cancer Research* 45(11 Part 2): 5895–5899.

Chen, C. J., Y. C. Chuang, S. L. You et al. 1986. A retrospective study on malignant neoplasms of bladder, lung and liver in blackfoot disease endemic area in Taiwan. *British Journal of Cancer* 53(3):399–405.

Chen, H. W., M. M. Frey, D. Clifford et al. 1999. Arsenic treatment considerations. *Journal of American Water Works Associations* 91(3):74–85.

Chen, Y., J. H. Graziano, F. Parvez et al. 2011a. Arsenic exposure from drinking water and mortality from cardiovascular disease in Bangladesh: Prospective cohort study. *British Medical Journal* 342:d2431–d2442.

Chen, Y. C., Y. L. Guo, H. J. Su et al. 2003a. Arsenic methylation and skin cancer risk in southwestern Taiwan. *Journal of Occupational and Environmental Medicine* 45:241–248.

Chen, T. C., Z. Y. Hseu, J. S. Jean et al. 2016. Association between arsenic and different-sized dissolved organic matter in the groundwater of black-foot disease area, Taiwan. *Chemosphere* 159:214–220.

Chen, C. L., L. I. Hsu, H. Y. Chiou et al. 2004a. Ingested arsenic, cigarette smoking, and lung cancer risk: A follow-up study in arseniasis-endemic areas in Taiwan. *Journal of American Medical Association* 292(24):2984–2990.

Chen, C. J., Y. M. Hsueh, M. S. Lai et al. 1995. Increased prevalence of hypertension and long-term arsenic exposure. *Hypertension* 25(1):53–60.

Chen, C. S. and S. J. Jiang. 1996. Determination of As, Sb, Bi and Hg in water samples by flow-injection inductively coupled plasma mass spectrometry with an in-situ nebulizer/hydride generator. *Spectrochimica Acta Part B: Atomic Spectroscopy* 51(14):1813–1821.

Chen, Y. and M. R. Karagas. 2013. Arsenic and cardiovascular disease: New evidence from the United States. *Annals of Internal Medicine* 159(10):713–714.

Chen, C. J., T. L. Kuo, M. M. Wu. 1988b. Arsenic and cancers. *Lancet* 331 (8582): 414–415.

Chen, B. C. and C. M. Liao. 2008. A body-weight-based method to estimate inorganic arsenic body burden through tilapia consumption in Taiwan. *Bulletin of Environmental Contamination and Toxicology* 80:289–293.

Chen, B., Q. Liu, A. Popowich et al. 2015. Therapeutic and analytical applications of arsenic binding to proteins. *Metallomics* 7(1):39–55.

Chen, M., L. Q. Ma, W. G. Harris. 2002. Arsenic concentrations in Florida surface soils. *Soil Science Society of America Journal* 66(2):632–640.

Chen, K., J. McAneney, R. Blong et al. 2004b. Defining area at risk and its effect in catastrophe loss estimation: A dasymetric mapping approach. *Applied Geography* 24:97–117.

Chen, Y., F. Parvez, M. Gamble et al. 2009. Arsenic exposure at low-to-moderate levels and skin lesions, arsenic metabolism, neurological functions, and biomarkers for respiratory and cardiovascular diseases: Review of recent findings from the Health Effects of Arsenic Longitudinal Study (HEALS) in Bangladesh. *Toxicology and Applied Pharmacology* 239(2):184–192.

Chen, J., H. Qian, H. Wu et al. 2017a. Assessment of arsenic and fluoride pollution in groundwater in Dawukou area, Northwest China, and the associated health risk for inhabitants. *Environmental Earth Sciences*, 76: 314 [doi:10.1007/s12665-017-6629-2].

Chen, Y. C., H. J. Su, Y. L. Guo et al. 2003b. Arsenic methylation and bladder cancer risk in Taiwan. *Cancer Causes and Control* 14(4):303–310.

Chen, C. J. and C. J. Wang. 1990. Ecological correlation between arsenic level in well water and age-adjusted mortality from malignant neoplasms. *Cancer Research* 50(17):5470–5474.

Chen, C. J., M. M. Wu, S. S. Lee et al. 1988a. Atherogenicity and carcinogenicity of high-arsenic artesian well water. Multiple risk factors and related malignant neoplasms of Blackfoot disease. *Arteriosclerosis* 8:452–460.

Chen, Y., F. Wu, M. Liu et al. 2013b. A prosepctive study of asernic exposure, arsenic methylation capacity, and risk of cardiovascular disease in Bangladesh. *Environmental Health Perspectives* 121(7):832–838.

Chen, Y., F. Wu, F. Parvez et al. 2013a. Arsenic exposure from drinking water and QT-interval prolongation: Results from the Health Effects of Arsenic Longitudinal Study. *Environmental Health Perspectives* 121:427–432.

Chen, K. P., H. Y. Wu, and T. C. Wu. 1962. Epidemiologic studies on blackfoot disease in Taiwan: III 3. Physicochemical characteristics of drinking water in endemic blackfoot disease areas. *Memoirs of the College of Medicine of the National Taiwan University* 8(1–2):115–129.

Chen, X., X. C. Zeng, J. Wang et al. 2017b. Microbial communities involved in arsenic mobilization and release from the deep sediments into groundwater in Jianghan plain, Central China. *Science of The Total Environment* 579:989–999.

Cheng, Y. Y., N. C. Huang, Y. T. Chang et al. 2017. Associations between arsenic in drinking water and the progression of chronic kidney disease: A nationwide study in Taiwan. *Journal of Hazardous Materials* 321:432–439.

Cheng, T. J., D. S. Ke, H. R. Guo. 2010. The association between arsenic exposure from drinking water and cerebrovascular disease mortality in Taiwan. *Water Research* 44(19):5770–5776.

Cheng, Z., A. van Geen, A. A. Seddique et al. 2005. Limited temporal variability of arsenic concentrations in 20 wells monitored for 3 years in Araihazar Bangladesh. *Environmental Science & Technology* 39(13):4759–4766.

Cheng, P. S., S. F. Weng, C. H. Chiang et al. 2016. Relationship between arsenic-containing drinking water and skin cancers in the arseniasis endemic areas in Taiwan. *The Journal of Dermatology* 43(2):181–186.

Chenini, I., A. B. Mammou, M. E. El May. 2010. Groundwater recharge zone mapping using GIS-based multi-criteria analysis: A case study in Central Tunisia (Maknassy Basin). *Water Resources Management* 24(5):921–939.

Cherry, N., K. Shaik, C. McDonald et al. 2008. Stillbirth in rural Bangladesh: Arsenic exposure and other etiological factors: A report from Gonoshasthaya Kendra. *Bulletin of the World Health Organization*, 86:172–177.

Cherry, N., K. Shaik, C. McDonald et al. 2010. Manganese, arsenic, and infant mortality in Bangladesh: An ecological analysis. *Archives of Environmental and Occupational Health*, 65(3):148–153.

Chesla, C. A. 1995. Hermeneutic phenomenology: An approach to understanding families. *Journal of Family Nursing*, 1:68–78.

Chiang, H. S., H. R. Guo, C. L. Hong et al. 1993. The incidence of bladder cancer in the black foot disease endemic area in Taiwan. *British Journal of Urology* 71(3):274–278.

Chiou, H. Y., S. T. Chion, Y. H. Hsu et al. 2001. Incidence of transitional cell carcinoma and arsenic in drinking water: A follow-up study of 8102 residents in a arseniasis endemic area in Northeastern Taiwan. *American Journal of Epidemiology*, 153(5):411–418.

Chiou, H. Y., Y. M. Hsueh, K. F. Liaw et al. 1995. Incidence of internal cancers and ingested inorganic arsenic: A seven-year follow-up study in Taiwan. *Cancer Research* 55(6):1296–1300.

Chiou, H. Y., W. I. Huang, C. L. Su et al. 1997. Dose-response relationship between prevalence of cerebrovascular disease and ingested inorganic arsenic. *Stroke* 28:1717–1723.

Chitsazan, M., M. S. Dorraninejad, A. Zarasvandi et al. 2009. Occurrence, distribution and source of arsenic in deep groundwater wells in Maydavood area, southwestern Iran. *Environmental Geology* 58:727–737.

Chiu, H. F., C. C. Chang, S. S. Tsai et al. 2006. Does arsenic exposure increase the risk for diabetes mellitus? *Journal of Occupational and Environmental Medicine* 48(1):63–67.

Chiu, H. F. and C. Y. Yang. 2005. Decreasing trend in renal disease mortality after cessation from arsenic exposure in a previous arseniasis-endemic area in southwestern Taiwan. *Journal of Toxicology and Environmental Health: Part A* 68(5):319–327.

Chiverton, A., J. Hannaford, I. Holman et al. 2015. Which catchment characteristics control the temporal dependence structure of daily river flows? *Hydrological Process* 29(6):1353–1369.

Choprapawon, C. P. and S. Ajjimangkul. 1999. Major interventions on chronic arsenic poisoning in Ronpibool district, Thailand-Review and long-term follow-up. In *Arsenic Exposure and Health Effects*, eds. W. R. Chappel, C.O. Abernathy, R.L. Calderon, 355–362. Amsterdam: Elsevier.

Choprapawon, C. and Y. Porapakkham. 2001. Occurrence of cancer in arsenic contaminated area, Ronpibool District, Nakorn Srithmmarat Province, Thailand. In *Arsenic Exposure and Health Effects IV*, eds. W. R. Chappell, C.O. Abernathy, R.L. Calderon, 201–206. New York: Elsevier.

Chou, W. C., C. Y. Chuang, P. C. Huang et al. 2012. Arsenic exposure in pregnancy increases the risk of adverse birth outcomes of newborns in Taiwan. In *Proceedings of the AACR Special Conference on Post-GWAS Horizons in Molecular Epidemiology: Digging Deeper into the Environment*, Nov 11–14. Philadelphia, USA: AACR.

Choudhury, S. 2015. Comparative study on linear and non-linear geostatistical estimation methods: A case study on iron deposit. *Procedia Earth and Planetary Science* 11:131–139.

Choudhury, R., C. Mahanta, S. Verma et al. 2017. Arsenic distribution along different hydrogeomorphic zones in parts of the Brahmaputra River Valley, Assam (India). *Hydrogeology Journal* 25(4):1153–1163.

Chow, W. S. 1986. Investigation of the presence of excessive arsenic and floride in well water in Kg. Sekolah, Ulu Kepong [abstract]. In *Geological Society of Malaysia Annual Conference*, Kuala Lumpur, 28–29 April. *Warta Geologi*, 12 (2): 94.

Chowdhury, A. M. R. 2004. Arsenic crisis in Bangladesh. *Scientific American* 291:86–91.

Chowdhury, U. K., B. K. Biswas, T. R. Chowdhury et al. 2000. Groundwater arsenic contamination in Bangladesh and West Bengal, India. *Environmental Health Perspectives*, 108: 393–397.

Chowdhury, M. A. I., M. T. Uddin, M. F. Ahmed et al. 2006. Collapse of socio-economic base of Bangladesh by arsenic contamination in groundwater. *Pakistan Journal of Biological Sciences* 9:1617–1627.

Christoplos, I., J. Mitchell, A. Liljelund. 2001. Re-framing risks: The changing context of disaster mitigation and preparedness. *Disasters* 23(3):185–198.

Chung, C. J., Y. L. Huang, Y. K. Huang et al. 2013b. Urinary arsenic profiles and the risks of cancer mortality: A population-based 20-year follow-up study in arseniasis-endemic areas in Taiwan. *Environmental Research* 122:25–30.

Chung, J. Y., B. G. Kim, B. K. Lee et al. 2016. Urinary arsenic species concentration in residents living near abandoned metal mines in South Korea. *Annals of Occupational and Environmental Medicine* 28:67.

Chung, Y. L., Y. P. Liaw, B. F. Hwang et al. 2013a. Arsenic in drinking and lung cancer mortality in Taiwan. *Journal of Asian Earth Sciences* 77:327–331.

Ciarrocca, M., F. Tomei, T. Caciari et al. 2012. Exposure to arsenic in urban and rural areas and effects on thyroid hormones. *Inhalation Toxicology* 24(9):589–598.

Cilesiz, S. 2011. A phenomenological approach to experiences with technology: Current state, promise, and future directions for research. *Educational Technology Research and Development* 59(4):487–510.

Ciminelli, V. S., M. Gasparon, J. C. Ng et al. 2017. Dietary arsenic exposure in Brazil: The contribution of rice and beans. *Chemosphere* 168:996–1003.

Cinderby, S. 1999. Geographic information systems (GIS) for participation: The future of environmental GIS? *International Journal of Environmental Pollution* 11(3):304–315.

Cinnirella, S., G. Buttafuoco, N. Pirrone. 2005. Stochastic analysis to assess the spatial distribution of groundwater nitrate concentrations in the Po catchment (Italy). *Environmental Pollution*, 133:569–580.

Cinti, D., P. P. Poncia, L. Brusca et al. 2015. Spatial distribution of arsenic, uranium and vanadium in the volcanic-sedimentary aquifers of the Vicano-Cimino Volcanic District (Central Italy). *Journal of Geochemical Exploration*, 152:123–133.

Cinti, D., P. P. Poncia, M. Procesi et al. 2013. Geostatistical techniques application to dissolved radon hazard mapping: An example from the western sector of the Sabatini Volcanic District and the Tolfa Mountains (central Italy). *Applied Geochemistry* 35:312–324.

Civita, M. V. 2010. The combined approach when assessing and mapping groundwater vulnerability to contamination. *Journal of Water Resource and Protection* 2:14–28.

Claesson, M. and J. Fagerberg. 2003. *Arsenic in groundwater of Santiago del Estero, Argentina: Sources, mobilization controls and remediation with natural materials*. MSc Dissertation, Royal Institute of Technology (KTH): Stockholm.

Clewell, H. J., P. R. Gentry, J. W. Yager. 2016. Considerations for a biologically based risk assessment for arsenic. In *Arsenic: Exposure Sources, Health Risks, and Mechanisms of Toxicity*, ed. J.C. States, 511–534. New Jersey: John Wiley & Sons.

Clifford, D. and Z. Zhang. 1994. Arsenic Chemistry and Speciation. *American Water Works Association Annual Conference*. New York, June 19–23.

Coe, E. and A. D. Schimmer. 2008. Catalase activity and arsenic sensitivity in acute leukemia. *Leukemia & Lymphoma*, 49(10):1976–1981.

Coelho, N. M. M., A. C. da Silva, C. M. da Silva. 2002. Determination of As(III) and total inorganic arsenic by flow injection hydride generation atomic absorption spectrometry. *Analytica Chimica Acta* 460(2):227–233.

Coetsiers, M. and K. Walraevens. 2006. Chemical characterization of the Neogene aquifer, Belgium. *Hydrogeological Journal* 14(8):1556–1568.

Cohen, S. M., L. L. Arnold, B. D. Beck et al. 2013. Evaluation of the carcinogenicity of inorganic arsenic. *Critical Review Toxicology* 43(9):711–752.

Cohen, S. M., L. L. Arnold, M. Eldan et al. 2006. Methylated arsenicals: The implications of metabolism and carcinogenicity studies in rodents to human risk assessment. *Critical Reviews in Toxicology*, 36:99–133.

Cohrssen, J. J. and V. T. Covello. 1989. *Risk Analysis: A Guide to Principles and Methods for Analyzing Health and Environmental Risks*. Washington DC: Council on Environmental Quality.

Çöl, M., C. Çöl, A. Soran et al. 1999. Arsenic-related Bowen's disease, palmar keratosis, and skin cancer. *Environmental Health Perspectives* 107(8):687–689.

Çolak, M., Ü. Gemici, G. Tarcan. 2003. The effects of colemanite deposits on the arsenic concentrations of soil, and groundwater in Iğdeköy-Emet-Kütahya-Turkey. *Water Air and Soil Pollution* 149:127–143.

Collotta, M., P. A. Bertazzi, V. Bollati. 2013. Epigenetics and pesticides. *Toxicology* 307:35–41.

Comar, C. L. 1979. Risk: A pragmatic de minimis approach. *Science* 203(4378):319.

Concha, G., K. Broberg, M. Grander et al. 2010. High-level exposure to lithium, boron, cesium, and arsenic via drinking water in the Andes of northern Argentina. *Environmental Science & Technology* 44(17):6875–6880.

Concha, G., B. Nermell, M. Vahter (1998): Metabolism of inorganic arsenic in children with chronic high arsenic exposure in Northern Argentina. *Environmental Health Perspectives* 106 (6): 355–359.

Connor, S. and F. Pearce. 2001. *British Scientists' failed to check for arsenic risk*. The Independent (UK), January 19.

Coppola, D. P. (2011) *Introduction to International Disaster Management*. Amsterdam: Elsevier.

Coronado-González, J. A., L. M. Del Razo, G. García-Vargas et al. 2007. Inorganic arsenic exposure and type 2 diabetes mellitus in Mexico. *Environmental Research* 104(3):383–389.

Corradini, F., F. Meza, R. Calderón. 2017. Trace element content in soil after a sediment-laden flood in northern Chile. *Journal of Soils and Sediments* 17(10):2500–2515.

Covello, V. T. 2006. Risk communication and message mapping: A new tool for communicating effectively in public health emergencies and disasters. *Journal of Emergency Management* 4(3):25–40.

Cressie, N. 1985. Fitting variogram models by weighted least square. *Mathematical Geology* 17(5):563–586.

Creswell, J. W. 2009. *Research Design: Qualitative, Quantitative, and Mixed Method Approaches*. Thousand Oaks: Sage.

Creswell, J. W. 2013. *Qualitative Inquiry and Research Design: Choosing Among five traditions*. Thousand Oaks: Sage.

Crichton, D. 1999. The risk triangle. In *Natural Disaster Management*, ed. J. Ingleton, 102–103. London: Tudor Rose.

Crighton, D., S. Wilkinson, J. O'Prey et al. 2006. DRAM, a p53- induced modulator of autophagy, is critical for apoptosis. *Cell* 126(1):121–134.

Crotty, M. 1998. *The Foundations of Social Research: Meaning and Perspective in the Research Process.* Sydney: Allen and Unwin.

Crow, B. and F. Sultana. 2002. Gender, class and access to water: Three cases in a poor and crowded delta. *Society and Natural Resources* 15:709–724.

Cubadda, F., B. P. Jackson, K. L. Cottingham et al. 2017. Human exposure to dietary inorganic arsenic and other arsenic species: State of knowledge, gaps and uncertainties. *Science of the Total Environment* 579:1228–1239.

Cui, X., T. Wakai, Y. Shirai et al. 2006. Arsenic trioxide inhibits DNA methyltransferase and restores methylation-silenced genes in human liver cancer cells. *Human Pathology* 37(3):298–311.

Cullen, W. R. 2008. *Is Arsenic an Aphrodisiac? The Sociochemistry of an Element.* Cambridge, UK: Royal Society of Chemistry.

Cullen, W. and K. J. Reimer. 1989. Arsenic speciation in the environment. *Chemistry Reviews* 89:713–764.

Cumbal, L. H., V. Aguirre, R. Tipan et al. 2006. Monitoring concentrations, speciation and mobility of arsenic in geothermal sources of Ecuador's North-Center Andean Region. In *Natural Arsenic in Groundwaters of Latin America*, eds. J. Bundschuh, M. A. Armienta, P. Bhattacharya et al. Freiberg: Bergakademie.

Cutter, G. A., L. S. Cutter, A. M. Featherstone et al. 2001. Antimony and arsenic biogeochemistry in the western Atlantic Ocean, Part II. *Topical Studies in Oceanography* 48:2895–2915.

Cuzick, J., P. Sasieni, S. Evans. 1992. Ingested arsenic, keratosis and bladder cancers. *American Journal of Epidemiology* 136(4):417–421.

D'Acqui, L. P., C. A. Santi, F. Maselli. 2007. Use of ecosystem information to improve soil organic carbon mapping of a Mediterranean island. *Journal of Environmental Quality* 36:262–271.

D'Ippoliti, D., E. Santelli, M. De Sario et al. 2015. Arsenic in drinking water and mortality for cancer and chronic diseases in central Italy, 1990–2010. *PLoS One* 10:e0138182.

D'Monte, D. 2004. Squabbling over arsenic poisoning. (https://groups.yahoo.com/neo/groups/arsenic-crisis/conversations/messages/262). Accessed 2 February 2006.

Dabeka, R. W. and G. M. A. Lacroix. 1987. Total arsenic in foods after sequential wet digestion, dry ashing, coprecipitation with ammonium pyrrolidine dithiocarbamate, and graphite-furnace atomic absorption spectrometry. *Journal of the Association of Official Analytical Chemists* 70(5):866–870.

Dahal, B. M., M. Fuerhacker, A. Mentler et al. 2008. Arsenic contamination of soils and agricultural plants through irrigation water in Nepal. *Environmental Pollution* 155(1):157–163.

Damodaran, A. 2008. *Strategic Risk Taking: A Framework for Risk Management.* New Jersey (USA): Pearson Education Inc.

Dana, P. H. 2010. Participatory mapping. In *Encyclopedia of Geography*, ed. B. Warf, 2124–2125. Thousand Oaks: Sage.

Dangleben, N. L., C. F. Skibola, M. T. Smith. 2013. Arsenic immunotoxicity: A review. *Environmental Health* 12:73. [http://www.ehjournal.net/content/12/1/73]

Dangles, O., F. C. Carpio, M. Villares et al. 2010. Community-based participatory research helps farmers and scientists to manage invasive pests in the Ecuadorian Andes. *AMBIO* 39(4):325–335.

Dani, S. U. 2013. Osteoresorptive arsenic intoxication. *Bone* 53(2):541–545.

Das, D., A. Chatterjee, B. K. Mandal et al. 1995. Arsenic in ground water in six districts of West Bengal, India: The biggest arsenic calamity in the world. Part-2. Arsenic concentration in drinking water, hair, nail, urine, skin-scale and liver tissue (biopsy) of the affected people. *The Analyst* 120(3):917–24.

Das, S., C. C. Liu, J. S. Jean et al. 2016. Dissimilatory arsenate reduction and in situ microbial activities and diversity in arsenic-rich groundwater of Chianan Plain, Southwestern Taiwan. *Microbial Ecology* 71(2):365–374.

Das, D. K. and L. N. Mandal. 1988. Effect of puddling and different times of organic matter application before puddling on the availability of applied iron, manganese, copper and zinc in rice soils. *Agrochimica* 32:327–336.

Das, H. K., A. K. Mitra, P. K. Sengupta et al. 2004. Arsenic concentrations in rice, vegetables, and fish in Bangladesh: A preliminary study. *Environment International* 30(3):383–387.

Das, N., S. Paul, D. Chatterjee et al. 2012. Arsenic exposure through drinking water increases the risk of liver and cardiovascular diseases in the population of West Bengal, India. *BMC Public Health* 12: 639 [www.biomedcentral.com/1471-2458/12/639].

Das B., M. M. Rahman, B. Nayak et al. 2009. Groundwater arsenic contamination, its health effects and approach for mitigation in West Bengal, India and Bangladesh. *Water Quality, Exposure and Health* 1:5–21.

Das, D., G. Samanta, B. K. Mandal et al. 1996. Arsenic in groundwater in six districts of West Bengal, India. *Environmental Geochemistry and Health* 18(1):5–15.

Das, N. K. and S. R. Sengupta. 2008. Arsenicosis: Diagnosis and treatment. *Indian Journal of Dermatology, Venereology and Leprology* 74(6):571–581.

Dash, J. P., A. Sarangi, D. K. Singh. 2010. Spatial variability of groundwater depth and quality parameters in the national capital territory of Delhi. *Environmental Management* 45:640–650.

Dastgiri, S., M. Mosaferi, M. A. H. Fizi et al. 2010. Arsenic exposure, dermatological lesions, hypertension, and chromosomal abnormalities among people in a rural community of Northwest Iran. *Journal of Health, Population, and Nutrition* 28(1):14–22.

Datta, D. V. 1976. Arsenic and non-cirrhotic portal hypertension (letter). *Lancet* 1:433.

Dauphiné, D. C., C. Ferreccio, S. Guntur et al. 2011. Lung function in adults following in utero and childhood exposure to arsenic in drinking water: Preliminary findings. *International Archives of Occupational and Environmental Health* 84:591–600.

Dauphiné, D. C., A.H. Smith, Y. Yuan et al. 2013. Case-control study of arsenic in drinking water and lung cancer in California and Nevada. *International Journal of Environmental Research and Public Health* 10(8):3310–3324.

Davey, J. C., A. P. Nomikos, M. Wungjiranirun et al. 2008. Arsenic as an endocrine disruptor: Arsenic disrupts retinoic acid receptor-and thyroid hormone receptor-mediated gene regulation and thyroid hormone-mediated amphibian tail metamorphosis. *Environmental Health Perspectives* 116:165–172.

David, M. 1977. *Geostatistical Ore Reserve Estimation*. Amsterdam: Elsevier.

Davidson, C. I., W. D. Goold, T. P. Mathison et al. 1985. Airborne trace elements in Great Smoky Mountains, Olympic, and Glacier National Parks. *Environmental Science and Technology* 19(1):27–35.

Davies, J. and C. Exley. 1992. *Short term BGS pilot project to assess the hydrogeological character of the main aquifer units of central and north-eastern Bangladesh and possible toxicity of groundwater to fish and humans.* BGS Technical Report (WD/92/43R), Keyworth: British Geological Survey.

Dávila-Esqueda, M. E., M. E. Jiménez-Capdeville, J. M. Delgado et al. 2012. Effects of arsenic exposure during the pre- and postnatal development on the puberty of female offspring. *Experimental and Toxicologic Pathology* 64(1):25–30.

Davis, M. A., A. J. Signes-Pastor, M. Argos et al. 2017. Assessment of human dietary exposure to arsenic through rice. *Science of The Total Environment* 586:1237–1244.

Davraz, A. 2015. Studies of geogenic groundwater contamination in Southwestern Anatolia, Turkey. *Procedia Earth and Planbetary Science* 15:435–441.

DCH/SOES. 1997. *Arsenic pollution in groundwater of Bangladesh.* Dhaka Community Hospital, Dhaka, and School of Environmental Studies, Jadavpur University, Kolkata.

De Chaudhuri, S., P. Ghosh, N. Sarma et al. 2008. Genetic variants associated with arsenic susceptibility: Study of purine nucleoside phosphorylase, arsenic (+3) methyl transferase and glutathione S-transferase omega genes. *Environmental Health Perspectives* 116(4):501–505.

De, B. K., D. Majumdar, S. Sen et al. 2004. Pulmonary involvement in chronic arsenic poisoning from drinking contaminated ground-water. *Journal of Association of Physicians of India* 52:395–400.

De Vizcaya-Ruiz, A., O. Barbier, R. Ruiz-Ramos. 2009. Biomarkers of oxidative stress and damage in human populations exposed to arsenic. *Mutation Research* 674:85–92.

Decrop, A. 1999. Triangulation in qualitative tourism research. *Tourism Management* 20:157–161.

Deetz, S. 1996. Describing differences in approaches to organization science: Rethinking Burrell and Morgan and their legacy. *Organization Science* 7(2):191–207.

Dehghanzadeh, R., N. S. Hir, J. S. Sis et al. 2015. Integrated assessment of spatial and temporal variations of groundwater quality in the Eastern Area of Urmia Salt Lake Basin using multivariate statistical analysis. *Water Resources Management* 29(4):1351–1364.

Del Razo, L. M., M. A. Arellano, M. E. Cebrián. 1990. The oxidation states of arsenic in well-water from a chronic arsenicism area of northern Mexico. *Environmental Pollution* 64:143–153.

Del Razo, L. M., G. G. García-Vargas, O. L. Valenzuela et al. 2011. Exposure to arsenic in drinking water is associated with increased prevalence of diabetes: A cross-sectional study in the Zimapán and Lagunera regions in Mexico. *Environmental Health* 10:73.

Del Razo, L. M., J. L. Hernández, G. García-Vargas et al. 1994. Urinary excretion of arsenic species in a human population chronically exposed to arsenic via drinking water: A pilot study. In *Arsenic Exposure and Health*, eds. W. R. Chappell, C. O. Abernathy, C. R. Cothern, 91–100. Middlesex (UK): Science Reviews Ltd.

Delbari, M., M. Amiri, M. B. Motlagh. 2016. Assessing groundwater quality for irrigation using indicator kriging method. *Applied Water Science* 6(4): 371–381.

Delgado-Andrade, C., M. Navarro, H. Lopez. 2003. Determination of total arsenic levels by hydride generation atomic absorption spectrometry in foods from south-east Spain: Estimation of daily dietary intake. *Food Additives & Contaminants* 20:923–932.

Demeritt, D. 1998. Science, social constructivism and nature. In *Remaking Reality: Nature at the Millennium*, eds. B. Braun and N. Castree, 173–193. London: Routledge.

DeNicola, E., O. S. Aburizaiza, A. Siddique et al. 2015. Climate change and water scarcity: The case of Saudi Arabia. *Annals of Global Health* 81(3):342–353.

Denzin, N. K. and Y. S. Lincoln. 2005. *The Sage Handbook of Qualitative Research*. Thousand Oaks: Sage.

Derby, S. L. and R. L. Keeney. 1981. Risk analysis: Understanding 'how safe is safe enough'? *Risk Analysis* 1(3):217–224.

Derr, R. 2016. What is risk? In *Threat Assessment and Risk Analysis: An Applied Approach*, eds. G. Allen and R. Derr, 9–24. Amsterdam: Elsevier and BH.

Dervin, B. 1994. Information-democracy: An examination of underlying assumptions. *Journal of the American Society for Information Science* 45(6):369–385.

Deschamps, E., V. S. T. Ciminelli, F. T. Lange et al. 2002. Soil and sediment geochemistry of the Iron Quadrangle, Brazil: The case of arsenic. *Journal Soils Sediments* 2:216–222.

DeSesso, J. M., C. F. Jacodson, A. R. Scialli et al. 1998. An assessment of the developmental toxicity of inorganic arsenic. *Reproductive Toxicology* 12(4):385–433.

Deshpande, L. S. and S. P. Pande. 2005. Development of arsenic testing field kit: A tool for rapid on-site screening of arsenic contaminated water sources. *Environmental Monitoring and Assessment* 101:93–101.

Deutsch, C. V. 2002. *Geostatistical Reservoir Modeling*. New York: Oxford University Press.

Deutsch, C. and A. Journel. 1998. *GSLIB: Geostatistical Software Library and User's Guide*. New York: Oxford University Press.

Dey, I. 2007. Grounded theory. In *Qualitative Research Practice*, eds. C. Seale, G. Gobo, J. F. Gubrium, and D. Silverman, 80–93. London: Sage.

Dhar, R. K., Y. Zheng, M. Stute et al. 2008. Temporal variability of groundwater chemistry in shallow and deep aquifers of Araihazar, Bangladesh. *Journal of Contaminant Hydrology* 99:97–111.

Dhar, R.K., B.K. Biswas, G. Samanta et al. 1997. Groundwater arsenic calamity in Bangladesh. *Current Science* 73:48–59.

Dheeman, D. S., C. Packianathan, J. K. Pillai et al. 2014. Pathway of human AS3MT arsenic methylation. *Chemical Research in Toxicology* 27(11):1979–89.

Díaz, O. P., R. Arcos, Y. Tapia et al. 2015. Estimation of arsenic intake from drinking water and food (raw and cooked) in a rural village of Northern Chile. Urine as a biomarker of recent exposure. *International Journal of Environmental Research and Public Health* 12(5):5614–5633.

Díaz, S. L., M. E. Espósito, M. de C. Blanco et al. 2016. Control factors of the spatial distribution of arsenic and other associated elements in loess soils and waters of the southern Pampa (Argentina). *Catena* 140:205–216.

Didar-Ul-Islam, S. M., R. K. Majumder, M. J. Uddin et al. 2017. Hydrochemical characteristics and quality assessment of groundwater in Patuakhali District, Southern Coastal Region of Bangladesh. *Exposure and Health* 9(1):43–50.

Diggle, P. J., J. A. Tawn, R. A. Moyeed. 1998. Model-based geostatistics. *Applied Statistics* 47:299–350.

Ding, W., L. G. Hudson, K. J. Liu. 2005. Inorganic arsenic compounds cause oxidative damage to DNA and protein by inducing ROS and RNS generation in human keratinocytes. *Molecular and Cellular Biochemistry* 279(1–2):105–112.

Ding, W. W. and R. E. Sturgeon. 1996. Evaluation of electrochemical hydride generation for the determination of arsenic and selenium in sea water by graphite furnace atomic absorption with in situ concentration. *Spectrochimica Acta Part B: Atomic Spectroscopy* 51(11):1325–1334.

Dittmar, T. 2004. Hydrochemical processes controlling arsenic and heavy metal contamination in the Elqui river system (Chile). *Science of the Total Environment* 325(1–3):193–207.

Dittmar, J., A. Voegelin, L. C. Roberts et al. 2010. Arsenic accumulation in a paddy field in Bangladesh: Seasonal dynamics and trends over a three-year monitoring period. *Environmental Science and Technology* 44:2925–2931.

Diwakar, J., S. G. Johnston, E.D. Burton et al. 2015. Arsenic mobilization in an alluvial aquifer of the Terai region, Nepal. *Journal of Hydrology: Regional Studies* 4(Part A):59–79.

Dixon-Woods, M., S. Bonas, A. Booth et al. 2006. How can systematic reviews incorporate qualitative research? A critical perspective. *Qualitative Research* 6:27–44.

Dobrzynski, D. 2007. Chemical diversity of groundwater in the Carboniferous-Permian aquifer in the Unislaw Slaski-Sokolowsko area (the Sudetes, Poland): A geochemical modelling approach. *Acta Geologica Polonica* 57(1):97–112.
DoE. 1994. *Environmental Quality Standards for Bangladesh*. Dhaka: Department of Environment, Government of Bangladesh.
Dogan, M., A. U. Dogan, C. Celebi et al. 2005. Geogenic arsenic and a survey of skin lesions in Emet Region of Kutahya, Turkey. *Indoor and Built Environment* 14(6):533–536.
Dokou, Z., N. N. Kourgialas, G. P. Karatzas. 2015. Assessing groundwater quality in Greece based on spatial and temporal analysis. *Environmental Monitoring and Assessment* 187(12):774.
Dong, J. and S. Y. Su. 2009. The association between arsenic and children's intelligence: A meta-analysis. *Biological Trace Element Research* 129(1–3):88–93.
Dong, F., R. Zheng, X. Chen et al. 2016. Caring for dying cancer patients in the Chinese cultural context: A qualitative study from the perspectives of physicians and nurses. *European Journal of Oncology Nursing* 21:189–196.
Dowling, M. 2007. From Husserl to van Manen. A review of different phenomenological approaches. *International Journal of Nursing Studies* 44:131–142.
Dowling, C. B., R. J. Poreda, A. R. Basu et al. 2002. Geochemical study of arsenic release mechanisms in the Bengal Basin groundwater. *Water Resources Research* 38:1173–1190.
DPHE. 1996. *Presence of Arsenic in Groundwater in the 18 District Towns Projects (DTP): Short Mission Report*. Dhaka: Department of Public Health and Engineering. Government of Bangladesh.
DPHE/JICA. 2010. *Situation Analysis of Arsenic Mitigation 2009*. Dhaka: Department of Public Health Engineering and Japan International Cooperation Agency.
DPHE/UNICEF. 2013. *SHEWA-B and Nationwide Water Point Mapping Programme (NWMP) Survey Results, 2012–2013*. Dhaka: Department of Public Health Engineering and UNICEF.
Drahota, P., T. Paces, Z. Pertold et al. 2006. Weathering and erosion fluxes of arsenic in watershed mass budgets. *Science of the Total Environment* 372(1):306–316.
Draoui, M., J. Vias, B. Andreo et al. 2008. A comparative study of four vulnerability mapping methods in a detritic aquifer under Mediterranean climatic conditions. *Environmental Geology* 54:455–463.
Drouhot, S., F. Raoul, N. Crini et al. 2014. Responses of wild small mammals to arsenic pollution at a partially remediated mining site in Southern France. *Science of the Total Environment* 470–471:1012–22.
Druwe, I. L. and R. R. Vaillancourt. 2010. Influence of arsenate and arsenite on signal transduction pathways: an update. *Archives of Toxicology* 84(8):585–596.
Duan, N. 1982. Models for human exposure to air pollution. *Environmental International* 8:305–309.
Duan, Y., Y. Gan, Y. Wang et al. 2015. Temporal variation of groundwater level and arsenic concentration at Jianghan Plain, central China. *Journal of Geochemical Exploration*, 149:106–119.
Duarte, F. A., J. S. F. Pereira, M. F. Mesco et al. 2007. Evaluation of liquid chromatography inductively coupled plasma mass spectrometry for arsenic speciation in water from industrial treatment of shale. *Spectrochimica Acta Part B: Atomic Spectroscopy* 62(9):978–984.
Dubner, S. J. and Levitt, S. D. (2006). *How many lives did Dale Earnhardt save?* The New York Times (February 19). [http://www.nytimes.com/2006/02/19/magazine/how-many-lives-did-dale-earnhardt-save.html].
Dubois, G. and S. Galmarini. 2004. Introduction to the Spatial Interpolation Comparison (SIC). *Applied GIS* 1(2):9–11.
Dummer, T. J. B., Z. M. Yu, L. Nauta et al. 2015. Geostatistical modelling of arsenic in drinking water wells and related toenail arsenic concentrations across Nova Scotia, Canada. *Science of the Total Environment* 505:1248–1258.
Durant, J. 1997. Scientific truth and political reality: Professional and public perceptions of risk. In *Science, Policy and Risk*, ed. J. Ashworth, 45–51. London: Royal Society.
Duxbury, J. M., A. B. Mayer, J. G. Lauren et al. 2003. Food chain aspects of arsenic contamination in Bangladesh: Effects on quality and productivity of rice. *Journal of Environmental Science and Health: Part A* 38(1):61–69.
Eary, L. E., D. Rai, S. V. Mattigod et al. 1990. Geochemical factors controlling the mobilization of inorganic constituents from fossil fuel combustion residues. I: Review of the major elements. *Journal of Environmental Quality* 19:202–214.
Eastman, J. R., W. Jin, P. A. K. Kyem et al. 1995. Raster procedures for multi-criteria/multi-objective decisions. *Photogrammetric Engineering and Remote Sensing* 61(5):539–547.
Ebi, K. L. 2009. Facilitating climate justice through community-based adaptation in the health sector. *Environmental Justice* 2(4):191–195.

Eblin, K. E., M. E. Bowen, D. W. Cromey et al. 2006. Arsenite and monomethylarsonous acid generate oxidative stress response in human bladder cell culture. *Toxicology and Applied Pharmacology* 217(1):7–14.

Edet, A. E. and O. E. Offiong. 2003. Evaluation of water quality pollution indices from heavy metal contamination monitoring. A case sudy from Akpabuyo-Odukpani area, Lower Cross River Basin (southeastern Nigeria). *GeoJournal* 57:295–304.

Edmond, G. and D. Mercer. 2004. Daubert and the exclusionary ethos: The convergence of corporate and judicial attitudes towards the admissibility of expert evidence in tort litigation. *Law and Policy* 26:231–57.

Edmunds, W. M., K. M. Ahmed, P. G. Whitehead. 2015. A review of arsenic and its impacts in groundwater of the Ganges-Brahmaputra-Meghna delta, Bangladesh. *Environmental Science: Processes & Impacts* 17(6):1032–46.

Edmunds, W. M., J. M. Cook, D. G. Kinniburgh et al. 1989. *Trace Element Occurrence in British Groundwaters.* Keyworth: British Geological Survey Research Report (SD/89/3).

Edwards, J. and G. Kaimal. 2016. Using meta-synthesis to support application of qualitative methods findings in practice: A discussion of meta-ethnography, narrative synthesis, and critical interpretive synthesis. *The Arts in Psychotherapy* 51:30–35.

Edwards, C. and A. Titchen. 2003. Research into patients' perspectives: Relevance and usefulness of phenomenological sociology. *Journal of Advanced Nursing* 44(5):450–460.

EFSA (European Food Safety Authority). 2014. Dietary exposure to inorganic arsenic in the European population. *EFSA Journal* 12(3):3597.

Eiche, E., M. Berg, S. M. Hönig et al. 2017. Origin and availability of organic matter leading to arsenic mobilisation in aquifers of the Red River Delta, Vietnam. *Applied Geochemistry* 77:184–193.

Eichstaedt, C. A., T. Antao, A. Cardona et al. 2015. Positive selection of AS3MT to arsenic water in Andean populations. *Mutation Research* 780:97–102.

Eisner, E. W. 2017. *The Enlightened Eye: Qualitative Inquiry and the Enhancement of Educational Practice.* New York: Teachers College Press.

El-Demerdash, F. M., M. I. Yousef, F. M. Radwan. 2009. Ameliorating effect of curcumin on sodium arsenite-induced oxidative damage and lipid peroxidation in different rat organs. *Food and Chemical Toxicology* 47(1):249–254.

Ellenhorn, M. J. 1997. Arsenic. In *Ellenhorn's Medical Toxicology: Diagnosis and Treatment of Human Poisoning*, eds. M. J. Ellenhorn, S. Schonwald, G. Ordog, J. Wasserberger, 1538–1542. Baltimore: Williams and Wilkins.

Elliot, J. A. and M. Campbell. 2001. The environmental imprints and complexes of social dynamics in rural Africa: cases from Zimbabwe and Ghana. *Geoforum* 33: 221–237.

El-Makky, A. M. 2011. Statistical analyses of La, Ce, Nd, Y, Nb, Ti, P, and Zr in bedrocks and their significance in geochemical exploration at the Um Garayat goldmine area, Eastern Desert, Egypt. *Natural Resources Research* 20:157–176.

Elumalai, V., K. Brindha, B. Sithole et al. 2017. Spatial interpolation methods and geostatistics for mapping groundwater contamination in a coastal area. *Environmental Science and Pollution Research* 24(12):11601–11617.

Emmi, P. C. and C. A. Horton. 1996. Seismic risk assessment, accuracy requirements, and GIS based sensitivity analysis. In *GIS and Environmental Modeling: Progress and Research Issues*, eds. M. F. Goodchild, L.T. Steyaert, B.O. Parks et al. 191–195. Fort Collins: GIS World Books.

Engel, B., K. Navulur, B. Cooper et al. 1996. Estimating groundwater vulnerability to nonpoint source pollution from nitrates and pesticides on a regional scale. In *Application of Geographic Information Systems in Hydrology, Water Resources Management*, eds. K. Kover and H. P. Nachtnebel, 521–526. Wallingford: IAHS.

Engel, R. R. and A. H. Smith. 1994. Arsenic in drinking water and mortality from vascular disease: An ecologic analysis in 30 counties in the United States. *Archives of Environmental Health* 49(5):418–27.

Englyst, V., N. G. Lundström, L. Gerhardsson et al. 2001. Lung cancer risks among lead smelter workers also exposed to arsenic. *The Science of the Total Environment* 273(1–3):77–82.

Engström, K. S., M. Vahter, T. Fletcher et al. 2015. Genetic variation in arsenic (+3 oxidation state) methyltransferase (AS3MT), arsenic metabolism and risk of basal cell carcinoma in a European population. *Environmental and Molecular Mutagenesis* 56(1):60–69.

Engström, K., M. Vahter, S. J. Mlakar et al. 2011. Polymorphisms in arsenic (+III oxidation state) methyltransferase (AS3MT) predict gene expression of AS3MT as well as arsenic metabolism. *Environmental Health Perspectives* 119(2):182–188.

EPA. 2012. *Arsenic: Rule-Making History.* Washington DC: U.S. Environmental Protection Agency.

EPA. 2013. *IRIS Toxicological Review of Inorganic Arsenic (Cancer)*. Washington, DC: U.S. Environmental Protection Agency.

Erickson, M. L., R. J. Barnes. 2006. Arsenic concentration variability in public water system wells in Minnesota, USA. *Applied Geochemistry* 21:305–317.

Ernest, K. O. and J. M. Christoper. 1995. An overview of arsenic removal processes. *Desalination* 103:79–88.

Espinosa, E., M. A. Armienta, O. Cruz et al. 2009. Geochemical distribution of arsenic, cadmium, lead and zinc in river sediments affected by tailings in Zimapán, a historical polymetalic mining zone of México. *Environmental Geology* 58:1467–1477.

Espinoza, M. A. and J. Bundschuh. 2008. Natural arsenic groundwater contamination of the sedimentary aquifers of Southwestern Sébaco valley, Nicaragua. In *Natural Arsenic in Groundwaters in Latin America*, eds. J. Bundschuh, M. A. Armienta, P. Birkle, P. Bhattacharya, J. Matschullat, A. B. Mukherjee, 109–122. Boca Raton: CRC Press.

Ettinger, A. S., T. E. Arbuckle, M. Fisher et al. 2017. Arsenic levels among pregnant women and newborns in Canada: Results from the Maternal-Infant Research on Environmental Chemicals (MIREC) cohort. *Environmental Research*, 153:8–16.

Evans, S. 1998. *The Ubiquitous Poison*. Liverpool: Liverpool Medical Institution. 6–18, [http://users.physics.harvard.edu/~wilson/arsenic/UbiquitousPoison.pdf].

Even, E., H. Masuda, T. Shibata et al. 2017. Geochemical distribution and fate of arsenic in water and sediments of rivers from the Hokusetsu area, Japan. *Journal of Hydrology: Regional Studies* 9:34–47.

Faita, F., L. Cori, F. Bianchi et al. 2013. Arsenic-induced genotoxicity and genetic susceptibility to arsenic-related pathologies. *International Journal of Environmental Research and Public Health* 10(4):1527–1546.

FAO/UNICEF/WHO/WSP. 2010. *Towards an arsenic safe environment in Bangladesh*. A report published on the World Water Day, 22 March 2010, Dhaka. [www.unicef.org/bangladesh/Towards_an_arsenic_safe_environ_report_22Mar2010.pdf]

FAO/WHO. 2011. Safety evaluation of certain contaminants in food, prepared by the seventy-second meeting of the joint FAO (Food and Agriculture Organization)/WHO (World Health Organization) expert committee on food additives. *WHO Food Additives Series*, 63:153–316.

Farías, S. S., V. A. Casa, C. Vázquez et al. 2003. Natural contamination with arsenic and other trace elements in ground waters of Argentine Pampean Plain. *The Science of the Total Environment* 309(1–3):187–199.

Farías, S. S., A. Londonio, C. Quintero et al. 2015. On-line speciation and quantification of four arsenical species in rice samples collected in Argentina using a HPLC-HG-AFS coupling. *Microchemical Journal* 120:34–39.

Farooq, S. H., D. Chandrasekharam, S. Norra et al. 2011. Temporal variations in arsenic concentration in the groundwater of Murshidabad District, West Bengal, India. *Environmental Earth Sciences*, 62(2):223–232.

Farooqi, A., H. Masuda, M. Kusakabe et al. 2007. Distribution of highly arsenic and fluoride contaminated groundwater from east Punjab, Pakistan, and the controlling role of anthropogenic pollutants in the natural hydrological cycle. *Geochemical Journal* 41(4):213–234.

Farzan, S. F., E. B. Brickley, Z. Li et al. 2017. Maternal and infant inflammatory markers in relation to prenatal arsenic exposure in a U.S. pregnancy cohort. *Environmental Research* 156:426–433.

Farzan, S. F., M. R. Karagas, Y. Chen. 2013a. In utero and early life arsenic exposure in relation to long-term health and disease. *Toxicology and Applied Pharmacology* 272:384–390.

Farzan, S. F., M. R. Karagas, J. Jiang et al. 2015. Gene-arsenic interaction in longitudinal changes of blood pressure: Findings from the Health Effects of Arsenic Longitudinal Study (HEALS) in Bangladesh. *Toxicology and Applied Pharmacology* 288(1):95–105.

Farzan, S. F., T. Waterboer, J. Gui et al. 2013b. Cutaneous alpha, beta and gamma human papillomaviruses in relation to squamous cell carcinoma of the skin: A population-based study. *International Journal of Cancer* 133(7):1713–1720.

Fatmi, Z., I. N. Abbasi, M. Ahmed et al. 2013. Burden of skin lesions of arsenicosis at higher exposure through groundwater of Taluka Gambat district Khairpur, Pakistan: A cross-sectional survey. *Environmental Geochemistry and Health* 35(3):341–346.

Fatmi, Z., I. Azam, F. Ahmed et al. 2009. Health burden of skin lesions at low arsenic exposure through groundwater in Pakistan. Is river the source? *Environmental Research* 109:575–581.

Faustman, E. M., and G. S. Omenn. 2008. Risk assessment. In *Casarett and Doull's Toxicology - The Basic Science of Poisons*, ed. C. D. Klaassen, 107–128. New York: McGraw-Hill Medical Publishing.

Fee, D. B. 2016. Neurological effects of arsenic exposure. In *Arsenic: Exposure Sources, Health Risks, and Mechanisms of Toxicity*, ed. J. C. States, 193–220. New Jersey: John Wiley & Sons.

Fei, D. L., D. C. Koestler, Z. Li et al. 2013. Association between in utero arsenic exposure, placental gene expression, and infant birth weight: A US birth cohort study. *Environmental Health* 12:58 [DOI: 10.1186/1476-069X-12-58].

Fenaux, P., C. Chomienne, L. Degos, 2001. Treatment of acute promyelocytic leukaemia. *Clinical Haematology*, 14:153–174.

Fendorf, S., H. A. Michael, A. van Geen. 2010. Spatial and temporal variations of groundwater arsenic in South and Southeast Asia. *Science* 328(5982):1123–1127.

Feng, H., Y. Gao, L. Zhao et al. 2013. Biomarkers of renal toxicity caused by exposure to arsenic in drinking water. *Environmental Toxicology and Pharmacology* 35(3):495–501.

Fengzhou, Q., G. Guoying, X. Chun. 1991. A compact, versatile, integrated nebulizer-hydride generator system for simultaneous determination of volatile elemental hydrides and other elements by ICP-AES. *Applied Spectroscopy* 45(2):287–292.

Ferguson H. F. and J. Gavis. 1972. A review of the arsenic cycle in natural waters. *Water Research* 6:1259–1274.

Fernández, M., P. Valdebenito, E. Chaparro et al. 2015. Epidemiological and clinicopathological characteristics of arsenic-related bladder cancer: A comparison between affected populations from contaminated and reference sites. *European Urology Supplements* 14(2):e835.

Ferral, A. E., E. Alaniz, M. S. Tagle et al. 2014. Hydrogeochemical characterization of the presence of arsenic in the Puelche aquifer in the area of Mataderos, Buenos Aires province, Argentina. In *One Century of the Discovery of Arsenicosis in Latin America (1914–2014)*, eds. M.I. Litter, H. B. Nicolli, M. Meichtry et al., 157–158. Boca Raton: CRC Press.

Ferreccio, C., C. González, M. Milosavjlevic et al. 2000. Lung cancer and arsenic concentrations in drinking water in Chile. *Epidemiology* 11(16):673–679.

Ferreccio, C., A. H. Smith, V. Durán et al. 2013. Case-control study of arsenic in drinking water and kidney cancer in uniquely exposed northern Chile. *American Journal of Epidemiology* 178(5):813–818.

Ferreira, M. A. and A. A. Barros. 2002. Determination of As(III) and arsenic(V) in natural waters by cathodic stripping voltammetry at a hanging mercury drop electrode. *Analytica Chimica Acta* 459(1):151–159.

Feussner, J. R., J. D. Shelburne, S. Bredehoeft et al. 1979. Arsenic-induced bone marrow toxicity: Ultrastructural and electron-probe analysis. *Blood* 53:820–827.

Fewtrell, L., R. Fuge, D. Kay. 2005. An estimation of the global burden of disease due to skin lesions caused by arsenic in drinking water. *Journal of Water and Health* 3(2):101–107.

Figueiredo, B. R., R. P. Borba, R. S. Angélica. 2007. Arsenic occurrence in Brazil and human exposure. *Environmental Geochemistry and Health* 29:109–118.

Fiksel, J. 1987. De minimis risk: From concept to practice. In *De Minimis Risk*, ed. C. Whipple, 3–7. New York: Plenum Press.

Finch, H., J. Lewis, and C. Turley. 2013. Focus groups. In *Qualitative Research Practice: A Guide for Social Science Students and Researchers*, eds. J. Ritchie, J. Lewis, C. M. Nicholls, and R. Ormston, 211–242. Los Angeles: Sage.

Fincher, R. M. and R. M. Koerker. 1987. Long-term survival in acute arsenic encephalopathy. Follow-up using newer measures of electrophysiologic parameters. *American Journal of Medicine* 82(3):549–552.

Finkelman, R. B., H. E. Belkin, B. Zheng. 1999. Health impacts of domestic coal use in China. *Proceedings of the National Academy of Sciences of the United States of America* 96:3427–3431.

Finlay, L. 2012. Debating phenomenological methods. In *Hermeneutic Phenomenology in Education: Method and Practice*, eds. N. Friesen, C. Henriksson, and T. Saevi, 17–37. Rotterdam: Sense Publishers.

Fitzmaurice, M. 2004. The International Court of Justice and the environment. *Non-State Actors and International Law* 4:173–197.

Fjeld, R. A., N. A. Eisenberg, K. L. Compton. 2007. *Quantitative Environmental Risk Analysis for Human Health*. New Jersey: John Wiley & Sons, Inc.

Flanagan, S. M, M. Belaval, J. D. Ayotte. 2014. *Arsenic, iron, lead, manganese, and uranium concentrations in private bedrock wells in southeastern New Hampshire, 2012–2013*. New Hampshire-Vermont: US Geological Survey.

Flanagan, S. V., R. B. Johnston, Y. Zheng. 2012. Arsenic in tubewell water in Bangladesh: Health and economic impacts and implications for arsenic mitigation. *Bulletin of the World Health Organization* 90 (11): 839–846.

Flanagan, S. V., S. E. Spayd, N. A. Procopio et al. 2016. Arsenic in private well water part 3 of 3: Socioeconomic vulnerability to exposure in Maine and New Jersey. *Science of the Total Environment* 562:1019–1030.

Fletcher, D., A. De Massis, M. Nordqvist. 2016. Qualitative research practices and family business scholarship: A review and future research agenda. *Journal of Family Business Strategy* 7:8–25.

Bibliography

Flora, S. J. S. 2011. Arsenic-induced oxidative stress and its reversibility. *Free Radical Biology and Medicine* 51(2):257–281.

Flora, S. J. S. 2015. Arsenic: Chemistry, occurrence, and exposure. In *Handbook of Arsenic Toxicity*, ed. S. J. S. Flora, 1–49. Amsterdam: Elsevier and Academic Press.

Flynn, H. C., V. M. Mahon, G. C. Diaz et al. 2002. Assessment of bioavailable arsenic and copper in soils and sediments from the Antofagasta region of northern Chile. *Science of the Total Environment* 286(1–3):51–59.

Focazio, M. J., A. H. Welch, S. A. Watkins et al. 2000. *A retrospective analysis on the occurrence of arsenic in ground-water resources in the United States and limitations in drinking-water-supply characterizations*. Water Resources Investigations Report 99–4279. US Geological Survey, Reston, VA, USA.

Ford, L. A., R. Crabtree, A. Hubbell. 2009. Crossing borders in health communication research: Toward an ecological understanding of context, complexity, and consequences in community-based health education in the US-Mexico borderlands. *Health Communication* 24(7):608–618.

Forkner, C. E. and T. F. M. Scott. 1931. Arsenic as a therapeutic agent in chronic myelogenous leukemia: Preliminary report. *The Journal of the American Medical Association* 97(1):3–5.

Forshufvud, S. 1962. *Who Killed Napoleon?* London: Hutchinson.

Foster, S. S. D. 1987. Fundamental concepts in aquifer vulnerability, pollution risk and protection strategy. In *Vulnerability of Soil and Groundwater to Pollutants. The Hague, Netherlands Organization for Applied Scientific Research*, eds. W. van Duijvenbooden and H. G. van Waegeningh, 69–86. International Conference, Noordwijk Aan Zee, the Netherlands.

Foster, F., G. Craun, K. G. Brown. 2002. Detection of excess arsenic-related cancer risks. *Environmental Health Perspectives* 110:12–13.

Foster, J. A. and A. T. McDonald. 2000. Assessing pollution risks to water supply intakes using geographical information systems (GIS). *Environmental Modelling & Software* 15:225–234.

Foster, S. and A. Tuinhof. 2004. *Arsenic in Groundwater of South and East Asia: Current Understanding as a Basis for Mitigation*. Washington DC: World Bank.

Foust Jr, R. D., P. Mohapatra, A. M. Compton-O'Brien, J. Reifel. 2004. Groundwater arsenic in the Verde Valley in central Arizona, USA. *Applied Geochemistry* 19:251–255.

Foy, H. M., S. Tarmapai, P. Eamchan et al. 1992. Chronic arsenic poisoning from well water in a mining area in Thailand. *Asia Pacific Journal of Public Health* 6(3):150–152.

Franzblau, A. and R. Lilis. 1989. Acute arsenic intoxication from environmental arsenic exposure. *Archives of Environmental Health* 44:385–390.

Fraser, B. 2012. Cancer cluster in Chile linked to arsenic contamination. *Lancet* 379(9816):603.

Freikowski, D., H. Neidhardt, J. Winter et al. 2013. Effect of carbon sources and of sulfate on microbial arsenicmobilization in sediments of West Bengal, India. *Ecotoxicology and Environmental Safety* 91:139–146.

Frengstad, B., A. K. M. Skrede, D. Banks et al. 2000. The chemistry of Norwegian groundwaters: III. The distribution of trace elements in 476 crystalline bedrock groundwaters, as analysed by ICP-MS techniques. *Science of the Total Environment* 246(1):21–40.

Freshwater, D. 2005. Writing, rigour and reflexivity in nursing research. *Journal of Research in Nursing* 10:311–315.

Frewer, L. J. 1999. Public risk perceptions and risk communications. In *Risk Communication and Public Health*, eds. P. Bennett and Sir K. Calman, 20–32. Oxford: Oxford University Press.

Frewer, L. J., C. Howard, R. Shepherd. 1997. Public concerns about general and specific applications of genetic engineering: Risk, benefit and ethics. *Science, Technology and Human Values* 22:98–124.

Fryer, M., C. D. Collins, H. Ferrier et al. 2006. Human exposure modelling for chemical risk assessment: A review of current approaches and research and policy implications. *Environmental Science and Policy* 9:261–274.

Fu, S., J. Wu, Y. Li et al. 2014. Urinary arsenic metabolism in a Western Chinese population exposed to high-dose inorganic arsenic in drinking water: Influence of ethnicity and genetic polymorphisms. *Toxicology and Applied Pharmacology* 274(1):117–23.

Fujii, R. and W. C. Swain. 1995. *Areal Distribution of Selected Trace Elements, Salinity, and Major Ions in Shallow Ground Water*, Tulare Basin, Southern San Joaquin Valley, California. US Geological Survey Water Resources Investig. Report, 95–4048.

Fytianos K. and C. Christophoridis. 2004. Nitrate, arsenic and chloride pollution of drinking water in northern Greece. Eleboration by Applying GIS. *Environmental Monitoring and Assessment* 93(1–3):55–67.

Gadamer, H. G. 1960. *Truth and Method*. New York: Continuum.

Gadamer, H. G. 1976. *Philosophical Heremeneutics*. Berkeley: University of California Press.

Gain, A. K. and C. Giupponi. 2015. A dynamic assessment of water scarcity risk in the Lower Brahmaputra River Basin: An integrated approach. *Ecological Indicators* 48:120–131.

Gaines, R. V., H. C. W. Skinner, E. E. Foord et al. 1997. *Dana's New Mineralogy* (8th edition). New York: John Wiley & Sons.

Gamboa-Loira, B., M. E. Cebrián, F. Franco-Marina et al. 2017. Arsenic metabolism and cancer risk: A meta-analysis. *Environmental Research* 156:551–558.

Gamhewage, G. 2016. Risk communication - A moving target in the fight against infectious hazards and epidemics. *WHO Weekly Epidemiological Record* 91(7):82–87.

Gao, L., B. Gao, D. Xu et al. 2017. Assessing remobilization characteristics of arsenic (As) in tributary sediment cores in the largest reservoir, China. *Ecotoxicology and Environmental Safety* 140:48–54.

Garai, R., A. K. Chakraborty, S. B. Dey et al. 1984. Chronic arsenic poisoning from tube-well water. *Journal of Indian Medical Association* 82(1):34–35.

García-Esquinas, E., M. Pollán, J. G. Umans et al. 2013. Arsenic exposure and cancer mortality in a US-based prospective cohort: The strong heart study. *Cancer Epidemiology, Biomarkers & Prevention* 22:1944–1953.

García-Sánchez, A. and E. Alvarez-Ayuso. 2003. Arsenic in soils and waters and its relation to geology and mining activities (Salamanca Province, Spain). *Journal of Geochemical Exploration* 80(1):69–79.

García-Sánchez, A., A. Moyano, P. Mayorga. 2005. High arsenic contents in groundwater of central Spain. *Environmental Geology* 47:847–854.

García-Vargas, G. G., L. M. Del Razo, M. E. Cebrián et al. 1994. Altered urinary porphyrin excretion in a human population chronically exposed to arsenic in Mexico. *Human & Experimental Toxicology* 13(2):839–847.

Gardner, R. M., M. Kippler, F. Tofail et al. 2013. Environmental exposure to metals and children's growth to age 5 years: A prospective cohort study. *American Journal of Epidemiology* 177(12):1356–67.

Garelick, H., H. Jones, A. Dybowska et al. 2008. Arsenic pollution sources. *Reviews of Environmental Contamination and Toxicology* 197:17–60.

Gault, A. G., H. A. L. Rowland, J. M. Charnock et al. 2008. Arsenic in hair and nails of individuals exposed to arsenic-rich groundwaters in Kandal Province, Cambodia. *Science of the Total Environment* 393:168–176.

Gay, J. R. and A. Korre. 2006. A spatially-evaluated methodology for assessing risk to a population from contaminated land. *Environmental Pollution* 142:227–234.

Gbadebo, A. M. and A. S. Mohammed. 2004. Arsenic pollution in aquifers located within the limestone areas of Ogunstate, South-Western Nigeria. *32nd International Geological Congress, Pre-Congress Workshop BWO 06: Natural Arsenic in Groundwater*, 18–19 August, Florence.

Ge, F., X. P. Lu, H. L. Zeng et al. 2009. Proteomic and functional analyses reveal a dual molecular mechanism underlying arsenic-induced apoptosis in human multiple myeloma cells. *Journal of Proteome Research* 8(6):3006–3019.

Geanellos, R. 1998. Hermeneutic philosophy. Part 1: Implications for its use as methodology in interpretive nursing research. *Nursing Inquiry* 5(3):154–163.

Geertz, C. 1973. *The Interpretation of Cultures*. New York: Basic Books.

Gelmann, E. R., E. Gurzau, A. Gurzau et al. 2013. A pilot study: The importance of inter-individual differences in inor-ganic arsenic metabolism for birth weight outcome. *Environmental Toxicology and Pharmacology* 36(3):1266–1275.

Gemici, Ü. and G. Tarcan. 2004. Hydrogeological and hydrogeochemical features of the Heybeli Spa, Afyon, Turkey: Arsenic and the other contaminants in the thermal waters. *Bulletin of Environmental Contamental and Toxicology* 72:1104–1114.

Gemici, Ü., G. Tarcan, C. Helvacı et al. 2008. High arsenic and boron concentrations in groundwaters related to mining activity in the Bigadiç borate deposits (Western Turkey). *Applied Geochemistry* 23:2462–2476.

Gentry, P. R., H. J. Clewell III, T. B. Greene et al. 2014. The impact of recent advances in research on arsenic cancer risk assessment. *Regulatory Toxicology and Pharmacology* 69(1):91–104.

Gentry, P. R., T. B. McDonald, D. E. Sullivan et al. 2010. Analysis of genomic dose-response information on arsenic to inform key events in a mode of action for carcinogenicity. *Environmental and Molecular Mutagenesis* 51(1):1–14.

Gerhardsson, L., D. Brune, G. F. Nordberg et al. 1988. Multielemental assay of tissues of deceased smelter workers and controls. *Science of the Total Environment* 74:97–110.

Gerlach, T. M., M. P. Doukas, K. A. McGee et al. 2001. Soil efflux and total emission rates of magmatic $CO2$ at the Horseshoe Lake tree kill, Mammoth Mountain, California, 1995–1999. *Chemical Geology* 177(1–2):101–16.

Bibliography

Gerrard, S. 2000. Environmental risk management. In *Environmental Science for Environmental Management*, ed. T. O'Riordan, 435–468. Harlow: Prentice Hall.

Gesler, W. 1986: The use of spatial analysis in medical geography: A review. *Social Science and Medicine* 23:963–973.

Ghariani, M., M. L. Adrien, M. Raucoules et al. 1991. Subacute arsenic poisoning. *Annales Françaises d'Anesthésie et de Réanimation* 10(3):304–307.

Ghose, R. 2001. Use of information technology for community empowerment: transforming geographic information systems into community information systems. *Transactions in GIS* 5(2):141–163.

Ghosh, A. 2013. Evaluation of chronic arsenic poisoning due to consumption of contaminated ground water in West Bengal, India. *International Journal of Preventive Medicine* 4(8):976–979.

Ghosh, P., M. Banerjee, S. De Chaudhuri et al. 2007. Comparison of health effects between individuals with and without skin lesions in the population exposed to arsenic through drinking water in West Bengal, India. *Journal of Exposure Science and Environmental Epidemiology* 17:215–223.

Ghosh, A. R. and K. Parial. 2014. Applications of remote sensing, geographic information system and geostatistics in the study of arsenic contamination in groundwater. In *Recent Trends in Modelling of Environmental Contaminants*, ed. D. Sengupta, 197–212. New Delhi: Springer.

Gibaud, S. and G. Jaouen. 2010. Arsenic-based drugs: From Fowler's solution to modern anticancer chemotherapy. *Topics in Organometallic Chemistry* 32:1–20.

Gibb, H. J. 1997. Epidemiology and cancer risk assessment. In *Fundamentals of Risk Analysis and Risk Management*, ed. V. Molak, 23–32. Boca Raton: CRC and Lewis Publishers.

Gibb, H., C. Haver, D. Gaylor et al. 2011. Utility of recent studies to assess the National Research Council 2001 estimates of cancer risk from ingested arsenic. *Environmental Health Perspectives* 119(3):284–290.

Gibbon, M. 2000. The Health Analysis and Action Cycle an Empowering Approach to Women's Health. *Sociological Research Online* 4(4). [http://www.socresonline.org.uk/4/4/gibbon.html].

Gilbert, S. G. 2004. *A Small Dose of Toxicology: The Health Effects of Common Chemicals*. London: CRC Press.

Gilbert-Diamond, D., J. A. Emond, E. R. Baker et al. 2016. Relation between in utero arsenic exposure and birth outcomes in a cohort of mothers and their newborns from New Hampshire. *Environmental Health Perspectives* 124:1299–1307.

Gilbert-Diamond, D., Z. Li, A. E. Perry et al. 2013. A population-based case-control study of urinary arsenic species and squamous cell carcinoma in New Hampshire, USA. *Environmental Health Perspectives* 121(10):1154–1160.

Giménez, M. C., R. M. Osicka, P. S. Blanes et al. 2014. Heavy metal concentration in arsenic contaminated groundwater of the Chaco Province, Argentina. In *One Century of the Discovery of Arsenicosis in Latin America (1914–2014)*, eds. M. I. Litter, H. B. Nicolli, M. Meichtry et al. 135–137. Boca Raton: CRC Press.

Giorgi, A. 1985. *Phenomenology and Psychological Research*. Pittsburgh: Duquesne University Press.

Giuliano, G., E. Preziosi, and R. Vivona. 2005. Valutazione della qualita dellae aquae sotterannee a scopi idroptabiliti: il caso del Lazio settentrionale. In *Presenza e diffusione dell'arsenico nel sottouolo e nelle risorse idriche italiane: nuovi strumenti di valutazione dinamiche di mobilizzazione*, ed. M. G. Scialoja, 97–106. Bologna: ARPA Emilia-Romagna.

Glaser, B. G. 1978. *Theoretical Sensitivity*. Mill Valley (CA): Sociology Press.

Glaser, B. G. and A. L. Strauss. 1967. *The Discovery of Grounded Theory: Strategies for Qualitative Research*. Chicago: Aldine.

Glesne, C. and A. Peshkin. 1992. *Becoming Qualitative Researchers: An Introduction*. New York: Longman.

Glik, D. C. 2007. Risk communication for public health emergencies. *Annual Review of Public Health* 28:33–54.

GOB. 1999. *National Water Policy 1999*. Ministry of Water Resources, Dhaka: Government of Bangladesh.

GOB. 2001. *National Water Management Plan 2001*. Ministry of Water Resources, Dhaka: Government of Bangladesh.

GOB. 2003. *Report of the Committee on Surface Water Development and Management for Drinking Water Supply in the Arsenic Affected Areas of Bangladesh*. Dhaka: Ministry of Local Government, Rural Development & Co-operatives, Government of Bangladesh.

GOB. 2004. *National Policy for Arsenic Mitigation 2004 and Implementation Plan for Arsenic Mitigation in Bangladesh*. Local Government Division; Ministry of Local Government, Rural Development & Cooperatives. Dhaka: Government of Bangladesh.

GOB. 2011. *Sector Development Plan (SDP) for Water Supply and Sanitation Sector in Bangladesh (FY 2011–25)*. Local Government Division; Ministry of Local Government, Rural Development & Cooperatives. Dhaka: Government of Bangladesh.

GOB. 2013. *Bangladesh Water Act 2013.* (Act No 14 of 2013). Article 3, Bangladesh Gazette, pp. 14277–14300, December 29, 2013, Dhaka: Government of Bangladesh.

GOB. 2014. *National Strategy for Water Supply and Sanitation.* (Draft Final). Policy Support Unit, Local Government Division; Ministry of Local Government, Rural Development & Cooperatives. Dhaka: Government of Bangladesh.

Gochfeld, M. 1995. Chemical agents. In *Environmental Medicine: Principles and Practice*, eds. S. M. Brooks, M. Gochfeld, J. Herzstein et al., 592–614. St. Louis: Mosby Publishing, pp.592–614.

Gogu, R. C. and A. Dassargues. 2000. Current trends and future challenges in groundwater vulnerability assessment using overlay and index methods. *Environmental Geology* 39(6):549–559.

Gogu, R. C., V. Hallet, A. Dassargues. 2003. Comparison of aquifer vulnerability assessment techniques. Application to the Néblon River basin (Belgium). *Environmental Geology* 44:881–892.

Goldsmith, J. R., M. Deane, J. Thom et al. 1972. Evaluation of health implications of elevated arsenic in well water. *Water Research* 6:1133–1136.

Golia, E. E., A. Dimirkou, St. A. Floras. 2015. Spatial monitoring of arsenic and heavy metals in the Almyros area, Central Greece. Statistical approach for assessing the sources of contamination. *Environmental Monitoring and Assessment*, 187 (7): 399.

Gómez, J. J., J. Lillo, B. Sahún. 2006. Naturally occurring arsenic in groundwater and identification of the geochemical sources in the Duero Cenozoic Basin, Spain. *Environmental Geology* 50(8):1151–1170.

Gómez-Arroyo, S., M. A. Armienta, J. Cortés-Eslava et al. 1997. Sister chromatid exchanges in Vicia faba induced by arsenic contaminated drinking water from Zimapan, Hidalgo, Mexico. *Mutation Research* 394:1–7.

Gomez-Rubio, P., J. Roberge, L. Arendell et al. 2011. Association between body mass index and arsenic methylation efficiency in adult women from southwest U.S. and northwest Mexico. *Toxicology and Applied Pharmacology* 252(2):176–182.

Gonçalves, J. A. C., J. C. de Lena, J. F. Paiva et al. 2007. Arsenic in the groundwater of Ouro Preto (Brazil): Its temporal behavior as influenced by the hydric regime and hydrogeology. *Environmental Geology* 53:785–793.

Gong, G., J. Basom, S. Mattevada et al. 2015. Association of hypothyroidism with low-level arsenic exposure in rural West Texas. *Environmental Research* 138:154–160.

Gong, Z., W. F. Chan, X. Wang et al. 2001. Determination of arsenic and antimony by microwave plasma atomic emission spectrometry coupled with hydride generation and a PTFE membrane separator. *Analytica Chimica Acta* 450(1–2):207–214.

Gong, G., K. A. Hargrave, V. Hobson et al. 2011. Low-level groundwater arsenic exposure impacts cognition: a project FRONTIER study. *Journal of Environmental Health* 74(2):16–22.

Gong, G., S. Mattevada, S. F. O'Bryant. 2014. Comparison of the accuracy of kriging and IDW interpolations in estimating groundwater arsenic concentrations in Texas. *Environmental Research* 130:59–69.

Gong, G. and S. E. O'Bryant. 2012. Low-level arsenic exposure, AS3MT gene polymorphism and cardiovascular diseases in rural Texas counties. *Environmental Research* 113:52–57.

Gonsebatt, M. E., L. Vega, A. M. Salazar et al. 1997. Cytogenetic effects in human exposure to arsenic. *Mutation Research* 386(3):219–228.

González-Horta, C., L. Ballinas-Casarrubias, B. Sánchez-Ramírez et al. 2015. A concurrent exposure to arsenic and fluoride from drinking water in Chihuahua, Mexico. *International Journal of Environmental Research and Public Health* 12(5):4587–4601.

Goodell, L. S., V. C. Stage, N. K. Cooke. 2016. Practical qualitative research strategies: Training interviewers and coders. *Journal of Nutrition Education and Behavior* 48(8):578–585.

Goodland, M. 2016. *Rainwater harvesting may no longer be outlawed in Colorado.* Colorado: The Colorado Independent. (http://www.coloradoindependent.com/158551/rainwater-harvesting-may-no-longer-be-outlawed-in-colorado).

Goodwin, T. A., L. Parker, B. E. Fisher et al. 2010. *Toenails, tap water and you: The arsenic connection.* Mineral Resources Branch report of activities 2009. Government of Nova Scotia, Canada. (No. ME 2010–1).

Goosen, M. F. A., E. N. Laboy-Nieves, E. Emmanuel. 2010. Environmental and human health risk management: An overview. In *Environmental and Human Health: Risk Management in Developing Countries*, eds. E. N. Laboy-Nieves, M. F. A. Goosen, E. Emmanuel, 3–11. Boca Raton: CRC Press.

Goovaerts, P. 1997. *Geostatistics for Natural Resources Evaluation.* Oxford: Oxford University Press.

Goovaerts, P. 1999. Geostatistics in soil science: State-of-the-art and perspectives. *Geoderma* 89:1–45.

Bibliography

Goovaerts, P., G. AvRuskin, J. Meliker et al. 2005. Geostatistical modeling of the spatial variability of arsenic in groundwater of southeast Michigan. *Water Resources Research* 41:1–19.

Gorby, M. S. 1988. Arsenic poisoning. *Western Journal of Medicine* 149:308–315.

Gorgij, A. D., O. Kisi, A. A. Moghaddam et al. 2017. Groundwater quality ranking for drinking purposes, using the entropy method and the spatial autocorrelation index. *Environmental Earth Sciences* 76:269 [doi:10.1007/s12665-017-6589-6].

Gough, D. 2007. Weight of evidence: a framework for the appraisal of the quality and relevance of evidence. *Education* 22(2):213–228.

Goyenechea, M. 1913. Sobre la nueva enfermedad descubierta en Bell Ville. *Revista medica de Rosario* 7:48.

Graham, J. D., L. C. Green, M. J. Roberts. 1988. *In Search of Safety: Chemicals and Cancer Risk*. Cambridge, MA: Harvard University Press.

Graham-Evans, B., H. H. Cohly, H. Yu et al. 2004. Arsenic-induced genotoxic and cytotoxic effects in human keratinocytes, melanocytes and dendritic cells. *International Journal of Environmental Research and Public Health* 1(2):83–89.

Gray, T. 2012. Arsenic. In *The Elements: A Visual Exploration of Every Known Atom in the Universe*, eds. T. Gray. and N. Mann, 86–87. New York: Black Dog & Leventhal Publishers, Inc.

Grégoire, D. C. and M. L. Ballinas. 1997. Direct determination of arsenic in fresh and saline waters by electrothermal vaporization inductively coupled plasma mass spectrometry. *Spectrochimica Acta - Part B* 52:75–82.

Gribble, M. O., B. V. Howard, J. G. Umans et al. 2012. Arsenic exposure, diabetes prevalence, and diabetes control in the Strong Heart Study. *American Journal of Epidemiology* 176(10):865–874.

Griffiths, J. K., P. Shand, J. Ingram. 2003. *Baseline Report Series 8: The Permo-Triassic Sandstones of Manchester and East Cheshire*. British Geological Survey Reprt (CR/03/265N).

Griffiths, J. K., P. Shand, J. Ingram. 2005. *Baseline Report Series 19: The Permo-Triassic Sandstones of Liverpool and Rufford*. British Geological Survey Reprt (CR/05/131N).

Griffiths, J. K., P. Shand, P. Marchant. 2006. *Baseline Report Series 23: The Lincolnshire Limestone*. British Geological Survey Report (CR/06/060N).

Grimmett, R. E. R. and I. G. McIntosh. 1939. Occurence of arsenic in soils and waters in the Waiotapu Valley, and its relation to stock health. *New Zealand Journal of Science and Technology* 21:138–150.

Grinspan, D. and R. Biagini. 1985. The manifestation of arsenic poisoning caused by drinking water. *Medicina cutanea Ibero-Latino-Americana* 13:85–109.

Grondin, J. 1994. *Introduction to Philosophical Hermeneutics*. New Haven, CT: Yale University Press.

Grondin, J. 1995. *Sources of Hermeneutics*. Albany: State University of New York Press.

Gross, E. L. and D. J. Low. 2013. *Arsenic Concentrations, Related Environmental Factors, and the Predicted Probability of Elevated Arsenic in Groundwater in Pennsylvania*. Virginia: US Geological Survey.

Grossier, P. and M. Ledrans. 1999. Arsenic in drinking water: A primary approach to assess exposure of the French population. *Genie Urbain-Genie Rural* 2:27–32.

Grund, S. C., K. Hanusch, H. W. Wolf. 2005. Arsenic and arsenic compounds. In *Ullmann's Encyclopedia of Industrial Chemistry*, ed. B. Elvers, 199–240. Vol 4. Weinheim: Wiley-VCH.

Guan, H., F. Piao, X. Zhang et al. 2012. Prenatal exposure to arsenic and its effects on fetal development in the general population of Dalian. *Biological Trace Element Research* 149(1):10–15.

Gude, J. C. J., L. C. Rietveld, D. van Halem. 2018. Biological As(III) oxidation in rapid sand filters. *Journal of Water Process Engineering* 21:107–115.

Gullick, J., M. Krivograd, S. Taggart et al. 2017. A phenomenological construct of caring among spouses following acute coronary syndrome. *Medicine, Health Care and Philosophy* 20(3):393–404.

Gunderson, E. L. 1995. FDA total diet study, July 1986-April 1991, dietary intakes of pesticides, selected elements and other chemicals. *Journal of AOAC International* 78:1353–1363.

Gundert-Remy, U., G. Damm, H. Foth et al. 2015. High exposure to inorganic arsenic by food: the need for risk reduction. *Archives of Toxicology* 89(12): 2219–2227.

Gunduz, O., C. Bakar, C. Simsek et al. 2015. Statistical analysis of causes of death (2005–2010) in villages of Simav Plain, Turkey, with high arsenic levels in drinking water supplies. *Archives of Environmental and Occupational Health* 70(1):35–46.

Gunduz, O., C. Simsek, A. Hasozbek. 2010. Arsenic pollution in the groundwater of Simav Plain, Turkey: Its impact on water quality and human health. *Water, Air, & Soil Pollution* 205(1–4):43–62.

Guo, H., Y. Zhang, L. Xing et al. 2012. Spatial variation in arsenic and fluoride concentrations of shallow groundwater from the town of Shahai in the Hetao basin, Inner Mongolia. *Applied Geochemistry* 27:2187–2196.

Guo, H. M., S. Z. Yang, X. H. Tang et al. 2008. Groundwater geochemistry and its implications for arsenic mobilization in shallow aquifers of the Hetao Basin, Inner Mongolia. *Science of the Total Environment* 393(1):131–144.

Guo, H. M., Y. Zhang, Y. F. Jia et al. 2013. Spatial and temporal evolutions of groundwater arsenic approximately along the flow path in the Hetao basin, Inner Mongolia. *Chinese Science Bulletin* 58(25):3070–3079.

Guo, H. R. 2004. Arsenic level in drinking water and mortality of lung cancer (Taiwan). *Cancer Causes and Control* 15(2):171–177.

Guo, H. R., H. Chiang, H. Hu et al. 1997. Arsenic in drinking water and incidence of urinary cancers. *Epidemiology* 8(5):545–550.

Guo, H. R. and Y. C. Tseng. 2000. Arsenic in drinking water and bladder cancer: comparison between studies based on cancer registry and death certificates. *Environmental Geochemistry and Health* 22:83–91.

Guo, J. X., L. Hu, P. Z. Yand et al. 2007b. Chronic arsenic poisoning in drinking water in Inner Mongolia and its associated health effects. *Journal of Environmental Science and Health: Part A* 42:1853–1858.

Guo, Q., H. Guo, Y. Yang et al. 2014. Hydrogeochemical contrasts between low and high arsenic groundwater and its implications for arsenic mobilization in shallow aquifers of the northern Yinchuan Basin, P.R. China. *Journal of Hydrology* 518:464–476.

Guo, Q., Y. Wang, X. Gao, T. Ma. 2007a. A new model (DRARCH) for assessing groundwater vulnerability to arsenic contamination at basin scale: A case study in Taiyuan basin, northern China. *Environmental Geology* 52: 923–932.

Guo, X., Y. Fujino, J. Chai et al. 2003. The prevalence of subjective symptoms after exposure to arsenic in drinking water in Inner Mongolia, China. *Journal of Epidemiology* 13(4):211–215.

Guo, X., Y. Fujino, S. Kaneko et al. 2001. Arsenic contamination of groundwater and prevalence of arsenical dermatosis in the Hetao plain area, Inner Mongolia, China. *Molecular and Cellular Biochemistry* 222(1–2):137–140.

Gurung, J. K., H. Ishiga, M. S. Khadka et al. 2007. The geochemical study of fluvio-lacustrine aquifers in the Kathmandu Basin (Nepal) and the implications for the mobilization of arsenic. *Environmental Geology* 52:503–517.

Gurzau, E. S. and A. E. Gurzau. 2001. Arsenic in drinking water from groundwater in Transylvania, Romania. In *Arsenic Exposure and Health Effects IV*, eds. W. R. Chapell, C. O. Abernathy, R. L. Calderon, 181–184. Amsterdam: Elsevier.

Gusenius, E. M. 1967. Beginnings of greatness in Swedish chemistry: Georg Brandt (1694–1768). *Transactions of the Kansas Academy of Science* 70(4):413–425.

Gutiérrez-Pizano, A., R. E. Rodríguez, G. J. Romero et al. 1996. Eliminación del arsénico en agua potable de pozos. *Actas INAGEQ* 2:319–322.

Gyorgyey, F. 1987. Arsenic and no lace. *Caduceus* 3:40–65.

Habuda-Stanić, M., M. Kuleš, B. Kalajdžić et al. 2007. Quality of groundwater in eastern Croatia - the problem of arsenic pollution. *Desalination* 210(1–3):157–162.

Hadi, A. 2003. Fighting arsenic at the grassroots: Experience of BRAC's community awareness initiative in Bangladesh. *Health Policy Plann* 18:83–100.

Hafeman, D. M., H. Ahsan, E. D. Louis et al. 2005. Association between arsenic exposure and a measure of subclinical sensory neuropathy in Bangladesh. *Journal of Occupational and Environmental Medicine* 47(8):778–784.

Hagiwara, K., T. Inui, Y. Koike et al. 2013. Determination of diphenylarsinic acid, phenylarsonic acid and inorganic arsenic in drinking water by graphite-furnace atomic-absorption spectrometry after simultaneous separation and preconcentration with solid-phase extraction disks. *Analytical Sciences* 29(12):1153–58.

Hagiwara, K., T. Inui, Y. Koike et al. 2015. Speciation of inorganic arsenic in drinking water by wavelength-dispersive X-ray fluorescence spectrometry after *in situ* preconcentration with miniature solid-phase extraction disks. *Talanta* 134:739–744.

Haining, R. P. 1996. Designing a health needs GIS with spatial analysis capability. In *Spatial Analytical Perspectives on GIS, (GISDATA 4)*, eds. M. Fischer, H. J. Scholten and D. Unwin, 60–74. London: Taylor & Francis.

Hakala, E. and A. Hallikainen. 2004. Exposure of the Finnish population to arsenic, effects and health risks. In *Arsenic in Finland: Distribution, Environmental Impacts and Risks*, eds. K. Loukola-Ruskeeniemi and P. Lahermo, 153–166. Geological Survey of Finland (in Finnish).

Halatek, T., H. Sinczuk-Walczak, S. Rabieh et al. 2009. Association between occupational exposure to arsenic and neurological, respiratory and renal effects. *Toxicology and Applied Pharmacology* 239(2):193–199.

Bibliography

Haldimann M., E. Pfammatter, P. Venetz et al. 2005. Occurrence of arsenic in drinking water of the canton Valais. *Mitteilungen aus dem Gebiete der Lebensmittel-untersuchung un Hygiene* 96:89–105.

Halim, M. A., R. K. Majumder, S. A. Nessa et al. 2010. Arsenic in shallow aquifer in the eastern region of Bangladesh: Insights from principal component analysis of groundwater compositions. *Environmental Monitoring and Assessment* 161:453–472.

Hall, A. H. 2002. Chronic arsenic poisoning. *Toxicology Letters* 128(1–3):69–72.

Hall, E. M., J. Acevedo, F. G. López et al. 2017. Hypertension among adults exposed to drinking water arsenic in Northern Chile. *Environmental Research* 153:99–105.

Hall, M. N. and M. V. Gamble. 2012. Nutritional manipulation of one-carbon metabolism: Effects on arsenic methylation and toxicity. *Journal of Toxicology* 2012:1–11. [http://dx.doi.org/10.1155/2012/595307].

Halperin, A. 2003. *Arsenic found in rural Mekong river wells*. The Cambodian Daily, Phnom Penh, Cambodia, Vol. 11, June 25.

Halsey, P. M. 2000. *Arsenic Contamination Study of Drinking Water in Nepal*. Unpublished Thesis, Master of Engineering, MIT: USA, May.

Hamadani, J. D., F. Tofail, B. Nermell et al. 2011. Critical windows of exposure for arsenic-associated impairment of cognitive function in pre-school girls and boys: A population-based cohort study. *International Journal of Epidemiology* 40(6):1593–1604.

Hamedani, M. L., I. R. Plimer, C. Xu. 2012. Orebody modelling for exploration: The western mineralisation, Broken Hill, NSW. *Natural Resources Research* 21(3):325–345.

Hammersley, M. 1992. *What's Wrong with Ethnography? Methodological Explorations*. London: Routledge.

Han, S., F. Zhang, H. Zhang et al. 2013. Spatial and temporal patterns of groundwater arsenic in shallow and deep groundwater of Yinchuan Plain, China. *Journal of Geochemical Exploration* 135:71–78.

Hanchett, S. 2004. *Social Aspects of the Arsenic Contamination of Drinking Water: A Review of Knowledge and Practice in Bangladesh and West Bengal*. Dhaka: Local Government Division, Bangladesh.

Hanchett, S., Q. Nahar, A. van Agthoven et al. 2000. *Arsenic Awareness in Six Bangladesh Towns*. Dhaka: The Royal Netherlands Embassy.

Haq, N. 2000. Many NGOs allegedly using arsenic patients. *The Daily Star*, July 13, Dhaka, Bangladesh: The Daily Star Centre. [http://www.eng-consult.com/arsenic/as170.txt].

Haq, N. 2001. Massive arsenic mitigation activities in Comilla. *The Daily Star*, September 17, Dhaka, Bangladesh: The Daily Star Centre. [http://www.eng-consult.com/arsenic/as219.txt].

Haque, R., D. N. G. Mazumder, S. Samanta et al. 2003. Arsenic in drinking water and skin lesions: Dose-response data from West Bengal, India. *Epidemiology* 14(2):174–182.

Harding, B. 1983. What is the status of arsenic as a human carcinogen. In *Arsenic: Industrial, Biomedical Environmental Perspectives*, eds. W. H. Lederer and R. J. Fensterheim, 203–209. London: Van Nostrand Reinhold.

Harrington, J. M., J. P. Middaugh, D. L. Morse et al. 1978. A survey of a population exposed to high concentrations of arsenic in well water in Fairbanks, Alaska. *American Journal of Epidemiology* 108(5):377–385.

Harris, T. and D. Weiner. 1998. Empowerment: Marginalization and community-integrated GIS. *Cartography and Geographic Information Systems* 25(2):67–76.

Harrison, P. T. C. and P. Holmes. 2006. Assessing risks to human health. *Issues in Environmental Science and Technology* 22:65–83.

Hartwig, A. 2013. Metal interaction with redox regulation: An integrating concept in metal carcinogenesis? *Free Radical Biology and Medicine* 55:63–72.

Hartwig, A., H. Blessing, T. Schwerdtle et al. 2003. Modulation of DNA repair process by arsenic and selenium compounds. *Toxicology* 193(1–2):161–169.

Harvey, C. F., K. N. Ashfaque, W. Yu et al. 2006. Groundwater dynamics and arsenic contamination in Bangladesh. *Chemical Geology* 228:112–136.

Harvey, F. and N. R. Chrisman. 1998. Boundary objects and the social construction of GIS technology. *Environment and Planning A* 30(9):1683–94.

Harvey, C. F., C. H. Swartz, A. B. M. Badruzzman et al. 2002. Arsenic mobility and groundwater extraction in Bangladesh. *Science* 298:1602–1606.

Hassan, M. M. 1997. Mapping the spatial distribution of irrigation demand areas during the dry season in Bangladesh: A GIS application. *Oriental Geographer* 41(1):48–63.

Hassan, M. M. 2000. Rational and approaches of the arsenic research in Bangladesh. *Journal of the Bangladesh National Geographical Association* 27 and 28 (1 and 2): 45–59.

Hassan, M. M. 2003. *Arsenic Toxicity in Bangladesh: Health and Social Hazards*. Unpublished PhD Thesis, Durham: Durham University, United Kingdom.

Hassan, M. M. 2005. Arsenic poisoning in Bangladesh: Spatial mitigation planning with GIS and public participation. *Health Policy* 74(3):247–260.

Hassan, M. M. 2015. *Scanning and Mapping the WASH Situation in Coastal Bangladesh: Problems and Potential*. Geo-Ecological Research Team (GeRT): Dhaka. Funded by the ICCO Cooperation, the Netherlands.

Hassan, M. M. and P. J. Atkins. 2006. Arsenic in Bangladesh. *Geography Review* 19(4):14–17.

Hassan, M. M. and P. J. Atkins. 2007. Arsenic risk mapping in Bangladesh: A simulation technique of cokriging estimation from regional count data. *Journal of Environmental Science and Health, Part A* 42:1719–28.

Hassan, M. M. and P. J. Atkins. 2011. Application of geostatistics with Indicator Kriging for analyzing spatial variability of groundwater arsenic concentrations in southwest Bangladesh. *Journal of Environmental Science and Health, Part A* 46(11):1185–1196.

Hassan, M. M. and R. Ahamed. 2017. Arsenic-safe aquifers in Coastal Bangladesh: An investigation with Ordinary Kriging Estimation. *The International Archives for Photogrammetry, Remote Sensing and Spatial Information Sciences* XLII-4/W5:97–105.

Hassan, M. M., P. J. Atkins, C. E. Dunn. 2003. The spatial pattern of risk from arsenic poisoning: A Bangladesh case study. *Journal of Environmental Science and Health, Part A* A38(1):1–24.

Hassan, M. M., P. J. Atkins, C. E. Dunn. 2004. Suitable arsenic mitigation options in Bangladesh: Voices of local people. *Indian Journal of Landscape Systems and Ecological Studies* 24(2):1–7.

Hassan, M. M., P. J. Atkins, C. E. Dunn. 2005. Social implications of arsenic poisoning in Bangladesh. *Social Science & Medicine* 61(10):2201–2211.

Hassan, M. M., P. J. Atkins, C. E. Dunn. 2006. Pattern of groundwater arsenic concentrations in different aquifers. *Oriental Geographers* 50(2):1–18.

Hathaway, G. J., N. H. Proctor, J. P. Hughes et al. 1991. Arsenic and arsine. In *Chemical Hazards of the Workplace*, eds. N. H. Proctor and J. P. Hughes, 92–96. New York: Van Nostrand Reinhold Co.

Haupert, T. A., J. H. Wiersma, J. M. Goldring. 1996. Health effects of ingesting arsenic-contaminated groundwater. *Wisconsin Medical Journal* 95(2):100–104.

Hawkesworth, S., Y. Wagatsuma, M. Kippler et al. 2013. Early exposure to toxic metals has a limited effect on blood pressure or kidney function in later childhood, rural Bangladesh. *International Journal of Epidemiology* 42(1):176–85.

Hay, I. 2000. *Qualitative Research Methods in Human Geography*. South Melbourne: Oxford University Press.

He, J., M. A. Teng, D. E. N. G. Yamin et al. 2009. Environmental geochemistry of high arsenic groundwater at western Hetao plain, Inner Mongolia. *Frontiers of Earth Science in China* 3(1):63–72.

He, X., G. Guoquan, Z. Hui et al. 1997. Fluorometric determination of arsenic(III) with fluorescein. *Microchemical Journal* 56:327–331.

Heaney, C. D., B. Kmush, A. Navas-Acien et al. 2015. Arsenic exposure and hepatitis E virus infection during pregnancy. *Environmental Research* 142:273–280.

Heck, J. E., Y. Chen, V. R. Grann et al. 2008. Arsenic exposure and anemia in Bangladesh: A population-based study. *Journal of Occupational and Environmental Medicine* 50(1):80–87.

Heck, J. E., J. W. Nieves, Y. Chen et al. 2009. Dietary intake of methionine, cysteine, and protein and urinary arsenic excretion in Bangladesh. *Environmental Health Perspectives* 117(1):99–104.

Heerboth, S., K. Lapinska, N. Snyder et al. 2014. Use of epigenetic drugs in disease: An overview. *Genetics and Epigenetics* 6:9–19.

Hei, T. K. and M. Filipic. 2004. Role of oxidative damage in the genotoxicity of arsenic. *Free Radical Biology and Medicine* 37:574–581.

Hei, T. K., S. X. Liu, C. Waldren. 1998. Mutagenicity of arsenic in mammalian cells: Role of reactive oxygen species. *Proceedings of the National Academy of Sciences of the United States of America* 95:8103–8107.

Heidegger, M. 1927. *Being and Time (Sein und Zeit)*. New York: Harper & Row.

Heidegger, M. 1993. Being and Time: Introduction. In *Basic Writings: Ten Key Essays, plus the Introduction to Being and Time*, ed. D. F. Krell, 37–88. San Francisco: Harper Collins.

Heinemeyer, G. 2008. Concepts of exposure analysis for consumer risk assessment. *Experimental and Toxicologic Pathology* 60:207–212.

Heinrichs, G. and P. Udluft. 1999. Natural arsenic in Triassic rocks: A source of drinking water contamination in Bavaria, Germany. *Hydrogeology Journal* 7:468–476.

Hengl, T. 2009. *A Practical Guide to Geostatistical Mapping*. Amsterdam: University of Amsterdam.

Henning, F. A. and D. E. Konasewich. 1984. *Characterization and assessment of wood preservation facilities in British Columbia*. West Vancouver, BC, Canada Environmental Protection Services, Pacific region, Environment Canada.

Hepburn, A. and J. Potter. 2007. Discourse analytic practice. In *Qualitative Research Practice*, eds. C. Seale, G. Gobo, J. F. Gubrium, and D. Silverman, 168–184. London: Sage.

Herath, I., M. Vithanage, J. Bundschuh et al. 2016. Natural arsenic in global groundwaters: Distribution and geochemical triggers for mobilization. *Current Pollution Reports* 2(1):68–89.

Hering, J. G., P. Y. Chen, J. A. Wilkie et al. 1996. Arsenic removal by ferric chloride. *Journal of American Water Works Associations* 88(4):155–167.

Hernández-García, M. E. and E. Custodio. 2004. Natural baseline quality of Madrid Tertiary Detrital Aquifer Groundwater (Spain): A basis for aquifer management. *Environmental Geology* 46(2):173–188.

Hernández-Zavala, A., L. M. Del Razo, C. Aguilar et al. 1998. Alteration in bilirubin excretion in individuals chronically exposed to arsenic in Mexico. *Toxicology Letters* 99(2):79–84.

Herrera, V., A. S. Amaro, C. Carrasco. 2014. Speciation of arsenic in a saline aquatic ecosystem in northern Chile. In *One Century of the Discovery of Arsenicosis in Latin America (1914–2014)*, eds. M. I. Litter, H. B. Nicolli, M. Meichtry et al. 58–59. Boca Raton: CRC Press.

Hershey, J. W., T. S. Oostdyk, P. N. Keliher. 1988. Determination of arsenic and selenium in environmental and agricultural samples by hydride generation atomic adsorption spectrometry. *Journal of the Association of Official Analytical Chemists* 71(6):1090–1093.

Higgs, J. 2001. Charting standpoints in qualitative research. In *Critical Moments in Qualitative Research*, eds. H. Byrne-Armstrong, J. Higgs, D. Horsfall, 44–67. Oxford (UK): Butterworth-Heinemann.

Hill, J., H. Faisal, A. C. Bagtzoglou. 2009. Zonal management of arsenic contaminated ground water in northwestern Bangladesh. *Journal of Environmental Management* 90:3721–3729.

Hindmarsh, J. T. and P. F. Corso. 2008. *The Death of Napoleon: The Last Campaign*. Philadelphia: Xlibris.

Hindmarsh, J. T., O. R. McLetchie, L. P. M. Heffernan et al. 1977. Electromyographic abnormalities in chronic environmental arsenicalism. *Journal of Analytical Toxicology* 1(6):270–276.

Hinwood, A. L., D. J. Jolley, M. R. Sim. 1999. Cancer incidence and high environmental arsenic concentrations in rural populations: Results of an ecological study. *International Journal of Environmental Health Research* 9(2):131–141.

Hiscock, K. M., A. A. Lovett, J. P. Parfitt. 1995. Groundwater vulnerability assessment: Two case studies using GIS methodology. *The Quarterly Journal of Engineering Geology* 28:179–194.

Hoekstra, A. Y. 2016. A critique on the water-scarcity weighted water footprint in LCA. *Ecological Indicators* 66: 564–573.

Holroyd, A. E. M. 2007. Interpretive hermeneutic phenomenology: Clarifying understanding. *The Indo-Pacific Journal of Phenomenology* 7(2):1–12.

Holsapple, M. P. and K. B. Wallace. 2008. Dose response considerations in risk assessment - An overview of recent ILSI activities. *Toxicology Letters* 180:85–92.

Holton, G. A. 2004. Defining risk. *Financial Analysts Journal* 60:19–25.

Honderich, T. 2005. *The Oxford Companion to Philosophy*. Oxford: Oxford University Press.

Hong, F., T. Y. Jin, A. H. Zhang. 2004. Risk assessment of renal dysfunction caused by co-exposure to arsenic, cadmium using benchmark dose calculation in a Chinese population. *Biometals* 17(5):573–580.

Hong, Q. N., P. Pluye, M. Bujold et al. 2017. Convergent and sequential synthesis designs: Implications for conducting and reporting systematic reviews of qualitative and quantitative evidence. *Systematic Reviews* 6:61.

Hopenhayn-Rich, C., M. L. Biggs, A. H. Smith. 1998. Lung and kidney cancer mortality associated with arsenic in drinking water in Córdoba, Argentina. *International Journal of Epidemiology* 27(4):561–569.

Hopenhayn-Rich, C., S. R. Browning, I. Hertz-Picciotto et al. 2000. Chronic arsenic exposure and risk of infant mortality in two areas of Chile. *Environmental Health Perspective* 108(7):667–673.

Hopenhayn-Rich, C., M. L. Biggs, A. Fuchs et al. 1996. Bladder cancer mortality associated with arsenic in drinking water in Argentina. *Epidemiology* 7:117–1124.

Hopenhayn-Rich, C., C. Ferreccio, S. R. Browning et al. 2003. Arsenic exposure from drinking water and birth weight. *Epidemiology* 14(5):593–602.

Hossain, M. 1991. *Agriculture in Bangladesh: Performance, Problem and Prospects*. Dhaka: University Press Limited.

Hossain, M. M. and J. Inauen. 2014. Differences in stakeholders' and end users' preferences of arsenic mitigation options in Bangladesh. *Journal of Public Health* 22(4):335–350.

Hossain, M. M. and M. Piantanakulchai. 2013. Groundwater arsenic contamination risk prediction using GIS and classification tree method. *Engineering Geology* 156:37–45.

Hossain, F. and B. Sivakumar. 2005. Spatial pattern of arsenic contamination in shallow wells of Bangladesh: Regional geology and nonlinear dynamics. *Stochastic Environmental Research and Risk Assessment* 20:66–76.

Howard, G. 2010. *Social Aspects of Access to Healthcare for Arsenicosis Patients*. Dhaka: Arsenic Policy Support Unit.

HRW. 2016. *Nepotism and Neglect: The Failing Response to Arsenic in the Drinking Water of Bangladesh's Rural Poor*. New York: Human Rights Watch.

Hsieh, Y. C., L. M. Lien, W. T. Chung et al. 2011. Significantly increased risk of carotid atherosclerosis with arsenic exposure and polymorphisms in arsenic metabolism genes. *Environmental Research* 111(6):804–810.

Hsu, K. H., J. R. Froines, C. J. Chen. 1997. Studies of arsenic ingestion from drinking water in northeastern Taiwan: chemical speciation and urinary metabolites. In *Arsenic Exposure and Health Effects*, eds. C. O. Abernathy, R. L. Calderon, W. R. Chappell, 190–209. London: Chapman Hall.

Hsu, L. I., Y. H. Wang, H. Y. Chiou et al. 2013. The association of diabetes mellitus with subsequent internal cancers in the arsenic-exposed area of Taiwan. *Journal of Asian Earth Sciences* 73:452–459.

Hsu, L. I., M. M. Wu, Y. H. Wang et al. 2015. Association of environmental arsenic exposure, genetic polymorphisms of susceptible genes, and skin cancers in Taiwan. *BioMed Research International* 2015: 892579 (http://dx.doi.org/10.1155/2015/892579).

Hsueh, Y. M., G. Cheng, M. Wu et al. 1995. Multiple risk factors associated with arsenic-induced skin cancer: Effects of chronic liver disease and malnutritional status. *British Journal of Cancer* 71(1):109–114.

Hsueh, Y. M., C. J. Chung, H. S. Shiue et al. 2009. Urinary arsenic species and CKD in a Taiwanese population: a case control study. *American Journal of Kidney Diseases* 54(5):859–870.

Hsueh, Y. M., H. Y. Chiou, Y. L. Huang et al. 1997. Serum beta-carotene level, arsenic methylation capability, and incidence of skin cancer. *Cancer Epidemiology, Biomarkers and Prevention* 6(8):589–596.

Hu, K., Y. Huang, H. Li et al. 2005. Spatial variability of shallow groundwater level, electrical conductivity and nitrate concentration, and risk assessment of nitrate contamination in north China plain. *Environmental International* 31:896–903.

Hu, K., B. Li, Y. Lu et al. 2004. Comparison of various spatial interpolation methods for non-stationary regional soil mercury content. *Environmental Science* 25(3):132–137.

Huang, C. F., Y. W. Chen, C. Y. Yang et al. 2011. Arsenic and diabetes: Current perspectives. *Kaohsiung Journal of Medical Sciences* 27:402–410.

Huang, M., S. J. Choi, D. W. Kim et al. 2009. Risk assessment of low-level cadmium and arsenic on the kidney. *Journal of Toxicology and Environmental Health: Part A*, 72(21–22):1493–1498.

Huang, Y. K., Y. L. Huang, Y. M. Hsueh et al. 2008. Arsenic exposure, urinary arsenic speciation, and the incidence of urothelial carcinoma: A twelve-year follow-up study. *Cancer Causes and Control* 19(8):829–839.

Huang, C., W. Y. Ma, J. Li et al. 1999. Arsenic induces apoptosis through a c-Jun NH2-terminal kinase-dependent, p53-independent pathway. *Cancer Research* 59(13):3053–3058.

Huang, Y. Z., X. C. Qian, G. Q. Wang et al. 1985. Endemic chronic arsenism in Xinjiang. *Chinease Medical Journal* 98:219–222.

Huang, C. Y., C. T. Su, C. J. Chung et al. 2012. Urinary total arsenic and 8-hydroxydeoxyguanosine are associated with renal cell carcinoma in an area without obvious arsenic exposure. *Toxicology and Applied Pharmacology* 262:349–354.

Huang, Y. K., C. H. Tseng, Y. L. Huang et al. 2007. Arsenic methylation capability and hypertension risk in subjects living in arseniasis hyperendemic areas in southwestern Taiwan. *Toxicology and Applied Pharmacology* 218(2):135–142.

Hubaux, R., D. D. Becker-Santos, K. S. Enfield et al. 2013. Molecular features in arsenic-induced lung tumors. *Molecular Cancer*, 12:20 (doi: 10.1186/1476-4598-12-20).

Huber, P. 1991. *Galileo's Revenge: Junk Science in the Courtroom*. New York: Basic Books.

Huda, N., S. Hossain, M. Rahman et al. 2014. Elevated levels of plasma uric acid and its relation to hypertension in arsenic-endemic human individuals in Bangladesh. *Toxicology and Applied Pharmacology* 281:11–18.

Hudler, P. 2012. Genetic aspects of gastric cancer instability. *Scientific World Journal* 2012:761909 (doi:10.1100/2012/761909).

Hug S. J., O. X. Leupin, M. Berg. 2008. Bangladesh and Vietnam: Different groundwater compositions require different approaches to arsenic mitigation. *Environmental Science and Technology* 42:6318–6323.

Hughes, M. F. 2002. Arsenic toxicity and potential mechanisms of action. *Toxicology Letters* 133(1):1–16.

Hughes, M. F. 2016. History of arsenic as a poison and a medicinal agent. In *Arsenic: Exposure Sources, Health Risks, and Mechanisms of Toxicity*, ed. J. C. States, 3–22. New Jersey: John Wiley & Sons.

Hughes, M. F., B. D. Beck, Y. Chen et al. 2011. Arsenic exposure and toxicology: A historical perspective. *Toxicological Sciences* 123(2):305–332.

Hunt, K. M., R. K. Srivastava, C. A. Elmets et al. 2014. The mechanistic basis of arsenicosis: Pathogenesis of skin cancer. *Cancer Letters* 354(2):211–219.

Huntsman-Mapila, P., T. Mapila, M. Letshwenyo, P. Wolski, C. Hemond. 2006. Characterization of arsenic occurrence in the water and sediments of the Okavango Delta, NW Botswana. *Applied Geochemistry* 21:1376–1391.

Huq, S. M., J. C. Joardar, S. Parvin et al. 2006. Arsenic contamination in food-chain: Transfer of arsenic into food materials through groundwater irrigation. *Journal of Health, Population, and Nutrition* 24(3):305–316.

Hurtado-Jiménez, R. and J. L. Gardea-Torresdey. 2006. Arsenic in drinking water in the Los Altos de Jalisco region of Mexico. *Pan American Journal of Public Health* 20(4):236–247.

Husserl, E. 1962. *Ideas: General Introduction to Pure Phenomenology*. New York: Collier.

Hutchinson, J. 1887. Arsenic cancer. *The British Medical Journal*, 2: 1280–1281.

Huyck, K. L., M. L. Kile, G. Mahiuddin et al. 2007. Maternal arsenic exposure associated with low birth weight in Bangladesh. *Journal of Occupational and Environmental Medicine* 49(10):1097–1104.

Hyson Jr, J. M. 2007. A history of arsenic in dentistry. *Journal of California Dental Association* 35:135–139.

IAEA. 1989. *Evaluating the Reliability of Predictions Made Using Environmental Transport Models*. Safety Series No. 100. Vienna: International Atomic Energy Agency.

IARC. 1980. *Some Metals and Metallic Compounds. IARC Monographs on the Evaluation of Carcinogenic Risks to Humans*. Lyon, France: International Agency for Research on Cancer.

IARC. 1990. *Arsenic and Arsenic Compounds In IARC Monographs on the Evaluation of the Carcinogenic Risk of Chemicals to Humans*. Geneva: World Health Organization.

IARC. 2004. *IARC Monographs on the Evaluation of Carcinogenic Risks to Humans*. Volume 84 (Some Drinking-Water Disinfectants and Contaminants, including Arsenic). Lyon, France: International Agency for Research on Cancer.

IARC. 2012. *A Review of Human Carcinogens: Arsenic, Metals, Fibres, and Dusts*. Lyon, France: International Agency for Research on Cancer.

IGME. 2015. *Levels and Trends in Child Mortality*. United Nations Inter-Agency Group for Child Mortality Estimation. New York: UNICEF.

Ilowite, J., P. Spiegler, S. Chawla. 2008. Bronchiectasis: New findings in the pathogenesis and treatment of this disease. *Current Opinion in Infectious Diseases*, 21(2):163–167.

Im, J. and J. Park. 2013. Stochastic structural optimization using particle swarm optimization, surrogate models and Bayesian statistics. *Chinese Journal of Aeronautics* 26(1):112–121.

Inauen, J., M. M. Hossain, R. B. Johnston et al., 2013. Acceptance and use of eight arsenic-safe drinking water options in Bangladesh. *PLoS ONE* 8(1):e53640. [doi: 10.1371/journal.pone.0053640].

Intamo, P., A. Suddhiprakarn, I. Kheoruenromne et al. 2016. Metals and arsenic concentrations of Ultisols adjacent to mine sites on limestone in Western Thailand. *Geoderma Regional*, 7(3):300–310.

IPCS. 2004. IPCS glossary of key exposure assessment terminology. In *IPCS Risk Assessment Terminology*. Geneva: World Health Organization.

Isaaks, E. H. and R. M. Srivastava. 1989. *An Introduction to Applied Geostatistics*. New York: Oxford University Press.

Islam, F. S., A. G. Gault, C. Boothman et al. 2004a. Role of metal-reducing bacteria in arsenic release from Bengal delta sediments. *Nature* 430(6995):68–71.

Islam, K., A. Haque, R. Karim et al. 2011. Dose-response relationship between arsenic exposure and the serum enzymes for liver function tests in the individuals exposed to arsenic: A cross sectional study in Bangladesh. *Environmental Health* 10:64 [doi:10.1186/1476-069X-10-64].

Islam, M. R., I. Khan, J. Attia et al. 2012a. Association between hypertension and chronic arsenic exposure in drinking water: A cross-sectional study in Bangladesh. *International Journal of Environmental Research and Public Health* 9(12):4522–4536.

Islam, M. R., I. P. Khan, S. M. N. Hassan et al. 2012b. Association between type 2 diabetes and chronic arsenic exposure in drinking water: A cross sectional study in Bangladesh. *Environmental Health* 11(1):38, [doi:10.1186/1476-069X-11-38].

Islam, L. N., A. H. M. N. Nabi, M. M. Rahman et al. 2004b. Association of clinical complications with nutritional status and the prevalence of leukopenia among arsenic patients in Bangladesh. *International Journal of Environmental Research and Public Health* 1(2):74–82.

Islam, L. N., A. H. M. N. Nabi, M. M. Rahman et al. 2007. Association of respiratory complications and elevated serum immunoglobulins with drinking water arsenic toxicity in human. *Journal of Environmental Science and Health: Part A* 42(12):1807–1814.

Islam, S., M. M. Rahman, M. R. Islam et al. 2016. Arsenic accumulation in rice: Consequences of rice genotypes and management practices to reduce human health risk. *Environment International* 96:139–155.

Islam, A. R. M. T., S. Shen, M. Bodrud-Doza et al. 2017. Assessment of trace elements of groundwater and their spatial distribution in Rangpur district, Bangladesh. *Arab Journal of Geosciences*, 10: 95, [doi:10.1007/s12517-017-2886-3].

Jaafar, R., I. Omar, A. J. Jidon et al. 1993. Skin cancer caused by chronic arsenical poisoning—a report of three cases. *Medical Journal of Malaysia*, 48(1):86–92.

Jabłońska-Czapla, M., S. Szopa, K. Grygoyć et al. 2014. Development and validation of HPLC-ICP-MS method for the determination inorganic Cr, As and Sb speciation forms and its application for Pławniowice reservoir (Poland) water and bottom sediments variability study. *Talanta* 120:475–483.

Jakariya, M., P. Bhattacharya. M. M. Hassan et al. 2009. Temporal variation of groundwater arsenic concentrations in Southwest Bangladesh. In *Natural Arsenic in Groundwater of Latin America - Occurrence, Health Impact and Remediation*, eds. J. Bundschuh, M. A. Armienta, P. Birkle et al. 225–233. Leiden: CRC Press.

Jalali, M., S. Karami, A. F. Marj. 2016. Geostatistical evaluation of spatial variation related to groundwater quality database: Case study for Arak Plain aquifer, Iran. *Environmental Modeling & Assessment*, 21(6):707–719.

James, K.A., T. Byers, J. E. Hokason et al. 2015. Association between lifetime exposure to inorganic arsenic in drinking water and coronary heart disease in Colorado residents. *Environmental Health Perspectives* 123:128–134.

James, K. A., J. A. Marshall, J. E. Hokanson et al. 2013. A case-cohort study examining life time exposure to inorganic arsenic in drinking water and diabetes mellitus. *Environmental Research* 123:33–38.

Jang, C. S., Y. T. Liou, C. P. Liang. 2010. Probabilistically determining roles of groundwater used in aquacultural fishponds. *Journal of Hydrology* 388:491–500.

Jarup, L. 1992. *Dose-response Relation for Occupational Exposure to Arsenic and Cadmium*. Sweden: National Institute for Occupational health.

Jarva, J., T. Tarvainen, J. Reinikainen. 2008. Application of arsenic baselines in the assessment of soil contamination in Finland. *Environmental Geochemistry and Health* 30:613–621.

Jasmin, I. and P. Mallikarjuna. 2015. Delineation of groundwater potential zones in Araniar River basin, Tamil Nadu, India: An integrated remote sensing and geographical information system approach. *Environmental Earth Sciences* 73(7):3833–3847.

Javed, S., A. Ali, S. Ullah. 2017. Spatial assessment of water quality parameters in Jhelum city (Pakistan). *Environmental Monitoring and Assessment* 189:119 [doi:10.1007/s10661-017-5822-9].

Jayasumana, M., P. Paranagama, M. Amarasinghe et al. 2013. Possible link of chronic arsenic toxicity with chronic kidney disease of unknown etiology in Sri Lanka. *Journal of Natural Sciences Research* 3:64–73.

Jekel, M. R. 1994. Removal of arsenic in drinking water treatment. In *Arsenic in the Environment, Part 1: Cycling and Characterization*, ed. J. O. Nriagu, 119–132. New York: John Wiley & Sons.

Jensen, L. A. and M. N. Allen. 1996. Meta-synthesis of qualitative findings. *Qualitative Health Research* 6(4):553–560.

Jessen, S., F. Larsen, C. B. Koch et al. 2005. Sorption and desorption of arsenic to ferrihydrite in a sand filter. *Environmental Science and Technology* 39(20):8045–8051.

Jha, S. K. and V. K. Mishra. 2016. Fluoride and arsenic in groundwater: Occurrence and geochemical processes controlling mobilisation. In *Innovative Saline Agriculture*, eds. J. C. Dagar, P. C. Sharma, D. K. Sharma et al. 351–370. India: Springer.

Jiang, Y. D., C. H. Chang, T. Y. Tai et al. 2012. Incidence and prevalence rates of diabetes mellitus in Taiwan: Analysis of the 2000–2009 Nationwide Health Insurance database. *Journal of the Formosan Medical Association* 111(11):599–604.

Jiang, J., M. Liu, F. Parvez et al. 2015. Association between arsenic exposure from drinking water and longitudinal change in blood pressure among HEALS cohort participants. *Environmental Health Perspectives* 123(8):806–812.

Jiang, H., Y. Ma, X. Chen et al. 2010. Genistein synergizes with arsenic trioxide to suppress human hepatocellular carcinoma. *Cancer Science* 101(4):975–983.

Jiang, Q. Q. and B. R. Singh. 1994. Effect of different forms and sources of arsenic on crop yield and arsenic concentration. *Water Air and Soil Pollution* 74:321–343.

JICA/ENPHO. 2005. *Arsenic vulnerability in groundwater resources in Kathmandu Valley*. Final Report, Japan International Cooperation Agency and Environment and Public Health Organization, Nepal.

Jin, M., Y. Fang, L. Zhao. 2005. Variable selection in generalized linear models with canonical link functions. *Statistics & Probability Letters* 71:371–382.

Jin, Y., C. Liang, G. He et al. 2003. Study on distribution of endemic arsenism in China. *Wei Sheng Yan Jiu* 32(6):519–540. (in Chinese and abstract in English).

Jindal, R. and P. Ratanamalaya. 2006. Investigations on the status of arsenic contamination in southern Thailand. In *Southeast Asian Water Environment 1*, eds. S. Ohgaki, K. Fukushi, H. Katayama et al. 223–231. London: IWA Publishing.

Jitmanee, K., M. Oshima, S. Motomizu. 2005. Speciation of arsenic (III) and arsenic (V) by inductively coupled plasma-atomic emission spectrometry coupled with preconcentration system. *Talanta* 66:529–533.

JMP. 2015. *25 Years Progress on Sanitation and Drinking Water: 2015 Update and MDG Assessment*. New York: UNICEF and World Health Organization.

Johnson, K. 2009. *It's Now Legal to Catch a Raindrop in Colorado*. The New York Times. (http://www.nytimes.com/2009/06/29/us/29rain.html?em)

Johnson, P., A. Buehring, G. Symon et al. 2007. Defining qualitative management research. *Qualitative Research in Organizations and Management: An International Journal* 3(1):23–43.

Johnston, R., H. Heijnen, P. Wurzel. 2001. Safe water technology. Chapter VI, Geneva: World Health Organization, [http://www.who.int/water_sanitation_health/dwq/arsenicun6.pdf].

Johnston, R. B. and M. H. Sarker. 2007. Arsenic mitigation in Bangladesh: National screening data and case studies in three upazilas. *Journal of Environmental Science and Health: Part A* 42(12):1889–1896.

Jomova, K., Z. Jenisova, M. Feszterova et al. 2011. Arsenic: Toxicity, oxidative stress and human disease. *Journal of Applied Toxicology* 31(2):95–107.

Jones, E. 2000. *Household Level Arsenic Removal Methodologies: Passive Sedimentation, Bucket Treatment Unit and Safi Filter*. (Preliminary Research Report: March), Dhaka: WaterAid Bangladesh.

Jordan, G. H. and B. Shrestha. 2000. A participatory GIS for community forestry user groups in Nepal: Putting people before the technology. *PLA Notes* 39:14–18.

Journel, A. G. 1983. Nonparametric estimation of spatial distributions. *Mathematical Geology* 15:445–468.

Journel, A. G. and C. J. Huijbregts. 1978. *Mining Geostatistics*. San Diego: Academic Press.

Jovanovic, D., Z. Rasic-Milutinovic, K. Paunovic et al. 2013. Low levels of arsenic in drinking water and type 2 diabetes in Middle Banat region, Serbia. *International Journal of Hygiene and Environmental Health* 216(1):50–55.

Júnior, F. W. A., E. M. de Oliveira Silveira, J. M. de Mello et al. 2015. Change detection in Brazilian savannas using semivariograms derived from NDVI images. *Ciência e Agrotecnologia* 39(2):103–109.

Kaivo-oja, J. 2016. Towards better participatory processes in technology foresight: How to link participatory foresight research to the methodological machinery of qualitative research and phenomenology? *Futures* 86:94–106.

Kalibo, H. and K. Medley. 2007. Participatory resource mapping for adaptive collaborative management at Mt. Kasigau, Kenya. *Landscape and Urban Planning* 82:145–158.

Kamijo, Y., K. Soma, Y. Asari et al. 1998. Survival after massive arsenic poisoning self-treated by high fluid intake. *Journal of Toxicology Clinical toxicology* 36(1–2):27–29.

Kaneto, H., N. Katakami, D. Kawamori et al. 2007. Involvement of oxidative stress in the pathogenesis of diabetes. *Antioxidants & Redox Signaling* 9(3):355–66.

Kanner, A. 2004. Equity in toxic tort litigation: Unjust enrichment and the poor. *Law and Policy* 26:209–230.

Kao, T. M. and S. R. Kao. 1954. Studies on the cause of a particular dry gangrene. *Journal of Formosan Medical Association* 53:272.

Kapaj, S., H. Peterson, K. Liber et al. 2006. Human health effects from chronic arsenic poisoning - a review. *Journal of Environmental Science and Health: Part A*, 41(10):2399–2428.

Kaplan, S. and B. J. Garrick. 1981. On the quantitative definition of risk. *Risk Analysis* 1(1):11–27.

Kaplan, B. and J. A. Maxell. 2005. Qualitative research methods for evaluating computer information systems. In *Evaluating Health Care Information Systems*, eds. J. G. Anderson and C. E. Aydin, 30–55. New York: Springer.

Karagas, M. R., A. S. Andrew, H. H. Nelson et al. 2012. SLC39A2 and FSIP1 polymorphisms as potential modifiers of arsenic-related bladder cancer. *Human Genetics* 131:453–461.

Karagas, M. R., A. Gossai, B. Pierce et al. 2015. Drinking water arsenic contamination, skin lesions, and malignancies: A systematic review of the global evidence. *Current Environmental Health Reports* 2(1):52–68.

Karagas, M. R., T. A. Stukel, J. S. Morris et al. 2001. Skin cancer risk in relation to toenail arsenic concentrations in a US population-based case-control study. *American Journal of Epidemiology* 153(6):559–65.

Karcher, S., L. Cáceres, M. Jekel et al. 1999. Arsenic removal from water supplies in Northern Chile using ferric chloride coagulation. *Water and Environment Journal* 13(3):164–169.

Kartinen, E. O. and C. J. Martin. 1995. An overview of arsenic removal processess. *Desalination* 103:79–88.

Kates, R. W. 1985. Hazard assessment: Art, science, and ideology. In *Perilous Progress: Managing the hazards of technology*, eds. R. W. Kates, C. Hohenemser, and J. X. Kasperson, 53–79. Boulder: Westview Press.

Katsoyiannis, I. A., S. J. Hug, A. Ammann et al. 2007. Arsenic speciation and uranium concentrations in drinking water supply wells in Northern Greece: Correlations with redox indicative parameters and implications for groundwater treatment. *Science of the Total Environment* 383(1–3):128–140.

Katsoyiannis, I. A. and A. A. Katsoyiannis. 2006. Arsenic and other metal contamination of groundwaters in the industrial area of Thessaloniki. *Environmental Monitoring and Assessment* 123:393–406.

Kavaf, N. and M. T. Nalbantcilar. 2007. Assessment of contamination characteristics in waters of the Kütahya Plain, Turkey. *CLEAN - Soil, Air, Water* 35(6):585–593.

Kavanagh, P., M. E. Farago, I. Thornton et al. 1998. Urinary arsenic species in Devon and Cornwall residents, UK: A pilot study. *Analyst* 123:27–29.

Keeney, R. L. 1995. Understanding life-threatening risks. *Risk Analysis* 15:627–637.

Keith, L. H. 1992. *EPA's IRIS Chemical Information Database.* Chelsea: Lewis Publishers.

Kelepertsis, A., D. Alexakis, K. Skordas. 2006. Arsenic, antimony and other toxic elements in the drinking water of eastern Thessaly in Greece and its possible effects on human health. *Environmental Geology* 50(1):76–84.

Kennedy, G. W. and D. S. Finlayson-Bourque. 2011. *Groundwater Chemistry from Bedrock Aquifers in Nova Scotia.* Nova Scotia Department of Natural Resources, Mineral Resources Branch, Nova Scotia, Canada.

Kerlin, B. A. 1998. Pursuit of the PhD: Is it good for your health? Paper presented at *the 4th International Multidisciplinary Qualitative Health Research Conference.* Vancouver, British Columbia, February 19–21.

Kevekordes, S., R. Suchenwirth, T. Gebel et al. 1998. Drinking water supply with reference to geogenic arsenic contamination. *Gesundheitswesen* 60(10):576–579 (in German).

Kew, J., C. Morris, A. Aihie et al. 1993. Arsenic and mercury intoxication due to Indian ethnic remedies. *British Medical Journal* 306:506–507.

Keya, M. K. 2004. Mental health of arsenic victims in Bangladesh. *South African Anthropology* 4:215–223.

Khan, M. M. H., K. Aklimunnessa, M. Kabir et al. 2006. Case-control study of arsenicosis in some arsenic contaminated villages of Bangladesh. *The Sapporo Medical Journal*, 75(4):51–61.

Khan, N. I., G. Owens, D. Bruce et al. 2009. Human arsenic exposure and risk assessment at the landscape level: A review. *Environmental Geochemistry and Health* 31:143–166.

Khan, M. H., S. Sarkar, N. Khan et al. 2010a. Assessment of low ABSPI among arsenic exposed and non-exposed populations: A pilot study. *Bangladesh Medical Research Council Bulletin* 36(1):23–26.

Khan, K. A., J. L. Stroud, Y. G. Zhu et al. 2010b. Arsenic bioavailability to rice is elevated in Bangladeshi paddy soils. *Environmental Science and Technology* 44(22):8515–8521.

Khan, N. I. and H. Yang. 2014. Arsenic mitigation in Bangladesh: An analysis of institutional Stakeholders' opinions. *Science of the Total Environment* 488–489:493–504.

Khandker, S., R. K. Dey, A. Z. M. M. Islam et al. 2006. Arsenic-safe drinking water and antioxidants for the management of arsenicosis patients. *Bangladesh Journal of Pharmacology* 1:42–50.

Khorasanipour, M. and E. Esmaeilzadeh. 2015. Geo-genic arsenic contamination in the Kerman Cenozoic Magmatic Arc, Kerman, Iran: Implications for the source identification and regional analysis. *Applied Geochemistry* 63:610–622.

Khuri, A. I. 2001. An overview of the use of generalized linear models in response surface methodology. *Nonlinear Analysis* 47:2023–2034.

Kienberger, S. 2014. Mapping environmental risks - Quantitative and spatialmodeling approaches. *Journal of Maps* 10(2):269–275.

Kile, M. L., A. Cardenas, E. Rodrigues et al. 2016b. Estimating effects of arsenic exposure during pregnancy on perinatal outcomes in a Bangladeshi cohort. *Epidemiology* 27(2):173–181.

Kile, M. L., J. M. Faraj, A. G. Ronnenberg et al. 2016a. A cross sectional study of anemia and iron deficiency as risk factors for arsenic-induced skin lesions in Bangladeshi women. *BMC Public Health* 16:158 (https://doi.org/10.1186/s12889-016-2824-4).

Kile, M. L., E. Hoffman, E. G. Rodrigues et al. 2011. A pathway-based analysis of urinary arsenic metabolites and skin lesions. *American Journal of Epidemiology* 173(7):778–786.

Kile, M. L., E. A. Houseman, A. A. Baccarelli et al. 2014. Effect of prenatal arsenic exposure on DNA methylation and leukocyte subpopulations in cord blood. *Epigenetics* 9(5):774–782.

Kim, K. W., P. Chanpiwat, H. T. Hanh et al. 2011. Arsenic geochemistry of groundwater in Southeast Asia. *Frontiers of Medicine* 5(4):420–433.

Kim, Y. and B. K. Lee. 2011. Association between urinary arsenic and diabetes mellitus in the Korean general population according to KNHANES 2008. *Science of the Total Environment* 409(19):4054–4062.

Kim, N. H., C. C. Mason, R. G. Nelson et al. 2013. Arsenic exposure and incidence of type 2 diabetes in Southwestern American Indians. *American Journal of Epidemiology* 177(9):962–969.

Kim, Y., H. Park, J. Kim et al. 2004. Health risk assessment for uranium in Korean groundwater. *Journal of Environmental Radioactivity* 77:77–85.

Kinn, L. G., H. Holgersen, T. J. Ekeland et al. 2013. Metasynthesis and bricolage: An artistic exercise of creating a collage of meaning. *Qualitative Health Research* 23(9):1285–1292.

Kinniburgh, D. G. and W. Kosmus. 2002. Arsenic contamination in groundwater: Some analytical considerations. *Talanta* 58(1):165–180.

Kippler, M., H. Skröder, S. M. Rahman et al. 2016. Elevated childhood exposure to arsenic despite reduced drinking water concentrations: A longitudinal cohort study in rural Bangladesh. *Environment International* 86:119–125.

Kitzinger, C. 2007. Feminist approaches. In *Qualitative Research Practice*, eds. C. Seale, G. Gobo, J. F. Gubrium, and D. Silverman, 113–128. London: Sage.

Klassen, R. A., S. L. Douma, A. Ford et al. 2009. *Geoscience modelling of relative variation in natural arsenic hazard potential in New Brunswick*. Geological Survey of Canada, Current Research 2009-7, pp. 1–9.

Kleefstra, T., A. Schenck, J. M. Kramer et al. 2014. The genetics of cognitive epigenetics. *Neuropharmacology* 80:83–94.

Kligerman, A. D., C. L. Doerr, A. H. Tennant et al. 2003. Methylated trivalent arsenicals as candidate ultimate genotoxic forms of arsenic: Induction of chromosomal mutations but not gene mutations. *Environmental and Molecular Mutagenesis* 42(3):192–205.

Kligerman, A. D. and A. H. Tennant. 2007. Insights into the carcinogenic mode of action of arsenic. *Toxicology and Applied Pharmacology* 222(3):281–288.

Knight, F. H. 1921. *Risk, Uncertainty, and Profit*. New York: Hart, Schaffner, and Marx.

Ko, R. J. 1999. Causes, epidemiology, and clinical evaluation of suspected herbal poisoning. *Clinical Toxicology* 37:697–708.

Kocar, B. D., M. L. Polizzotto, S. G. Benner et al. 2008. Integrated biogeochemical and hydrologic process driving arsenic release from shallow sediments to groundwater of the Mekong delta. *Applied Geochemistry* 23(11):3059–3071.

Koch, T. 1996. Implementation of a hermeneutic inquiry in nursing: Philosophy, rigour, and representation. *Journal of Advanced Nursing* 24(1):174–184.

Koch, T. 1999. An interpretive research process: Revisiting phenomenological and hermeneutical approaches. *Nurse Researcher* 6:20–34.

Koch, T. K. and A. Harrington. 1998. Reconceptualizing rigour: The case for reflexivity. *Journal of Advanced Nursing* 28(6):882–91.

Koenig, T. H. and M. L. Rustad. 2004. Toxic torts, politics, and environmental justice: The case for crimtorts. *Law and Policy* 26:189–207.

Kohnhorst, R., T. Kunito, T. Agusa et al. 2005. Arsenic in groundwater in selected countries in South and Southeast Asia: A review. *Journal of Tropical Medicine and Parasitology* 28:73–82.

Kokilavani, V., M. A. Devi, K. Sivarajan et al. 2005. Combined efficacies of dl-alpha-lipoic acid and meso 2,3 dimercaptosuccinic acid against arsenic induced toxicity in antioxidant systems of rats. *Toxicology Letters* 160:1–7.

Komorowicz, I. and D. Barałkiewicz. 2011. Arsenic and its speciation in water samples by high performance liquid chromatography inductively coupled plasma mass spectrometry - Last decade review. *Talanta* 84(2):247–261.

Komorowicz, I. and D. Baralkiewicz. 2014. Arsenic speciation in water by high-performance liquid chromatography/ inductively coupled plasma mass spectrometry-method validation and uncertainty estimation. *Rapid Communications in Mass Spectrometry* 28:159–168.

Komorowicz, I. and D. Barałkiewicz. 2016. Determination of total arsenic and arsenic species in drinking water, surface water, wastewater, and snow from Wielkopolska, Kujawy-Pomerania, and Lower Silesia provinces, Poland. *Environmental Monitoring and Assessment* 188(9):504.

Kondo, H., Y. Ishiguro, K. Ohno et al. 1999. Naturally occurring arsenic in the groundwaters in the southern region of Fukuoka Prefecture, Japan. *Water Research* 33(8):1967–1972.

Kong, A. P., K. Xiao, K. C. Choi et al. 2012. Associations between microRNA (miR-21, 126, 155 and 221), albuminuria and heavy metals in Hong Kong Chinese adolescents. *Clinica Chimica Acta* 413(13–14):1053–1057.

Korngold, E., N. Belayev, L. Aronov. 2001. Removal of arsenic from drinking water by anion exchangers. *Desalination* 141(1):81–84.

Kortatsi, B. K., J. Asigbe, G. A. Dartey et al. 2008. Reconnaissance survey of arsenic concentration in groundwater in south-eastern Ghana. *West African Journal of Applied Ecology* 13:16–26.

Kortsenshteyn, V. N., A. P. Karaseva, A. K. Aleshina. 1973. Distribution of arsenic in deep groundwater of the Middle Caspian Artesian Basin. *Geokhimiya* 4:612–17. (In Russian).

Kouras, A., I. Katsoyiannis, D. Voutsa. 2007. Distribution of arsenic in groundwater in the area of Chalkidiki, Northern Greece. *Journal of Hazardous Materials* 147(3):890–899.

Kourgialas, N. N. and G. P. Karatzas. 2015. An integrated approach for the assessment of groundwater contamination risk/vulnerability using analytical and numerical tools within a GIS framework. *Hydrological Sciences Journal* 60(1):111–132.

Kralj, P. 2004. Chemical composition of low temperature (<20–40°C) thermal waters in Slovenia. *Environmental Geology* 46(5):635–642.

Krefting, L. 1991. Rigor in qualitative research: The assessment of trustworthiness. *American Journal of Occupational Therapy* 45:214–222.

Krige, D. G. 1951. A statistical approach to some basic mine valuation problems on the Witwatersrand. *Journal of the Chemical, Metallurgical and Mining Society of South Africa* 52(6):119–139.

Krishna, M. V. B., K. Chandrasekaran, D. Karunasagar et al. 2001. A combined treatment approach using Fenton's reagent and zero valent iron for the removal of arsenic from drinking water. *Journal of Hazardous Materials* 84(2–3):229–240.

Kuhn, T. S. 1962. *The Structure of the Scientific Revolutions*. Chicago: The University of Chicago Press.

Kuiper, K. 2010. *Mesopotamia: The World's Earliest Civilization*. New York: Britannica Educational Publishing in association with Rosen Education Service.

Kulkarni, H. V., N. Mladenov, K. H. Johannesson et al. 2017. Contrasting dissolved organic matter quality in groundwater in Holocene and Pleistocene aquifers and implications for influencing arsenic mobility. *Applied Geochemistry* 77:194–205.

Kullo, I. J. and K. Ding. 2007. Mechanisms of disease: The genetic basis of coronary heart disease. *Nature Clinical Practice. Cardiovascular Medicine* 4(10):558–569.

Kumar, A., M. S. Rahman, M. A. Iqubal et al. 2016b. Ground water arsenic contamination: A local survey in India. *International Journal of Preventive Medicine*, 7:100 (doi:10.4103/2008-7802.188085).

Kumar, M., M. M. Rahman, A. L. Ramanathan et al. 2016a. Arsenic and other elements in drinking water and dietary components from the middle Gangetic plain of Bihar, India: Health risk index. *Science of the Total Environment* 539:125–134.

Kumari, S., A. K. Singh, A. K. Verma et al. 2014. Assessment and spatial distribution of groundwater quality in industrial areas of Ghaziabad, India. *Environmental Monitoring and Assessment*, 186(1):501–514.

Kuo, T. L. 1968. Arsenic content of artesian well water in endemic area of chronic arsenic poisoning. *Rep. Inst. Pathol. National Taiwan University* 20:7–13.

Kuo, C. C., B. V. Howard, J. G. Umans et al. 2015. Arsenic exposure, arsenic metabolism, and incident diabetes in the strong heart study. *Diabetes Care* 38(4):620–627.

Kuroda, K., T. Hayashi, A. Funabiki et al. 2017. Holocene estuarine sediments as a source of arsenic in Pleistocene groundwater in suburbs of Hanoi, Vietnam. *Hydrology Journal*, 25(4):1137–1152.

Kurttio, P., E. Pukkala, H. Kahelin et al. 1999. Arsenic concentrations in well water and risk of bladder and kidney cancer in Finland. *Environmental Health Perspectives* 107(9):705–710.

Kwok, R. K., R. B. Kaufmann, M. Jakariya. 2006. Arsenic in drinking-water and reproductive health outcomes: a study of participants in the Bangladesh Integrated Nutrition Programme. *Journal of Health, Population, and Nutrition* 24(2):190–205.

Kwok, R. K., P. Mendola, Z. Y. Liu et al. 2007. Drinking water arsenic exposure and blood pressure in healthy women of reproductive age in Inner Mongolia, China. *Toxicology and Applied Pharmacology* 222(3):337–343.

Kwong, Y. T. J., S. Beauchemin, M. F. Hossain et al. 2007. Transformation and mobilization of arsenic in the historic Cobalt mining camp, Ontario, Canada. *Journal of Geochemical Exploration* 92:133–150.

Kwong, Y. L. and D. Todd. 1997. Delicious poison: Arsenic trioxide for the treatment of leukemia. *Blood* 89:3487–3488.

Kyem, P. A. K. 2002. Promoting local community participation in forest management through a PPGIS application in southern Ghana. In *Community Participation and Geographical Information System*, eds. W. J. Craig, T. M. Harris and D. Weiner, 218–231. London: Taylor & Francis.

Lahermo, P., G. Alfthan, D. Wang. 1998. Selenium and arsenic in the environment in Finland. *Journal of Environmental Pathology, Toxicology and Oncology* 17:205–216.

Lai, M. S., Y. M. Hsueh, C. J. Chen et al. 1994. Ingested inorganic arsenic and prevalence of diabetes mellitus. *American Journal of Epidemiology* 139(5):484–492.

Lalor, G., R. Rattray, P. Simpson et al. 1999. Geochemistry of an arsenic anomaly in St. Elizabeth, Jamaica. *Environmental Chemistry and Health* 21(1):3–11.

Lamm, S. H., A. Engel, M. B. Kruse et al. 2004. Arsenic in drinking water and bladder cancer mortality in the United States: an analysis based on 133 U.S. counties and 30 years of observation. *Journal of Occupational and Environmental Medicine* 46(3):298–306.

Lamm, S. H., S. A. Robbins, C. Zhou et al. 2013. Bladder/lung cancer mortality in Blackfoot-disease (BFD)-endemic area villages with low (<150 µg/L) well water arsenic levels-an exploration of the dose-response Poisson analysis. *Regulatory Toxicology and Pharmacology* 65(1):147–156.

Lander, J. J., R. J. Stanley, H. W. Sumner et al. 1975. Angiosarcoma of the liver associated with Fowler's solution (potassium arsenite). *Gastroenterology* 68(6):1582–1586.

Lark, R. M. 2000. A comparison of some robust estimators of the variogram for use in soil survey. *European Journal of Soil Science* 51(1):137–157.

Larsen, E. H., L. Moseholm, M. M. Nielsen. 1992. Atmospheric deposition of trace elements around point sources and human health risk assessment. II: Uptake of arsenic and chromium by vegetables grown near a wood preservation factory. *Science of The Total Environment* 126(3):263–275.

Larsen, F., N. Q. Pham, N. D. Dang et al. 2008. Controlling geological and hydrogeological processes in an arsenic contaminated aquifer on the Red River flood plain, Vietnam. *Applied Geochemistry* 23:3099–3115.

Lasky, T., W. Sun, A. Kadry. 2004. Mean total arsenic concentrations in chicken 1989–2000 and estimated exposures for consumers of chicken. *Environmental Health Perspectives* 112:18–21.

Last, J. M. 1995. *A Dictionary of Epidemiology*. New York: Oxford University Press.

Latour, B. 2000. When things strike back: A possible contribution of 'science studies' to the social sciences. *British Journal of Sociology* 51:107–123.

Lawrence, J. M., R. Contreras, W. Chen et al. 2008. Trends in the prevalence of preexisting diabetes and gestational diabetes mellitus among a racially/ethnically diverse population of pregnant women, 1999–2005. *Diabetes Care* 31:899–904.

Lawrence, A. R., D. C. Gooddy, M. Kanatharan et al. 2000. Groundwater evolution beneath Hat Yai, a rapidly developing city in Thailand. *Hydrogeology Journal* 8:564–575.

Lawson, A. 2003. *Bangladesh Arsenic Case Begins*. London: British Broadcasting Corporation (BBC). [http://news.bbc.co.uk/2/hi/south_asia/2886079.stm].

Lawson, M., D. A. Polya, A. J. Boyce et al. 2016. Tracing organic matter composition and distribution and its role on arsenic release in shallow Cambodian groundwaters. *Geochimica et Cosmochimica Acta* 178:160–177.

Le, X. C., W. R. Cullen, K. J. Reimer et al. 1992. A new continuous hydride generator for the determination of arsenic, antimony and tin by hybride generation atomic absorption spectrometry. *Analytica Chimica Acta* 258:307–315.

Lee, T. R. 1981. Perception of risk: The public's perception of risk and the question of irrationality. *Proceedings of the Royal Society of London* 376(1764):5–16.

Lee, J. J., C. S. Jang, S. W. Wang et al. 2007. Evaluation of potential health risk of arsenic-affected groundwater using indicator kriging and dose response model. *Science of the Total Environment* 384:151–162.

Lee, J. C., H. Y. Lee, C. H. Moon et al. 2013. Arsenic trioxide as a vascular disrupting agent: Synergistic effect with irinotecan on tumor growth delay in a CT26 allograft Model. *Translational Oncology* 6(1):83–91.

Leininger, M. 1994. Evaluation criteria and critique of qualitative research studies. In *Critical Issues in Qualitative Research Methods*, ed. J. M. Morse, 95–115. Thousand Oaks: Sage.

Leonardi, G., M. Vahter, F. Clemens et al. 2012. Inorganic arsenic and basal cell carcinoma in areas of Hungary, Romania, and Slovakia: A case-control study. *Environmental Health Perspectives* 120(5):721–726.

Lerman, B. B., N. Ali, D. Green. 1980. Megaloblastic, dyserythropoietic anemia following arsenic ingestion. *Annals of Clinical and Laboratory Science* 10:515–517.

Lesseur, C., D. Gilbert-Diamonda, A. S. Andrewa et al. 2012. A case-control study of polymorphisms in xenobiotic and arsenic metabolism genes and arsenic-related bladder cancer in New Hampshire. *Toxicology Letters* 210:100–106.

Lester, S. 1999. *An Introduction to Phenomenological Research*. Taunton, UK: Stan Lester Developments.

Lewis, D. R., J. W. Southwick, R. Ouellet-Hellstrom et al. 1999. Drinking water arsenic in Utah: A cohort mortality study. *Environmental Health Perspectives* 107(5):359–65.

Leybourne, M. I. and E. M. Cameron. 2008. Source, transport, and fate of rhenium, selenium, molybdenum, arsenic, and copper in groundwater associated with porphyry-Cu deposits, Atacama Desert, Chile. *Chemical Geology* 247:208–228.

Li, Y., H. Guo, C. Hao. 2014a. Arsenic release from shallow aquifers of the Hetao basin, Inner Mongolia: evidence from bacterial community in aquifer sediments and groundwater. *Ecotoxicology* 23(10):1900–14.

Li, H., S. Horke, U. Forstermann. 2014b. Vascular oxidative stress, nitric oxide and atherosclerosis. *Atherosclerosis* 237:208–219.

Li, J., Z. Y. Huang, Y. Hu et al. 2013b. Potential risk assessment of trace metals by consuming shellfish collected from Xiamen, China. *Environmental Science and Pollution Research* 20(5):2937–2947.

Li, X., B. Li, S. Xi et al. 2013a. Prolonged environmental exposure of arsenic through drinking water on the risk of hypertension and type 2 diabetes. *Environmental Science and Pollution Research* 20(11):8151–8161.

Li, Y., D. Wang, Y. Liu et al. 2017. A predictive risk model of groundwater arsenic contamination in China applied to the Huai River Basin, with a focus on the region's cluster of elevated cancer mortalities. *Applied Geochemistry* 77:178–183.

Li, H., T. Xu, W. Tong et al. 2008. Comparison of cardiovascular risk factors between prehypertension and hypertension in a Mongolian population, Inner Mongolia, China. *Circulation Journal* 72(10):1666–1673.

Liang, C. P., Y. C. Chien, C. H. Jang et al. 2017. Spatial analysis of human health risk due to arsenic exposure through drinking groundwater in Taiwan's Pingtung Plain. *International Journal of Environmental Research and Public Health*, 14:81 (doi:10.3390/ijerph14010081).

Liao, Y. T., C. J. Chen, W. F. Li et al. 2012. Elevated lactate dehydrogenase activity and increased cardiovascular mortality in the arsenic-endemic areas of Southwestern Taiwan. *Toxicology and Applied Pharmacology* 262(3):232–237.

Liao, C. M., H. H. Shen, T. L. Lin et al. 2008. Arsenic cancer risk posed to human health from tilapia consumption in Taiwan. *Ecotoxicology and Environmental Safety* 70:27–37.

Liaw, J., G. Marshall, Y. Yuan et al. 2008. Increased childhood liver cancer mortality and arsenic in drinking water in northern Chile. *Cancer Epidemiology, Biomarkers and Prevention* 17(8):1982–1987.

Liévremont, D., P. N. Bertin, M. C. Lett. 2009. Arsenic in contaminated waters: Biogeochemical cycle, microbial metabolism and biotreatment processes. *Biochimie* 91:1229–1237.

Lin, Y. B., Y. P. Lin, C. W. Liu et al. 2006. Mapping of spatial multi-scale sources of arsenic variation in groundwater on Chianan floodplain of Taiwan. *Science of the Total Environment* 370:168–181.

Lin, H. J., T. I. Sung, C. Y. Chen et al. 2013. Arsenic levels in drinking water and mortality of liver cancer in Taiwan. *Journal of Hazardous Materials* 262:1132–1138.

Lincoln, Y. S. and E. G. Guba. 1985. *Naturalistic Inquiry*. Newbury Park: Sage.

Lincoln, Y. S. and E. G. Guba. 2000. Paradigmatic controversies, contradictions, and emerging confluences. In *Handbook of Qualitative Research*, eds. N. K. Denzin and Y. S. Lincoln, 163–188. Thousand Oaks: Sage.

Lindberg, A. L., W. Goessler, E. Gurzau et al. 2006. Arsenic exposure in Hungary, Romania and Slovakia. *Journal of Environmental Monitoring* 8:203–208.

Lindberg, A. L., M. Rahman, L. A. Persson et al. 2008. The risk of arsenic induced skin lesions in Bangladeshi men and women is affected by arsenic metabolism and the age at first exposure. *Toxicology and Applied Pharmacology* 230(1):9–16.

Lioy, P. 1990. Assessing total human exposure to contaminants. *Environmental Science & Technology* 24:938–945.

Lisabeth, L. D., H. J. Ahn, J. J. Chen et al. 2010. Arsenic in drinking water and stroke hospitalizations in Michigan. *Stroke* 41(11):2499–2504.

Liu, C. W., C. S. Jang, C. M. Liao. 2004. Evaluation of arsenic contamination potential using indicator kriging in the Yun-Lin Aquifer (Taiwan). *Science of the Total Environment* 321:173–188.

Liu, C. W., K. H. Lin, Y. M. Kuo. 2003. Application of factor analysis in the assessment of groundwater quality in a black foot disease area in Taiwan. *Science of the Total Environment* 313(743):77–89.

Liu, J., Q. Liu, H. Yang. 2016. Assessing water scarcity by simultaneously considering environmental flow requirements, water quantity, and water quality. *Ecological Indicators* 60:434–441.

Liu, J. and M. P. Waalkes. 2008. Liver is a target of arsenic carcinogenesis. *Toxicological Sciences* 105(1):24–32.

Liu, F. F., J. P. Wang, Y. J. Zheng et al. 2013. Biomarkers for the evaluation of population health status 16 years after the intervention of arsenic-contaminated groundwater in Xinjiang, China. *Journal of Hazardous Materials* 262:1159–1166.

Liu, X., J. Wu, J. Xu. 2006. Characterizing the risk assessment of heavy metals and sampling uncertainty analysis in paddy field by geostatistics and GIS. *Environmental Pollution* 141:257–264.

Livengood, A. and K. Kunte. 2012. Enabling participatory planning with GIS: A case study of settlement mapping in Cuttack, India. *Environment & Urbanization* 24(1):77–97.

Lo, M. C. 1975. *Report on the Investigation of Arsenic Content of Well Water in the Province of Taiwan*. Taiwan Provincial Institute of Sanitary Department, Taiwan.

Lofland, J., D. A. Snow, L. Anderson et al. 2006. *Analyzing Social Settings: A Guide to Qualitative Observation and Analysis*. Belmont: Wadsworth.

Lokuge, K. M., W. Smith, B. Caldwell et al. 2004. The effect of arsenic mitigation interventions on disease burden in Bangladesh. *Environmental Health Perspectives* 112:1172–1177.

Lomboy, M., J. Riego de Dios, B. Magtibay et al. 2017. Updating national standards for drinking-water: A Philippine experience. *Journal of Water and Health* 15(2):288–295.

Longley, P.A., M. F. Goodchild, D. J. Maguire et al. 2001. *Geographic Information Systems & Science*. Chichester: John Wiley & Sons.

López, D. L., J. Bundschuh, P. Birkle et al. 2012. Arsenic in volcanic geothermal fluids of Latin America. *Science of the Total Environment* 429:57–75.

López-Carrillo, L., R. U. Hernández-Ramírez, A. J. Gandolfi et al. 2014. Arsenic methylation capacity is associated with breast cancer in northern Mexico. *Toxicology and Applied Pharmacology* 280:53–59.

Lovertt, A. A., J. P. Parfitt, J. S. Brainard. 1997. Using GIS in risk analysis: A case study of hazardous waste transport. *Risk Analysis* 17:625–633.

Low, H. 2016. *Why Houses in Bermuda Have White Stepped Roofs?* London: British Broadcasting Corporation (BBC). [www.bbc.com/news/magazine-38222271].

Lowrance, W. W. 1976. *Of Acceptable Risk: Science and the Determination of Safety*. Los Altos, CA: William Kaufmann.

Lu, F. J. 1990. Blackfoot disease: Arsenic or humic acid? *The Lancet* 336(8707):115–116.

Lu, T. H., C. C. Su, Y. W. Chen et al. 2011. Arsenic induces pancreatic β-cell apoptosis via the oxidative stress-regulated mitochondria-dependent and endoplasmic reticulum stress-triggered signaling pathways. *Toxicology Letters* 201(1):15–26.

Lu, P., Y. Su, Z. Niu et al. 2007. Geostatistical analysis and risk assessment on soil total nitrogen and total soil phosphorus in the Dongting Lake Plain Area, China. *Journal of Environmental Quality* 36:935–942.

Luansang, C., S. Boonmahathanakorn, M. L. Domingo-Price. 2012. The role of community architects in upgrading; reflecting on the experience in Asia. *Environment and Urbanization* 24(2):497–512.

Lucanie, R. 1998. Papal poisoning: Preferred pastime of the Pontiffs? *Mithridata* 8(1):4–7 (Issue: 15).

Lucas, P. J., J. Baird, L. Arai et al. 2007. Worked examples of alternative methods for the synthesis of qualitative and quantitative research in systematic reviews. *BMC Medical Research Methodology* 7:4 [doi:10.1186/1471-2288-7-4].

Luczaj, J. and K. Masarik. 2015. Groundwater quantity and quality issues in a water-rich region: Examples from Wisconsin, USA. *Resources* 4:323–357.

Luna, A. L., L. C. Acosta-Saavedra, L. Lopez-Carrillo et al. 2010. Arsenic alters monocyte superoxide anion and nitric oxide production in environmentally exposed children. *Toxicology and Applied Pharmacology* 245(2):244–251.

Luo, Y. L., P. A. Jiang, Y. H. Yu et al. 2006. Investigation and assessment on arsenic pollution of groundwater and soil in Kuitu, Xinjiang province. *Arid Land Geography* 29:705–709 (In Chinese with English Abstract).

Luong, J. H. T., E. Majid, K. B. Male. 2007. Analytical tools for monitoring arsenic in the environment. *The Open Analytical Chemistry Journal* 1:7–14.

Luu, T. T. G., S. Sthiannopkao, K. W. Kim. 2009. Arsenic and other trace elements contamination in groundwater and a risk assessment study for the residents in the Kandal Province of Cambodia. *Environmental International* 35(3):455–460.

Luz, F., A. Mateus, J. X. Matos et al. 2014. Cu- and Zn-soil anomalies in the NE border of the South Portuguese Zone (Iberian Variscades, Portugal) identified by multifractal and geostatistical analyses. *Natural Resources Research* 23:195–215.

Ma, Y., Z. Ma, S. Yin et al. 2017. Arsenic and fluoride induce apoptosis, inflammation and oxidative stress in cultured human umbilical vein endothelial cells. *Chemosphere* 167:454–461.

Ma, J., M. K. Sengupta, D. Yuan et al. 2014. Speciation and detection of arsenic in aqueous samples: A review of recent progress in non-atomic spectrometric methods. *Analytica Chimica Acta* 831:1–23.

Ma, L., L. Wang, Y. Jia et al. 2016. Arsenic speciation in locally grown rice grains from Hunan Province, China: Spatial distribution and potential health risk. *Science of the Total Environment* 557–558:438–444.

Ma, H. Z., Y. J. Xia, K. G. Wu et al. 1999. Arsenic exposure and health effects in Bayingnormen, Inner Mongolia. In *Arsenic Exposure and Health Effects*, eds. W. R. Chappell, C. O. Abernathy, R. L. Calderon, 127–131. Amsterdam: Elsevier.

Mabit, L. and C. Bernard. 2007. Assessment of spatial distribution of fallout radionuclides through geostatistics concept. *Journal of Environmental Radioactivity* 97:206–219.

Machiwal, D., A. Mishra, M. K. Jha et al. 2012. Modeling short-term spatial and temporal variability of groundwater level using geostatistics and GIS. *Natural Resources Research* 21(1):117–136.

MacIntosh, D. L., P. L. Williams, D. J. Hunter et al. 1997. Evaluation of a food frequency questionnaire - Food composition approach for estimating dietary intake of inorganic arsenic and methylmercury. *Cancer Epidemiology, Biomarkers and Prevention* 6:1043–1050.

Maclean, K. S. and W. M. Langille. 1981. Arsenic in orchard and potato soils and its relation to selected chemical properties and anions. *Plant and Soil* 61(3):413–418.

Macnaghten, P. and G. Myers. 2007. Focus groups. In *Qualitative Research Practice*, eds. C. Seale, G. Gobo, J. F. Gubrium, and D. Silverman, 65–79. London: Sage.

Madajewicz, M., A. Pfaff, A. van Geen et al. 2007. Can information alone change behavior? Response to arsenic contamination of groundwater in Bangladesh. *Journal of Development Economics* 84:731–754.

Madden, E. F. and B. A. Fowler. 2000. Mechanisms of nephrotoxicity from metal combinations: A review. *Drug and Chemical Toxicology* 23(1):1–12.

Maguire, D. J. 1991. An overview and definition of GIS. In *Geographical Information Systems: Principles and Applications*, eds. D. J. Maguire, M. F. Goodchild, D. W. Rhind, 9–20. Vol 1: Principles. London: Longman.

Maharjan, M., R. R. Shrestha, S. A. Ahmad et al. 2006. Prevalence of arsenicosis in Terai, Nepal. *Journal of Health, Population and Nutrition* 24(2):246–252.

Maharjan, M., C. Watanabe, S. A. Ahmad et al. 2005. Short report: Arsenic contamination in drinking water and skin manifestations in lowland Nepal: The first community-based survey. *American Journal of Tropical Medicine and Hygiene* 73(2):477–479.

Mahmud, A. and P. Capella. 1999. *Poisoned Villagers to sue Unicef*. London: The Guardian. [https://www.theguardian.com/world/1999/jul/22/2].

Mailloux, S. J. and F. J. Mootz. 2017. From Hermeneutics to Rhetoric, and back again. *Social Science Research Network* 2017:1–32. [http://dx.doi.org/10.2139/ssrn.2895218].

Maity, J. P., B. Nath, C. Y. Chen et al. 2011. Arsenic-enriched groundwaters of India, Bangladesh and Taiwan - comparison of hydrochemical characteristics and mobility constraints. *Journal of Environmental Science and Health: Part A*, 46(11):1163–76.

Majumder, S., B. Nath, S. Sarkar et al. 2014. Size-fractionation of groundwater arsenic in alluvial aquifers of West Bengal, India: the role of organic and inorganic colloids. *Science of the Total Environment* 468–469:804–812.

Malavé-Llamas, K. and M. del C. Cotto-Maldonado. 2010. Communicating environmental risks in developing countries. In *Environmental and Human Health: Risk Management in Developing Countries*, eds. E. N. Laboy-Nieves, M. F. A. Goosen, E. Emmanuel, 13–28. Boca Raton: CRC Press.

Maltby, H. J., J. M. de Vries-Erich, K. Lund. 2016. Being the stranger: Comparing study abroad experiences of nursing students in low and high income countries through hermeneutical phenomenology. *Nurse Education Today* 45:114–119.

Malve, O., T. Hjerppe, S. Tattari et al. 2016. Participatory operations model for cost-efficient monitoring and modeling of river basins - A systematic approach. *Science of the Total Environment* 540:79–89.

Mandal, P. 2017. Molecular insight of arsenic-induced carcinogenesis and its prevention. *Naunyn-Schmiedeberg's Archives of Pharmacology* 390(5):443–455.

Mandal, B. K., B. K. Biswas, R. K. Dhar et al. 1999. Groundwater arsenic contamination and sufferings of people in West Bengal, India and Bangladesh. Status report up to March 1998. In *Metals and Genetics*, ed. B. Sarkar, 41–65. New York: Kluwer Publishers.

Mandal, B. K., T. R. Chowdhury, G. Samanta et al. 1996. Arsenic in groundwater in seven districts of West Bengal, India - the biggest arsenic calamity in the world. *Current Science* 70(11):976–987.

Mandal, B. K. and K. T. Suzuki. 2002. Arsenic round the world: A review. *Talanta* 58:201–235.

Manna, P., M. Sinha, P. Sil. 2008. Arsenic-induced oxidative myocardial injury: Protective role of arjunolic acid. *Archives of Toxicology* 82(3):137–149.

Mapa, R. B. and D. Kumaragamage. 1996. Variability of soil properties in a tropical Alfisol used for shifting cultivation. *Soil Technology* 9:187–197.

Mapedza, E., J. Wright, R. Fawcett. 2003. An investigation of land cover change in Mafungautsi Forest, Zimbabwe, using GIS and participatory mapping. *Applied Geography* 23(1):1–21.

Marchiset-Ferlay, N., C. Savanovitch, M. P. Sauvant-Rochat. 2012. What is the best biomarker to assess arsenic exposure via drinking water? *Environment International* 39:150–71.

Margat, J. 1968. *Vulnerabilite des nappes d'eau souterraine a la pollution. Bases de la cartographic*. BRGM, 68 SGL198 HYD. Orleans, France.

Mariner, P. E., F. J. Holzmer, R. E. Jackson et al. 1996. Effects of high pH on arsenic mobility in shallow sand aquifer and on aquiferpermeability along the adjacent shoreline. *Environtal Science and Technology* 30:1645–1651.

Markley, C. T. and B. E. Herbert. 2009. Arsenic risk assessment: The importance of speciation in different hydrologic systems. *Water, Air, & Soil Pollution* 204:385–398.

Marko, K., N. S. Al-Amri, A. M. M. Elfeki. 2014. Geostatistical analysis using GIS for mapping groundwater quality: Case study in the recharge area of Wadi Usfan, western Saudi Arabia. *Arabian Journal of Geosciences* 7(12):5239–5252.

Markoulakis, R. and B. Kirsh. 2013. Difficulties for university students with mental health problems: A critical interpretive synthesis. *The Review of Higher Education* 37(1):77–100.

Marsh, J. 1837. Separation of arsenic. In *The American Journal of Pharmacy*, ed. J. Carson, 307–314. Vol II. Philadelphia: Merriew and Gunn.

Marshall, G., C. Ferreccio, Y. Yuan et al. 2007. Fifty-year study of lung and bladder cancer mortality in Chile related to arsenic in drinking water. *Journal of National Cancer Institute* 99(12):920–28.

Marsit, C. J. 2015. Influence of environmental exposure on human epigenetic regulation. *The Journal of Experimental Biology* 218(Part 1):71–79.

Martin, D. 1991. *Geographic Information Systems and their Socioeconomic Applications*. London: Routledge.

Martin, J. D. and T. W. Simpson. 2005. Use of Kriging models approximate deterministic computer models. *AIAA Journal* 43(4):853–863.

Martinez, V. D., E. A. Vucic, D. D. Becker-Santos et al. 2011. Arsenic exposure and the induction of human cancers. *Journal of Toxicology* 431287:1–13 (doi:10.1155/2011/431287).

Martínez-Graña, A. M., J. Goy, C. Zazo et al. 2014. Characterization of environmental impact on resources, using strategic assessment of environmental impact and management of natural spaces of "Las Batuecas-Sierra de Francia" and "Quilamas" (Salamanca, Spain). *Environmental Earth Sciences* 71(1):39–51.

Martyn, C. N., D. J. Barker, C. Osmond et al. 1989. Geographical relation between Alzheimer's disease and aluminum in drinking water. *Lancet* 1(8629):59–62.

Mason, B. and L. G. Berry. 1978. *Elements of Mineralogy*. New York: Freeman.

Masoud, A. A. 2014. Groundwater quality assessment of the shallow aquifers west of the Nile Delta (Egypt) using multivariate statistical and geostatistical techniques. *Journal of African Earth Sciences* 95:123–137.

Mass, M. J., A. Tennant, B. C. Roop et al. 2001. Methylated trivalent arsenic species are genotoxic. *Chemical Research in Toxicology* 14(4):355–361.

Massey, E. W., D. Wold, A. Heyman. 1984. Arsenic: Homicidal intoxication. *Southern Medical Journal* 77:848–851.

Matera, V. and I. Le Hecho. 2001. Arsenic behavior in contaminated soils: Mobility and speciation. In *Heavy Metals Release in Soils*, eds. H. M. Selim and D. M. Sparks, 207–235. Boca Raton: CRC Press.

Matheron, G. 1963. Principles of geostatistics. *Economic Geology* 58:1246–1266.

Mathew, L., A. Vale, J. E. Adcock. 2010. Arsenical peripheral neuropathy. *Practical Neurology* 10(1):34–38.

Matschullat, J., R. P. Borba, E. Deschamps et al. 2000. Human and environmental contamination in the Iron Quadrangle, Brazil. *Applied Geochemistry* 15:181–190.

Mauro, C. D., M. Hoogerwerf, A. J. C. Sinke. 2000. A GIS Based risk assessment model: Application on sites contaminated by chlorinated solvents and petroleum hydrocarbons (BTEX). In *Fourth International Conference on Integrating GIS and Environmental Modelling (GIS/EM4): Problems, Prospects and Research Needs*. Alberta, Canada.

Maxwell, A. J. 1998. Designing a qualitative study. In *Handbook of Applied Social Science Research Methods*, eds. L. Bickman and D. J. Rog, 214–253. Thousand Oaks: Sage.

Mayorga, P., A. Moyano, H. M. Anawar et al. 2013. Temporal variation of arsenic and nitrate content in groundwater of the Duero River Basin (Spain). *Physics and Chemistry of the Earth* 58–60:22–27.

Mazumder, D. N. G. 2005. Effect of chronic intake of arsenic-contaminated water on liver. *Toxicology and Applied Pharmacology* 206(2):169–175.

Mazumder, D. N. G. 2007a. Arsenic and non-malignant lung disease. *Journal of Environmental Science and Health: Part A* 42(12):1859–1867.

Mazumder, D. N. G. 2007b. Hepatology in Kolkata. *Indian Journal of Gastroenterology* 26:S18–S23.

Mazumder, D. N. G. 2008. Chronic arsenic toxicity and human health. *Indian Journal of Medical Research* 128:436–447.

Mazumder, D. N. G., A. K. Chakraborty, A. Ghose et al. 1988. Chronic arsenic toxicity from drinking tube-well water in rural West Bengal. *Bulletin World Health Organization* 66(4):499–506.

Mazumder, D. N. G. and U. B. Dasgupta. 2011. Chronic arsenic toxicity: Studies in West Bengal, India. *Kaohsiung Journal of Medical Sciences* 27:360–370.

Mazumder, D. N. G., J. Das Gupta, A. Santra et al. 1998b. Chronic arsenic toxicity in West Bengal - The worst calamity in the world. *Journal of Indian Medical Association* 96(1):4–7.

Mazumder, D. N. G., A. Ghose, K. K. Majumdar et al. 2010. Arsenic contamination of ground water and its health impact on population of District of Nadia, West Bengal, India. *Indian Journal of Community Medicine* 35:331–338.

Mazumder, D. N. G., R. Haque, N. Ghosh et al. 1998a. Arsenic levels in drinking water and the prevalence of skin lesions in West Bengal, India. *International Journal of Epidemiology* 27(5):871–877.

Mazumder, D. N. G., R. Haque, N. Ghosh et al. 2000. Arsenic in drinking water and the prevalence of respiratory effects in West Bengal, India. *International Journal of Epidemiology* 29:1047–1052.

Mazumder, D. N. G., I. Purkayastha, A. Ghose et al. 2012. Hypertension in chronic arsenic exposure: A case control study in West Bengal. *Journal of Environmental Science and Health: Part A*, 47(11): 1514–1520.

Mazumder, D. N. G., C. Steinmaus, P. Bhattacharya et al. 2005. Bronchiectasis in persons with skin lesions resulting from arsenic in drinking water. *Epidemiology* 16(6):760–765.

Mazzella, A. and A. Mazzella. 2013. The importance of the model choice for experimental semivariogram modeling and its consequence in evaluation process. *Journal of Engineering*, 2013:1–10. [http://dx.doi.org/10.1155/2013/960105].

Mbotake, I. T. 2006. A priliminary study of sources of arsenic contamination in southwest Cameroon. *Journal of Environmental Hydrology* 14:1–11.

McArthur, J. M., D. M. Banerjee, K. A. Hudson-Edwards et al. 2004. Natural organic matter in sedimentary basins and its relation to arsenic in anoxic ground water: The example of West Bengal and its worldwide implications. *Applied Geochemistry* 19:1255–1293.

McBean, E. A. 2013. Risk characterization for arsenic-impacted water sources, including ground-truthing. *Stochastic Environmental Research and Risk Assessment* 27:705–711.

McCall, M. and P. Minang. 2005. Assessing participatory GIS for community-based natural resource management: Claiming community forests in Cameroon. *The Geographical Journal* 171(4):340–356.

McClelland, G. H., W. D. Schulze, B. Hurd. 1990. The effect of risk beliefs on property values: A case study of hazardous waste site. *Risk Analysis* 10(4):485–497.

McClintock, T. R., Y. Chen, F. Parvez et al. 2014. Association between arsenic exposure from drinking water and hematuria: Results from the Health Effects of Arsenic Longitudinal Study. *Toxicology and Applied Pharmacology* 276(1):21–27.

McClintock, T. R., F. Parvez, F. Wu et al. 2016. Major dietary patterns and carotid intima-media thickness in Bangladesh. *Public Health Nutrition* 19(2):218–229.

McCullagh, P. and J. A. Nelder. 1989. *Generalized Linear Models*. London: Chapman and Hall.

McDonald, C., R. Hoque, N. Huda. 2007. Risk of arsenic-related skin lesions in Bangladeshi villages at relatively low exposure: A report from Gonoshasthaya Kendra. *Bulletin of the World Health Organization* 85(9):668–673.

McGrath, D., C. Zhang, O. T. Carton. 2004. Geostatistical analyses and hazard assessment on soil lead in Silvermines area, Ireland. *Environmental Pollution* 127:239–248.

McKenzie, R. C., J. R. Arthur, G. J. Beckett. 2002. Selenium and the regulation of cell signaling, growth, and survival: Molecular and mechanistic aspects. *Antioxidants & Redox Signaling* 4(2):339–351.

McLaren, S. J. and N. D. Kim. 1995. Evidence for a seasonal fluctuation of arsenic in New Zeland's longest river and the effect of treatment on concentrations in drinking water. *Environmental Pollution* 90(1):67–73.

McWilliam, C. L. 2010. Phenomenology. In *The SAGE Handbook of Qualitative Methods in Health Research*, eds. I. Bourgeault, R. Dingwall, and R. de Vries, 229–248. London: Sage.

Medeiros, M.K. and A.J. Gandolfi. 2016. Bladder cancer and arsenic. In *Arsenic: Exposure Sources, Health Risks, and Mechanisms of Toxicity*, ed. J. C. States, 163–192. New Jersey: John Wiley & Sons.

Medrano, M. A., R. Boix, R. Pastor-Barriuso et al. 2010. Arsenic in public water supplies and cardiovascular mortality in Spain. *Environmental Research* 110(5):448–454.

Meek, W. J. 1955. The gentle art of poisoning. *Journal of the American Medical Association* 158:335–339.

Meharg, A. A. 2004a. Arsenic in rice - understanding a new disaster for South-East Asia. *Trends Plant Science* 9:415–417.

Meharg, A. A. 2004b. *The Venomous Earth: How Arsenic Caused the World's Worst Poisoning*. Macmillan, Basingstoke: Palgrave.

Meharg, A. A. 2005. *Venomous Earth: How Arsenic Caused the World's Worst Mass Poisoning*. Hampshire: Macmillan.

Meharg, A. A., E. Lomb, P. N. Williams, K. G. Scheckel, J. Feldmann, A. Raab, Y. Zhu, R. Islam. 2008. Speciation and localization of arsenic in white and brown rice grains. *Environmental Science and Technology* 42:1051–1057.

Meharg, A. A. and M. Rahman. 2003. Arsenic contamination of Bangladesh paddy field soils: Implications for rice contribution to arsenic consumption. *Environmental Science and Technology* 37:229–234.

Mehrjardi, R. T., M. Z. Jahromi, S. Mahmodi et al. 2008. Spatial distribution of groundwater quality with geostatistics, case study: Yazd-Ardakan plain. *World Applied Sciences Journal* 4(1):9–17.

Melak, D., C. Ferreccio, D. Kalman et al. 2014. Arsenic methylation and lung and bladder cancer in a case-control study in northern Chile. *Toxicology and Applied Pharmacology* 274:225–231.

Meliker, J. R., M. J. Slotnick, G. A. AvRuskin et al. 2010. Lifetime exposure to arsenic in drinking water and bladder cancer: A population-based case-control study in Michigan, USA. *Cancer Causes and Control* 21(5):745–757.

Meliker, J. R., R. L. Wahl, L. L. Cameron et al. 2007. Arsenic in drinking water and cerebrovascular disease, diabetes mellitus, and kidney disease in Michigan: A standardized mortality ratio analysis. *Environmental Health* 6:4 (https://doi.org/10.1186/1476-069X-6-4).

Melkonian, S., M. Argos, Y. Chen et al. 2012. Intakes of several nutrients are associated with incidence of arsenic-related keratotic skin lesions in Bangladesh. *The Journal of Nutrition* 142(12):2128–2134.

Mendez, M. A., C. González-Horta, B. Sánchez-Ramírez et al. 2016. Chronic exposure to arsenic and markers of cardiometabolic risk: A cross-sectional study in Chihuahua, Mexico. *Environmental Health Perspectives* 124(1):104–111.

Menoni, A. 2004. Land use planning in hazard mitigation: Intervening in social and systematic vulnerabilities - an application to seismic risk prevention. In *Natural Disasters and Sustainable Development*, eds. R. Casale and C. Margottini, 165–182. Berlin: Springer.

Meranger, J. C., K. S. Subramanian, R. F. McCurdy. 1999. Arsenic in Nova Scotian groundwater. *Science of the Total Environment* 39:49–55.

Merola, R. B., T. T. Hien, D. T. T. Quyen et al. 2015. Arsenic exposure to drinking water in the Mekong Delta. *Science of the Total Environment*, 511:544–552.

Merriam, S. B. 2009. *Qualitative Research: A Guide to Design and Implementation*. San Francisco, CA: John Wiley & Sons.

Merry, R. H., K. G. Tiller, A. M. Alston. 1983. Accumulation of copper, lead, and arsenic in some Australian orchard soils. *Australian Journal of Soil Research* 21:549–561.

Michalowska, B. 2004. *Bringing aid agencies to account*. (http://lawzone.thelawyer.com/cgi-bin/item.cgi?id=110514&d=205&h=207&f=259). Accessed 2 February 2006.

Middleton, D. R. S., M. J. Watts, E. M. Hamilton et al. 2016. Urinary arsenic profiles reveal exposures to inorganic arsenic from private drinking water supplies in Cornwall, UK. *Scientific Reports* 6:25656. (doi:10.1038/srep25656).

Miles, M. and A. Huberman. 1994. *Qualitative Data Analysis*. London: Sage.

Milintawisamai, M., S. Boonchalermkit, M. Fukuda et al. 1997. Application of isotope technique to study groundwater pollution by arsenic in Nakhon Si Thammarat Province, Thailand. The paper presented at the *Conference on Toxic Metal Studies of Pak Panang and Pattani River Basins*, Research and Development office, Prince of Songkla University, Thailand, 17 October.

Millward, G. E., H. J. Kitts, L. Ebdon et al. 1997. Arsenic in the Thames Plume, UK. *Marine Environmental Research* 44(1):51–67.

Milton, A. H., Z. Hasan, A. Rahman et al. 2001. Chronic arsenic poisoning and respiratory effects in Bangladesh. *Journal of Occupational Health* 43:136–140.

Milton, A. H., W. Smith, B. Rahman et al. 2005. Chronic arsenic exposure and adverse pregnancy outcomes in Bangladesh. *Epidemiology* 16(1):82–86.

Mimi, Z. A. and A. Assi. 2009. Intrinsic vulnerability, hazard and risk mapping for karst aquifers: A case study. *Journal of Hydrology* 364:298–310.

Minatel, B. C., A. P. Sage, C. Anderson et al. 2018. Environmental arsenic exposure: From genetic susceptibility to pathogenesis. *Environment International* 112:183–197.

Mintzberg, H. 1988. Opening up the definition of strategy. In *The Strategy Process: Concept, Context and Cases*, eds. J. B. Quinn, H. Hintzberg, and R. M. James, 13–20. Englewood Cliffs, NJ: Prentice Hall.

Mirlean, N., P. Baisch, D. Diniz. 2014. Arsenic in groundwater of the Paraiba do Sul delta, Brazil: An atmospheric source? *Science of the Total Environment* 482–483:148–156.

Mitchell, J. K. 1990. Human dimensions of environmental hazards, complexity, disparity, and the search for guidance. In *Nothing to Fear*, ed. A. Kirby, 131–175. Tucson: University of Arizona Press.

Mitchell-Heggs, C. A., M. Conway, J. Cassar. 1990. Herbal medicine as a cause of combined lead and arsenic poisoning. *Human & Experimental Toxicology* 9:195–196.

Mitsunobu, S., N. Hamanura, T. Kataoka et al. 2013. Arsenic attenuation in geothermal streamwater coupled with biogenic arsenic(III) oxidation. *Applied Geochemistry* 35:154–160.

Mizan, S. A., A. Chatterjee, S. Ahmed. 2017. Arsenic enrichment in groundwater in southern flood plain of Ganga-Son interfluves. *Arabian Journal of Geosciences* 10:100. (doi:10.1007/s12517-017-2880-9).

Mjørud, M., K. Engedal, J. Røsvik et al. 2017. Living with dementia in a nursing home, as described by persons with dementia: A phenomenological hermeneutic study. *BMC Health Services Research* 17:93. (doi:10.1186/s12913-017-2053-2).

Moeller, K., H. Copes, A. Hochstetler. 2016. Advancing restrictive deterrence: A qualitative meta-synthesis. *Journal of Criminal Justice* 46:82–93.

MoH. 2004. *Survey Report on Arsenic Determination in Mongolia*. Ullanbaatar: Ministry of Health, Public Health Institute.

Mok, W. M. and C. M. Wai. 1994. Mobilisation of arsenic in contaminated river waters. In *Arsenic in the Environment, Part 1: Cycling and Characterization*, ed. J. O. Nriagu, 99–117. New York: John Wiley & Sons.

Molak, V. 1997. Introduction and overview. In *Fundamentals of Risk Analysis and Risk Management*, ed. V. Molak, 1–10. Boca Raton: CRC and Lewis Publishers.

Molina, R., C. Schulz, J. Bernardos et al. 2014. Association between arsenic in groundwater and malignant tumors in La Pampa, Argentina. In *One Century of the Discovery of Arsenicosis in Latin America (1914–2014)*, eds. M. I. Litter, H. B. Nicolli, M. Meichtry et al. 644–645. *Proceedings of the 5th International Congress on Arsenic in the Environment*, May 11–16, Buenos Aires, Argentina.

Mondal, P., C. B. Majumder, B. Mohanty. 2006. Laboratory based approaches for arsenic remediation from contaminated water: Recent developments. *Journal of Hazardous Materials* B137:464–479.

Mondal, D. and D. A. Polya. 2008. Rice is a major exposure route for arsenic in Chakdaha Block, Nadia district, West Bengal, India: A probabilistic risk assessment. *Applied Geochemistry* 23:2987–2998.

Monn, C. 2001. Exposure assessment of air pollutants: A review on spatial heterogeneity and indoor/outdoor/personal exposure to suspended particulate matter, nitrogen dioxide and ozone. *Atmospheric Environment* 35:1–32.

Montes, R. V., A. M. Martinez-Graña, J. R. Martínez Catalán et al. 2016. Vulnerability to groundwater contamination, SW salamanca, Spain. *Journal of Maps* 12(issue sup1):147–155.

Moon, K., E. Guallar, A. Navas-Acien. 2012. Arsenic exposure and cardiovascular disease: An updated systematic review. *Current Atherosclerosis Reports* 14(6):542–555.

Moon, K. A., E. Guallar, J. G. Umans et al. 2013. Association between exposure to low to moderate arsenic levels and incident cardiovascular disease. A prospective cohort study. *Annals of Internal Medicine* 159(10):649–659.

Moore, L. E., M. Lu, A. H. Smith. 2002. Childhood cancer incidence and arsenic exposure in drinking water in Nevada. *Archives of Environmental Health* 57:201–206.

Moore, L. E., A. H. Smith, C. Hopenhayn-Rich et al. 1997. Micronuclei in exfoliated bladder cells among individuals chronically exposed to arsenic in drinking water. *Cancer Epidemiology, Biomarkers & Prevention* 6(1):31–36.

Morales, K. H., L. Ryan, T. L. Kuo et al. 2000. Risk of internal cancers from arsenic in drinking water. *Environmental Health Perspectives* 108:655–661.

Mordukhovich, I., R. O. Wright, C. Amarasiriwardena et al. 2009. Association between low-level environmental arsenic exposure and QT interval duration in a general population study. *American Journal of Epidemiology* 170:739–746.

Morell, I., M. V. Esteller, E. Gimenez. 2006. The presence of arsenic in arenaceous rocks: A case study of an aquifer in the Spanish Mediterranean. In *Natural Arsenic in Grondwaters of Latin America*, eds. J. Bundschuh, M. A. Armienta, P. Bhattacharya et al. Freiberg: Bergakademie (Abstract).

Morgan, B. 2005. Social protest against privatization of water: Forging cosmopolitan citizenship? In *Sustainable Justice: Reconciling International Economic, Environmental and Social Law*, eds. M. C. Cordonier-Segger and C. G. Weeramantry, 339–354. Leiden: Martinus Nijhoff.

Morra, P., S. Bagli, G. Spadoni. 2006. The analysis of human healthy risk with a detailed procedure operating in a GIS environment. *Environment International* 32:444–454.

Morris, J. S., M. Schmid, S. Newman et al. 1974. Arsenic and noncirrhotic portal hypertension. *Gastroenterology* 66:86–94.

Morrison, L. G., L. Yardley, J. Powell et al. 2012. What design features are used in effective e-health interventions? A review using techniques from critical interpretive synthesis. *Telemedicine and e-Health* 18(2):137–144.

Morse, J. M. and P. A. Field. 1995. *Qualitative Research Methods for Health Professionals*. Thousand Oaks: Sage.

Mortenson, S., M. Liu, B. R. Burleson et al. 2006. Exploring cultural and individual differences (and similarities) related to skilled emotional support. *Journal of Cross Cultural Psychology* 3:366–385.

Morton, W., G. Starr, D. Pohl et al. 1976. Skin cancer and water arsenic in Lane County, Oregon. *Cancer* 37:2523–2532.

Mortoza, S. 2003. *Case against British Geological Survey (BGS) underway in London*. Dhaka: Arsenic Crisis Info Centre. [www.bicn.com/acic/resources/infobank/mortoza/mortoza10.htm].

Mosaferi, M., M. Yunesian, S. Dastgiri et al. 2008. Prevalence of skin lesions and exposure to arsenic in drinking water in Iran. *Science of the Total Environment* 390:69–76.

Moschione, E., G. Cuzzio, M. Campins et al. 2014. Water quality data and presence of arsenic in groundwater in Trenque Lauquen County, province of Buenos Aires (Argentina). In *One Century of the Discovery of Arsenicosis in Latin America (1914–2014)*, eds. M. I. Litter, H. B. Nicolli, M. Meichtry et al. 129–131. Boca Raton: CRC Press.

Mosler, H. J., O. R. Blöchliger, J. Inauen. 2010. Personal, social, and situational factors influencing the consumption of drinking water from arsenic-safe deep tubewells in Bangladesh. *Journal of Environmental Management* 91:1316–1323.

Mostafa, M. G. and N. Cherry. 2013. Arsenic in drinking water and renal cancers in rural Bangladesh. *Occupational and Environmental Medicine* 70(11):768–773.

Moules, N. J. 2002. Hermeneutic inquiry: Paying heed to history and Hermes—An ancestral, substantive, and methodological tale. *International Journal of Qualitative Methods* 1(3):1–21. [https://doi.org/10.1177/160940690200100301].

Moules, N. J., G. McCaffery, J. C. Field et al. 2015. *Conducting Hermeneutic Research: From Philosophy to Practice*. New York: Peter Lang.

Mroczek, E. K. 2005. Contributions of arsenic and chloride from the Kaweraw geothermal field to the Tarawera River, New Zealand. *Geothermics* 34(2):218–233.

Mueller, S. H. 2002. *A geochemical characterization of groundwater near Fairbanks, Alaska, with emphasis on arsenic hydrogeochemistry*. MS thesis (unpublished), Boulder (Colorado): University of Colorado.

Muenyi, C. S., M. Ljungman, J. C. States. 2015. Arsenic disruption of DNA damage responses–potential role in carcinogenesis and chemotherapy. *Biomolecules* 5(4):2184–2193.

Mukherjee, M. 1995. Micro-level planning for rural development and participatory rural appraisal techniques: similarities and potential. In *Participatory Rural Appraisal: Methods and Applications in Rural Planning*, ed. M. Mukherjee. New Delhi: Vikas Publishing House.

Mukherjee, A. B., P. Bhattacharya, G. Jacks et al. 2006b. Groundwater arsenic contamination in India. In *Managing Arsenic in the Environment: From Soil to Human Health*, eds. R. Naidu, G. Owens, E. Smith et al. 553–593. Melbourne: CSRIO.

Mukherjee, A. and A. E. Fryar. 2008. Deeper groundwater chemistry and geochemical modeling of the arsenic affected western Bengal basin, West Bengal, India. *Applied Geochemistry* 23:863–894.

Mukherjee, A., M. Kundu, B. Basu et al. 2017. Arsenic load in rice ecosystem and its mitigation through deficit irrigation. *Journal of Environmental Management* 197:89–95.

Mukherjee, S. C., M. M. Rahman, U. K. Chowdhury et al. 2003. Neuropathy in arsenic toxicity from groundwater arsenic contamination in West Bengal, India. *Journal of Environmental Science and Health: Part A* 38:165–183.

Mukherjee, S. C., K. C. Saha, S. Pati et al. 2005. Murshidabad—one of the nine groundwater arsenic-affected districts of West Bengal, India. Part II: Dermatological, neurological, and obstetric findings. *Clinical Toxicology* 43(7):835–848.

Mukherjee, A., M. K. Sengupta, M. A. Hossain et al. 2006a. Arsenic contamination in groundwater: A global perspective with emphasis on the Asian scenario. *The Journal of Health, Population and Nutrition* 24(2):142–163.

Mumford, A. C., J. L. Barringer, W. M. Benzel et al. 2012. Microbial transformations of arsenic: Mobilization from glauconitic sediments to water. *Water Research* 46(9):2859–68.

Mumford, J. L., K. Wu, Y. Xia et al. 2007. Chronic arsenic exposure and cardiac repolarization abnormalities with QT interval prolongation in a population-based study. *Environmental Health Perspectives* 115:690–694.

Mumpower, J. 1986. An analysis of the *de minimis* strategy for risk management. *Risk Analysis* 6(4):437–446.

Mundey, M. K., M. Roy, S. Roy et al. 2013. Antioxidant potential of Ocimum sanctum in arsenic induced nervous tissue damage. *Brazilian Journal of Veterinary Pathology* 6:95–101.

Munhall, P. L. 2007. *Nursing Research: A Qualitative Perspective*. Sudbury: Jones and Bartlett.

Muñoz, M. O., J. L. G. Aróstegui, P. Bhattacharya et al. 2016. Geochemistry of naturally occurring arsenic in groundwater and surface-water in the southern part of the Poopó Lake basin, Bolivian Altiplano. *Groundwater for Sustainable Development* 2–3:104–116.

Muñoz, M.O., L. Huallpara, E. B. Coariti et al. 2014. Natural arsenic occurrence in drinking water and assessment of water quality in the southern part of the Poopó lake basin, Bolivian Altiplano. In *One Century of the Discovery of Arsenicosis in Latin America (1914–2014)*, eds. M. I. Litter, H. B. Nicolli, M. Meichtry et al. 154–156. Boca Raton: CRC Press.

Murcott, S. 2012. *Arsenic Contamination in the World: An International Source Book.* London: IWA Publishing.

Murphy, E. A. and M. Aucott. 1998. An assessment of the amounts of arsenical pesticides used historically in a geographical area. *Science of the Total Environment* 218(2–3):89–101.

Murshed, R., R. M. Douglas, G. Ranmuthugala et al. 2004. Clinicians' roles in the management of arsenicosis in Bangladesh: Interview study. *British Medical Journal* 328:493–494.

Murshid, S. 2004. Arsenic victims vs the British Geological Survey. *The Daily Star*, September 04, Dhaka, Bangladesh: The Daily Star Centre. [http://www.thedailystar.net/law/2004/09/04/right.htm].

Musil, S., T. Matoušek, J. M. Currier et al. 2014. Speciation analysis of arsenic by selective hydride generation-cryotrapping-atomic fluorescence spectrometry with flame-in-gas-shield atomizer: Achieving extremely low detection limits with inexpensive instrumentation. *Analytical Chemistry* 86:10422–10428.

Myers, M. D. 2009. *Qualitative Research in Business and Management.* London: Sage.

Myers, S. L., D. T. Lobdell, Z. Liu et al. 2010. Maternal drinking water arsenic exposure and perinatal outcomes in Inner Mongolia, China. *Journal of Epidemiology and Community Health* 64(4):325–329.

Nabi, A. H., M. M. Rahman, L. N. Islam. 2005. Evaluation of biochemical changes in chronic arsenic poisoning among Bangladeshi patients. *International Journal of Environmental Research and Public Health* 2(3–4):385–93.

Nachman, K. E., J. P. Graham, L. B. Price et al. 2005. Arsenic: A roadblock to potential animal waste management solutions. *Environmental Health Perspectives* 113(9):1123–1124.

Nagy, G. and L. Korom. 1983. Late skin symptoms of arsenic poisoning in the arsenic endemy in Bugac-Alsómonostor. *Zeitschrift Fur Hautkrankheiten* 58(13):961–964.

Nakadaira, H., K. Endoh, M. Katagiri et al. 2002. Elevated mortality from lung cancer associated with arsenic exposure for a limited duration. *Journal of Occupational and Environmental Medicine* 44(3):291–299.

Nakahara, H., M. Yanokura, Y. Murakami. 1978. Environmental effects of geothermal waste water on the nearby river system. *Journal of Radioanalytical Chemistry* 45(1):25–36.

Narany, T. S., M. F. Ramli, A. Z. Aris et al. 2014. Spatiotemporal variation of groundwater quality using integrated multivariate statistical and geostatistical approaches in Amol-Babol Plain, Iran. *Environmental Monitoring and Assessment* 186(9):5797–5815.

Nas, B. 2009. Geostatistical approach to assessment of spatial distribution of groundwater quality. *Polish Journal of Environmental Studies* 18:1073–1082.

NASC/ENPHO. 2004. *The State of Arsenic in Nepal - 2003.* Kathmandu: National Arsenic Steering Committee and Environment and Public Health Organization.

Nasreen, M. 2003. Social impacts of arsenicosis. In *Arsenic Contamination: Bangladesh Perspective*, ed. M. F. Ahmed, 340–353. Dhaka: ITN-Bangladesh.

Nath, B., S. Chakraborty, A. Burnol et al. 2009. Mobility of arsenic in the sub-surface environment: An integrated hydrogeochemical study and sorption model of the sandy aquifer materials. *Journal of Hydrology* 364:236–248.

Naujokas, M. F., B. Anderson, H. Ahsan et al. 2013. The broad scope of health effects from chronic arsenic exposure: Update on a worldwide public health problem. *Environmental Health Perspectives* 121(3):295–302.

Navas, A. and J. Machín. 2002. Spatial distribution of heavy metals and arsenic in soils of Aragón (northeast Spain): controlling factors and environmental implications. *Applied Geochemistry* 17:961–973.

Navas-Acien, A., A. R. Sharrett, E. K. Silbergeld et al. 2005. Arsenic exposure and cardiovascular disease: a systematic review of the epidemiologic evidence. *American Journal of Epidemiology* 162(11): 1037–1049.

Navas-Acien, A., E. K. Silbergeld, R. Pastor-Barriuso et al. 2008. Arsenic exposure and prevalence of type 2 diabetes in US adults. *Journal of American Medical Association* 300(7):814–822.

Navon, L. and A. Morag. 2004. Liminality as biographical disruption: Unclassifiability following hormonal therapy for advanced prostate cancer. *Social Science & Medicine* 58:2337–2347.

Näykki, T., P. Perämäki, J. Kujala et al. 2001. Optimization of a flow injection hydride generation atomic absorption spectrometric method for the determination of arsenic, antimony and selenium in iron chloride/sulfate-based water treatment chemical. *Analytica Chimica Acta* 439(2):229–238.

Neamtiu, I., M. S. Bloom, G. Gati et al. 2015. Pregnant women in Timis County, Romania are exposed primarily to low-level (<10μg/L) arsenic through residential drinking water consumption. *International Journal of Hygiene and Environmental Health* 218(4):371–379.

Neku, A. and N. Tandulkar. 2003. An overview of arsenic contamination in groundwater of Nepal and its removal at household level. *Journal of Physics* IV(107):941–944.

Nelder, J. A. and R. W. M. Wedderburn. 1972. Generalized linear models. *Journal of the Royal Statistical Society: Series A* 135:370–384.

NERC. 2005. *NERC Briefing*. Swindon: Natural Environment Research Council. 1–4. [http://www.bgs.ac.uk/news/news/BangladeshUpdate.pdf].

Neshiwat, L. F., M. L. Friedland, B. Schorr-Lesnick et al. 1992. Hepatic angiosarcoma. *The American Journal of Medicine* 93(2):219–222.

Neuman, W. L. and L. W. Kreuger. 2003. *Social Work Research Methods: Qualitative and Quantitative Approaches*. Boston: Allyn and Bacon.

Neumann, R. B., K. N. Ashfaque, A. B. M. Badruzzaman et al. 2010. Anthropogenic influences on groundwater arsenic concentrations in Bangladesh. *Nature Geoscience* 3:46–52.

Newton, D. E. 2007. *Forensic Chemistry*. New York: Infobase Publishing.

Ngai, T. K. K., R. R. Shrestha, B. Dangol et al. 2007. Design for sustainable development - Household drinking water filter for arsenic and pathogen treatment in Nepal. *Journal of Environ Science and Health, Part A* 42(12):1879–1888.

Nguyen, V. A., S. Bang, P. H. Viet et al. 2009. Contamination of groundwater and risk assessment for arsenic exposure in Ha Nam province, Vietnam. *Environment International* 35:466–472.

Nguyen, K. P. and R. Itoi. 2009. Source and release mechanism of arsenic in aquifers of the Mekong Delta, Vietnam. *Journal of Contaminant Hydrology* 103:58–69.

Nickson, R. T., J. M. McArthur, P. Ravenscroft et al. 2000. Mechanism of arsenic release to groundwater, Bangladesh and West Bengal. *Applied Geochemistry* 15:403–13.

Nickson, R. T., J. M. McArthur, B. Shrestha et al. 2005. Arsenic and other drinking water quality issues, Muzaffargarh district, Pakistan. *Applied Geochemistry* 20:55–68.

Nickson, R., J. McArther, W. Burgees et al. 1998. Arsenic poisoning of Bangladesh groundwater. *Nature* 395(6700):338.

Nickson, R., C. S. Sengupta, P. Mitra et al. 2007. Current knowledge on the distribution of arsenic in ground water in five states of India. *Journal of Environmental Science and Health: Part A* 42(12):1707–1718.

Nicolli, H. B., J. W. García, C. M. Falcón et al. 2012. Mobilization of arsenic and other trace elements of health concern in groundwater from the Salí River Basin, Tucumán Province, Argentina. *Environmental Geochemistry and Health* 34:251–62.

Nicolli, H. B., J. M. Suriano, M. A. G. Peral et al. 1989. Groundwater contamination with arsenic and other trace-elements in an area of the Pampa, province of Córdoba, Argentina. *Environmental Geology and Water Science* 14:3–16.

Nicolli, H. B., A. Tineo, J. W. García et al. 2001. Trace elements quality problem in groundwater from Tucumán, Argentina. In *Water Rock Interaction*, ed. R. Cidu, 993–996. Lisse, South Holland: Swets and Zeitlinger.

Nieuwenhuijsen, M., D. Paustenbach, R. Duarte-Davidson. 2006. New developments in exposure assessment: The impact on the practice of health risk assessment and epidemiological studies. *Environment International* 32:996–1009.

Nikolaidis, C., M. Orfanidis, D. Hauri et al. 2013. Public health risk assessment associated with heavy metal and arsenic exposure near an abandoned mine (Kirki, Greece). *International Journal of Environmental Health Research* 23(6):507–519.

NJGS. 2006. *Arsenic in New Jersey Ground Water*. New Jersey: New Jersey Geological Survey.

Nordberg, G. F. 2010. Biomarkers of exposure, effects and susceptibility in humans and their application in studies of interactions among metals in China. *Toxicology Letters* 192(1):45–49.

Norman, D. I., G. P. Miller, L. Branvold et al. 2001. Arsenic in Ghana, West Africa, groundwaters. *US Geological Survey Workshop on Arsenic in the Environment*. [http://wwwbrr.cr.usgs.gov/arsenic/FinalAbsPDF/Norman.pdf].

Norra, S., Z. A. Berner, P. Agarwala et al. 2005. Impact of irrigation with arsenic rich groundwater on soil and crops: A geochemical case study in West Bengal delta plain, India. *Applied Geochemistry* 20:1890–1906.

Nosek, C. M., M. M. Scheckel, T. Waterbury et al. 2016. The Collaborative Improvement Model: An interpretive study of revising a curriculum. *Journal of Professional Nursing* 33(1):38–50.

NRC. 1983. *Risk Assessment in the Federal Government: Managing the Process.* Washington, DC: National Research Council.
NRC. 1999a. *Risk Assessment of Radon in Drinking Water.* Washington, DC: National Research Council.
NRC. 1999b. *Arsenic in Drinking Water.* Washington, DC: The National Research Council.
NRC. 2001. *Arsenic in Drinking Water: 2001 update.* Washington, DC: National Research Council.
NRC. 2014. *Critical Aspects of EPA's IRIS Assessment of Inorganic Arsenic: Interim Report.* Washington, DC: National Research Council.
NRCS. 2000. *Research on Arsenic Contamination in the Groundwater of Terai Nepal.* Final Report, Nepal Red Cross Society, Nepal.
NRDC. 2000. *Arsenic and Old Laws.* New York: Natural Resources Defense Council.
Nriagu, J. O. 2002. Arsenic poisoning through the ages. In *Environmental Chemistry of Arsenic*, ed. W. T. Frankenberger Jr, 1–26. New York: Marcel Dekker.
Nriagu, J. O., P. Bhattacharya, A. Mukherjee et al. 2007. Arsenic in soil and groundwater: An overview. In *Arsenic in Soil and Groundwater Environment*, eds. P. Bhattacharya, A. Mukherjee, J. Bundschuh et al. 3–60. Amsterdam: Elsevier.
Nriagu, J. O., T. S. Lin, D. N. G. Mazumder et al. 2012. E-cadherin polymorphisms and susceptibility to arsenic-related skin lesions in West Bengal, India. *Science of the Total Environment* 420:65–72.
O'Bryant, S. E., M. Edwards, C. V. Menon et al. 2011. Long-term low-level arsenic exposure is associated with poorer neuropsychological functioning: A Project FRONTIER study. *International Journal of Environmental Research and Public Health* 8(3):861–874.
O'Neill, P. 1990. *Heavy Metals in Soils.* London: Black and son Ltd.
Obiri, S. 2007. Determination of heavy metals in water from boreholes in Dumasi in the Wassa West District of Western Region of Republic of Ghana. *Environmental Monitoring and Assessment* 130(1–3):455–463.
Ogawa, Y., D. Ishiyama, N. Shikazono et al. 2012. The role of hydrous ferric oxide precipitation in the fractionation of arsenic, gallium, and indium during the neutralization of acidic hot spring water by river water in the Tama River watershed, Japan. *Geochimica et Cosmochimica Acta* 86:367–383.
Ojeda, G., A. Duran, E. Duran et al. 2014. Study of the concentration of arsenic in drinking water from the Province of Tucuman, Argentina. In *One Century of the Discovery of Arsenicosis in Latin America (1914–2014)*, eds. M. I. Litter, H. B. Nicolli, M. Meichtry et al., 138–139. Boca Raton: CRC Press.
Oke, S. A. 2003. Preliminary assessment of the impact of petroleum refinery, Kaduna, Northern Nigeria on the environment and human health. *Medical Geology Newsletter* 7:11–13.
Okonkwo, J. O. 2007. Arsenic status and distribution in soils at disused cattle dip in South Africa. *Bulletin of Environmental Contamination and Toxicology* 79(4):380–383.
Okrent, D. 1980. Comment on societal risk. *Science* 208:372–375.
Olea, R. A. 1994. Fundamentals of semivariogram estimation, modeling, and usage. In *Stochastic Modeling and Geostatistics: Principles, Methods, and Case Studies*, eds. J. M. Yarus and R. L. Chambers, 27–36. Oklahoma: The American Association of Petroleum Geologists.
Olea, R. A., N. J. Raju, J. J. Egozcue et al. 2017. Advancements in hydrochemistry mapping: Methods and application to groundwater arsenic and iron concentrations in Varanasi, Uttar Pradesh, India. *Stochastic Environmental Research and Risk Assessment*, 2017:1–19. [https://doi.org/10.1007/s00477-017-1390-3].
Oliver, M. A. and A. L. Khayrat. 2001. A geostatistical investigation of the spatial variation of radon in soil. *Computers and Geosciences* 27:939–957.
Oliver, M. A. and R. Webster. 2014. A tutorial guide to geostatistics: Computing and modelling variograms and kriging. *Catena* 113:56–69.
Oliver, M. A. and R. Webster. 2015. *Basic Steps in Geostatistics: The Variogram and Kriging.* Heidelberg: Springer.
Omenn, G. S. 2007. The Risk Assessment - Risk Management Paradigm. In *Risk Assessment for Environmental Health*, eds. M. G. Robson and W. A. Toscano, 11–30. San Francisco: John Wiley & Sons.
Omenn, G. S. and E. M. Faustman. 2002. Risk assessment and risk management. In *Oxford Textbook of Public Health*, eds. R. Detels, J. McEwen, R. Beaglehole et al. 1083–1104. New York: Oxford University Press.
Ongley, L. K., L. Sherman, A. Armienta et al. 2007. Arsenic in the soils of Zimapán, Mexico. *Environmental Pollution* 145:793–799.
Oono, M., H. Masuda, M. Kusakabe. 2002. *Seasonal Change of Arsenic Concentrations in Groundwater at the south of Osaka Prefecture.* 49th Annual Meeting of the Geochemical Society of Japan, p.12, Tokyo: The Geochemical Society of Japan.
Opresko, D. M. 1992. Risk Assessment Information System Database. In *Toxicity Profile: Arsenic*, ed. Oak Ridge Reservation Environmental Restoration Program, Biomedical and Environmental Information Analysis Section, Health and Safety Research Division, Tennessee, USA.

Oremland, R. S. and J. F. Stolz. 2003. The ecology of arsenic. *Science* 300(5621):939–944.

Ortega-Guerrero, A. 2016. Evaporative concentration of arsenic in groundwater: Health and environmental implications, La Laguna Region, Mexico. *Environmental Geochemistry and Health* 39(5):987–1003.

Osorio-Yáñez, C., J. C. Ayllon-Vergara, G. Aguilar-Madrid et al. 2013. Carotid intima-media thickness and plasma asymmetric dimethylarginine in Mexican children exposed to inorganic arsenic. *Environmental Health Perspectives* 121(9):1090–1096.

Ott, W. R. 1982. Concepts of human exposure to air pollution. *Environmental International* 7:179–196.

Ott, W. R. 1990. Total human exposure: Basic concepts, EPA field studies and future research needs. *Journal of Air and Waste Management Association* 40:966–975.

Ott, W. R., A. C. Steinemann, L. A. Wallace. 2007. *Exposure Analysis*. Boca Raton: Taylor & Francis Group (CRC Press).

Ou, C., A. St-Hilaire, T. B. Ouarda et al. 2012. Coupling geostatistical approaches with PCA and fuzzy optimal model (FOM) for the integrated assessment of sampling locations of water quality monitoring networks (WQMNs). *Journal of Environmental Monitoring* 14(12):3118–3128.

Oyarzun, R., J. Lillo, P. Higueras et al. 2004. Strong arsenic enrichment in sediments from the Elqui watershed, Northern Chile: Industrial (gold mining at El Indio - Tambo district) vs. geologic processes. *Journal of Geochemical Exploration* 84(2):53–64.

Özdemir, A. 2013. *Hydrogeological Investigation of Çeltikçi (Burdur) Plain*. MSc Thesis (unpublished). Suleyman Demirel University, Turkey.

Pachauri, V., A. Mehta, D. Mishra et al. 2013. Arsenic induced neuronal apoptosis in guinea pigs is Ca^{2+} dependent and abrogated by chelation therapy: Role of voltage gated calcium channels. *Neuro Toxicology* 35:137–145.

Padawangi, R., E. Turpin, Herlily et al. 2016. Mapping an alternative community river: The case of the Ciliwung. *Sustainable Cities and Society* 20:147–157.

Paijitprapapon, A. and V. Ramnarong. 1994. Groundwater contamination by arsenic from mining industry in Ron Phibun, Thailand. Paper presented at the *2nd Asian Regional Scope Workshop on Groundwater Contamination*, Adelaide, South Australia.

Pakulska, D. and S. Czerczak. 2006. Hazardous effects of arsine: A short review. *International Journal of Occupational Medicine and Environmental Health* 19(1):36–44.

Palaneeswari, M. S., P. M. Rajan, S. Silambanan et al. 2013. Blood arsenic and cadmium concentrations in end-stage renal disease patients who were on maintenance haemodialysis. *Journal of Clinical and Diagnostic Research* 7(5):809–813.

Palm, R. 1998. Urban earthquake hazards: The impacts of culture on perceived risk and response in the USA and Japan. *Applied Geography* 18(1):35–46.

Palmer, R. E. 1969. *Hermeneutics: Interpretation Theory in Schleiermacher, Dilthey, Heidegger, and Gadamer*. Evanston: Northwestern University Press.

Pan, W. C., W. J. Seow, M. L. Kile et al. 2013. Association of low to moderate levels of arsenic exposure with risk of type 2 diabetes in Bangladesh. *American Journal of Epidemiology* 178(10):1563–1570.

Pande, S. P., L. S. Deshpande, S. N. Kaul. 2001. Laboratory and field assessment of arsenic testing field kits in Bangladesh and West Bengal, India. *Environmental Monitoring and Assessment* 68:1–18.

Panek, J. 2015. How participatory mapping can drive community empowerment - A case study of Koffiekraal, South Africa. *South African Geographical Journal* 97(1):18–30.

Panigatti, M. C., R. M. Boglione, C. A. Griffa et al. 2014. Groundwater arsenic in the central-west of the Santa Fe Province, Argentine. In *One Century of the Discovery of Arsenicosis in Latin America (1914–2014)*, eds. M. I. Litter, H. B. Nicolli, M. Meichtry et al., 159–161. Boca Raton: CRC Press.

Panthi, S. R., S. Sharma, A. K. Mishra. 2006. Recent status of arsenic contamination in groundwater of Nepal - A review. *Kathmandu University Journal of Science, Engineering and Technology* 2(1):1–11.

Pardo-Igúzquiza, E., M. Chica-Olmo, J. A. Luque-Espinar et al. 2015. Compositional cokriging formapping the probability risk of groundwater contamination by nitrates. *Science of the Total Environment* 532:162–175.

Parris, G. E. 2005. The possible use of arsine (AsH3) as therapy for malaria. *Medical Hypotheses* 64(6):1100–1101.

Parrish, A. R., X. H. Zheng, K. D. Turney et al. 1999. Enhanced transcription factor DNA binding and gene expression induced by arsenite or arsenate in renal slices. *Toxicological Sciences* 50(1):98–105.

Partington, J. R. 1962. *A Short History of Chemistry*. New York: Dover Publications, Inc.

Parvez, F., Y. Chen, P. W. Brandt-Rauf et al. 2010. A prospective study of respiratory symptoms associated with chronic arsenic exposure in Bangladesh: Findings from the Health Effects of Arsenic Longitudinal Study (HEALS). *Thorax* 65:528–533.

Parvez, F., Y. Chen, M. Argos et al. 2006. Prevalence of arsenic exposure from drinking water and awareness of its health risks in a Bangladeshi population: Results from a large population-based study. *Environmental Health Perspectives* 114:355–359.

Parvez, F., Y. Chen, P. W. Brandt-Rauf et al. 2008. Nonmalignant respiratory effects of chronic arsenic exposure from drinking water among never-smokers in Bangladesh. *Environmental Health Perspectives* 116:190–195.

Parvez, F., Y. Chen, M. Yunus et al. 2011b. Associations of arsenic exposure with impaired lung function and mortality from diseases of the respiratory system: Findings from the health effects of Arsenic Longitudinal Study (HEALS). *Epidemiology*, 22: S179.

Parvez, F., Y. Chen, M. Yunus et al. 2013. Arsenic exposure and impaired lung function: Findings from a large population-based prospective cohort study. *American Journal of Respiratory and Critical Care Medicine* 188(7):813–819.

Parvez, F., G. A. Wasserman, P. Factor-Litvak et al. 2011a. Arsenic exposure and motor function among children in Bangladesh. *Environmental Health Perspectives* 119(11):1665–1670.

Parviainen, A., K. Vaajasaari, K. Loukola-Ruskeeniemi et al. 2006. *Anthropogenic sSources in the Pirkanmaa Region in Finland*. Espoo: Geological Survey of Finland (GTK).

Patel, P. H. 2001. The crisis of capacity: A case study of arsenic contamination in Bangladesh and West Bengal. *Unpublished Honors Thesis*, Department of Environmental Science and Public Policy, Harvard University, USA.

Paterson, B. L., C. J. Dubouloz, J. Chevrier et al. 2009. Conducting qualitative metasynthesis research: Insights from a metasynthesis project. *International Journal of Qualitative Methods* 8(3):22–33.

Patgiri, D. K. and T. C. Baruah. 1995. Spatial variability of total porosity, air entry potential and saturation water content in a cultivated inceptisol II: Estimation through kriging. *Agricultural Water Management*, 27:11–16.

Patton, M. Q. 2002. *Qualitative Research and Education Methods*. Thousand Oaks: Sage.

Paul, B. K. 2004. Arsenic contamination awareness among the rural residents in Bangladesh. *Social Science and Medicine* 59:1741–1755.

Paul, B. K. 2011. *Environmental Hazards and Disasters: Contexts, Perspectives and Management*. Oxford: John Wiley & Sons.

Paul, D. S., A. W. Harmon, V. Devesa et al. 2007. Molecular mechanisms of the diabetogenic effects of arsenic: Inhibition of insulin signaling by arsenite and methlarsonous acid. *Environmental Health Perspectives* 115(5):734–742.

Pavittranon, S., K. Sripaoraya, S. Ramchuen et al. 2003. Laboratory case identification of arsenic in Ronpibul Village, Thailand (2000–2002). *Journal of Environmental Science and Health: Part A* 38(1):213–221.

Pawlowsky-Glahn, V. and J. J. Egozcue. 2016. Spatial analysis of compositional data: A historical review. *Journal of Geochemical Exploration* 164:28–32.

Pazirandeh, A., A. H. Brati, M. G. Marageh. 1998. Determination of arsenic in hair using neutron activation. *Applied Radiatiation and Isotopes* 49(7):753–59.

Pearce, F. 2001. Bangladesh's arsenic poisoning: Who is to blame? *The UNESCO Courier* 2001(January):10–13.

Pearce, F. and J. Hecht. 2002. Flawed water tests put millions at risk. *New Scientist* 2369(2002):521. [https://www.newscientist.com/issue/2369/].

Pechrada, J., K. Sajjaphan, M. J. Sadowsky. 2010. Structure and diversity of arsenic-resistant bacteria in an old tin mine area of Thailand. *Journal of Microbiology and Biotechnology* 20(1):169–178.

Pei, Q., N. Ma, J. Zhang et al. 2013. Oxidative DNA damage of peripheral blood polymorphonuclear leukocytes, selectively induced by chronic arsenic exposure, is associated with extent of arsenic-related skin lesions. *Toxicology and Applied Pharmacology* 266:143–149.

Peng, H., N. Zhang, M. He et al. 2015. Simultaneous speciation analysis of inorganic arsenic, chromium and selenium in environmental waters by 3-(2-aminoethylamino) propyltrimethoxysilane modified multi-wall carbon nanotubes packed microcolumn solid phase extraction and ICP-MS. *Talanta* 131:266–272.

Peräkylä, A. 2007. Conversation analysis. In *Qualitative Research Practice*, eds. C. Seale, G. Gobo, J. F. Gubrium, and D. Silverman, 153–167. London: Sage.

Perkins, C. 2007. Community mapping. *The Cartographic Journal* 44(2):127–137.

Peryea, F. J. 1998. Historical use of lead arsenate insecticides, resulting soil contamination and implications for soil remediation. *16th World Congress of Soil Science Proceedings*, 20–26 August 1998, Montpelllier, France (http://soils.tfrec.wsu.edu/leadhistory.htm).

Pesic, D. and V. Srdanov. 1977. Determination of arsenic premises by atomic-absorption spectrometry. *Proceedings of VI Yugoslav Conference on General and Applied Spectroscopy*, Belgrade.

Pesola, G. R., M. Argos, Y. Chen et al. 2015. Dipstick proteinuria as a predictor of all-cause and cardiovascular disease mortality in Bangladesh: A prospective cohort study. *Preventive Medicine* 78:72–77.

Peters, S. C., J. D. Blum, M. R. Karagas et al. 2006. Sources and exposure of the New Hampshire population to arsenic in public and private drinking water supplies. *Chemical Geology* 228:72–84.

Peters, S. C. and L. Burkert. 2008. The occurrence and geochemistry of arsenic in groundwaters of the Newark basin of Pennsylvania. *Applied Geochemistry* 23:85–98.

Peters, R., L. Stocken, R. Thompson. 1945. British Anti-Lewisite (BAL). *Nature* 156(3969):616–619.

Peterson, M. 2002. What is a de minimis risk? *Risk Management* 4(2):47–55.

Peterson, P. J., L. M. Benson, R. Zeive. 1981. Metalloids. In *Arsenic and Effect of Heavy Metal Pollution on Plants*, ed. M. W. Lepp. Vol 1. London: Applied Science Publications.

Peterson, M. L. and R. Carpenter. 1983. Biogeochemical processes affecting total arsenic and arsenic species distributions in an intermittently anoxic Fjord. *Marine Chemistry* 12(4):295–321.

Pettine, M., M. Camusso, W. Martinotti. 1992. Dissolved and particulate transport of arsenic and chromium in the Po River, Italy. *The Science of Total Environment* 119:253–280.

Pfeifer, H. R., A. Häussermann, J. C. Lavanchy et al. 2007. Distribution and behavior of arsenic in soils and waters in the vicinity of the former gold-arsenic mine of Salanfe, Western Switzerland. *Journal of Geochemical Exploration* 93(3):121–134.

Pfeifer, H. R., M. Hassouna, N. Plata. 2010. Arsenic in the different environmental compartments of Switzerland: An updated inventory. *COST Action 637-Meteau: 4th International Conference Proceedings 2010*, Kristianstad, Sweden.

Pfeifer, H. R. and J. Zobrist. 2002. Arsenic in drinking water - also a problem in Switzerland? *EAWAG News* 53e:15–17.

Pham, L. H., H. T. Nguyen, C. V. Tran et al. 2017. Arsenic and other trace elements in groundwater and human urine in Ha Nam province, the Northern Vietnam: Contamination characteristics and risk assessment. *Environmental Geochemistry and Health* 39(3):517–529.

Phan, K., K. W. Kim, J. H. Hashim. 2014. Environmental arsenic epidemiology in the Mekong river basin of Cambodia. *Environmental Research* 135:37–41.

Phan, K., K. W. Kim, L. Huoy et al. 2016. Current status of arsenic exposure and social implication in the Mekong River basin of Cambodia. *Environmental Geochemistry and Health* 38(3):763–772.

Phan, K., S. Sthiannopkao, K. W. Kim et al. 2010. Health risk assessment of inorganic arsenic intake of Cambodia residents through groundwater drinking pathway. *Water Research* 44(19): 5777–5788.

Phung, D., D. Connell, S. Rutherford et al. 2017. Cardiovascular risk from water arsenic exposure in Vietnam: Application of systematic review and meta-regression analysis in chemical health risk assessment. *Chemosphere* 177:167–175.

Pi, K., Y. Wang, X. Xie et al. 2015. Hydrogeochemistry of co-occurring geogenic arsenic, fluoride and iodine in groundwater at Datong Basin, northern China. *Journal of Hazardous Materials* 300:652–661.

Pi, K., Y. Wang, X. Xie et al. 2016. Multilevel hydrogeochemical monitoring of spatial distribution of arsenic: A case study at Datong Basin, northern China. *Journal of Geochemical Exploration* 161:16–26.

Piccini, C., A. Marchetti, R. Farina et al. 2012. Application of indicator kriging to evaluate the probability of exceeding nitrate contamination thresholds. *International Journal of Environmental Research* 6(4):853–862.

Pierce, B. L., M. G. Kibriya, L. Tong et al. 2012. Genome-wide association study identifies chromosome 10q24.32 variants associated with arsenic metabolism and toxicity phenotypes in Bangladesh. *PLoS Genetics* 8(2):e1002522.

Pierce, B. L., L. Tong, M. Argos et al. 2013. Arsenic metabolism efficiency has a causal role in arsenic toxicity: Mendelian randomization and gene-environment interaction. *International Journal of Epidemiology* 42(6):1862–1871.

Pilsner, J. R., X. Liu, H. Ahsan et al. 2009. Folate deficiency, hyperhomocysteinemia, low urinary creatinine, and hypomethylation of leukocyte DNA are risk factors for arsenic-induced skin lesions. *Environmental Health Perspectives* 117(2):254–60.

Pimparkar, B. D. and A. Bhave. 2010. Arsenicosis: Review of recent advances. *Journal of Association of Physicians of India* 58:617–624.

Piñeiro, A. M., J. Moreda-Piñeiro, E. Alonso-Rodríguez et al. 2013. Arsenic species determination in human scalp hair by pressurized hot water extraction and high performance liquid chromatography-inductively coupled plasma-mass spectrometry. *Talanta* 105:422–428.

Planer-Friedrich, B., C. Härtig, H. Lissner et al. 2012. Organic carbon mobilization in a Bangladesh aquifer explained by seasonal monsoon-driven storativity changes. *Applied Geochemistry* 27(12):2324–2334.

Plantamura, J., F. Dorandeu, P. Burnat et al. 2011. Arsine: An unknown industrial chemical toxic. *Annales Pharmaceutiques Françaises* 69(4):196–200.

Platanias, L. C. 2009. Biological responses to arsenic compounds. *Journal of Biological Chemistry* 284(28):18583–18587.

Plate, E. J. 2002. Risk management for hydraulic systems under hydrological loads. In *Risk, Reliability, Uncertainty, and Robustness of Water Resources Systems*, eds. J. J. Bogardi and Z. W. Kundewicz, 209–220. Cambridge: Cambridge University Press.

Pohl, P. and B. Pruisisz. 2004. Ion-exchange column chromatography - An attempt to speciate arsenic. *Trends in Analytical Chemistry* 23(1):63–69.

Pokhrel, D., B. S. Bhandari, T. Viraraghavan. 2009. Arsenic contamination of groundwater in the Terai region of Nepal: An overview of health concerns and treatment options. *Environment International* 35:157–161.

Polit, D. F. and C. T. Beck. 2012. *Nursing Research: Generating and Assessing Evidence for Nursing Practice*. Philadelphia: Wolters Kluwer Health/Lippincott Williams & Wilkins.

Polizzotto, M. L., B. D. Kocar, S. G. Benner et al. 2008. Nearsurface wetland sediments as a source of arsenic release to ground water in Asia. *Nature* 454:505–508.

Polya, D. A., D. Mondal, A. K. Giri et al. 2009. Probabilistic risk assessment of groundwater arsenic. *Geological Society of America Abstracts with Programs* 41:344.

Pontes, B. M. S., E. de A. Menor, J. A. Figueiredo. 2014. Arsenic, selenium and lead contamination from the waters in surface Itapessoca catchment, northeastern Brazil. In *One Century of the Discovery of Arsenicosis in Latin America (1914–2014)*, eds. M. I. Litter, H. B. Nicolli, M. Meichtry et al. 125–128. Boca Raton: CRC Press.

Pories, M. L., J. Hodgson, M. A. Rose et al. 2016. Following bariatric surgery: An exploration of the couples' experience. *Obesity Surgery* 26(1):54–60.

Porter, R. 1997. *The Greatest Benefit to Mankind: A Medical History of Humanity*. London: Harper Collins.

Postma, D., F. Larsen, N. T. Thai et al. 2012. Groundwater arsenic concentration in Vietnam controlled by sediment age. *Nature Geoscience* 5:656–661.

Powell, R. A. and H. M. Single. 1996. Methodology matters - V: Focus groups. *International Journal for Quality in Health Care* 8(5):499–504.

Pradhan, B. 2006. Arsenic contaminated drinking water and nutrition status of the rural communities in Bagahi village, Rautahat district, Nepal. *Journal of Institute of Medicine* 28(2):47–51.

Prakash, V. B. 1994. Arsenic and Ayurveda. *Leukemia & Lymphoma* 16(1–2):189–190.

Prasad, P. and D. Sinha. 2017. Low-level arsenic causes chronic inflammation and suppresses expression of phagocytic receptors. *Environmental Science and Pollution Research* 24(12):11708–11721.

Pu, Y. S., S. M. Yang, Y. K. Huang et al. 2007. Urinary arsenic profile affects the risk of urothelial carcinoma even at low arsenic exposure. *Toxicology and Applied Pharmacology* 218(2):99–106.

Pugh, C. and B. Criddle. 2004. Environmental claims and personal injury - an overview. (http://studylib.net/doc/8902130/environmental-claims-and-personal). Accessed 2 February 2006.

Puginier, O. 2001. Can participatory land use plan at community level in the highlands of northern Thailand use geographic information systems (GIS) as a communication tool? Paper presented at the *International Workshop on "Participatory Technology Development and Local Knowledge for Sustainable Land Use in South East Asia"*. Chaing Mai: Thailand, 6–7:6.

Pullen-James, S. and S. E. Woods. 2006. Occupational arsine gas exposure. *Journal of the National Medical Association* 98:1998–2001.

Puntoriero, M., A. V. Volpedo, A. Fernández Cirelli. 2014. Arsenic, fluoride, and vanadium in surfacewater (Chasicó Lake, Argentina). *Frontiers in Environmental Science* 2:1–5.

Putnam, L. L. and S. Banghart. 2017. Interpretive approaches. In *The International Encyclopedia of Organizational Communication*, eds. C. R. Scott and L. Lewis, 1–17. Chennai: Wiley Blackwell.

Pyrcz, M. J. and C. V. Deutsch. 2014. *Geostatistical Reservoir Modeling*. New York: Oxford University Press.

Quamruzzaman, Q., M. Rahman, M. A. Salam et al. 2003. Painting tubewells red or green alone does not help arsenicosos patients. In *Arsenic Exposure and Health Effects IV: Proceedings of the Fourth International Conference on Arsenic Exposure and Health Effects*, eds. W. R. Chappell, C. O. Abernaty, R. L. Calderon et al. 415–419. Amsterdam: Elsevier.

Quansah, R., F. A. Armah, D. K. Essumang et al. 2015. Association of arsenic with adverse pregnancy outcomes/infant mortality: A systematic review and meta-analysis. *Environmental Health Perspectives* 123(5):412–421.

Radloff, K. A., Y. Zheng, M. Stute et al. 2017. Reversible adsorption and flushing of arsenic in a shallow, Holocene aquifer of Bangladesh. *Applied Geochemistry* 77:142–157.

Radosavljević, V. and B. Jakovljević. 2008. Arsenic and bladder cancer: Observations and suggestions. *Journal of Environmental Health* 71(3):40–42.

Rahman, F. A., D. L. Allan, C. J. Rosen et al. 2004. Arsenic availability from chromated copper arsenate (CCA)-treated wood. *Journal of Environmental Quality* 33(1):173–180.

Rahman, M. A. and H. Hasegawa. 2011. High levels of inorganic arsenic in rice in areas where arsenic-contaminated water is used for irrigation and cooking. *Science of the Total Environment* 409(22):4645–4655.

Rahman, M. M., D. Mondal, B. Das et al. 2014. Status of groundwater arsenic contamination in all 17 blocks of Nadia district in the state of West Bengal, India: A 23-year study report. *Journal of Hydrology: Part C* 518:363–372.

Rahman, M., D. Mukherjee, M. Sengupta et al. 2002. Effectiveness and reliability of arsenic field testing kits: are the million-dollar screening projects effective or not? *Journal of Environmental Science and Technology* 36(24):5385–94.

Rahman, A., L. Å. Persson, B. Nermell et al. 2010. Arsenic exposure and risk of spontaneous abortion, stillbirth, and infant mortality. *Epidemiology* 21(6):797–804.

Rahman, M., N. Sohel, S. K. Hore et al. 2015. Prenatal arsenic exposure and drowning among children in Bangladesh. *Global Health Action* 8:28702.

Rahman, M., M. Tondel, S. A. Ahmad et al. 1998. Diabetes mellitus associated with arsenic exposure in Bangladesh. *American Journal of Epidemiology* 148:198–203.

Rahman, M., M. Tondel, S. A. Ahmad et al. 1999. Hypertension and arsenic exposure in Bangladesh. *Hypertension* 33:74–78.

Rahman, A., M. Vahter, E. C. Ekström et al. 2007. Association of arsenic exposure during pregnancy with fetal loss and infant death: a cohort study in Bangladesh. *American Journal of Epidemiology* 165:1389–1396.

Rahman, A., M. Vahter, A. H. Smith et al. 2009. Arsenic exposure during pregnancy and size at birth: a prospective cohort study in Bangladesh. *American Journal of Epidemiology* 169(3):304–312.

Rahman, M., M. Vahter, N. Sohel et al. 2006. Arsenic exposure and age- and sex-specific risk for skin lesions: A population-based case-referent study in Bangladesh. *Environmental Health Perspectives* 114(12):1847–1852.

Rajaković, L. V., Ž. N. Todorović, V. N. Rajaković-Ognjanović et al. 2013. Analytical methods for arsenic speciation analysis. *Journal of the Serbian Chemical Society* 78(10):1461–1479.

Ranft, U., P. Miskovic, B. Pesch et al. 2003. Association between arsenic exposure from a coal-burning power plant and urinary arsenic concentrations in Prievidza District, Slovakia. *Environmental Health Perspectives* 111(7):889–894.

Rao, P. S. 2004. *Second report on the legal regimes for the allocation of loss in case of transboundary harm arising out of hazardous activities*. United Nations General Assembly 56th Session A/CN.4/540, United Nations, New York.

Rapant, S. and K. Krčmová. 2007. Health risk assessment maps for arsenic groundwater content: Application of national geochemical databases. *Environmental Geochemistry and Health* 29(2):131–144.

Rapley, T. 2007. Interviews. In *Qualitative Research Practice*, eds. C. Seale, G. Gobo, J. F. Gubrium, and D. Silverman, 15–33. London: Sage.

Raqib, R., S. Ahmed, R. Sultana et al. 2009. Effects of in utero arsenic exposure on child immunity and morbidity in rural Bangladesh. *Toxicology Letters* 185:197–202.

Rasheed, H., P. Kay, R. Slack et al. 2017. Human exposure assessment of different arsenic species in household water sources in a high risk arsenic area. *Science of the Total Environment* 584–585:631–641.

Rashid, H. E. 1991. *Geography of Bangladesh*. Dhaka: University Press Limited.

Rasool, A., T. Xiao, Z. T. Baig et al. 2015. Co-occurrence of arsenic and fluoride in the groundwater of Punjab, Pakistan: source discrimination and health risk assessment. *Environmental Science and Pollution Research International* 22(24):19729–746.

Rasool, A., T. Xiao, A. Farooqi et al. 2016. Arsenic and heavy metal contaminations in the tube well water of Punjab, Pakistan and risk assessment: A case study. *Ecological Engineering* 95:90–100.

Rasul, S. B., A. K. M. Munir, Z. A. Hossain et al. 2002. Electrochemical measurement and speciation of inorganic arsenic in groundwater of Bangladesh. *Talanta* 58(1):33–43.

Ratnaike, R. N. 2003. Acute and chronic arsenic toxicity. *Postgraduate Medical Journal* 79(933):391–396.

RATSC. 1999. *Risk Assessment Approaches Used by UK Government for Evaluating Human Health Effects of Chemicals*. Risk Assessment, Toxicology Steering Committee, Institute for Environment and Health, UK.

Ravbar, N. and N. Goldscheider. 2007. Proposed methodology of vulnerability and contamination risk mapping for the protection of karst aquifers in Slovenia. *Acta Carsologica* 36(3):397–411.

Ravenscroft, P., H. Brammer, and K. Richards. 2009. *Arsenic Pollution: A Global Synthesis*. Singapore: Wiley-Blackwell.

Ravenscroft, P., J. M. McArthur, M. A. Hoque. 2013. Stable groundwater quality in deep aquifers of Southern Bangladesh: The case against sustainable abstraction. *Science of the Total Environment* 454–455:627–638.

Reay, T. and Z. Zhang. 2014. Qualitative methods in family business research. In *The Sage Handbook of Family Business*, eds. L. Melin, M. Nordqvist, P. Sharma, 573–593. London: Sage.

Rebolledo, B., A. Gil, X. Flotats et al. 2016. Assessment of groundwater vulnerability to nitrates from agricultural sources using a GIS-compatible logic multicriteria model. *Journal of Environmental Management* 171:70–80.

Recio-Vega, R., T. Gonzalez-Cortes, E. Olivas-Calderon et al. 2014. In utero and early childhood exposure to arsenic decreases lung function in children. *Journal of Applied Toxicology* 35(4):358–66.

Regelson, W., U. Kim, J. Ospina et al. 1968. Hemangioendothelial sarcoma of liver from chronic arsenic intoxication by Fowler's solution. *Cancer* 21:514–522.

Rehman F., M. R. Jan, F. Bibi. 2009. *Determination of Lead and Arsenic in Drinking Water of the Selected Localities of Pakistan. Strengthening the Collaboration between the AASA Clean Water Programme and the IAP Water Programme: Proceedings of the Regional Workshop*. October 20–23. Barnual, Russia.

Rehman, Z. U., S. Khan, K. Qin et al. 2016. Quantification of inorganic arsenic exposure and cancer risk via consumption of vegetables in southern selected districts of Pakistan. *Science of the Total Environment* 550:321–329.

Reichard, J. F. and A. Puga. 2010. Effects of arsenic exposure on DNA methylation and epigenetic gene regulation. *Epigenomics* 2(1):87–104.

Reichard, J. F., M. Schnekenburger, A. Puga. 2007. Long term low-dose arsenic exposure induces loss of DNA methylation. *Biochemical and Biophysical Research Communications* 352(1):188–192.

Reimann, C., K. Bjorvatn, B. Frengstad et al. 2003. Drinking water quality in the Ethiopian section of the East African Rift Valley I - data and health aspects. *Science of the Total Environment* 311:65–80.

Reiners, G. M. 2012. Understanding the difference between Husserl's (descriptive) and Heidegger's (interpretive) phenomenological research. *Journal of Nursing Care* 1(5):1–3. [http://dx.doi.org/10.4172/2167-1168.1000119].

Ren, X., C. M. McHale, C. F. Skibola et al. 2011. An emerging role for epigenetic dysregulation in arsenic toxicity and carcinogenesis. *Environmental Health Perspectives* 119(1):11–19.

Ren, J. L., X. Z. Zhang, Y. X. Sun et al. 2016. Antimony and arsenic biogeochemistry in the East China Sea. *Deep Sea Research Part II: Topical Studies in Oceanography* 124:29–42.

Renwick, A. G., Barlow, S. M., I. Hertz-Picciotto et al. 2003. Risk characterization of chemicals in food and diet. *Food and Chemical Toxicology* 41(9):1211–1271.

Rest, A. 1998. The indispensability of an International Environmental Court. *Review of European Community and International Environmental Law* 7:63–67.

Rezaei, M., M. J. Khodayar, E. Seydi et al. 2017. Acute, but not chronic, exposure to arsenic provokes glucose intolerance in rats: Possible roles for oxidative stress and the adrenergic pathway. *Canadian Journal of Diabetes* 41(3):273–280.

Rezende, P. S., L. M. Costa, C. C. Windmöller. 2015. Arsenic mobility in sediments from Paracatu River Basin, MG, Brazil. *Archives of Environmental Contamination and Toxicology* 68(3):588–602.

Rich, M. and K. R. Ginsburg. 1999. The reason and rhyme of qualitative research: Why, when, and how to use qualitative methods in the study of adolescent health. *Journal of Adolescent health* 25(6):371–378.

Richards, L. A., J. Sültenfuß, C. J. Ballentine et al. 2017. Tritium tracers of rapid surface water ingression into arsenic-bearing aquifers in the Lower Mekong Basin, Cambodia. *Procedia Earth and Planetary Science* 17:845–848.

Richardson, L. and St. E. A. Pierre. 2005. Writing: A method of inquiry. In *Handbook of Qualitative Research*, eds. N. K. Denzin and Y. S. Lincoln, 959–978. Thousand Oaks: Sage.

Ricoeur, P. 1981. *Hermeneutics and the Human Sciences: Essays on Language, Action and Interpretation*. Cambridge: Cambridge University Press.

Riedel, F. N. and T. Eikmann. 1986. Natural occurrence of arsenic and its compounds in soils and rocks. *Wissenschaftszentrum Umwelt* 3–4:108–117.

Riethmiller, S. 2005. From Atoxyl to Salvarsan: Searching for the magic bullet. *Chemotherapy* 51:235–242.

Rimal, R. N., M. Lapinski, R. Cook et al. 2005. Moving toward a theory of normative influences: How perceived benefits and similarity moderate the impact of descriptive norms on behaviors. *Journal of Health Communication* 10:433–450.

Rivara, M. I., M. Cebrián, G. Corey et al. 1997. Cancer risk in an arsenic-contaminated area of Chile. *Toxicology and Industrial Health* 13(2–3):321–338.

Rizzo, D. M. and D. E. Dougherty. 1994. Characterization of aquifer properties using artificial neural networks: Neural Kriging. *Water Resource Research* 30:483–497.

Robbins, P. 2003. Beyond ground truth: GIS and the environment knowledge of herders, professional foresters, and other traditional communities. *Human Ecology* 31(2):233–253.

Roberts, L. C., S. J. Hug, A. Voegelin et al. 2011. Arsenic dynamics in porewater of an intermittently irrigated paddy field in Bangladesh. *Environmental Science and Technology* 45(3):971–976.

Robinson, G. R. Jr. and J. D. Ayotte. 2006. The influence of geology and land use on arsenic in stream sediments and ground waters in New England, USA. *Applied Geochemistry* 21: 1482–1497.

Robinson, D. A., I. Lebron, B. Kocar et al. 2009. Time-lapse geophysical imaging of soil moisture dynamics in tropical deltaic soils: An aid to interpreting hydrological and geochemical processes. *Water Resources Research* 45(1):W00D32.

Robles, A. D., P. Polizzi, M. B. Romero et al. 2016. Geochemical mobility of arsenic in the surficial waters from Argentina. *Environmental Earth Sciences* 75:1479.

Robles-Osorio, M. L., I. N. Pérez-Maldonado, D. Martín delCampo et al. 2012. Urinary arsenic levels and risk of renal injury in a cross-sectional study in open population. *Revistade Investigación Clínica* 64:609–614.

Robson, M. 2003. Methodologies for assessing exposures to metals: Human host factor. *Ecotoxicology and Environmental Safety* 56:104–109.

Robson, M. and F. Ellerbusch. 2007. Introduction to risk assessment in public health. In *Risk Assessment for Environmental Health*, eds. M. Robson and W. Toscano, 1–10. San Francisco: John Wiley & Sons.

Rodricks, J. and M. R. Taylor. 1983. Application of risk assessment to food safety decision making. *Regulatory Toxicology and Pharmacology* 3:275–307.

Rodríguez, V. M., M. E. Jiménez-Capdeville, M. Giordano. 2003. The effects of arsenic exposure on the nervous system. *Toxicology Letters* 145(1):1–18.

Rodríguez, R., J. A. Ramos, A. Armienta. 2004. Groundwater arsenic variations: The role of local geology and rainfall. *Applied Geochemistry* 19(2):245–250.

Rodríguez-Lado, L., G. Sun, M. Berg et al. 2013. Groundwater arsenic contamination throughout China. *Science* 341(6148):866–868.

Rohe, G. H. 1897. Arsenic. In *Reference-Book of Practical Therapeutics*, ed. F. P. Foster. Volume 1, New York: D. Appleton and Company.

Roig-Navarro, A. F., Y. Martinez-Bravo, F. J. López et al. 2001. Simultaneous determination of arsenic species and chromium(VI) by high-performance liquid chromatography-inductively coupled plasma-mass spectrometry. *Journal of Chromatography A* 912(2):319–327.

Rojewski, M. T., C. Baldus, W. Knauf et al. 2002. Dual effects of arsenic trioxide (As2O3) on non-acute promyelocytic leukaemia myeloid cell lines: Induction of apoptosis and inhibition of proliferation. *British Journal of Haematology* 116(3):555–563.

Röllin, H. B., K. Channa, B. G. Olutola et al. 2017. Evaluation of in utero exposure to arsenic in South Africa. *Science of the Total Environment* 575:338–346.

Roman, S. and V. Peuraniemi. 1999. The effect of bedrock, glacial deposits and a waste disposal site on groundwater quality in the Haukipudas area, Northern Finland. In *Groundwater in the Urban Environment*, ed. J. Chilton, 329–334. Rotterdam: Balkema.

Romero, L., H. Alonso, P. Campano et al. 2003. Arsenic enrichment in waters and sediments of the Rio Loa (Second Region, Chile). *Applied Geochemistry* 18(9):1399–1416.

Romero-Schmidt, H., A. Naranjo-Pulido, L. Méndez-Rodríguez et al. 2001. Environmental health risks by arsenic consumption in water wells in the Cape region, Mexico. In *Environmental Health Risk*, eds. C. A. Brebbia and D. Fajzieva, 131–138. Southampton: WIT Press.

Rondeau, V., D. Commenges, H. Jacqmin-Gadda et al. 2009. Aluminum and silica in drinking water and the risk of Alzheimer's disease or cognitive decline: Findings from 15-year follow-up of the PAQUID cohort. *American Journal of Epidemiology* 169(4):489–496.

Rosado, J. L., D. Ronquillo, K. Kordas et al. 2007. Arsenic exposure and cognitive performance in Mexican school children. *Environmental Health Perspectives* 115(9):1371–1375.

Rosales-Castillo, J. A., L. C. Acosta-Saavedra, R. Torres et al. 2004. Arsenic exposure and human papillomavirus response in non-melanoma skin cancer Mexican patients: A pilot study. *International Archives of Occupational and Environmental Health* 77(6):418–423.

Rosenberg, H. G. 1974. Systemic arterial disease and chronic arsenicism in infants. *Archives of Pathology* 97(6):360–365.

Rosenberg, D. 1995. *The Hidden Holmes: His Theory of Torts in History.* Cambridge, MA: Harvard University Press.

Rosenboom, J. W. 2004. *Not Just Red or Green: An Analysis of Arsenic Data from 15 Upazilas.* Dhaka: Arsenic Policy Support Unit (APSU), Bangladesh.

Rosenthal, G. 2007. Biographical research. In *Qualitative Research Practice*, eds. C. Seale, G. Gobo, J. F. Gubrium, and D. Silverman, 48–64. London: Sage.

Rossman, T. G. 2003. Mechanisms of arsenic carcinogenesis: An integrated approach. *Mutation Research* 533(1–2):37–65.

Rossman, T. G., A. N. Uddin, F. J. Burns. 2004. Evidence that arsenite acts as a cocarcinogen in skin cancer. *Toxicology and Applied Pharmacology* 198(3):394–404.

Rott, U. and M. Friedle. 1999. Subterranean removal of arsenic from groundwater. In *Arsenic Exposure and Health Effects III*, eds. W. R. Chappell, C. O. Abernathy, R. L. Calderon, 389–396. Oxford: Elsevier.

Rouhani, S. and D. E. Mayers. 1990. Problems in space-time kriging of geohydrological data. *Mathematical Geology* 22(5):611–623.

Routh, J. and O. Hjelmquist. 2011. Distribution of arsenic and its mobility in shallow aquifer sediments from Ambikanagar, West Bengal, India. *Applied Geochemistry* 26(4):505–515.

Rowe, W. D. 1977. *An Anatomy of Risk.* New York: Wiley.

Rowland, H. A. L., A. G. Gault, P. Lythgoe et al. 2008. Geochemistry of aquifer sediments and arsenic-rich groundwaters from Kandal Province, Cambodia. *Applied Geochemistry* 23(11):3029–3046.

Rowland, H. A. L., D. A. Polya, J. R. Lloyd et al. 2006. Characterisation of organic matter in a shallow, reducing, arsenic-rich aquifer, West Bengal. *Organic Geochemistry* 37(9):1101–1114.

Roy, A., K. Kordas, P. Lopez et al. 2011. Association between arsenic exposure and behavior among first-graders from Torreón, Mexico. *Environmental Research* 111(5):670–676.

Roy, R. V., Y. O. Son, P. Pratheeshkumar et al. 2015. Epigenetic targets of arsenic: Emphasis on epigenetic modifications during carcinogenesis. *Journal of Environmental Pathology, Toxicology and Oncology* 34(1):63–84.

Rudenko, A. and L. H. Tsai. 2014. Epigenetic modifications in the nervous system and their impact upon cognitive impairments. *Neuropharmacology* 80:70–82.

Rudman, D. L., M. Y. Egan, C. E. McGrath et al. 2016. Low vision rehabilitation, age-related vision loss, and risk: A critical interpretive synthesis. *The Gerontologist* 56(3):e32–e45.

Rudnai, T., J. Sándor, M. Kádár et al. 2014. Arsenic in drinking water and congenital heart anomalies in Hungary. *International Journal of Hygiene and Environmental Health* 217(8):813–818.

Ruiz-Navarro, M. L., M. Navarro-Alarcón, H. L. González-de la Serrana et al. 1998. Urine arsenic concentrations in healthy adults as indicators of environmental contamination: Relation with some pathologies. *Science of the Total Environment* 216(1–2):55–61.

Ryan, K. M., A. C. Phillips, K. H. Vousden. 2001. Regulation and function of the p53 tumor suppressor protein. *Current Opinion in Cell Biology* 13(3):332–337.

Ryan, P. C., D. P. West, K. Hattori et al. 2015. The influence of metamorphic grade on arsenic in metasedimentary bedrock aquifers: a case study from Western New England, USA. *Science of the Total Environment* 505:1320–1330.

Saad, A. and M. A. Hassanien. 2001. Assessment of arsenic level in the hair of the nonoccupational Egyptian population: Pilot study. *Environment International* 27(6):471–478.

Saady, J. J., R. V. Blanke, A. Poklis. 1989. Estimation of the body burden of arsenic in a child fatally poisoned by arsenite weedkiller. *Journal of Analytical Toxicology* 13:310–312.

Sabel, C. E., A. C. Gatrell, M. Loytonen et al. 2000. Modelling exposure opportunities: Estimating relative risk for motor neuron disease in Finland. *Social Science Medicine* 50:1121–1137.

Saha, K. 1995. Chronic arsenical dermatoses from tube-well water in West Bengal during 1983–87. *Indian Journal of Dermatology* 40:1–12.

Saha, K. C. 2003. Diagnosis of arsenicosis. *Journal of Environmental Science and Health: Part A* 38:255–272.

Saha, J., A. Dikshit, M. Bandyopadhyay et al. 1999. A review of arsenic poisoning and its effects on human health. *Critical Reviews in Environmental Science and Technology* 29:281–313.

Saltori, R. 2004. *Arsenic contamination in Afganistan preliminary findings.* Conference on Water Quality - Arsenic Mitigation, Taiyuan, 23–26 November, UNICEF and Water & Sanitation Group (Afganistan).

Samal, A. C., S. Kar, J. P. Maity et al. 2013. Arsenicosis and its relationship with nutritional status in two arsenic affected areas of West Bengal, India. *Journal of Asian Earth Sciences* 77:303–310.

Samanta, G. and D. Chakraborti. 1997. Flow injection atomic absorption spectrometry for the standardization of arsenic, lead and mercury in environmental and biological standard reference materials. *Fresenius' Journal of Analytical Chemistry* 357:827–832.

Samanta, G., T. R. Chowdhury, B. K. Mandal et al. 1999. Flow injection hydride generation atomic absorption spectrometry for determination of arsenic in water and biological samples from arsenic-affected districts of West Bengal, India, and Bangladesh. *Microchemical Journal* 62(1):174–191.

Sancha, A. M. and M. L. Castro. 2001. Arsenic in Latin America: Occurrence, exposure, health effects and remediation. In *Arsenic Exposure and Health Effects IV*, eds. W. R. Chapell, C. O. Abernathy, R. L. Calderon, 87–96. Amsterdam: Elsevier.

Sandman, P. M. 1987. Risk communication: Facing public outrage. *EPA Journal* November 1987:21–22.

Sandoval, J. and M. Esteller. 2012. Cancer epigenomics: Beyond genomics. *Current Opinion in Genetics and Development* 22(1):50–55.

Sandström, P., T. Pahlén, L. Edenius et al. 2003. Conflict resolution by participatory management: Remote sensing and GIS as tools for communicating land use needs for reindeer herding in northern Sweden. *AMIBO* 32(8):557–567.

Santolaya, B., C. L. Salazar, C. R. Santolaya et al. 1995. *Arsénico. Impacto sobre el hombre y su entorno. II Región de Chile (Antofagasta)*. Centro de Investigaciones Ecobiológicas y Médicas de Altura (CIEMA) División Chuquicamata-Codelco, Chile, pp69–87.

Santra, A., J. Das Gupta, B. K. De et al. 1999. Hepatic manifestations in chronic arsenic toxicity. *Indian Journal of Gastroenterology* 18:152–155.

Sanz, N.A., M. P. Diez, A. V. de Miguel et al. 2001. Nivel de arsenico en abastecimientos de agua de consumo de origen subterraneo en la comunidad de Madrid. *Revista Española de Salud Pública* 75:421–432 (in Spanish).

Saoudi, A., A. Zeghnoun, M. L. Bidondo et al. 2012. Urinary arsenic levels in the French adult population: The French National Nutrition and Health Study, 2006–2007. *Science of the Total Environment* 433:206–15.

Saposnik, G. 2010. Drinking water and risk of stroke: The hidden element. *Stroke* 41(11):2451–2452.

Sarkar, A. 2010. Ecosystem perspective of groundwater arsenic contamination in India and relevance in policy. *EcoHealth* 7(1):114–126.

Sarkar, S., G. Horn, K. Moulton et al. 2013. Cancer development, progression, and therapy: An epigenetic overview. *International Journal of Molecular Sciences* 14(10):21087–21113.

Sarker, M. M. R. 2008. Determinates of arsenicosis patients treatment cost in rural Bangladesh. *Bangladesh Journal of Environmental Science* 14: 80–83.

Sarker, M. M. R. 2010. Determinants of arsenicosis patients' perception and social implications of arsenic poisoning through groundwater in Bangladesh. *International Journal of Environmental Research and Public Health* 7(10):3644–3656.

Sarma, N. 2016. Skin manifestations of chronic arsenicosis. In *Arsenic: Exposure Sources, Health Risks, and Mechanisms of Toxicity*, ed. J. C. States, 127–136. New Jersey: John Wiley & Sons.

Sasaki, A., Y. Oshima, A. Fujimura. 2007. An approach to elucidate potential mechanism of renal toxicity of arsenic trioxide. *Experimental Hematology* 35(2):252–262.

Sato, H. 2010. The policies and politics of health risk management. In *Management of Health Risks from Environment and Food*, ed. H. Sato, 3–26. Amsterdam: Springer.

Saunders, S. G., D. J. Barrington, S. Sridharan et al. 2016. Addressing WaSH challenges in Pacific Island Countries: A participatory marketing systems mapping approach to empower informal settlement community action. *Habitat International* 55:159–166.

Saunders, J. R., L. D. Knopper, I. Koch et al. 2010. Arsenic transformations and biomarkers in meadow voles (Microtus pennsylvanicus) living on an abandoned gold mine site in Montague, Nova Scotia, Canada. *Science of the Total Environment* 408:829–835.

Saunders, J. A., M. K. Lee, A. Uddin et al. 2005. Natural arsenic contamination of Holocene alluvial aquifers by linked glaciation, weathering and microbial processes. *Geochemistry Geophysics Geosystems* 6(4):1–7.

Sawada, N., M. Iwasaki, M. Inoue et al. 2013. Dietary arsenic intake and subsequent risk of cancer: The Japan Public Health Center-based (JPHC) Prospective Study. *Cancer Causes Control* 24(7):1403–1415.

Saxe, J. K., T. S. Bowers, K. R. Reid. 2006. Arsenic. In *Environmental Forensics: Contaminant Specific Guide*, eds. R. D. Morrison and B. L. Murphy, 279–292. Burlington, MA: Academic Press.

Sbarato, V. M. and H. J. Sánchez. 2001. Analysis of arsenic pollution in groundwater aquifers by X-ray fluorescence. *Applied Radiation and Isotopes* 54(5):737–740.

Scandura, T. A. and E. Williams. 2000. Research methodology in management: Current practices, trends, and implications for future research. *Academy of Management Journal* 43:1248–64.

Scheindlin, S. 2005. The duplicitous nature of inorganic arsenic. *Molecular Interventions* 5:60–64.

Schiappa, E. 1996. Towards a pragmatic approach to definition: 'wetlands' and the politics of meaning. In *Environmental Pragmatism*, eds. A. Light and E. Katz, 209–230. London: Routledge.

Schleiermacher, F. D. E. 1985. General hermeneutics. In *The Hermeneutics Reader: Texts of the German Tradition from the Enlightenment to the Present*, ed. K. Mueller-Vollmer, 73–85. New York: Continuum Publishing Company.

Schmidt, C. W. 2014. Low-dose arsenic: In search of a risk threshold. *Environmental Health Perspectives* 22(5):A131–A134.

Schoen, A., B. Beck, R. Jharma et al. 2004. Arsenic toxicity of low doses: Epidemiology and mode of action consideration. *Toxicology and Applied Pharmacology* 198(3):253–267.

Schoof, R. A., L. J. Yost, E. Crecelius et al. 1998. Dietary arsenic intake in Taiwanese districts with elevated arsenic in drinking water. *Human and Ecological Risk Assessment: An International Journal* 4(1):117–135.

Schoof, R. A., L. J. Yost, J. Eickhoff. 1999. A market basket survey of inorganic arsenic in food. *Food and Chemical Toxicology* 37:839–846.

Schwenzer, S. P., C. E. Tommasco, M. Kerstn et al. 2001. Speciation and oxidation kinetics of arsenic in the thermal springs of Wiesbaden spa, Germay. *Fresenius Journal of Analytical Chemistry* 371:927–933.

Sciandrello, G., F. Caradonna, M. Mauro et al. 2004. Arsenic-induced DNA hypomethylation affects chromosomal instability in mammalian cells. *Carcinogenesis* 25(3):413–417.

Segura, M., J. Muñoz, Y. Madrid et al. 2002. Stability study of As(III), As(V), MMA and DMA by anion exchange chromatography and HG-AFS in wastewater samples. *Analytical and Bioanalytical Chemistry* 374(3):513–519.

Seip, H. M. and A. B. Heiberg. 1989. Pilot study on risk management of chemicals in the environment: An introduction. In *Risk Management of Chemicals in the Environment*, eds. H. M. Seip and A. B. Heiberg, 1–10. New York: Plenum Press (Published in cooperation with NATO Committee on the Challenges of Modern Society).

Selvaraj, V., J. Tomblin, M. Y. Armistead et al. 2013. Selenium (sodium selenite) causes cytotoxicity and apoptotic mediated cell death in PLHC-1 fish cell line through DNA and mitochondrial membrane potential damage. *Ecotoxicology and Environmental Safety* 87:80–88.

Selzer, P. M. and M. A. Ancel. 1983. Chronic arsenic poisoning masquerading as pernicious anemia. *The Western Journal of Medicine* 139(2):219–220.

Semenova, N. V., L. O. Leal, R. Forteza et al. 2002. Multisyringe flow-injection system for total inorganic arsenic determination by hydride generation-atomic fluorescence spectrometry. *Analytica Chimica Acta* 455(2):277–285.

Sen, D. and P. S. Biswas. 2012. Arsenicosis: Is it a protective or predisposing factor for mental illness? *Iranian Journal of Psychiatry* 7(4):180–183.

Sengupta, S. R., N. K. Das, P. K. Datta. 2008. Pathogenesis, clinical features and pathology of chronic arsenicosis. *Indian Journal of Dermatology, Venereology and Leprology* 74(6):559–570.

Sengupta, S., O. Sracek, J. S. Jean et al. 2014. Spatial variation of groundwater arsenic distribution in the Chianan Plain, SW Taiwan: Role of local hydrogeological factors and geothermal Sources. *Journal of Hydrology* 518:393–409.

Seow, W. J., M. L. Kile, A. A. Baccarelli et al. 2014. Epigenome-wide DNA methylation changes with development of arsenic-induced skin lesions in Bangladesh: A case-control follow-up study. *Environmental and Molecular Mutagenesis* 55(6):449–456.

Serfes, M. E., S. E. Spayd, G. C. Herman. 2005. Arsenic occurrence, sources, mobilization, and transport in groundwater in the Newark Basin of New Jersey. In *Advances in Arsenic Research: Integration of Experimental and Observational Studies and Implications for Mitigation*, eds. P. A. Oday, D. Vlassopoulos, Z. Meng, L. G. Benning, 175–190. Washington, DC: American Chemical Society.

Serfor-Armah, Y., B. V. Samlafo, P. O. Yeboah. 2009. Arsenic and mercury levels in human hairs and nails from gold mining areas in Wassa West district of Ghana. *Journal of Applied Science and Technology* 14(1–2):117–122.

Serón, F. J., J. I. Badal, F. J. Sabadell. 2001. Spatial prediction procedures for regionalization and 3-D imaging of Earth structures. *Physics of the Earth and Planetary Interiors* 123(2–4):149–168.

Serre, M. L., A. Kolovos, G. Christakos et al. 2003. An application of the holistochastic human exposure methodology to naturally occurring arsenic in Bangladesh Drinking Water. *Risk Analysis* 23:515–528.

Sexton, K., M. A. Callahan, E. F. Ryan et al. 1995b. Informed decisions about protecting and promoting public health: rationale for a national human exposure assessment survey. *Journal of Exposure Analysis and Environmental Epidemiology* 5:233–256.

Sexton, K., M. A. Callahan, E. F. Ryan. 1995c. Estimating exposure and dose to characterize health risks: the role of human tissue monitoring in exposure assessment. *Environmental Health Perspectives* 103 (supplementary 3):13–29.

Bibliography

Sexton, K., D. Kleffman, M. Callahan. 1995a. An introduction to the national human exposure assessment survey and related phase I field studies. *Journal of Exposure Analysis and Environmental Epidemiology* 5:229–232.

Sexton, K., K. Olden, B. L. Johnson. 1993. Environmental justice: The central role of research in establishing a credible scientific basis for informed decision making. *Toxicology and Industrial Health* 9:685–727.

Shady, A. M. 2008. Water scarcity: Can we live with it? *Resource* 15(3):9–11.

Shaffer, P. 1996. Beneath the poverty debate. *IDS Bulletin* 27:23–35.

Shakil, A. R. and W. B. Martin. 2000. Design improvement for pond sand filter. *Proceedings 26th WEDC Conference*, Dhaka, pp80–83.

Shamsipur, M., N. Fattahi, Y. Assadi et al. 2014. Speciation of As(III) and As(V) in water samples by graphite furnace atomic absorption spectrometry after solid phase extraction combined with dispersive liquid-liquid microextraction based on the solidification of floating organic drop. *Talanta* 130:26–32.

Shand, P., J. Cobbing, R. Tyler-White et al. 2003. *Baseline Report Series 9: The Lower Greensand of Southern England*. British Geological Survey Reprt (CR/03/273N).

Shand, P., R. Hargreaves, L. J. Brewerton. 1997. *The Natural (Baseline) Quality of Groundwaters in England and Wales: The Permo-Triassic Sandstones of Cumbria, North-West England*. Bristol (UK): Environment Agency.

Shankar, S., U. Shanker, Shikha. 2014. Arsenic contamination of groundwater: A review of sources, prevalence, health risks, and strategies for mitigation. *The Scientific World Journal* 2014:1–18. [http://dx.doi.org/10.1155/2014/304524].

Sharif, M. U., R. K. Davis, K. F. Steele et al. 2008. Distribution and variability of redox zones controlling spatial variability of arsenic in the Mississippi River Valley alluvial aquifer, southeastern Arkansas. *Journal of Contaminant Hydrology* 99(1–4):49–67.

Shearer, J. J., E. A. Wold, C. S. Umbaugh et al. 2015. Inorganic arsenic related changes in the stromal tumor microenvironment in a prostate cancer cell-conditioned media model. *Environmental Health Perspectives* 124(7):1009–1015.

Shek, D. L. and F. K. Y. Wu. 2017. The social indicators movement: Progress, paradigms, puzzles, promise and potential research directions. *Social Indicator Research* 2017:1–16. [https://doi.org/10.1007/s11205-017-1552-1].

Shen, H., W. Xu, J. Zhang et al. 2013. Urinary metabolic biomarkers link oxidative stress indicators associated with general arsenic exposure to male infertility in a Han Chinese population. *Environmental Science and Technology* 47(15):8843–8851.

Shepard, K. F., G. M. Jensen, B. J. Schmoll et al. 1993. Alternative approaches to research in physical therapy: Positivism and phenomenology. *Physical Therapy* 73(2):88–101.

Sherman, R. R. and R. B. Webb. 1988. *Qualitative Research in Education: Focus and Methods*. New York: Falmer Press.

Sherwood, C.L. and R.C. Lantz. 2016. Lung cancer and other pulmonary diseases. In *Arsenic: Exposure Sources, Health Risks, and Mechanisms of Toxicity*, ed. J. C. States, 137–162. New Jersey: John Wiley & Sons.

Shi, H., X. Shi, K. J. Liu. 2004. Oxidative mechanism of arsenic toxicity and carcinogenesis. *Molecular and Cellular Biochemistry* 255(1–2):67–78.

Shiber, J. G. 2005. Arsenic in domestic well water and health in Central Appalachia, USA. *Water, Air, and Soil Pollution* 160:327–341.

Shimada, N. 1996. Geochemical conditions enhancing the solubilization of arsenic into groundwater in Japan. *Applied Organometallic Chemistry* 10:667–674.

Shinkai, Y., D. V. Truc, D. Sumi et al. 2007. Arsenic and other metal contamination of groundwater in the Mekong River Delta, Vietnam. *Journal of Health Science* 53(3):344–346.

Shirai, S., Y. Suzuki, J. Yoshinaga et al. 2010. Maternal exposure to low-level heavy metals during pregnancy and birth size. *Journal of Environmental Science and Health: Part A* 45(11):1468–1474.

Shrestha, R. R. and A. Maskey. 2005. *Groundwater Arsenic Contamination in Terai Region of Nepal and Its Mitigation*. ENPHO, Kathmandu, Nepal.

Shrestha, R. R., M. P. Shrestha, N. P. Upadhyay et al. 2003a. Groundwater arsenic contamination, its health impact and mitigation program in Nepal. *Journal of Environmental Science and Health: Part A* 38:185–200.

Shrestha, R. R., M. P. Shrestha, N. P. Upadhyay et al. 2003b. Groundwater arsenic contamination in Nepal: A new challenge for water supply sector. In *Arsenic Exposure and Health Effects V*, eds. W. R. Chappel, C. O. Abernathy, R. I. Calderon, D. J. Thomas, 25–38. Amsterdam: Elsevier.

Shrivastava, A., A. Barla, S. Singh et al. 2017. Arsenic contamination in agricultural soils of Bengal deltaic region of West Bengal and its higher assimilation in monsoon rice. *Journal of Hazardous Materials-Part B* 15(324):526–534.

Sidhu, M. S., K. P. Desai, H. N. Lynch et al. 2015. Mechanisms of action for arsenic in cardiovascular toxicity and implications for risk assessment. *Toxicology* 331(1):78–99.

Sikdar, P. K. and S. Chakraborty. 2017. Numerical modelling of groundwater flow to understand the impacts of pumping on arsenic migration in the aquifer of North Bengal Plain. *Journal of Earth System Science*, 126:29 (doi:10.1007/s12040-017-0799-x).

Simeonova, P. P. and M. I. Luster. 2004. Arsenic and atherosclerosis. *Toxicology and Applied Pharmacology* 198(3):444–449.

Simeonova, P. P., S. Wang, W. Toriuma et al. 2000. Arsenic mediates cell proliferation and gene expression in the bladder epithelium: Association with AP-1 transactivation. *Cancer Research* 60:3445–3453.

Simsek, C., A. Elci, O. Gunduz et al. 2008. Hydrogeological and Hydrogeochemical Characterization of a Karstic Mountain Region. *Environmental Geology* 54(2):291–308.

Sinan, M. and M. Razack. 2009. An extension to the DRASTIC model to assess groundwater vulnerability to pollution: application to the Haouz aquifer of Marrakech (Morocco). *Environmental Geology* 57:349–363.

Singh, B. K. 2014. Flood hazard mapping with participatory GIS: The case of Gorakhpur. *Environment and Urbanization Asia* 5(1):161–173.

Singh, S. K. 2017. Conceptual framework of a cloud-based decision support system for arsenic health risk assessment. *Environmental Systems and Decisions* 37(4):435–450.

Singh, S. K. and A. K. Ghosh. 2012. Health risk assessment due to groundwater arsenic contamination: children are at high risk. *Human and Ecological Risk Assessment: An International Journal* 18(4):751–766.

Singh, A. P., R. K. Goel, T. Kaur. 2011. Mechanisms pertaining to arsenic toxicity. *Toxicology International* 18(2):87–93.

Singh, S. K. and E. A. Stern. 2017. Global arsenic contamination: Living with the poison nectar. *Environment: Science and Policy for Sustainable Development* 59(2):24–28.

Singh, S. K. and N. Vedwan. 2015. Mapping composite vulnerability to groundwater arsenic contamination: an analytical framework and a case study in India. *Natural Hazards* 75(2):1883–1908.

Sinha, D., J. Biswas, A. Bishayee. 2013. Nrf2-mediated redox signaling in arsenic carcinogenesis: a review. *Archives of Toxicology* 87(2):383–396.

Siripitaykunkit, U. 2000. Survey of chronic arsenic poisoning in Rhonpiboon, Nakkon Si Thammarat, Thailand. Proeedings of *the 6th International Conference on the Biochemistry of Trace Elements*, Guelph, Canada.

Smedley P. L. 1996. Arsenic in rural groundwater in Ghana. *Journal of African Earth Sciences* 22(4): 459–470.

Smedley, P. L., W. M. Edmunds, K. B. Pelig-Ba. 1996. Mobility of arsenic in groundwater in the Obuasi goldmining area of Ghana: Some implications for human health. *Environmental Geochemistry and Health* 113:163–181.

Smedley, P. L. and D. G. Kinniburgh. 2002. A review of the source, behaviour and distribution of arsenic in natural waters. *Applied Geochemistry* 17(5):517–568.

Smedley P. L., D. G. Kinniburgh, D. M. J. Macdonald et al. 2005. Arsenic associations in sediments from the loess aquifer of La Pampa, Argentina. *Applied Geochemistry* 20:989–1016.

Smedley P. L., J. Knudsen, D. Maiga. 2007. Arsenic in groundwater from mineralised Proterozoic basement rocks of Burkina Faso. *Applied Geochemistry* 22:1074–1092.

Smedley P. L., I. Neumann, R. Farrell. 2004. *Baseline Report Series 10: The Chalk of Yorkshire and North Humberside*. British Geological Survey Report (CR/04/128).

Smedley, P. L., H. B. Nicolli, D. M. J. MacDonald et al. 2002. Hydrogeochemistry of arsenic and other inorganic constituents in groundwaters from La Pampa, Argentina. *Applied Geochemistry* 17:259–284.

Smedley, P. L., H. B. Nicolli, D. M. J. Macdonald et al. 2008. Arsenic in groundwater and sediments from La Pampa Province, Argentina. In *Natural Arsenic in Groundwaters of Latin America*, eds. J. Bundschuh, M. A. Armienta, P. Birkle et al. 35–45. Boca Raton: CRC Press.

Smedley P. L., M. Zhang, G. Zhang et al. 2003. Mobilisation of arsenic and other trace elements in fluviolacustrine aquifers of the Huhhot Basin, Inner Mongolia. *Applied Geochemistry* 18:1453–1477.

Smith, D. 1997. Phenomenology: Methodology and method. In *Qualitative Research: Discourse on Methodologies*, ed. J. Higgs. Sydney: Hampden Press.

Smith, D. P. 2013. Relationships between the health of Alaska Native communities and our environment - Phase 1, Exploring and communicating. US Geological Survey Fact Sheet: 2013-3066 (https://pubs.usgs.gov/fs/2013/3066/).

Smith, J. K. 1983. Quantitative versus qualitative research: An attempt to clarify the issue. *Educational Researcher* 12(1):6–13.

Smith, R. C. 1995. GIS and long range economic planning for indigenous territories. *Cultural Survival Quarterly* 18(4):43–48.

Smith, A. H., A. P. Arroyo, D. N. G. Mazumdar et al. 2000b. Arsenic induced skin lesions among Atacameno people in northern Chile despite good nutrition and centuries of exposure. *Environmental Health Perspectives* 108:617–620.

Smith, A. H., M. Goycolea, R. Haque et al. 1998b. Marked increase in bladder and lung cancer mortality in a region of Northern Chile due to arsenic in drinking water. *American Journal of Epidemiology* 147(7):660–669.

Smith, J. L., J. J. Halvorson, R. I. Papendick. 1993. Using multiple-variable indicator kriging for evaluating soil quality. *Soil Science Society of America Journal* 57:743–749.

Smith, A. H., C. Hopenhayn-Rich, M. N. Bates et al. 1992. Cancer risks from arsenic in drinking water. *Environmental Health Perspectives* 97:259–67.

Smith J. V. S., J. Jankowski, J. Sammut. 2003. Vertical distribution of As(III) and As(V) in a coastal sandy aquifer: factors controlling the concentration and speciation of arsenic in the Stuarts Point groundwater system, northern New South Wales, Australia. *Applied Geochemistry* 18(9):1479–1496.

Smith, N. M., R. Lee, D. T. Heitkemper et al. 2006. Inorganic As in cooked rice and vegetables from Bangladeshi households. *Science of the Total Environment* 370(2–3):294–301.

Smith, A. H., E. O. Lingas, M. Rahman. 2000a. Contamination of drinking-water by arsenic in Bangladesh: A public health emergency. *Bulletin of the World Health Organization* 78:1093–1103.

Smith, A. H., P. A. Lopipero, M. N. Bates et al. 2002. Arsenic epidemiology and drinking water standards. *Science* 296:2145–2146.

Smith, A. H., G. Marshall, J. Liaw et al. 2012. Mortality in young adults following in utero and childhood exposure to arsenic in drinking water. *Environmental Health Perspectives* 120(11):1527–1531.

Smith, A. H., G. Marshall, Y. Yuan et al. 2006. Increased mortality from lung cancer and bronchiectasis in young adults after exposure to arsenic in utero and in early childhood. *Environmental Health Perspectives* 114:1293–1296.

Smith, A. H., G. Marshall, Y. Yuan et al. 2011. Evidence from Chile that arsenic in drinking water may increase mortality from pulmonary tuberculosis. *American Journal of Epidemiology* 173(4):414–420.

Smith, E., R. Naidu, A. M. Alston. 1998a. Arsenic in the soil environment: A review. *Advances in Agronomy*, 64:149–195.

Smith, K. and D. N. Petley. 2009. *Environmental Hazards: Assessing Risk and Reducing Disaster*. London: Routledge.

Smith, A. H. and M. M. Smith. 2004. Arsenic drinking water regulations in developing countries with extensive exposure. *Toxicology* 198:39–44.

Smith, A. H. and C. M. Steinmaus. 2009. Health effects of arsenic and chromium in drinking water: recent human findings. *Annual Review of Public Health* 30:107–122.

Sø, H. U., D. Postma, R. Jakobsen. 2017. Do Fe-oxides control the adsorption of arsenic in aquifers of the Red River floodplain, Vietnam? *Procedia Earth and Planetary Science* 17:300–303.

Sober, A. J. and J. M. Burstein. 1995. Precursors to skin cancer. *Cancer* 75(suppl S2):645–650.

Söderström, M. and B. Magnusson. 1995. Assessment of local agroclimatological conditions-a methodology. *Agricultural and Forest Meteorology* 72:243–260.

Sohel, N., L. A. Persson, M. Rahman et al. 2009. Arsenic in drinking water and adult mortality: A population-based cohort study in rural Bangladesh. *Epidemiology* 20(6):824–830.

Sohel, N., M. Vahter, M. Ali et al. 2010. Spatial patterns of fetal loss and infant death in an arsenic-affected area in Bangladesh. *International Journal of Health Geographics* 9:53.

Soignet, S. L., P. Maslak, Z. G. Wang et al. 1998. Complete remission after treatment of acute promyelocytic leukemia with arsenic trioxide. *The New England Journal of Medicine* 339(19):1341–1348.

Son, Y. O., P. Pratheeshkumar, R. V. Roy et al. 2015. Antioncogenic and oncogenic properties of Nrf2 in arsenic-induced carcinogenesis. *Journal of Biological Chemistry* 290(45):27090–27100.

Sorg, T. J., A. S. C. Chen, L. Wang. 2014. Arsenic species in drinking water wells in the USA with high arsenic concentrations. *Water Research* 48:156–169.

Sorvari, J., E. Schultz, E. Rossi et al. 2007. *Risk Assessment of Natural and anthropogenic Arsenic in Pirkanmaa region, Finland*. Finnish Environment Institute, Esko Rossi Oy, Pirkanmaa Regional Environment Centre, Geological Survey of Finland (GTK). Finland.

Southwick, J. W., A. E. Western, M. M. Beck et al. 1983. An epidemiological study of arsenic in drinking water in Millard County, Utah. In *Arsenic: Industrial, Biomedical, Environmental Perspectives*, eds. W. H. Lederer and R. J. Fensterheim, 210–225. New York: Van Nostrand Reinhold Company.

Spaulding, A. 2000. Micropolitical Behavior of Second Graders: A Qualitative Study of Student Resistance in the Classroom. *The Qualitative Report* 4 (1/2). (http://www.nova.edu/ssss/QR/QR4-1/spaulding.html).

Spayd, S. E., M. G. Robson, B. T. Buckley. 2015. Whole-house arsenic water treatment provided more effective arsenic exposure reduction than point-of-usewater treatment at NewJersey homes with arsenic in well water. *Science of the Total Environment* 505:1361–1369.

Spencer, L., J. Ritchie, R. Ormston et al. 2013. Analysis: Principles and processes. In *Qualitative Research Practice: A Guide for Social Science Students and Researchers*, eds. J. Ritchie, J. Lewis, C. M. Nicholls, and R. Ormston, 269–294. Los Angeles: Sage.

Srivastava, R. M. 1996. Describing spatial variability using geostatistical analysis. In *Geostatistics for Environmental and Geotechnical Applications*, eds. S. Rouhani, R. M. Srivastava, A. J. Desbarats et al. 13–19. Ann Arbor: American Society for Testing and Materials.

Sriwana, T., M. J. van Bergen, S. Sumarti et al. 1998. Volcanogenic pollution by acid water discharges along Ciwidey River, West Java Indonesia. *Journal of Geochemical Exploration* 62:161–182.

Stanger, G., T. Van Truong, K. S. L. T. M. Ngoc, T. V. Luyen, T. T. Thanh. 2005. Arsenic in groundwaters of the Lower Mekong. *Environmental Geochemistry and Health* 27(4):341–57.

States, J. C., S. Srivastava, Y. Chen et al. 2009. Arsenic and cardiovascular disease. *Toxicological Sciences* 107(2):312–323.

Stea, F., F. Bianchi, L. Cori et al. 2014. Cardiovascular effects of arsenic: clinical and epidemiological findings. *Environmental Science and Pollution Research International* 21(1):244–251.

Steinmaus, C., C. Ferreccio, J. Acevedo et al. 2014. Increased lung and bladder cancer incidence in adults after in utero and early-life arsenic exposure. *Cancer Epidemiology, Biomarkers and Prevention* 23(8):1529–1538.

Steinmaus, C., C. Ferreccio, J. Acevedo et al. 2016. High risks of lung disease associated with early-life and moderate lifetime arsenic exposure in northern Chile. *Toxicology and Applied Pharmacology* 313:10–15.

Steinmaus, C. M., C. Ferreccio, J. A. Romo et al. 2013. Drinking water arsenic in northern Chile: High cancer risks 40 years after exposure cessation. *Cancer Epidemiology, Biomarkers and Prevention* 22(4):623–630.

Steinmaus, C., Y. Yuan, M. N. Bates et al. 2003. Case-control study of bladder cancer and drinking water arsenic in the western United States. *American Journal of Epidemiology* 158:1193–2001.

Steinmaus, C., Y. Yuan, D. Kalman et al. 2010. Individual differences in arsenic metabolism and lung cancer in a case-control study in Cordoba, Argentina. *Toxicology and Applied Pharmacology* 247:138–145.

Steinmaus, C., Y. Yuan, J. Liaw et al. 2009. Low-level population exposure to inorganic arsenic in the United States and diabetes mellitus: a reanalysis. *Epidemiology* 20:807–815.

Steinmaus, C. M., Y. Yuan, A. H. Smith. 2005. The temporal stability of arsenic concentrations in well water in western Nevada. *Environmental Research* 99(2):164–168.

Stenner, R., T. Mitchell, S. Palmer. 2016. The role of philosophical hermeneutics in contributing to an understanding of physiotherapy practice: A reflexive illustration. *Physiotherapy* 103(3):330–334.

Sterling, R. O. and J. J. Helble. 2003. Gas-solid reaction of arsenic metal vapors with fly ash. *Chemosphere* 51(10):1111–19.

Sterns, J. D., C. B. Smith, J. R. Steele et al. 2014. Epigenetics and type II diabetes mellitus: underlying mechanisms of prenatal predisposition. *Frontiers in Cell and Developmental Biology* 2:15.

Stewart, A. 2014. The anthropology of family business: An imagined ideal. In *Sage Handbook of Family Business*, eds. L. Melin, M. Nordqvist, P. Sharma, 66–82. Thousand Oaks: Sage.

Sthiannopkao, S., K. W. Kim, K. H. Cho et al. 2010. Arsenic levels in human hair, Kandal Province, Cambodia: The influences of groundwater arsenic, consumption period, age and gender. *Applied Geochemistry* 25(1):81–90.

Sthiannopkao, S., K. W. Kim, S. Sotham et al. 2008. Arsenic and manganese in tube well waters of Prey Veng and Kandal Provinces, Cambodia. *Applied Geochemistry* 23(5):1086–1093.

Stiles, K. 2002. International support for NGOs in Bangladesh: Some unintended consequences. *World Development* 30:835–846.

Store, R. and J. Kangas. 2001. Integrating spatial multi-criteria evaluation and expert knowledge for GIS-based habitat suitability modelling. *Landscape and Urban Planning* 55:79–93.

Strauss, A. L. and J. Corbin. 1990. *Basics of Qualitative Research: Grounded Theory Procedures and Techniques*. Thousand Oaks: Sage.

Strauss, A. L. and J. Corbin. 1998. *Basics of Qualitative Research: Techniques and Procedures for Developing Grounded Theory*. Thousand Oaks: Sage.

Stryker, J. E., C. M. Moriarty, J. D. Jensen. 2008. Effects of newspaper coverage on public knowledge about modifiable cancer risks. *Health Communication* 23(4):380–390.

Stüben, D., Z. Berner, D. Chandrasekharam et al. 2003. Arsenic enrichment in groundwater of West Bengal, India: geochemical evidence for mobilization of arsenic under reducing conditions. *Applied Geochemistry* 18(9):1417–1434.

Su, C. C., J. L. Lu, K. Y. Tsai et al. 2011. Reduction in arsenic intake from water has different impacts on lung cancer and bladder cancer in an arseniasis endemic area in Taiwan. *Cancer Causes & Control* 22(1):101–108.

Sugár, É., E. Tatár, G. Záray et al. 2013. Field separation-based speciation analysis of inorganic arsenic in public well water in Hungary. *Microchemical Journal* 107:131–135.

Sultan, K. 2007. Distribution of metals and arsenic in soils of Central Victoria (Creswick-Ballarat), Australia. *Archives of Environmental Contamination and Toxicology* 52:339–346.

Sun, G. 2004. Arsenic contamination and arsenicosis in China. *Toxicology and Applied Pharmacology* 198:268–271.

Sun, G., X. Li, J. Pi et al. 2006. Current research problems of chronic arsenicosis in China. *Journal of Health, Population and Nutrition* 24(2):176–181.

Surdu, S. 2014. Non-melanoma skin cancer: Occupational risk from UV light and arsenic exposure. *Reviews of Environmental Health* 29(3):255–264.

Susko, M. L., M. S. Bloom, I. A. Neamtiu et al. 2017. Low-level arsenic exposure via drinking water consumption and female fecundity - A preliminary investigation. *Environmental Research* 154:120–125.

Suter, G. W., L. W. Barnthouse, R. V. O'Neill. 1987. Treatment of risk in environmental impact assessment. *Environmental Management* 11(3):295–303.

Suzuki, K. T., K. Kurasaki, N. Suzuki. 2007. Selenocysteine β-lyase and methylselenol demethylase in the metabolism of Se-methylated selenocompounds into selenide. *Biochimica et Biophysica Acta - General Subjects* 1770(7):1053–1061.

Suzuki, T. and I. Tsukamoto. 2006. Arsenite induces apoptosis in hepatocytes through an enhancement of the activation of Jun N-terminal kinase and p38 mitogen-activated protein kinase caused by partial hepatectomy. *Toxicology Letters* 165(3):257–264.

Suzuki, T., S. Yamashita, T. Ushijima et al. 2013. Genome-wide analysis of DNA methylation changes induced by gestational arsenic exposure in liver tumors. *Cancer Science* 104(12):1575–1585.

Svensson, M. 2007. *Mobilisation of geogenic arsenic into groundwater in Västerbotten County, Sweden.* Unpublished dissertation, Department of Earth Sciences, Uppsala University, Sweden.

Swami, M., B. Soni, H. R. Shah. 2014. Need of arsenic monitoring: A review. *International Journal of Research and Scientific Innovation* 1(7):253–258.

Swanson-Kauffman, K. and E. Schonwald. 1988. Phenomenology. In *Paths to Knowledge: Innovative Research Methods for Nursing*, ed. B. Sarter, 97–105. New York: National League for Nursing.

Swartz, C. H., N. K. Blute, B. Badruzzman et al. 2004. Mobility of arsenic in a Bangladesh aquifer: Inferences from geochemical profiles, leaching data, and mineralogical characterization. *Geochimica Acta* 68(22):4539–4557.

Sweeney, C. J., C. Takimoto, L. Wood et al. 2010. A pharmacokinetic and safety study of intravenous arsenic trioxide in adult cancer patients with renal impairment. *Cancer Chemotherapy and Pharmacology* 66:345–356.

Sy, S. M. T., C. M. Salud-Gnilo, C. Yap-Silva et al. 2017. A retrospective review of the dermatologic manifestations of chronic arsenic poisoning in the Philippines. *International Journal of Dermatology* 56(7):721–725.

Sydelko, P. J., J. E. Dolph, K. A. Majerus et al. 2000. An advance object-based software framework for complex ecosystem modeling and simulation. In *Fourth International Conference on Integrating GIS and Environmental Modeling (GIS/EM4): Problems, Prospects and Research Needs*, 1–9. Alberta, Canada.

Sykora, P. and E. T. Snow. 2008. Modulation of DNA polymerase beta-dependent base excision repair in cultured human cells after low dose exposure to arsenite. *Toxicology and Applied Pharmacology* 228(3):385–394.

Szuler, I. M., C. N. Williams, J. T. Hindmarsh et al. 1979. Massive variceal hemorrhage secondary to presinusoidal portal hypertension due to arsenic poisoning. *Journal of Canadian Medical Association* 120:168–171.

Taheri, M., M. H. M. Gharaie, J. Mehrzad et al. 2017. Hydrogeochemical and isotopic evaluation of arsenic contaminated waters in an argillic alteration zone. *Journal of Geochemical Exploration* 175:1–10.

Tamasi, G. and R. Cini. 2004. Heavy metals in drinking waters from Mount Amiata (Tuscany, Italy). Possible risk from arsenic for public health in the Province of Siena. *Science of the Total Environment* 327(1–3):41–51.

Tanaka, T. 1988. Distribution of arsenic in the natural environment with emphasis on rocks and soils. *Applied Organometallic Chemistry* 2(4):283–295.

Tanga, S., V. R. Goel, S. Patil. 2016. Chronic arsenic poisoning leading to skin malignancy in a community. *Clinical Skin Cancer* 1(1):15–19.

Tapio, S. and B. Grosche. 2006. Arsenic in the aetiology of cancer. *Mutation Research-Genetic Toxicology and Environmental Mutagenesis* 612:215–246.

Taylor, S., C. Papadopoulos, S. Vieillet et al. 2009. Assessing community health risks: Proactive vs reactive sampling. *American Journal of Environmental Science* (6):695–696.

Teegavarapu, R. S. V. and V. Chandramouli. 2005. Improved weighting methods, deterministic and stochastic data-driven models for estimation of missing precipitation records. *Journal of Hydrology* 312(1–4):191–206.

ten Have, P. 2007. Ethnomethodology. In *Qualitative Research Practice*, eds. C. Seale, G. Gobo, J. F. Gubrium, and D. Silverman, 139–152. London: Sage.

Terada, H., T. Sasagawa, H. Saito et al. 1962. Chronic arsenical poisoning and hematopoietic organs. *Acta Medca et Biologica* 9:279–92.

Thapinta, A. and P. Hudak. 2003. Use of geographic information systems for assessing groundwater pollution potential by pesticides in Central Thailand. *Environmental International* 29:87–93.

Thomas, D. J., M. Styblo, S. Lin. 2001. The cellular metabolism and systemic toxicity of arsenic. *Toxicology and Applied Pharmacology* 176(2):127–144.

Thornton, I. 1996. Sources and pathways of arsenic in the geochemical environment: health implication. *Environmental Geochemistry and Health*, 113: 153–61 (Special Publications).

Thornton, I. and M. Farago. 1997. The geochemistry of arsenic. In *Arsenic Exposure and Health Effects*, eds. C. O. Abernathy, R. L. Calderon, W. R. Chappell, 1–16. London: Chapman Hall.

Thucydides. 1954. *History of the Peloponnesian War.* New York: Penguin.

Thundiyil, J. G., Y. Yuan, A. H. Smith et al. 2007. Seasonal variation of arsenic concentration in wells in Nevada. *Environmental Research* 104(3):367–373.

Thywissen, K. 2006. *Components of Risk: A Comparative Glossary.* Bonn: United Nations University, Institute for Environment and Human Security (UNU-EHS).

Timmerman, P. 1981. Vulnerability, Resilience and the Collapse of Society. *Environmental Monograph 1.* Toronto: Institute for Environmental Studies, University of Toronto, Canada.

Titchen, A. 2000. *Professional Craft Knowledge in Patient-Centred Nursing and the Facilitation of its Development.* Oxford (UK): Ashdale Press.

Titchen, A. and D. McIntyre. 1993. A phenomenological approach to qualitative data analysis in nursing research. In *Changing Nursing Practice through Action Research*, ed. A. Titchen, 29–48. Oxford (UK): National Institute for Nursing, Centre for Practice Development and Research.

Tobin, G. A. and B. E. Montz. 1997. *Natural Hazards: Explanation and Integration.* New York: The Guilford Press.

Tokar, E. J., B. A. Diwan, J. M. Ward et al. 2011. Carcinogenic effects of "whole-life" exposure to inorganic arsenic in CD1 mice. *Toxicological Sciences* 119(1):73–83.

Tollestrup, K., F. J. Frost, M. Cristiani et al. 2005. Arsenic-induced skin conditions identified in southwest dermatology practices: An epidemiologic tool. *Environmental Geochemistry and Health* 27:47–53.

Tondel, M., M. Rahman, A. Magnuson et al. 1999. The relationship of arsenic levels in drinking water and the prevalence rate of skin lesions in Bangladesh. *Environmental Health Perspectives* 107(9):727–729.

Toner, P., J. Bowman, K. Clabby et al. 2005. *Water Quality in Ireland 2001–2003.* Wexford: Environmental Protection Agency.

Torres, I. S. I. and H. Ishiga. 2003. Assessment of geochemical conditions for the release of arsenic, iron, and copper into groundwater in the coastal aquifer at Yumigahama, Western Japan. In *Water Pollution VII: Modelling, Measuring and Prediction*, eds. C. A. Brebbia, D. Almorza, D. Sales, 147–157. Southampton: WIT Press.

Toujague, R. D. R. 1999. Arsênio e metais associados na região aurífera do Piririca, Vale do Ribeira, São Paulo, Brasil. PhD thesis, Instituto de Geociências, São Paulo: Universidade Estadual de Campinas.

Trang, N. T. 2004. *PGIS's relevance, applicability and conditions in local rural development: A case study with Village Development Planning in Bach Ma National Park buffer zone, Vietnam.* Unpublished MSc Thesis. The Netherlands: ITC; 2004.

Tripathi, N. and S. Bhattarya. 2004. Integrating indigenous knowledge and GIS for participatory natural resource management: state-of-the-practice. *The Electronic Journal on Information Systems in Developing Countries*, 17(3):1–13.

Tristán, E., A. Demetriades, M. H. Ramsey et al. 2000. Spatially resolved hazard and exposure assessments: An example of lead in soil at Lavrion, Greece. *Environmental Research Section A*, 82:33–45.

Trønnes, D. H. and A. B. Heiberg, 1989. Quantification of health risk due to chemicals: Methods and uncertainties. In *Risk Management of Chemicals in the Environment*, eds. H. M. Seip and A. B. Heiberg, 11–24. New York: Plenum Press (Published in cooperation with NATO Committee on the Challenges of Modern Society).

Tsai, S. Y., H. Y. Chou, H. W. The et al. 2003. The effects of chronic arsenic exposure from drinking water on neurobehavioral development in adolescence. *Neurotoxicology* 24:747–753.

Tsai, S. M., T. N. Wang, Y. C. Ko. 1999. Mortality for certain diseases in areas with high levels of arsenic in drinking water. *Archives of Environmental Health* 54(3):186–193.

Tsanis, I. K. and M. A. Gad. 2001. A GIS precipitation method for analysis of storm kinematics. *Environmental Modelling and Software*, 16(3):273–281.

Tseng, W. P. 1977. Effects and dose-response relationships of skin cancer and blackfoot disease with arsenic. *Environmental Health Perspectives* 19:109–119.

Tseng, C. H. 2002. An overview on peripheral vascular disease in blackfoot disease-hyperendemic villages in Taiwan. *Angiology* 53:529–537.

Tseng, C. H. 2003. Abnormal current perception thresholds measured by neurometer among residents in blackfoot disease-hyperendemic villages in Taiwan. *Toxicology Letters* 146(1):27–36.

Tseng, C. N. 2004. The potential biological mechanisms of arsenic-induced diabetes mellitus (Review). *Toxicology and Applied Pharmacology* 197:67–83.

Tseng, C. H. 2007. Metabolism of inorganic arsenic and non-cancerous health hazards associated with chronic exposure in humans. *Journal of Environmental Biology* 28(2):349–357.

Tseng, C. H. 2008. Cardiovascular disease in arsenic-exposed subjects living in the arseniasis hyperendemic areas in Taiwan. *Atherosclerosis* 199(1):12–18.

Tseng, C. H., C. K. Chong, C. J. Chen et al. 1996. Dose-response relationship between peripheral vascular disease and ingested inorganic arsenic among residents in blackfoot disease endemic villages in Taiwan. *Atherosclerosis* 120(1–2):125–33.

Tseng, C. H., C. K. Chong, C. P. Tseng et al. 2003. Long-term arsenic exposure and ischemic heart disease in arseniasis-hyperendemic villages in Taiwan. *Toxicology Letters* 137(1–2):15–21.

Tseng, W. P., H. M. Chu, S. W. How et al. 1968. Prevalence of skin cancer in an endemic area of chronic arsenism in Taiwan. *Journal of the National Cancer Institute* 40(3):453–463.

Tseng, C. H., Y. K. Huang, Y. L. Huang et al. 2005. Arsenic exposure, urinary arsenic speciation, and peripheral vascular disease in blackfoot disease-hyperendemic villages in Taiwan. *Toxicology and Applied Pharmacology* 206(3):299–308.

Tseng, C. H., T. Y. Tai, C. K. Chong et al. 2000. Long-term arsenic exposure and incidence of non-insulin-dependent diabetes mellitus: a cohort study in arseniasis-hyperendemic villages in Taiwan. *Environmental Health Perspectives* 108(9):847–851.

Tsuda, T., A. Babazono, E. Yamamoto et al. 1995. Ingested arsenic and internal cancer: a historical cohort study followed for 33 years. *American Journal of Epidemiology* 141(3):198–209.

Tsuji, J. S., V. Perez, M. R. Garry et al. 2014. Association of low-level arsenic exposure in drinking water with cardiovascular disease: a systematic review and risk assessment. *Toxicology* 323:78–94.

Tsutsumi, A., T. Izutsu, M. D. A. Islam et al. 2004. Depressive status of leprosy patients in Bangladesh: Association with self-perception of stigma. *Leprosy Review* 75(1):57–66.

Tuli, R., D. Chakrabarty, P. K. Trivedi et al. 2010. Recent advances in arsenic accumulation and metabolism in rice. *Molecular Breeding* 26(2):307–323.

Tun, M. K. 2002. *Report on Assessment of arsenic content in groundwater and the prevalence of arsenicosis in Thabaung and Kyonpyaw townships in Ayeyarwaddy division, Myanmar*. Yangoon: Yangoon Department of Medical Research.

Twigg, J. 1998. Understanding vulnerability: An introduction. In *Understanding Vulnerability: South Asian Perspectives*, eds. J. Twigg and M. Bhatt, 1–11. Colombo: Intermediate Technology Publications.

Tyler, C. R. and A. M. Allan. 2014. The effects of arsenic exposure on neurological and cognitive dysfunction in human and rodent studies: A review. *Current Environmental Health Reports* 1(2):132–147.

Ul-Haque, I., M. Baig, D. Nabi et al. 2007. Groundwater arsenic contamination - a multidirectional emerging threat to water scarce areas of Pakistan. *GQ07: Securing Groundwater Quality in Urban and Industrial Environments*. Proceedings of the 6th International Water Conference held in Fremantle, Western Australia, 2–7 December.

Umitsu, M. 1993. Late Quaternary sedimentary environments and landforms in the Ganges Delta. *Sediment Geology* 83:177–86.

UN. 2010. *The Human Right to Water and Sanitation*. The UN General Assembly Resolution (A/RES/64/292). July 28.
UN. 2015. *Sustainable Development Goals*. (http://www.un.org/sustainabledevelopment/water-and-sanitation/). Accessed on 17 April 2017.
UNDP/UNCHS. 2001. *Water Quality Testing in 11 Project Townships*. UNDP/UNCHS (United Nations Development Programme and United Nations Centre for Human Settlements).
UNHRC. 2012. *The human right to safe drinking water and sanitation*. UN Human Rights Council Resolution 21/2 of September 2012 (A/HRC/21/L.1).
UNICEF. 2000a. *Arsenic Mitigation in Bangladesh*. Dhaka: UNICEF.
UNICEF. 2000b. *Learning from Experience: Evaluation of UNICEF's Water and Environmental Sanitation Programme in India, 1966–1998*. New York: UNICEF.
UNICEF. 2008. *Arsenic Mitigation in Bangladesh*. [www.unicef.org/bangladesh/Arsenic.pdf] accessed on April 16, 2016.
UNICEF. 2010. *Arsenic Mitigation in Bangladesh*. Dhaka: UNICEF.
UNICEF. 2011. *Bangladesh national drinking water quality survey of 2009*. Dhaka: UNICEF.
UNICEF. 2013. *Project Completion Review Report: Sanitation, Hygiene Education and Water Supply in Bangladesh (SHEWA-B)*. Dhaka: UNICEF.
Urmson, J. O. and R. Jonathan. 2005. *The Concise Encyclopaedia of Western Philosophy*. New York: Routledge.
USEPA. 1988. *Special Report on Ingested Inorganic Arsenic: Skin Cancer, Nutritional Essentiality*. Washington, DC: US Environmental Protection Agency. (EPA 625/3-87/013: Risk Assessment Forum).
USEPA. 1989. *Risk Assessment Guidance for Superfund. Human Health Evaluation Manual Part A, Interim Final*. Washington, DC: US Environmental Protection Agency. (EPA/540/1-89/002).
USEPA. 1991. *Technical Support Document for Water Quality-Based Toxics Control*. Washington, DC: US Environmental Protection Agency. (EPA/505/2-90-001).
USEPA. 1992. Guidelines for Exposure Assessment. *EPA/600/Z-92/001, Federal Register* 57 (104): 22888–22938.
USEPA. 1998. *Integrated Risk Information System (IRIS): Arsenic, inorganic*, CASRN 7440-38-2.
USEPA. 2000. *Implementation Guidance for Arsenic*. Washington, DC: US Environmental Protection Agency.
USEPA. 2001a. *Technical Fact Sheet: Final Rule for Arsenic in Drinking Water*. 66: FR6976 (815-F-00-016), 22 January 2001. Washington, DC: US Environmental Protection Agency.
USEPA. 2001b. National primary drinking water regulations: Arsenic and clarifications to compliance and new source contaminants monitoring. *Federal Register*, 66: 6976–7066 (January 22, 2001).
USEPA. 2001c. *The integrated risk information system (IRIS)*. Cincinnati (OH), USA, U.S. Environmental Protection Agency (USPA), Environmental Criteria and Assessment Office, Washington, DC.
USEPA. 2002. *A Review of the Reference Dose and Reference Concentration Processes*. Washington, DC: US Environmental Protection Agency. (EPA/630/P-02/002F).
USEPA. 2004. *Monitoring Arsenic in the Environment: A Review of Science and Technologies for Field Measurements and Sensors*. Washington, DC: United States Environmental Protection Agency. (EPA 542/R-04/002: April 2004).
USEPA. 2005. *Guidelines for Carcinogen Risk Assessment Risk Assessment Forum*. Washington, DC: U.S. Environmental Protection Agency.
USEPA. 2009. *Exposure Factors Handbook 2009 Update* (External Review Draft). (EPA/600/R-09/052A). National Center for Environmental Assessment, US Environmental Protection Agency, Washington, DC.
USPHS. 1943. Public Health Service drinking water standards. *Public Health Reports*, US Public Health Service, 58:69–111.
Uyan, M. 2016. Determination of agricultural soil index using geostatistical analysis and GIS on land consolidation projects: A case study in Konya/Turkey. *Computers and Electronics in Agriculture* 123:402–409.
Uyan, M. and T. Cay. 2013. Spatial analyses of groundwater level differences using geostatistical modeling. *Environment and Ecological Statistics* 20(4):633–646.
Uyan, M., T. Cay, Y. Inceyol et al. 2015. Comparison of designed different land reallocation models in land consolidation: A case study in Konya/Turkey. *Computers and Electronics in Agriculture* 110:249–258.
Vahidnia, A., G. B. van der Voet, F. A. de Wolff. 2007. Arsenic neurotoxicity - a review. *Human and Experimental Toxicology* 26:823–832.
Vahter, M. 2008. Health effects of early life exposure to arsenic. *Basic & Clinical Pharmacology & Toxicology* 102:204–211.
Vahter, M. 2009. Effects of arsenic on maternal and fetal health. *Annual Review of Nutrition* 29:381–399.

Valenzuela, O. L., Z. Drobná, E. Hernández-Castellanos et al. 2009. Association of AS3MT polymorphisms and the risk of premalignant arsenic skin lesions. *Toxicology and Applied Pharmacology* 239(2):200–207.

Vall, O., M. Gómez-Culebras, O. Garcia-Algar et al. 2012. Assessment of prenatal exposure to arsenic in Tenerife Island. *PLoS One* 7(11):e50463.

Van Breda, S. G., S. M. Claessen, K. Lo et al. 2015. Epigenetic mechanisms underlying arsenic-associated lung carcinogenesis. *Archives of Toxicology* 89(11):1959–1969.

Van Dissen, R. and G. McVerry. 1994. Earthquake hazard and risk in New Zealand. In *Proceedings of the Natural Hazards Management Workshop*, eds. A. G. Hull and R. Coory, 67–71. Lower Hutt, New Zealand: Institute of geological and Nuclear Sciences.

Van Elteren, J. T., V. Stibilj, Z. Šlejkovec. 2002. Speciation of inorganic arsenic in some bottled Slovene mineral waters using HPLC-HGAFS and selective coprecipitation combined with FI-HGAFS. *Water Research* 36(12):2967–2974.

Van Geen, A., K. M. Ahmed, E. B. Ahmed et al. 2016. Inequitable allocation of deep community wells for reducing arsenic exposure in Bangladesh. *Journal of Water, Sanitation, and Hygiene for Development* 6(1):142–150.

Van Geen, A., K. M. Ahmed, A. A. Seddique et al. 2003b. Community wells to mitigate the arsenic crisis in Bangladesh. *Bulletin of the World Health Organization* 81:632–38.

Van Geen, A., H. Ahsan, A. H. Horneman et al. 2002. Promotion of well-switching to mitigate the arsenic crisis in Bangladesh. *Bulletin of the World Health Organization* 80:732–37.

Van Geen, A., Z. Cheng, Q. Jia. 2007. Monitoring 51 community wells in Araihazar, Bangladesh, for up to 5 years: implications for arsenic mitigation. *Journal of Environmental Science and Health: Part A* 42(12):1729–1740.

Van Geen, A., Z. Cheng, A. A. Seddique et al. 2005. Reliability of a commercial kit to test groundwater for arsenic in Bangladesh. *Environmental Science and Technology* 39:299–303.

Van Geen, A., K. H. Win, T. Zaw et al. 2014. Confirmation of elevated arsenic levels in groundwater of Myanmar. *Science of the Total Environment* 478:21–24.

Van Geen, A., Y. Zheng, R. Vesteeg et al. 2003a. Spatial variability of arsenic in 6000 tubewells in a 25 km^2 area of Bangladesh. *Water Resources Research* 39(5):1140.

Van Manen, M. 2003. *Researching Lived Experience. Human Science for an Action Sensitive Pedagogy*. Ontario: Althouse Press.

Van Manen, M. 2007. Phenomenology of practice. *Phenomenology and Practice*, 1(1):11–30.

Van Manen, M. and C. A. Adams. 2010. Qualitative research: Phenomenology. In *International Encyclopedia of Education*, eds. E. Baker, P. Peterson, B. McGaw, 449–455. Oxford: Elsevier.

Vanaei, M., D. Fathijoo, H. Rahimimoghaddam. 2006. *Study of environmental effect of arsenic in Divandareh area*. First conference on environmental and medicine geology. Tehran, Iran.

Varouchakis, E. A., D. T. Hristopulos, G. P. Karatzas. 2012. Improving kriging of groundwater level data using nonlinear normalizing transformations-a field application. *Hydrological Sciences Journal* 57(7):1404–1419.

Varpio, L., E. Paradis, C. Watling. 2017. Introducing a qualitative space. *Perspectives on Medical Education* 6(2):63–64.

Varsányi, I. 1989. Arsenic in deep groundwater. In *Proceedings of the 6th International Symposium on Water-rock Interaction (WRI–6)*, ed. D. L. Miles, 715–718. Malvern: AA Balkema, Rotterdam/Brookfield.

Varsányi, I., Z. Fodré, A. Bartha. 1991. Arsenic in drinking water and mortality in the Southern Great Plain, Hungary. *Environmental Geochemistry and Health* 13(1):14–22.

Varsányi, I. and L. O. Kovács. 2006. Arsenic, iron and organic matter in sediments and groundwater in the Pannonian Basin, Hungary. *Applied Geochemistry* 21:949–963.

Vassileva, E., A. Becker, J. A. C. Broekaert. 2001. Determination of arsenic and selenium species in groundwater and soil extracts by ion chromatography coupled to inductively coupled plasma mass spectrometry. *Analytica Chimica Acta* 441(1):135–146.

Vaughan, G. T. 1993. *The environmental chemistry and fate of arsenical pesticides in cattle tick dip sites and banana plantations*. Investigation Report CET/LHIR 148, CSIRO, Division of Coal and Energy Technology, Centre for Advanced Analytical Chemistry, Sydney, Australia.

Venteris, E. R., N. T. Basta, J. M. Bigham et al. 2014. Modeling spatial patterns in soil arsenic to estimate natural baseline concentrations. *Journal of Environmental Quality* 43:936–946.

Verplanck, P. L., S. H. Mueller, R. J. Goldfarb et al. 2008. Geochemical controls of elevated arsenic concentrations in groundwater, Ester Dome, Fairbanks district, Alaska. *Chemical Geology* 255:160–172.

Verret, W. J., Y. Chen, A. Ahmed et al. 2005. A randomized, double-blind placebo-controlled trial evaluating the effects of vitamin E and selenium on arsenic-induced skin lesions in Bangladesh. *Journal of Occupational & Environmental Medicine* 47:1026–1035.

Vine, M. F., D. Degnan, C. Hanchette. 1997. Geographic Information Systems: Their use in environmental epidemiologic research. *Environmental Health Perspectives* 1:598–605.

Violante, A., M. Ricciardella, S. D. Gaudio et al. 2006. Coprecipitation of arsenate with metal oxides: nature, mineralogy, and reactivity of aluminum precipitates. *Environmental Science and Technology* 40:4961–4967.

Vivona, R., E. Presosi, G. Giuliano et al. 2005. Geochemical characterization of a volcanic-sedimentary aquifer in Central Italy. In *Water-Rock Interaction: Proceedings of the 11th International Symposium on Water-Rock Interaction (WRI - 11)*, eds. R. B. Wanty and I. Seal, 513–517. Amsterdam: A.A. Balkema.

Vogelstein, B., D. Lane, A. J. Levine. 2000. Surfing the p53 network. *Nature* 408(6810):307–310.

Von Brumssen, M. 1999. *Genesis of high arsenic groundwater in the Bengal Delta Plains, West Bengal and Bangladesh*. Unpublished M.Sc. thesis, Department of Environmental Engineering, Stockholm: Royal School of Technology.

Von Ehrenstein, O. S., D. N. G. Mazumder, M. Hira-Smith et al. 2006. Pregnancy outcomes, infant mortality, and arsenic in drinking water in West Bengal, India. *American Journal of Epidemiology* 163(7):662–669.

Von Ehrenstein, O. S., D. N. G. Mazumder, Y. Yuan et al. 2005. Decrements in lung function related to arsenic in drinking water in West Bengal, India. *American Journal of Epidemiology* 162:533–541.

Von Ehrenstein, O. S., S. Poddar, Y. Yuan et al. 2007. Children's intellectual function in relation to arsenic exposure. *Epidemiology* 18(1):44–51.

Vörösmarty, C. J., P. B. McIntyre, M. O. Gessner et al. 2010. Global threats to human water security and river biodiversity. *Nature* 467:555–561.

Vukašinović-Pešić, V. L., N. Z. Blagojević, L. V. Rajaković. 2009. Comparative analysis of methods for determination of arsenic in coal and coal ash. *Instrumentation Science and Technology* 37(4):482–498.

Waalkes, M. P., J. Liu, B. A. Diwan. 2007. Transplacental arsenic carcinogenesis in mice. *Toxicological and Applied Pharmacology* 222(3):271–280.

Wade, T. J., Y. Xia, J. Mumford et al. 2015. Cardiovascular disease and arsenic exposure in Inner Mongolia, China: A case control study. *Environmental Health* 14: 35.

Wadge, G., A. P. Wislocki, E. J. Pearson. 1993. Spatial analysis in GIS for natural hazards assessment. In *Environmental Modeling with GIS*, eds. M. F. Goodchild, B. O. Parks, and L. T. Steyaert, 332–338. USA: Oxford University Press.

Wainwright, D. 1997. Can sociological research be qualitative, critical and valid? *The Qualitative Report* 3(2):1–12. (http://www.nova.edu/ssss/QR/QR3-2/wain.html).

Walker, I. F., N. Leigh-Hunt, A. C. K. Lee. 2016. Redesign and commissioning of sexual health services in England: A qualitative study. *Public Health* 139:134–140.

Wang, C. H., C. L. Chen, C. K. Hsiao et al. 2009. Increased risk of QT prolongation associated with atherosclerotic diseases in arseniasis-endemic area in Southwestern coast of Taiwan. *Toxicology and Applied Pharmacology* 239:320–324.

Wang, C. H., C. L. Chen, C. K. Hsiao et al. 2010. Arsenic-induced QT dispersion is associated with atherosclerotic diseases and predicts long-term cardiovascular mortality in subjects with previous exposure to arsenic: A 17-year follow-up study. *Cardiovascular Toxicology* 10(1):17–26.

Wang, Z. H., X. T. Cheng, J. Li et al. 2003. Investigation on arsenic concentration in drinking water and disease state of arsenism in Shanyin County. *Chinese Journal of Control of Endemic Disease* 18(5):293–295 (In Chinese with English abstract).

Wang, W., S. Cheng, D. Zhang. 2014b. Association of inorganic arsenic exposure with liver cancer mortality: A meta-analysis. *Environmental Research* 135:120–125.

Wang, H., Q. Cheng, R. Zuo. 2015. Spatial characteristics of geochemical patterns related to Fe mineralization in the southwestern Fujian province (China). *Journal of Geochemical Exploration* 148:259–269.

Wang, C. H., C. K. Hsiao, C. L. Chen et al. 2007b. A review of the epidemiologic literature on the role of environmental arsenic exposure and cardiovascular diseases. *Toxicology and Applied Pharmacology* 222(3):315–326.

Wang, T. S., T. Y. Hsu, C. H. Chung et al. 2001a. Arsenite induces oxidative DNA adducts and DNA-protein cross-links in mammalian cells. *Free Radical Biology and Medicine* 31(3):321–330.

Wang L. and J. Huang. 1994. Chronic arsenism from drinking water in some areas of Xinjiang, China. In *Arsenic in the Environment, Part II: Human Health and Ecosystem Effects*, ed. J. O. Nriagu, 159–172. New York: John Wiley & Sons.

Bibliography

Wang, C. H., J. S. Jeng, P. K. Yip et al. 2002. Biological gradient between long-term arsenic exposure and carotid atherosclerosis. *Circulation* 105:1804–1809.

Wang, L., M. C. Kou, C. Y. Weng et al. 2012. Arsenic modulates heme oxygenase-1, interleukin-6, and vascular endothelial growth factor expression in endothelial cells: roles of ROS, NF -κB, and MAPK pathways. *Archives of Toxicology* 86(6):879–896.

Wang, S. L., W. F. Li, C. J. Chen et al. 2011. Hypertension incidence after tap-water implementation: a 13-year follow-up study in the arseniasis-endemic area of Southwestern Taiwan. *Science of the Total Environment* 409(21):4528–4535.

Wang, Y., T. Ma, Z. Luo. 2001b. Geostatistical and geochemical analysis of surface water leakage into groundwater on a regional scale: a case study in the Liulin karst system, northwestern China. *Journal of Hydrology* 246:223–234.

Wang, S. and C. N. Mulligan. 2006. Occurrence of arsenic contamination in Canada: Sources, behavior and distribution. *Science of the Total Environment* 366:701–721.

Wang, S. X., Z. H. Wang, X. T. Cheng et al. 2007a. Arsenic and fluoride exposure in drinking water: Children's IQ and growth in Shanyin county, Shanxi province, China. *Environmental Health Perspectives* 115(4):643–647.

Wang, L., R. Wang, L. Fan et al. 2017. Arsenic trioxide is an immune adjuvant in liver cancer treatment. *Molecular Immunology* 81:118–126.

Wang, Y. H., M. M. Wu, C. T. Hong et al. 2007c. Effects of arsenic exposure and genetic polymorphisms of p53, glutathione S-transferase M1, T1, and P1 on the risk of carotid atherosclerosis in Taiwan. *Atherosclerosis* 192:305–312.

Wang, W., Z. Xie, Y. Lin et al. 2014a. Association of inorganic arsenic exposure with type 2 diabetes mellitus: A meta-analysis. *Journal of Epidemiology and Community Health* 68(2):176–184.

Wang, J., L. Zhao, Y. Wu. 1998. Environmental geochemical study on arsenic in arseniasis areas in Shanyin and Yingxian, Shanxi Province. *Geoscience* 12:243–248 (English abstract).

Wasserman, G. A., X. Liu, F. Parvez et al. 2004. Water arsenic exposure and children's intellectual function in Araihazar, Bangladesh. *Environmental Health Perspectives* 112(13):1329–1333.

Wasserman, G. A., X. Liu, F. Parvez et al. 2007. Water arsenic exposure and intellectual function in 6-year-old children in Araihazar, Bangladesh. *Environmental Health Perspectives* 115(2):285–289.

Wasserman, G. A., X. Liu, F. Parvez et al. 2011. Arsenic and manganese exposure and children's intellectual function. *Neurotoxicology* 32(4):450–457.

Watanabe, C. 2001. Environmental arsenic exposure in Bangladesh: Water versus extra-water intake of arsenic. *Environmental Science* 8:458–466.

Watanabe, T. and S. Hirano. 2013. Metabolism of arsenic and its toxicological relevance. *Archives of Toxicology* 87(6):969–979.

Wattanasen, K., S. Å. Elming, W. Lohawijarn et al. 2006. An integrated geophysical study of arsenic contaminated area in the peninsular Thailand. *Environmental Geology* 51:595–608.

Wauchope, R. D. and L. L. McDowell. 1984. Adsorption of phosphate, arsenate, methanearsonate, and cacodylate by lake and stream sediments: Comparisons with soils. *Journal of Environmental Quality* 13:499–504.

Webster, J. G. and D. K. Nordstrom. 2003. Geothermal arsenic: The sources, transport and fate of arsenic in geothermal systems. In *Arsenic in Ground Water*, eds. A. H. Welch and K. G. Stollenwerk, 101–126. Boston: Kluwer Academic.

Webster, R. and M. A. Oliver. 2007. *Geostatistics for Environmental Scientists*. West Sussex: John Wiley & Sons.

Wegelin, M. 1996. *Surface Water Treatment by Roughing Filters: A Design Construction and Operation Manual*. SANDEC Report No. 02/96.

Wei, B., J. Yu, H. Li et al. 2016. Arsenic metabolites and methylation capacity among individuals living in a rural area with endemic arseniasis in Inner Mongolia, China. *Biological Trace Element Research* 170(2):300–308.

Wei, B., J. Yu, L. Yang et al. 2017. Arsenic methylation and skin lesions in migrant and native adult women with chronic exposure to arsenic from drinking groundwater. *Environmental Geochemistry and Health* 39(1):89–98.

Wei, F. S., C. J. Zheng, J. S. Chen et al. 1991. Study on the background contents on 61 elements of soils in China. *Huaanjing Kexue* 12(14):12–19.

Wei, Y., J. Zhang, D. Zhang et al. 2014. Metal concentrations in various fish organs of different fish species from Poyang Lake, China. *Ecotoxicology and Environmental Safety* 104:182–188.

Weider, B. and S. Forshufvud. 1995. *Assassination on St Helena Revisited*. New York: John Wiley & Sons.

Weiner, D. and T. Harris. 2003. Community integrated GIS for land reform in South Africa. *Urban and Regional Information Systems Association (URISA) Journal* 15:61–73.

Weiner, D., T. Warner, T. M. Harris et al. 1995. Apartheid representations in a digital landscape: GIS, remote sensing, and local knowledge in Kiepersol, South Africa. *Cartography and Geographic Information Systems* 22(1):30–44.

Welch, A. H., D. R. Helsel, M. J. Focazio et al. 1999. Arsenic in ground water supplies of the United States. In *Arsenic Exposure and Health Effects*, eds. W. R. Chappell, C. O. Abernathy, R. L. Calderon, 9–17. Oxford: Elsevier.

Welch, A. H. and M. S. Lico. 1998. Factors controlling As and U in shallow groundwater, Southern Carson Desert, Nevada. *Applied geochemistry* 13:521–539.

Welch, A. H., M. S. Lico, J. L. Hughes. 1988. Arsenic in ground water of the Western United States. *Ground Water* 26(3):334–347.

Welch, A. H., D. B. Westjohn, D. R. Helsel et al. 2000. Arsenic in ground water of the United States: Occurrence and geochemistry. *Ground Water* 38(4):589–604.

Wen, C. P., T. Y. Cheng, M. K. Tsai et al. 2008. All-cause mortality attributable to chronic kidney disease: a prospective cohort study based on 462 293 adults in Taiwan. *Lancet* 371:2173–2182.

Wen, T. H., N. H. Lin, C. H. Lin et al. 2006. Spatial mapping of temporal risk characteristics to improve environmental health risk identification: A case study of a dengue epidemic in Taiwan. *Science of the Total Environment* 367:631–640.

Whanger, P. D., P. H. Weswig, J. C. Stoner. 1977. Arsenic levels in Oregon waters. *Environmental Health Perspectives* 19:139–143.

White, D. E. 1981. Active geothermal systems and hydrothermal ore deposits. In *Economic Geology: Seventy-Fifth Anniversary* Volume 1905–1980, ed. B. J. Skinner, 392–423. El Paso, Texas: Economic Geology Publishing Company.

White, M. J. 2004. Asbestos and the future of mass torts. *Journal of Economic Perspectives* 18:183–204.

White, D. E., J. D. Hem, G. A. Waring. 1963. Chemical composition of subsurface waters. In *Data of Geochemistry*, ed. M. Fleisher, 1–64. 6th ed. Washington, DC: US Geological Survey.

White, L. and A. Taket. 1997. Beyond Appraisal: Participatory Appraisal of Needs and the Development of Action (PANDA). *Omega International Journal of Management Science* 25(5):523–534.

WHO. 1958. *International Standards for Drinking-Water.* Geneva: World Health Organization.

WHO. 1963. *International Standards for Drinking-Water.* Geneva: World Health Organization, Second edition.

WHO. 1981. *Arsenic* (Environmental Health Criteria 18). Geneva: International Programme on Chemical Safety, World Health Organization.

WHO. 1993. *Guidelines for Drinking Water Quality: Recommendations.* Vol. 1, Geneva: World Health Organization.

WHO. 1996. *Arsenic in Drinking Water in Bangladesh: A Challenge in Near Future.* Country situation report, Dhaka: World Health Organisation.

WHO. 2000a. *Methods of Assessing Risk to Health from Exposure to Hazards Released from Landfills.* Copenhagen: World Health Organization.

WHO. 2000b. *Towards an Assessment of the Socioeconomic Impact of Arsenic Poisoning in Bangladesh.* Geneva: World Health Organization.

WHO. 2001. *Arsenic and Arsenic Compounds.* Environmental Health Criteria 224. Geneva: World Health Organisation.

WHO. 2003. *The Right to Water.* Geneva: World Health Organization.

WHO. 2004. *WHO Guidelines for Drinking-Water Quality.* Geneva: World Health Organization.

WHO. 2015. *Drinking Water.* Fact Sheet No 391. Geneva: World Health Organization. Accessed on May 23, 2016. [http://www.who.int/mediacentre/factsheets/fs391/en/].

Wilbers, G. J., M. Becker, L. T. Nga et al. 2014. Spatial and temporal variability of surface water pollution in theMekong Delta, Vietnam. *Science of the Total Environment* 485–486:653–665.

Wild, S., G. Roglic, A. Green et al. 2004. Global prevalence of diabetes: Estimates for the year 2000 and projections for 2030. *Diabetes Care* 27:1047–1053.

Wildavsky, A. 1988. *Searching for Safety.* New Brunswick (USA): Transaction Publishers.

Wilkinson, S. 2005. *A hydrogeological study of the Rarangi area, Marlborough (New Zealand).* Unpublished MSc thesis, University of Canterbury, New Zealand.

Williams, S., E. Annandale, J. Tritter. 1998b. The sociology of health and illness at the turn of the century: Back to the future? *Sociological Research Online* 3(4). [www.socresonline.org.uk/socresonline/3/4/1.html].

Bibliography

Williams, P. N., M. R. Islam, E. E. Adomako et al. 2006. Increase in rice grain arsenic for regions of Bangladesh irrigating paddies with elevated arsenic in groundwaters. *Environmental Science and Technology* 40:4903–4908.

Williams, T. M., B. G. Rawlins, B. Smith et al. 1998a. In-vitro determination of arsenic bioavailability in contaminated soil and mineral beneficiation waste from Rob Phibun, Southern Thailand: A basis for improved human risk assessment. *Environmental Geochemistry and Health* 20:169–177.

Wilson, H. L. 1997. *Simple Environmental Exposure Models in a GIS Framework*. Term project report, Austin: University of Texas.

Wilson, R. and E. A. C. Crouch. 1987. Risk assessment and comparisons: an introduction. *Science* 236(4799):267–270.

Wilson, F. H. and D. M. Hawkins. 1978. Arsenic in streams, stream sediments and groundwater, Fairbanks area, Alaska. *Environmental Geology* 2(4):195–202.

Winters, C. A. 1997. Living with chronic heart disease: A pilot study. *The Qualitative Report* 3(4). [www.nova.edu/ssss/QR/QR3-4/winters.html].

Withington, S. G., S. Joha, D. Baird et al. 2003. Assessing socio-economic factors in relation to stigmatization, impairment status, and selection for socioeconomic rehabilitation: A 1-year cohort of new leprosy cases in north Bangladesh. *Leprosy Review* 74(2):120–132.

Wodak, R. 2007. Critical discourse analysis. In *Qualitative Research Practice*, eds. C. Seale, G. Gobo, J. F. Gubrium, and D. Silverman, 185–201. London: Sage.

Wolcott, H. F. 1994. *Transforming Qualitative Data: Description, Analysis and Interpretation*. Thousand Oaks: Sage.

Wong, N. 2009. Investigating the effects of cancer risk and efficacy perceptions on cancer prevention adherence and intentions. *Health Communication* 24(2):95–105.

Wong, S. S., K. C. Tan, C. L. Goh. 1998. Cutaneous manifestations of chronic arsenicism: review of seventeen cases. *Journal of the American Academy of Dermatology* 38(2 pt 1):179–185.

Wong, H. H. and J. Wang. 2010. Merkel cell carcinoma. *Archives of Pathology and Laboratory Medicine* 134(11):1711–1716.

Wood, G. 1997. States without citizens: The problem of the franchise state. In *NGOs, States and Donors: Too Close for Comfort?* Eds. D. Hulme and M. Edwards, 79–92. Basingstoke: Macmillan.

Wood, G. 1999. Contesting water in Bangladesh: Knowledge, rights and governance. *Journal of International Development* 11:731–754.

Woolgar, J. A. and A. Triantafyllou. 2011. Squamous cell carcinoma and precursor lesions: Clinical pathology. *Periodontology 2000* 57(1):51–72.

Woolson, E. A. 1983. Man's perturbation on the arsenic cycle. In *Arsenic: Industrial, Biomedical, Environmental Perspectives*, eds. W. H. Lederer and R. J. Fensterheim, 393–408. New York: Van Nostrand Reinhold Company.

World Bank. 1999. *The Bangladesh Arsenic Mitigation Water Supply Project: Addressing a Massive Public Health Crisis*. Dhaka: The World Bank.

World Bank. 2005. *Towards a More Effective Operational response: Arsenic Contamination of Groundwater in South and East Asian Countries*. Volume I and II, Policy Report No. 31303, World Bank and Water and Sanitation Program, USA.

World Bank. 2007. *Arsenic Mitigation Water Supply*. Report No.: ICR000028. Washington, DC: The World Bank.

World Bank. 2011. *Implementation Completion and Results Report*. Bangladesh Water Supply Program Project. Report No.: ICR507. Washington DC: The World Bank.

World Bank. 2016. *Fact Sheet. Safe Water for Rural Populations of Bangladesh: Bangladesh Rural Water Supply and Sanitation Project (BRWSSP)*. Washington, DC: The World Bank.

WPSC. 2008. *The Safety of Chromated Copper Arsenate (CCA)-Treated Wood*. Wood Preservative Science Council, Manakin-Sabot, Varginia, USA. [http://www.woodpreservativescience.org/safety.shtml].

WRUD. 2001. *Preliminary Study on Arsenic Contamination in Selected Areas of Myanmar*. Report of the Water Resources Utilization Department, Ministry of Agriculture and Irrigation, Myanmar.

Wu, J., G. Chen, Y. Liao et al. 2011. Arsenic levels in the soil and risk of birth defects: A population-based case-control study using GIS technology. *Journal of Environmental Health* 74:20–25.

Wu, M. M., H. Y. Chiou, C. L. Chen et al. 2010. GT-repeat polymorphism in the heme oxygenase-1 gene promoter is associated with cardiovascular mortality risk in an arsenic-exposed population in Northeastern Taiwan. *Toxicology and Applied Pharmacology* 248(3):226–233.

Wu, M. M., H. Y. Chiou, T. W. Wang et al. 2001. Association of blood arsenic levels with increased reactive oxidants and decreased antioxidant capacity in a human population of Northeastern Taiwan. *Environmental Health Perspectives* 109(10):1011–1017.

Wu, F. and W. H. Farland. 2007. Risk assessment and regulatory decision making in environmental health. In *Risk Assessment for Environmental Health*, eds. M. Robson and W. Toscano, 31–54. San Francisco: John Wiley & Sons.

Wu, Y., R. Huxley, L. Li et al. 2008. Prevalence, awareness, treatment, and control of hypertension in China: data from the China National Nutrition and Health Survey 2002. *Circulation* 118:2679–2686.

Wu, F., F. Jasmine, M. G. Kibriya et al. 2012. Association between arsenic exposure from drinking water and plasma levels of cardiovascular markers. *American Journal of Epidemiology* 175(2):1252–1261.

Wu, M. M., T. L. Kuo, Y. H. Hwang et al. 1989. Dose-response relation between arsenic concentration in well water and mortality from cancers and vascular diseases. *American Journal of Epidemiology* 130(6):1123–1132.

Wu, W., S. Yin, H. Liu et al. 2014. The geostatistic-based spatial distribution variations of soil salts under long-term wastewater irrigation. *Environmental Monitoring and Assessment* 186(10):6747–6756.

Wyatt, C. J., C. Fimbres, L. Romo et al. 1998a. Incidence of heavy metal contamination in water supplies in northern Mexico. *Environmental Research* 76:114–119.

Wyatt, C. J., V. L. Quiroga, R. T. O. Acosta et al. 1998b. Excretion of arsenic (As) in urine of children, 7–11 years, exposed to elevated levels of As in the city water supply in Hermosillo, Sonora, México. *Environmental Research* 78:19–21.

Wyllie, J. 1937. An investigation of the source of arsenic in a well water. *Canadian Public Health Journal* 28:128–135.

Xia, Y. and J. Liu. 2004. An overview on chronic arsenism via drinking water in PR China. *Toxicology* 198(1–3):25–29.

Xu, C., H. S. He, Y. Hu et al. 2005. Latin hypercube sampling and geostatistical modeling of spatial uncertainty in a spatially explicit forest landscape model simulation. *Ecological Modelling* 185:255–269.

Xu, L., K. Yokoyama, Y. Tian et al. 2011. Decrease in birth weight and gestational age by arsenic among the newborn in Shanghai, China. *Japanese Journal of Public Health* 58(2):89–95.

Xue, H. 2003. *Transboundary Damage in International Law*. Cambridge: Cambridge University Press.

Yadav, I. C., S. Singh, N. L. Devi et al. 2012. Spatial distribution of arsenic in groundwater of southern Nepal. *Reviews of Environmental Contamination and Toxicology* 218:125–40.

Yalçin, S. and X. C. Le. 1998. Low pressure chromatographic separation of inorganic arsenic species using solid phase extraction cartridges. *Talanta* 47(3):787–96.

Yamaguchi, Y., H. Madhyastha, R. Madhyastha et al. 2016. Arsenic acid inhibits proliferation of skin fibroblasts, and increases cellular senescence through ROS mediated MST1-FOXO signaling pathway. *The Journal of Toxicological Sciences* 41(1):105–113.

Yamauchi, H. and B. A. Fowler. 1994. Toxicity and metabolism of inorganic and methylated arsenicals. In *Arsenic in the Environment, Part II: Human Health and Ecosystem Effects*, ed. J. O. Nriagu, 35–54. New York: John Wiley & Sons.

Yan, W., Y. Zhang, J. Zhang et al. 2011. Mutant p53 protein is targeted by arsenic for degradation and plays a role in arsenic-mediated growth suppression. *The Journal of Biological Chemistry* 286(20): 17478–86.

Yan-Chu, H. 1994. Arsenic distribution in soils. In *Arsenic in the Environment, Part 1: Cycling and Characterization*, ed. J. O. Nriagu, 17–47. New York: John Wiley & Sons.

Yang, C. Y. 2006. Does arsenic exposure increase the risk of development of peripheral vascular diseases in humans? *Journal of Toxicology and Environmental Health: Part A* 69(19):1797–1804.

Yang, C. Y., C. C. Chang, H. F. Chiu. 2008. Does arsenic exposure increase the risk for prostate cancer? *Journal of Toxicology and Environmental Health: Part A* 71(23):1559–1563.

Yang, C. Y., C. C. Chang, S. S. Tsai et al. 2003. Arsenic in drinking water and adverse pregnancy outcome in an arseniasis-endemic area in northeastern Taiwan. *Environmental Research* 91:29–34.

Yang, Q., H. B. Jung, R. G. Marvinney et al. 2012. Can arsenic occurrence rates in bedrock aquifers be predicted? *Environmental Science & Technology* 46(4):2080–2087.

Yang, H. J., C. Y. Lee, Y. J. Chiang et al. 2016. Distribution and hosts of arsenic in a sediment core from the Chianan Plain in SW Taiwan: Implications on arsenic primary source and release mechanisms. *Science of The Total Environment* 569–570:212–222.

Yang, Y., J. Pankow, H. Swan et al. 2017. Preparing for analysis: A practical guide for a critical step for procedural rigor in large-scale multisite qualitative research studies. *Quality & Quantity* 2017:1–14. [https://doi.org/10.1007/s11135-017-0490-y].

Yang, Y. S. and L. Wang. 2010. Catchment-scale vulnerability assessment of groundwater pollution from diffuse sources using the DRASTIC method: A case study. *Hydrological Sciences Journal* 55(7):1206–1216.

Yassi, A., T. Kjellström, T. de Kok et al. 2001. *Basic Environmental Health*. New York: Oxford University Press.

Yazdanpanah, N. 2016. Spatiotemporal mapping of groundwater quality for irrigation using geostatistical analysis combined with a linear regression method. *Modeling Earth System and Environment* 2:18.

Ye, H., Z. Yang, X. Wu et al. 2017. Sediment biomarker, bacterial community characterization of high arsenic aquifers in Jianghan Plain, China. *Scientific Report* 7:42037.

Yeh, T. C., Y. S. Tai, Y. S. Pu et al. 2015. Characteristics of arsenic-related bladder cancer: A study from Nationwide Cancer Registry Database in Taiwan. *Urological Science* 26(2):103–108.

Yeşilkanat, C. M., Y. Kobya, H. Taşkin et al. 2015. Dose rate estimates and spatial interpolation maps of outdoor gamma dose rate with geostatistical methods: A case study from Artvin, Turkey. *Journal of Environmental Radioactivity* 150:132–144.

Yih, L. H. and T. C. Lee. 2000. Arsenite induces p53 accumulation through an ATM-dependent pathway in human fibroblasts. *Cancer Research* 60(22):6346–6352.

Yin, R. K. 2003. *Case Study Research: Design and Methods*. Thousand Oaks: Sage.

Yoon, Y., S. Kim, Y. Chae et al. 2016. Evaluation of bioavailable arsenic and remediation performance using a whole-cell bioreporter. *Science of the Total Environment* 547:125–131.

Yoshida, T., H. Yamauchi, G. F. Sun. 2004. Chronic health effects in people exposed to arsenic via the drinking water: dose-response relationships in review. *Toxicology and Applied Pharmacology* 198(3):243–252.

Yu, X., T. Deng, Y. Guo et al. 2014. Arsenic species analysis in freshwater using liquid chromatography combined to hydride generation atomic fluorescence spectrometry. *Journal of Analytical Chemistry* 69:83–88.

Yu, W. H., C. M. Harvey, C. F. Harvey. 2003. Arsenic in groundwater in Bangladesh: A geostatistical and epidemiological framework for evaluating health effects, and potential remedies. *Water Resources Research* 39(6):1146–1163.

Yu, R. C., K. H. Hsu, C. J. Chen et al. 2000. Arsenic methylation capacity and skin cancer. *Cancer Epidemiology, Biomarkers and Prevention* 9(11):1259–1262.

Yu, H. S., W. T. Liao, C. Y. Chai. 2006. Arsenic carcinogenesis in the skin. *Journal of Biomedical Science* 13(5):657–666.

Yu, G., D. Sun, Y. Zheng. 2007. Health effects of exposure to natural arsenic in groundwater and coal in China: An overview of occurrence. *Environmental Health Perspectives* 115:636–642.

Yuan, Y., G. Marshall, C. Ferreccio et al. 2010. Kidney cancer mortality: Fifty-year latency patterns related to arsenic exposure. *Epidemiology* 21:103–108.

Yudovich, Y. E. and M. P. Ketris. 2005. Arsenic in coal: A review. *International Journal of Coal Geology* 61:141–196.

Yunus, F. M., S. Khan, P. Chowdhury et al. 2016. A review of groundwater arsenic contamination in Bangladesh: The millennium development goal era and beyond. *International Journal of Environmental Research and Public Health* 13(2):215.

Yunus, F. M., M. J. Rahman, M. Z. Alam et al. 2014. Relationship between arsenic skin lesions and the age of natural menopause. *BMC Public Health* 14:419 (doi: 10.1186/1471-2458-14-419).

Zakharova, T., F. Tatàno, V. Menshikov. 2002. Health cancer risk assessment for arsenic exposure in potentially contaminated areas by fertilizer plants: A possible regulatory approach applied to a case study in Moscow region-Russia. *Regulatory Toxicology and Pharmacology* 36(1):22–33.

Zaldívar, R. and G. L. Ghai. 1980. Clinical epidemiological studies on endemic chronic arsenic poisoning in children and adults, including observations on children with high- and low-intake of dietary arsenic. *Zentralbl Bakteriol B* 170:409–421.

Zaldívar, R. and A. Guillier. 1977. Environmental and clinical investigations on endemic chronic arsenic poisoning in infants and children. *Zentralbl Bakteriol Hyg* 165:226–234.

Zaldívar, R., L. Prunés, G. L. Ghai. 1981. Arsenic dose in patients with cutaneous carcinomata and hepatic haemangio-endothelioma after environmental and occupational exposure. *Archives of Toxicology* 47(2):145–154.

Zaman, A. 2001. Poison in the well. *New Internationalist*, 332:16–17 (https://newint.org/features/2001/03/05/poison/).

Zandbergen, P. A. 1998. Urban watershed ecological risk assessment using GIS: A case study of the Brunette River watershed in British Columbia, Canada. *Journal of Hazardous Materials* 61:163–173.

Zartarian, V., T. Bahadori, T. Mckone. 2005. Adoption of an official ISEA glossary. *Journal of Exposure Analysis and Environmental Epidemiology* 15:1–5.

Zartarian, V. G., W. R. Ott, N. Duan. 2006. Basic concepts and definition of exposure and dose. In *Exposure Analysis*, eds. W. R. Ott, A. C. Steinemann, L. A. Wallace, 33–64. Boca Raton: CRC Press.

Zaw, M. and M. T. Emett. 2002. Arsenic removal from water using advanced oxidation processes. *Toxicology Letters* 133(1):113–118.

Zayre, I., A. Gonzáleza, M. Krachler et al. 2006. Spatial distribution of natural enrichments of arsenic, selenium, and uranium in a minerotrophic peatland, Gola di Lago, Canton Ticino, Switzerland. *Environmental Science and Technology* 40:6568–6574.

Zernike, K. 2003. *Arsenic Poisoning at Church Mystifies a Maine Town.* The New York Times. (http://www.nytimes.com/2003/05/01/us/arsenic-poisoning-at-church-mystifies-a-maine-town.html).

Zhai, B., X. Jiang, C. He et al. 2015. Arsenic trioxide potentiates the anti-cancer activities of sorafenib against hepatocellular carcinoma by inhibiting Akt activation. *Tumour Biology* 36(4):2323–2334.

Zhang, J., X. Chen, P. Parkpian et al. 2001b. GIS Application on arsenic contamination and its risk assessment in Ronphibun, Nakhorn Si Thammarat, Thailand. *Geographical Information Sciences* 7(2):69–78.

Zhang, T. D., C. Q. Chen, Z. G. Wang et al. 2001a. Arsenic trioxide, a therapeutic agent for APL. *Oncogene* 20:7146–7153.

Zhang, Y. K., C. Dai, C. G. Yuan et al. 2017a. Establishment and characterization of arsenic trioxide resistant KB/ATO cells. *Acta Pharmaceutica Sinica B* 7(5):564–570.

Zhang, X., S. Jia, S. Yang et al. 2012. Arsenic trioxide induces G2/M arrest in hepatocellular carcinoma cells by increasing the tumor suppressor PTEN expression. *Journal of Cellular Biochemistry* 113(11):3528–3535.

Zhang, J., T. Ma, L. Feng et al. 2017b. Arsenic behavior in different biogeochemical zonations approximately along the groundwater flow path in Datong Basin, northern China. *Science of the Total Environment* 584–585:458–468.

Zhang, H., D. Ma, X. Hu. 2002. Arsenic pollution in groundwater from Hetao Area, China. *Environmental Geology* 41:638–643.

Zhang, C., G. Mao, S. He et al. 2013. Relationship between long-term exposure to low-level arsenic in drinking water and the prevalence of abnormal blood pressure. *Journal of Hazardous Materials* 262:1154–1158.

Zhang, J., X. Mu, W. Xu et al. 2014a. Exposure to arsenic via drinking water induces 5-hydroxymethylcytosine alteration in rat. *The Science of the Total Environment* 497–498:618–625.

Zhang, L., X. Qin, J. Tang et al. 2017c. Review of arsenic geochemical characteristics and its significance on arsenic pollution studies in karst groundwater, Southwest China. *Applied Geochemistry* 77:80–88.

Zhang, T. C., M. T. Schmitt, J. L. Mumford. 2003. Effects of arsenic on telomerase and telomeres in relation to cell proliferation and apoptosis in human keratinocytes and leukemia cells in vitro. *Carcinogenesis* 24(11):1811–1817.

Zhang, Q., D. Wang, Q. Zheng et al. 2014b. Joint effects of urinary arsenic methylation capacity with potential modifiers on arsenicosis: a cross-sectional study from an endemic arsenism area in Huhhot Basin, Northern China. *Environmental Research* 132:281–289.

Zhang, T., B. X. Zhang, P. Ye et al. 2006. The changes of trace protein in urine collected from 145 cases of arsenic poisoning patients caused by coal burning. *Journal of Chinese Microcirculation* 10:134–135.

Zheng, L. Y., J. G. Umans, M. Tellez-Plaza et al. 2013. Urine arsenic and prevalent albuminuria: Evidence from a population-based study. *American Journal of Kidney Diseases* 61(3):385–394.

Zheng, L., C. C. Kuo, J. Fadrowski et al. 2014. Arsenic and chronic kidney disease: A systematic review. *Current Environmental Health Reports* 1(3):192–207.

Zheng, Y., A. van Geen, M. Stute et al. 2005. Geochemical and hydrogeological contrasts between shallow and deeper aquifers in two villages of Araihazar, Bangladesh: Implications for deeper aquifers as drinking water sources. *Geochimica et Cosmochimica Acta* 69(22):5203–5218.

Zhou, Y., Y. Zeng, J. Zhou et al. 2017. Distribution of groundwater arsenic in Xinjiang, P.R. China. *Applied Geochemistry* 77:116–125.

Zhu, J., M. H. M. Koken, F. Quignon et al. 1997. Arsenic-induced PML targeting onto nuclear bodies: Implications for the treatment of acute promyelocytic leukemia. *Proceedings of the National Academy of Sciences of the United States of America* 94: 3978–3983.

Zierold, K. M., L. Knobeloch, H. Anderson. 2004. Prevalence of chronic diseases in adults exposed to arsenic-contaminated drinking water. *American Journal of Public Health* 94(11):1936–1937.

Zuo, R., J. Wang, G. Chen et al. 2015. Identification of weak anomalies: A multifractal perspective. *Journal of Geochemical Exploration* 148:12–24.

Zuo, R., Q. Xia, H. Wang. 2013. Compositional data analysis in the study of integrated geochemical anomalies associated with mineralization. *Applied Geochemistry* 28:202–211.

Index

A

Abnormal Electromyography (EMG), 129–130
Acceptable Daily Intake (ADI), 172–173
Acceptable Risk, 160–161
Acute Myeloid Leukaemia (AML), 10
Acute Promyelocytic Leukaemia (APL), 10
Acute Respiratory Infection (ARI), 230
Adaptation Strategies, 196, 198, 214–215, 216–217
Adsorption, 15, 18, 20, 58, 66, 67, 84, 264
Aeolian sediments, 57
Alanine Aminotransferase (ALT), 131, 132
Alanine Transaminase, 131
Albuminuria, 134
ALCAN, 231
Alkali desorption, 66, 264
Alkaline phosphatase, 131
Alluvial and Deltaic Aquifers, 2, 26, 33–38
Altered Cell Proliferation, 136–137, 149
Altered Cell Signaling,, 149
Altered DNA Methylation, 22, 149
Altered DNA Repair, 149
Altered Growth Factors, 149, 264–265
Aluminum Gallium Arsenide Crystals, 10
Alzheimer Disease, 131
Ammoniac Copper Arsenate, 11
Anemia, 132, 133, 147
Anisotropic Model, 75, 76
Anodic Stripping Voltammetric (ASV), 8
Aquifer Vulnerability, 72
Argentina, 2, 7, 25, 55–57, 72, 99, 103, 144, 146, 148, 184, 263–264
Arsanilic acid, 11
Arsenate ions, 4
Arsenate, 3–4, 7–8, 15, 17, 25, 43, 55, 99–100, 151, 153, 247
Arsenic Methylation Index (SMI), 147
Arsenic Mitigation, 70, 74, 93, 94, 221–255
Arsenic sulfides, 2, 10, 12
Arsenic Trioxide, 10, 11, 16, 20, 51–52, 100
Arsenic, 1, 2–3, 3–7, 8–12
Arsenic-Iron Removal Plants (AIRP), 38, 230, 236, 251
Arsenicosis, 2, 20–21, 37, 40, 44, 46, 65, 103, 126, 186, 189, 196, 201, 211, 214, 217, 219, 251, 259, 265–266
Arsenite, 1, 4, 7, 14, 15, 17, 46, 99–100, 151
Arsenopyrite, 3, 5, 13, 17, 41, 49, 54, 59, 65, 67
Arsonic Acid, 12
Arsphenamine, 9, 10
Aspartate Aminotransferase (SAT), 131, 132
Aspartate Transaminase, 131
Atomic Absorption Spectrometry (AAS), 7, 45
Australia, 11–12, 24, 64, 164
Average Daily Dose (ADD), 149, 176, 177, 186
Average Standard Error (ASE), 82

B

BAMWSP, 37, 223, 228, 229, 241, 250, 259
Bangladesh, 26, 33–38, 44, 69, 72, 86, 103–104, 124, 127, 132–133, 135, 176, 186, 203, 207, 209, 215, 221–255
Basal Cell Carcinoma (BCC), 48, 103, 137, 144–145, 183
Bell Ville Disease, 21, 26, 56, 263–264
Benchmark Concentration (BMC), 172
Benchmark Dose (BMD), 133, 172
Benchmark Dose Lower-confidence Limit (BMDL), 172, 173
Benchmark Response (BMR), 172
Best Guess, 188
Best Linear Unbiased Estimator (BLUE), 75, 79
β2-microglobulin (β2-MG), 75
Black Foot Disease, 21, 26, 48, 245, 263–264
Bladder Cancer, 48, 56, 61, 126, 147, 148, 151, 183, 184, 185, 189
Body Mass Index (BMI), 128, 154
Bolivia, 65
Botswana, 65
Bottom-up Approach, 93, 266
Bovine Spongiform Encephalopathy (BSE), 162
Bowen's Disease (BD), 44, 50, 103, 137, 144, 183
Brazil, 2, 26, 57, 74, 238
British Columbia, 51–52
Buffer Generation, 190, 191, 192
Burkina Faso, 62
Burma, 26, 38–39, 238

C

Cambodia, 2, 26, 39–40
Cameroon, 62–63
Canada, 24, 51–52, 72, 73, 129–130, 263–264
Cancer Slope Factor (CSF), 170, 172
Carcinogen, 1, 4, 20, 99, 137, 151, 157, 170, 172, 245, 264
Cardiac Ischemic Disease, 123
Cardiovascular Disease (CVD), 41, 51, 99, 123, 125, 149, 183, 223, 253, 264
Chile, 18–19, 21, 26, 57–58, 126, 146, 183, 254
China, 3, 15, 24, 28, 40–41, 73, 121, 238, 263–264
Chromated Copper Arsenate (CCA), 11
Chromosomal Aberrations, 21–22, 149
Chromosome Instability (CIN), 151
Chronic Daily Intake (CDI), 148, 153
Chronic Kidney Disease (CKD), 48, 121, 133, 147, 184
Chronic Obstructive Pulmonary Disease (COPD), 125, 126, 127
Cirrhosis, 131
Clara Cell Protein (CC16), 126
Cognitive Impairments, 130
Colombia, 65
Communication Process, 163
Community Integrated GIS, 93
Congenital Malformations, 134
Contaminant of Concern (CoC), 167, 169, 179
Coping Strategies, 196, 215–216, 217, 265
Correlogram, 76
Costa Rica, 65

Counter Mapping GIS, 93
Creutzfeldt Jakob Disease (CJD), 162
Critical Cartography, 93
Critical Interpretative Synthesis, 200
Croatia, 60
Cross-Validation, 82, 83
Cuba, 65
Cumulative Arsenic Exposure (CAE), 124, 128, 144, 147, 151, 179–180
Cytotoxicity, 12, 20, 22, 133, 245

D

De manifestis risk, 178, 265
De minimis risk, 178, 265
Deep Tubewell (DTW), 94, 96, 210, 217, 219, 234, 237, 240, 242, 250, 254
Denmark, 65, 145
Dermal effects, 102–122
Deterministic methods, 71, 72
Diabetes Mellitus (Type 1 and Type 2), 54, 102, 127, 129, 183
Dichlorodiphenyltrichloroethane (DDT), 11
Dimethyl Arsenic (DMA), 3–4, 12, 14, 121, 144, 145, 149
Dimethyl Arsenic Acid (DMAA), 3–4, 12, 14
Disability Adjusted Life Years (DALY), 46
Disodium Methyl Arsenate (DSMA), 12
Dissolution-precipitation, 20, 66
Dissolved Organic Carbon (DOC), 74
DNA damage, 21–22, 149, 151, 152–153
DNA Methyltransferase, 151
DNA repair, 22, 149
Dose-response, 23–24, 100, 122, 127, 144, 149, 167, 170–173
Dug well, 37, 41, 61, 62, 63, 224, 239–240, 251
Duty of Care, 257, 260, 267

E

Ecuador, 65
El Salvador, 65, 67
Endocrine System, 127–129
Estimated Exposure Dose (EED), 177
Estimated Glomerular Filtration Rate (eGFR), 133, 134, 147
Ethiopia, 63
Experimental Semivariogram, 75–78, 81
Exposure Assessment, 156, 166–167, 173–176, 185, 188, 265
Exposure Duration, 149, 150, 174, 176, 186, 190
Exposure Frequency, 176
Exposure-response, 171–172, 183, 189
Extrapolation, 70, 75, 172

F

Ferric Oxyhydroxides, 17
FI-HG-AAS, 7–8, 76, 84, 148, 186
FI-HG-AFS, 8
Finland, 30, 60–61, 183
Fowler's Solution, 9–10
France, 9, 26, 58–59
Full-Scale IQ (FSIQ), 130

G

Ganges Alluvial Plain, 41, 84
Ganges-Brahmaputra Plain, 2
Ganges-Brahmaputra-Meghna River Systems, 38
Gastrointestinal Disturbances, 54
Generalized Linear Models (GLM), 83
Genetically Modified Foods (GMF), 162, 263
Genotoxicity, 20, 21, 22, 149
Geographical Information Systems (GIS), 69, 70–71, 83, 93, 94–97, 156, 179–181, 190, 230
Geometric Anisotropy, 75
Geostatistical Logistic Regression Models, 73
Geostatistics, 69, 70–73, 74–75, 180
Geothermal Activity, 12, 13, 18
Germany, 59
GF-AAS, 7
Ghana, 63–64
Global Goals, 222
Global Positioning Systems (GPS), 84
Great Hungarian Plain, 2, 59–60
Greece, 59
Grey arsenic, 6
Ground Mapping, 93
Guatemala, 65

H

Hanging Mercury Drop Electrode (HMDE), 8
Hard-to-Reach (H2R), 231
Hazard Assessment, 166
Hazard Characterization, 171
Hazard Identification, 167–170
Hazard Quotient (HQ), 47, 177, 186, 265
Hazard, 155, 158, 164
Hematological System, 132–133
Hepatic Manifestation, 131–132
Herbicides Monosodium Methyl Arsenate (MSMA), 12
Hermeneutic Phenomenology, 198–200
Hermeneutics, 198, 199
High Performance Liquid Chromatography (HPLC), 7
Holocene Alluvial-lacustrine Aquifers, 41
Holocene Sediments, 40, 51
Honduras, 65
Human Papillomavirus (HPV), 144
Hungary, 59
Hydrated Ferric Oxides, 39
Hydride Generation Atomic Absorption Spectrometry (HG-AAS), 7
Hydride Generation-Atomic Fluorescence Spectrometry (HG-AFS), 7
Hyperkeratosis, 1, 20, 41, 49, 50, 54, 59, 103, 264
Hyperpigmentation, 1, 20, 41, 49, 50, 54, 60, 99, 103
Hypertension, 102, 123, 124, 183
Hypertensive Heart Disease, 55
Hypopigmentation, 54

I

ICCO Cooperation, 84
India, 10, 41–44, 126, 222
Indicator Kriging, 71, 72–73, 80
Inductively Coupled Plasma-Atomic Emission Spectrometry (ICP-AES), 7

Index

Inductively Coupled Plasma-Mass Spectrometry (ICP-MS), 7
Information-democracy, 70, 266
Inorganic Arsenic, 1, 4
Insulin-dependent DM (IDDM or T1DM), 127
Integumentary System, 102–122
Interpretive Research Paradigm, 198
Interpretivism, 198
Interstitial Lung Disease, 125
Inverse Distance Weighted (IDW), 72, 81–82
Inverse Squared Distance (ISD) Interpolation, 81
Ion Chromatography (IC), 7
Iran, 44–45
Ireland, 65
Iron oxyhydroxides, 46
Ischemic Heart Disease (ISHD), 102, 123
Italy, 60

J

James Marsh, 9
Japan, 45

K

Kai Dam, 21
Kathmandu Valley, 46
Kazakhstan, 65
Keratosis, 20, 21, 22, 37, 49, 50
Kidney Cancer, 21, 58, 102, 147–148, 149, 184
King of Poisons, 8, 263
Kriging Estimation, 72, 73, 74, 77–82, 192

L

Lactate Dehydrogenase (LDH), 124
Lag distance, 76
Lead hydrogen arsenate, 12
Least Developed Countries (LDC), 221
Leukopenia, 115, 132
Lifetime Carcinogenic Risk (LCR), 178, 186, 265
Lincolnshire limestone, 62
Linear Dose-response Assessment (LDRA), 172
Liver Cancer, 20, 21, 48, 50, 146–147, 149, 183, 189
Löllingite, 3, 14
Low Birth Weight (LBW), 116, 134, 136
Low Lift Pumps (LLP), 242
Lowest-Observable-Adverse-Effect Level (LOAEL), 172, 183
Lung Cancer, 20, 21, 45, 56, 58, 126, 145–146, 149, 172, 184
Lymphocytosis, 132

M

Malaysia, 65
Map interpolation, 71
Maximum Contamination Level (MCL), 22, 55, 181
Maximum likelihood estimation technique, 84
Mean Arterial Blood Pressure (MAP), 123, 124
Mean Error (ME), 82
Mean Squared Deviation Ratio (MSDR), 82
Mean Standardized Error (MSE), 82
Megaloblastic Erythropoiesis, 132

Melanosis, 20, 103, 122
Merkel Cell Carcinoma (MCC), 137
Metabolic Effects, 127–129
Meta-Synthesis, 200–202
Mexico, 26, 52–54
Micro Albumin (mALB), 133
Microsatellite Instability (MIN), 151
Microwave Plasma Torch (MPT), 8
Millennium Development Goals (MDG), 222, 229
Minimum Detection Limit (MDL), 186
Mitogen Activated Protein Kinase (MAPK), 134
Model-based Bayesian Ordinary Kriging, 73
Model-based geostatistics (MBG), 73
Model-based Ordinary Kriging, 73
Modification Factor (MF), 173
Monomethyl Arsenic (MMA), 3, 14, 121, 145, 149
Monomethyl Arsenic Acid (MMAA), 3, 14, 121
Monosodium Methanoarsonate (MSMA), 14
Motor Neuropathy, 129

N

N-acetyl-β_2-glucosaminidase (NAG), 133
Narrative Synthesis, 200–201
Neonatal, 134
Nepal, 45–46
Nephritis, 55
Nervous system, 20, 129–131
Neuropathy, 44
Neurotoxicity, 102
New Zealand, 64–65
Newton-Raphson Optimization Technique, 84
Nicaragua, 65
Niccolite, 3
Nigeria, 64
Noncirrhotic Portal Fibrosis (NCPF), 131
Nonflame Atomic Spectrometry (NAS), 7
Noninsulin-dependent DM (NIDDM or T2DM), 127
No-Observable-Adverse-Effect Level (NOAEL), 169, 172
Normocytic anemia, 132
Norway, 65
Nugget effect, 76, 77, 81, 85, 89

O

Obnoxious Risk, 178
Odds Ratio (OR), 120, 123, 143, 144
Ordinary Kriging (OK), 72, 79–80, 192
Organic arsenic, 4
Orpiment, 3, 13
Overlay Operation, 180, 190
Oxidation States, 3, 25
Oxidation-reduction, 66
Oxidative Stress, 125
Oxyhydroxide Reduction Hypothesis, 17–18

P

Pakistan, 46–47
Palmar hyperkeratosis, 102
Paris Green, 11, 12
Participant Observation, 200
Participatory Maps, 93
Participatory Remote Sensing, 93

Participatory Rural Appraisal (PRA), 71, 93, 201, 266
Participatory Sketch Mapping, 93
Participatory Stone Mapping, 93
Pentavalent Arsenic, 4, 14, 25
Peripheral Arterial Disease (PAD), 123
Peripheral Neuropathy, 129
Peripheral Vascular Disease (PVD), 47, 48, 54, 123, 223
Phenomenology, 199–200
Pigmentation, 222
Pipedwater Systems, 38, 241–242
Plantar Hyperkeratosis, 102
Plasma Uric Acid (PUA), 124
Poland, 65
Pond-Sand-Filters (PSF), 28, 232, 239, 251
Prediction error, 74, 82
Preterm birth/delivery (PD), 134, 136
Prevalence Odds Ratio (POR), 126
Prevalence Proportion Ratio (PPR), 135
Probabilistic Scale, 85
Probability, 70, 85, 158
Promyelocytic Leukemia Protein (PML), 10
Prostate cancer, 55, 148–149, 208
Proximity, 257
Public Forum GIS, 93
Public participation GIS (PPGIS), 70, 71, 93, 94–97, 266
Pulmonary System, 125–126
Pulse Pressure (PP), 123, 124
Pyrite oxidation hypothesis, 17

Q

Qualitative Enquiry, 196, 197
Qualitative Transcripts, 201–203

R

Radial Basis Function, 72
Rain Water Harvesting (RWH), 38, 224, 230, 232, 235, 238
Range, 76, 77, 78, 82, 85
Reactive Nitrogen Species (RNS), 152
Reactive Oxygen Species (ROS), 10, 129
READ-F, 231
Realgar, 3, 10
Redox Condition, 4, 15, 46
Reductive Dissolution, 66
Reference Dose (RfD), 177
Regionalized Variables, 74, 79
Relative Risk (RR), 122, 146
Remote Sensing (RS), 180
Renal System, 133–134
Reproductive System, 134–137
Residual Risk, 155
Resilience, 156, 159, 265
Respiratory Effects, 125, 264
Respiratory Tract Infections (RTI), 125
Retinol Binding Protein (RBP), 133
Rich Descriptive Narratives, 203
Risk Analysis, 156, 157, 161–164, 167
Risk Assessment, 155–156, 160, 161–164, 166–179, 265
Risk Aversion, 155
Risk Characterization, 156, 167, 177, 185, 188
Risk communication, 163, 164
Risk Management, 155, 161–164, 167, 173–174, 266

Risk Mapping, 156, 180
Risk Perception, 160, 163
Risk Reduction, 93, 163, 167
Risk, 155–193, 265
Romania, 60, 136
Root Mean Square Error (RMSE), 82
Root Mean Square Standardized Error (RMSSE), 82
Root-Mean-Square Prediction Error, 81
Russia, 24, 65

S

Semivariogram, 70, 75–77, 79, 81, 85
Serum Alkaline Phosphatase (S-ALP), 131, 132
Shobuz Biplab, 242
SIDKO, 231
Sill, 75, 76–77, 85, 192
Silver Diethyldithiocarbamate (Ag-DDTC) method, 41–42
Skin Cancer, 1–2, 21, 37–38, 41, 56, 101, 103, 137, 144–145, 170, 185, 189
Skin Lesions, 20–21, 43, 44, 50, 58, 101, 103, 121, 122, 126, 132, 144, 154, 206, 211, 253
Slovakia, 60, 137, 144–145, 183
SONO Filter, 231
Spain, 26, 61, 67, 74, 129, 136
Spatial Anisotropy, 76
Spatial Autocorrelation, 70, 75, 77, 82, 266
Spatial Continuity, 70, 74–82, 85
Spatial Decision-Support System (SDSS), 70, 94
Spatial discontinuity, 70, 72, 76, 85, 266
Spatial Interpolation, 75, 81, 192
Spatial prediction, 75
Spatial Risk Mapping, 156, 191, 266
Spatial Variability, 70, 72–74, 77, 92
Spherical Model, 77, 85
Splenomegaly, 131, 132
Spontaneous Abortion, 134, 136
Spontaneous Pregnancy Loss (SPL), 134
Squamous Cell Carcinoma (SCC), 48, 103
Square Wave Cathodic Stripping Voltammetry (SWCSV), 8
Stillbirth, 134–136
Stochastic Interpolation, 72
Stroke, 123, 246
Subclinical Sensory Neuropathy, 130
Sustainable Development Goals (SDG), 222
Sweden, 24, 61
Switzerland, 61

T

Taiwan, 26, 47–48, 99, 103, 127, 147, 223, 245
Tanalith, 12
Tennantite, 3
Thailand, 21, 26, 48–49, 144, 238
Threshold of Toxicity, 171
Thrombocytopenia, 132
Thyroid hormone, 127
Thyroxine (T4) Hormones, 127
Tien Giang Province, 51
Tolerable Daily Intake (TDI), 173
Tort Law, 257
Total Dissolved Arsenic, 51
Total Hazard Score (THS), 170

Index

Toxicity Assessment, 167, 170–173, 265
Toxicity-Concentration Screen (TCS), 169
Tri-iodothyronine (T3) Hormones, 127
Trivalent Arsenic, 137
Tucumán Province, 56
Tumor-suppressor Protein, 22, 53
Turkey, 49, 73

U

Uncertainty Factor (UF), 172, 173
United Kingdom, 62
Universal Kriging, 72
Uruguay, 65
USA, 22, 54, 154, 245, 258

V

Variogram, 75, 76, 81, 85, 192
Venezuela, 65
Verbal IQ (VIQ), 120
Verbatim Data, 196, 204
Vietnam, 26, 50, 66, 67, 263

Volcanic Activity, 13, 15, 58
Vulcano, 60
Vulnerability, 156, 158, 159, 165, 167, 265

W

Waiotapu Valley, 64
West Bengal, 18, 25, 41, 44, 70, 72, 74, 103, 121, 126, 127, 129, 130, 131, 135, 183, 222, 223
White arsenic, 2

X

X-ray Fluorescence (XRF), 7

Y

Yellow Orpiment, 2

Z

Zonal Anisotropy, 75
Zosimos of Panopolis, 2